THE DISTRIBUTION OF THE GALAXIES

This topical volume examines one of the leading problems in astronomy today – how galaxies cluster in our Universe. Many observational surveys and theoretical projects are currently underway to understand the distribution of galaxies. This is the first book to describe gravitational theory, computer simulations, and observations related to galaxy distribution functions (a general method for measuring the clustering and velocities of galaxies). It embeds distribution functions in a broader astronomical context, including other exciting contemporary topics such as correlation functions, fractals, bound clusters, topology, percolation, and minimal spanning trees.

Key results are derived and the necessary gravitational physics provided to ensure the book is self-contained. And throughout the book, theory, computer simulation, and observation are carefully interwoven and critically compared. The book also shows how future observations can test the theoretical models for the evolution of galaxy clustering at earlier times in our Universe.

This clear and authoritative volume is written at a level suitable for graduate students and will be of key interest to astronomers, cosmologists, physicists, and applied statisticians.

William C. Saslaw is professor of astronomy at the University of Virginia, Charlottesville and also does research at the National Radio Astronomy Observatory and at the University of Cambridge. He received his Ph.D. in applied mathematics and theoretical physics from Cambridge. He is also the author of *Gravitational Physics of Stellar and Galactic Systems*.

THE DISTRIBUTION OF THE GALAXIES

Gravitational Clustering in Cosmology

WILLIAM C. SASLAW

University of Cambridge
University of Virginia
National Radio Astronomy Observatory, U.S.A.

CAMBRIDGE
UNIVERSITY PRESS

CAMBRIDGE UNIVERSITY PRESS
Cambridge, New York, Melbourne, Madrid, Cape Town, Singapore, São Paulo

Cambridge University Press
The Edinburgh Building, Cambridge CB2 8RU, UK

Published in the United States of America by Cambridge University Press, New York

www.cambridge.org
Information on this title: www.cambridge.org/9780521394260

First published 2000
This digitally printed version 2008

A catalogue record for this publication is available from the British Library

Library of Congress Cataloguing in Publication data
Saslaw, William C.
The distribution of the galaxies : gravitational clustering in
cosmology / William C. Saslaw.
p.   cm.
Includes bibliographical references.
ISBN 0-521-39426-0
1. Galaxies – Clusters.   2. Gravitation.   I. Title.
QB858.7.S28   1999
523.1′12 – dc21
99-14191
CIP

ISBN 978-0-521-39426-0 hardback
ISBN 978-0-521-05092-0 paperback

Contents

Prologue

Despite appearances, it is not the Epilogue, but the Prologue that is often left for last. Only after seeing what is done, can one acknowledge and apologize. My main acknowledgments are to many students and collaborators, for they have taught me much. My apologies are to those colleagues who may not find enough of their own results in the pages still ahead. For them I can only echo Longfellow that "Art is long and Time is fleeting." The subject of large-scale structure in the universe, of which the distribution of the galaxies represents only a part, has burgeoned beyond all previous bounds as the new millennium approaches. Driven as much by the scope and depth of its questions as by new streams of data from the depths of time, there is an increasing excitement that fundamental answers are almost in reach. And there will be no stopping until they are found.

On the timescales of the physical processes we are about to consider, millennia count for very little. But on the timescale of our own understanding, years, decades, and certainly centuries have changed the whole conceptual structure surrounding our views. This may happen again when the role of dark matter becomes more transparent.

Meanwhile, this monograph is really no more than an extended essay on aspects of galaxy clustering that I've found especially interesting. It emphasizes galaxy distribution and correlation functions – but mostly distribution functions because correlations have already been discussed widely elsewhere. Besides, distribution functions contain more convenient descriptions of more information.

Both these statistics are embedded here in a broader context, but their main virtue is that, of all descriptions so far, correlations and distributions can be related most directly to physical theories of gravitational clustering. Even for gravitational clustering, I have emphasized the simplest and most fundamental problem: how gravitating point masses cluster self-consistently within the background of an expanding universe. Three general approaches – statistical thermodynamics, computer N-body experiments, and astronomical observations – all give consistent results, and all are discussed here in detail. The observational agreement suggests that this cosmological many-body problem will remain a useful and basic aspect for understanding galaxy clustering, whatever detailed scenario produced it. And we really need to understand clustering by the known gravitational forces of the galaxies alone, before adding more speculative contributions.

ix

The cosmological many-body problem, moreover, is endlessly fascinating in its own right. Much has been learned about it since Newton's first qualitative discussion, but much more remains for future discovery. Part of this fascination comes from its inherent and essential nonlinearity. Its statistical thermodynamics hints at a deeper theory. Computer simulations can check these theories and prevent them from going too far astray. We have just begun to explore the theory of systems with ranges of masses and different initial conditions. Important new problems are easy to find.

To begin, I have sketched the historical background, from Babylonian myths until 1970. Our subject started slowly and for centuries lay well outside the mainstreams of astronomy. Nevertheless it made quiet progress, and I've selected some milestones with the hindsights of history. These lead, in Part II, to a brief general review of the main descriptions of galaxy clustering; each has its weakness and all have some virtue. Thus the first third of the book provides an overall introduction.

Next, the general theme of gravity takes over. Part III discusses its relation to correlation functions in the context of the cosmological many-body problem and reviews several topics of recent interest along with a sketch of computer simulation techniques and results. This ends with a brief description of observations. Naturally there is some repetition of the earlier introduction. Although most discussions are self-contained, my earlier book *Gravitational Physics of Stellar and Galactic Systems* (GPSGS) sometimes extends them.

In the book's second half, I discuss distribution functions for galaxy positions and peculiar velocities and how they evolve. These are the generalizations, for the cosmological many-body system, of the Poisson and Maxwell–Boltzmann distributions in a perfect gas.

Distribution functions may be less familiar than correlations and other descriptions of clustering. So I've started by summarizing their mathematical properties that are especially useful for our exploration. Then the cosmic energy equation provides a dynamical link to the cosmological distributions. Like most complex dynamics, this link is easier to follow in the linear regime of fairly weak clustering. To examine the observed range of nonlinear clustering, it helps to develop the statistical thermodynamic theory. After reviewing thermodynamics, we apply it to derive the spatial and velocity distribution functions of the cosmological many-body problem. Then we follow their quasi-equilibrium evolution as the universe expands. There are no free parameters in this theory – after all it's just gravity.

Initially the applicability of gravitational thermodynamics to the cosmological many-body problem was rather surprising. To paraphrase Mark Twain, it gratified some astrophysicists and astonished the rest. This apparent impasse arose because thermodynamics is essentially an equilibrium theory, whereas gravitational clustering is a manifestly nonequilibrium phenomenon. What this seeming contradiction failed to appreciate, however, is that under a wide range of conditions, cosmological many-body clustering can evolve through a sequence of equilibrium states. This quasi-equilibrium evolution enables thermodynamics to provide a very good approximation.

Computer N-body experiments, which directly integrate the mutual orbits of many galaxies in an expanding universe, show that gravitational thermodynamics does indeed apply for a variety of initial conditions in different universes. Part V describes these tests. They also determine the conditions under which gravitational thermodynamics fails to give a good description of clustering and the reasons for this failure. We still need more diverse computer experiments to explore the whole range of the theory.

Naturally, the grandest experiment of all is the analog computer in the sky. In our own Universe, the initial conditions and detailed evolution of galaxy clustering remain controversial and highly uncertain. Here the cosmological many-body problem is perhaps the simplest model of this process. It assumes, in this application, that when the galaxies clustered most of the dark matter that affected their orbits was associated with the individual galaxies, either inside them or in halos around them. Thus galaxies acting effectively as point masses dominated the clustering. Other types of models are dominated by vast quantities of unobserved dark matter whose growing large-scale inhomogeneities determine the clustering, and galaxies merely go along for the ride. More than thirty years of increasingly ingenious and sensitive searching have failed to reveal the specific forms of dark matter these other models require.

As a model for galaxy clustering, the cosmological many-body theory describes the observed galaxy distribution functions remarkably well, as Part VI discusses. This suggests that the clustering effects of intergalactic dark matter are small, or that they have contrived to mimic many-body clustering. Consistency between models and observations is a good sign, but it is not proof. History has shown that our Universe is complex, and so a final decision here will have to await further developments.

Part VII introduces some aspects of clustering that may unfold in the future. These generally involve more complicated modeling than simple gravitational clustering, but such models are necessary to understand many detailed astrophysical consequences, including galaxy formation itself. No one doubts a connection between galaxy formation and clustering, but the nature and strength of this link is still so uncertain that at present I think it wise – or at least expedient – to consider clustering as a separate problem. The formation of galaxies (not a particularly well-defined process) sets the initial conditions for their clustering, which I assume are subsumed by those studied here. If not, we will have to modify them. Optimistically there is hope that eventually the fluctuations of the cosmic microwave background will lead to a clear solution of the origins of clustering.

The work of determining galaxy distribution functions from observations, numerical simulations, and gravitational theory in which I've participated has benefitted greatly from discussions with dozens of astronomers, physicists, and mathematicians. Most of the results have come from many collaborations over the years and over the world. Although some collaborators were officially called students, they quickly outgrew any secondary role and became equal participants. In a young subject, everyone rapidly reaches frontiers. It is adventurous and all great fun. For these collaborations, I especially thank Sverre Aarseth, H. M. Antia, Kumar Chitre, Paul Coleman, Phil Crane, Naresh Dadhich, James Edgar, Fan Fang, Andrew Hamilton, Shirin Haque-Copliah, Shogo Inagaki, Makoto Itoh, Sanjay Jain, Leo Krzewina, Ofer Lahav, Sunil Maharaj, Hojung Mo, Somak Raychaudhury, Yoel Rephaeli, Ravi Sheth, Trinh Thuan, and David Valls-Gabaud. In addition, I am happy to thank Ian Du Quesnay for helpful advice on the Greek philosophers; Arthur Haberman, E. R. Harrison, Ofer Lahav, Morton Roberts, and Mark Whittle for their detailed comments on substantial parts of the manuscript; and especially David Valls-Gabaud for reading and commenting on the entire manuscript. In the University of Cambridge I am very grateful to Pat Cassidy at Jesus College and to Judith Moss at the Institute of Astronomy for their exceptionally fine work, which converted the manuscript with its occasionally complicated equations into a form directly suitable for the printers. And both Jesus College and the Institute of Astronomy provided, as they have for many years, conditions especially conducive to creative research. I am glad to thank George Kessler at NRAO for producing many diagrams with an ideal blend of modern electronics and old-fashioned skill. Richard Sword's enthusiastic and interactive artistry designed the book's cover. It has also been a pleasure, as I expected it would be, to work with Simon Mitton, Adam Black, and the staff of the Cambridge University Press.

PART I

Historical

they ate the oxen of the sun,
the herd of Hélios Hypérion
Homer,   The Odyssey
(translation by Mandelbaum)

The history of understanding structure in our Universe is older than the story of Odysseus, and has as many twists and turns. Few of these paths remain familiar to most astronomers today, so in the early chapters I have simply collected some essential developments along the way. They are not without surprises. One of which is that many ideas now thought to be novel have really been known for tens, or hundreds, of years. Often they were no more than speculations, but sometimes they captured reality's core.

1

Cosmogony Myths and Primitive Notions

To every thing there is a season,
and a time to every purpose under heaven.
Ecclesiastes 3:1

Structure may be the most surprising property of our Universe. Structure differ-
entiates things, and the changing of things introduces the notion of time. We can
easily imagine universes without macroscopic structure, filled by an equilibrium
distribution of matter and radiation: homogeneous, isotropic, and dull. Indeed
such a state may have described our own Universe until fairly recently. The oldest
structures we know are the high-redshift galaxies and quasars, dating from an epoch
when the Universe was only about ten times smaller than it is now. Compare this
with the epoch of decoupling of matter and radiation when the scale of the Universe
was a thousand times smaller than now, with the epoch of nucleosynthesis at which
helium formed when the Universe was a billion times smaller, and with the almost
mythical quantum epoch when the Universe was more than 10^{50} times smaller.

Although the structure we now see may be considered fairly recent when viewed
in terms of the expanding scale of the Universe, it is very old in terms of total time
elapsed since the expansion began. This is because the relative rate of expansion of
the scale of the Universe, $\dot{R}(t)/R(t)$, slows down as the Universe expands. In typical
conventional cosmological models, the distance of the farthest galaxies and quasars
corresponds to more than 90% of the age of the Universe since the big bang. Earlier
structure no doubt existed, but our knowledge of it is much more uncertain.

Long before people knew the age and size of any astronomical structure, civi-
lizations imagined many origins for the Earth and the Moon, the Sun and the stars.
These cosmogony myths varied widely among different cultures, and sometimes rival
myths arose within the same culture. Most of these myths begin with a magnified
version of the locally familiar – rocks, oceans, personified gods with human vices,
a giant clam, even a mirror world beyond the sky. Such basic elements, like some
aspects of modern cosmology, had no origin themselves and were subject to no
further questioning. Eventually there occurred a split, disruption, battle, or sexual
union from which the Earth, Moon, Sun, oceans, sky, remaining gods, first people,
plants, and animals emerge in assorted ways and combinations for each different
myth. In several myths a cosmic egg forms or is produced. Something breaks it, or it
opens spontaneously, and the structure of the universe emerges. Putting everything
in its place, however, may require further work by the gods.

Cosmogony myths are usually grouped into historical or anthropological categories. But here, to emphasize their range, I suggest a thematic classification whose
sequence tends toward increasing abstraction. Some examples represent each type

3

and illustrate how many types contain common features. Naturally we view these myths and early cosmogonies through the lens of modern understanding. Any claim to decant the detailed attitudes of their original believers from our own concepts of the Universe would be largely illusory.

In the first group of myths, battles among the gods generate structure. Among the oldest is the Babylonian Epic of the Creation whose earliest known written tablets go back three thousand years and are probably based on still older versions. Not surprisingly, the Universe was organized by the god of Babylon himself, Marduk. He was the grandson of Apsu and Tiamet who personified the sweet and salt waters, from whose mingling was born a family of gods. They squabbled long and noisily, giving each other insomnia and indigestion, until Apsu and Tiamet disagreed over whether they should destroy their own progeny. Battle lines were drawn. Marduk, potentially the bravest, was selected king of the gods. He killed Tiamet by commanding the tempests and hurricanes to blow her full of wind until she was so puffed up she could not close her mouth. Then he shot an arrow into her heart and split her body into two halves, one forming the heavens and the other forming the Earth. After that it was relatively easy to organize the world, install the stars, fix the planetary orbits, and create humanity to do the hard work and leave the gods free to amuse themselves.

The old Greek version of the separation of Heaven and Earth is told by Hesoid (ca. 800 B.C.; see also Lang, 1884). His *Theogony*, an extensive geneology of three hundred gods, may derive its style partly from the earlier Babylonian epic, although this is unclear. It starts with Chaos, not in the sense of turbulence and erratic unpredictable motion we use today, but in the earlier sense of a vast formless dark chasm or void. Chaos was not a god, but a sort of principle, which produced the Earth (Gaia), underworld (Tartara), and love (Eros) gods. Earth bore Heaven (Uranus), then together they produced a bunch of nasty children, including Kronos whom Gaia encouraged to castrate Uranus. Thus Earth and Heaven were separated and a slew of other gods, nymphs, and giants developed from the bits and pieces of the wound.

Alternatively, according to the myths of Iceland, ice played a major role in the formation of the world from the void. One part of the void developed a region of clouds and shadows, another a region of fire, a third a fountain from which flowed rivers of ice. As some of this ice melted, Ymir, a giant who looked like a man, emerged, and then a giant cow to feed him along with other giants and giant men and women. These bore three gods who battled the giants, killing Ymir. His flesh became the land, his blood the sea, his bones the mountains, his hair the trees, and his empty skull the sky. Stray sparks from the region of the fire were placed in his skull by the gods to form the Moon, Sun, and stars, which were then set in motion. Maggots forming in Ymir's rotting corpse were made into the dwarfs of the underworld. Eventually the gods made humans from tree trunks.

A more abstract battle cosmogony was developed by the ancient Persians, especially the Zoroastrians. Two personifications, good (Ormazd) and evil (Ahriman), alternately created the opposing aspects of the world: light – darkness, life – death, truth – falsehood, summer – winter, pretty birds – biting insects, and so on for three thousand years. After another six thousand years Ormazd would win and the world would be purified. Observations can therefore put an upper limit on the age of this cosmogony.

In the second group of myths, the gods generate structure much more peacefully. In Genesis this is done by monotheistic decree rather than by magnifying tribal or family strife. First, light is created in the dark void, then the heaven and Earth are separated and heaven supplied with the Sun, Moon, and stars. In contrast, Japanese creation myths (perhaps partly derived from Chinese sources) began with three gods forming spontaneously in heaven who then hid themselves away. The Earth, which was first like a huge globule of oil floating on a vast sea, gave rise to two further hiding gods, and then seven generations of more normal gods. Izanagi and his wife Izanami, the last of these generations, caused the Earth to solidify by poking it with a spear given to them by other gods. They stood on the floating bridge of heaven during this process and a drop falling as they lifted up the spear formed the island of Onokoro, on which they landed. Their subsequent union gave birth to the rest of the Japanese islands (after some misbirths caused by the woman speaking first to the man, rather than waiting until she was spoken to) and then to a new range of nature gods. Izanami died in childbirth when the fire god was born and went to hell, where the distraught Izanagi followed trying to convince her to return. He failed, and after several misadventures returned to the surface where he cleansed himself by bathing in the sea and producing more gods. The greatest of these was the Sun goddess, Amaterasu, who resulted from Izanagi washing his left eye; the moon goddess, Tsukiyomi, came from washing his right eye. Thus were the Earth, Sun, and Moon created without the need for great battles.

The Pawnee Indians of Nebraska also had a peaceful cosmogony generated by decree. Their chief god, Tirawa, first placed the Sun in the east, and the Moon and the evening star in the west, the pole star in the north, and the Star of Death in the south. Four other stars between each of these supported the sky. He assembled the lightning, thunder, clouds, and winds creating a great dark storm. Dropping a pebble he caused the thick clouds to open revealing vast waters. Then Tirawa ordered the four star gods holding up heaven to smite the waters with maces, and lo the waters parted and Earth became manifest. Next he ordered the four star gods to sing and praise the creation of Earth, whence another huge storm rose up gouging out mountains and valleys. Three times more the gods sang and there came forth forests and prairies, flowing rivers, and seeds that grow. The first human chief was born to the Sun and the Moon, the first woman to the morning and evening star. The rest of humanity was created by the union of stars.

The third group of myths are less anthropomorphic animal cosmogonies. Islanders on Nauru, in the South Pacific, say that in the beginning only an Old-Spider floated above the primordial sea. One day she discovered a giant clam and looked in vain for a hole to enter by. So she tapped the clam. It sounded hollow and empty. With the aid of a magic incantation she opened the shell a bit and crawled inside where it was cramped and dark. Eventually, after a long search throughout the clam, Old-Spider found a snail. Although the snail was useless in its original form, she give it great power by sleeping with it under one of her arms for three days. Releasing the snail, she explored some more and found another, bigger snail, which got the same treatment. Politely, Old-Spider asked the newly energized first snail to open the clam a little wider to give them more room. As its reward, she turned the snail into the Moon. By the feeble light of the Moon, Old-Spider next found an enormous worm. Stronger than the snail, the worm raised the shell a little higher as the salty sweat poured from its body to form the sea. Higher still and higher he raised the upper half-shell until it became the sky. But the effort was too much, and the worm died of exhaustion. Old-Spider turned the second snail into the Sun near the lower half-shell, which formed the Earth.

Egg cosmogonies are the fourth group of myths. Present in the early Egyptian and Phoenician creation stories, they became especially prominent in the Greek Orphic and Indian Vedic cosmogonies. Followers of Orpheus believed that Ether, the spirit of the finite, gradually organized the cosmic matter into an enormous egg. Its shell was night, its upper part was the sky, and its lower part the Earth. Light, the first being, was born in its center and, combining with Night, created Heaven and Earth. In the Vedic cosmogonies, a golden egg arose and the Universe and all its contents were nascent within. After a thousand years the egg opened and Brahma emerged to begin organizing the Universe, first becoming a wild boar to raise the Earth above the waters.

A Finno–Ugric multi-egg cosmogony begins with the Daughter of Nature, Luonnotar, who grew bored of floating through the sky and descended to the sea where she lay amidst the waves for seven centuries. A large bird wanting to build a nest spied her knee sticking out of the sea. So the bird laid eggs there and sat on them for three days, until Luonnotar became irritated and flung them off. When they cracked open their lower parts combined to form the Earth, their upper parts became heaven, their yolks became the Sun, their whites the Moon, and the spotted bits formed the stars.

The fifth group of myths have inanimate origins. Sometimes, as with the primordial Egyptian concept of Nun it is the ocean that gives rise even to the gods. Elsewhere, as in Samoa, the Universe began as a series of rocks, first the rocks of heaven and then those of Earth. The Hawaiians had a myth in which the Universe was cyclic, each world emerging from the wreck of a previous one.

The sixth, most abstract, group of myths do not involve gods at all. Structure in these universes is completely self-generating. On the Marquesas Islands in the South Pacific is a myth that describes how "the primeval void started a swelling, a whirling, a vague growth, a boiling, a swallowing; there came out an infinite number of supports or posts, the big and the little, the long and the short, the hooked and the curved, and above all there emerged the solid Foundation, space and light, and innumerable rocks" (Luquet, 1959). This is not far from some modern cosmologies if we substitute "vacuum" for "void," "perturbations" for "supports or posts," and "galaxies" for "rocks." This last substitution signifies the scale over which people are aware of their surroundings. Essentially this is an evolutionary, rather than a theistic or geneological, view of the organization of the Universe. In common with almost all creation myths, darkness precedes light.

Gradually the early Greek philosopher-physicists replaced earlier creation myths with speculative attempts at rational explanations for the structure they knew. Ionians, typified by Thales of Miletus (about the sixth century B.C.), began this process by claiming that all natural things came into being from other natural things. The substance persists but its qualities change – a sort of conservation law – according to Aristotle's later description (*Metaphysics*). For Thales, water was the primitive element and he may have thought (though this is uncertain) that the Earth originated from water and continued to float on it. Suggestions (Kirk et al., 1983) for Thales's choice of water range from the influence of earlier Egyptian and Babylonian myths to the observation that corpses dry out.

Anaximander favored fire. Writing about the same time as Thales, he states (in the description of Hippolytus, the Roman theologian; Kirk et al., p. 135): "The heavenly bodies came into being as a circle of fire separated off from the fire in the world, and enclosed by air. There are breathing-holes, certain pipe-like passages, at which the heavenly bodies show themselves; accordingly eclipses occur when the breathing holes are blocked up." Thus cosmology continued its tradition of being often in error, but never in doubt.

More modern views emerged a century later as Leucippus and Democritus began to develop the implications of their atomic ideas. Since Leucippus thought there were an infinite number of atoms, it followed that there could be an infinite number of worlds.

The worlds come into being as follows: many bodies of all sorts of shapes move "by abscission from the infinite" into a great void; they come together there and produce a single whirl, in which, colliding with one another and revolving in all manner of ways, they begin to separate apart, like to like. But when their multitude prevents them from rotating any longer in equilibrium, those that are fine go out towards the surrounding void, as if sifted, while the rest "abide together" and, becoming entangled, unite their motions and make a first spherical structure.

So far, this sounds a bit like gravitational clustering, but then he continues more fancifully:

This structure stands apart like a "membrane" which contains in itself all kinds of bodies; and as they whirl around owing to the resistance of the middle, the surrounding membrane becomes thin, while contiguous atoms keep flowing together owing to contact with the whirl. So the earth came into being, the atoms that had been borne to the middle abiding together there. Again the containing membrane is itself increased, owing to the attraction of bodies outside; as it moves around in the whirl it takes in anything it touches. Some of these bodies that get entangled form a structure that is at first moist and muddy, but as they revolve with the whirl of the whole they dry out and then ignite to form the substance of the heavenly bodies. (Leucippus' theory described by Diogenes Laertius, a Roman biographer of the third century A.D., in Kirk et al., p. 417)

Thus the Universe is structured by natural, but ad hoc and unexplained causes. That the atomists thought in terms of short-range, rather than long-range, causes is suggested by Simplicius, the Roman Neoplatonist commentator describing their views in the sixth century A.D., a thousand years later: ". . . these atoms move in the infinite void, separate one from the other and differing in shapes, sizes, position and arrangement; overtaking each other they collide, and some are shaken away in any chance direction, while others, becoming intertwined one with another according to the congruity of their shapes, sizes, positions and arrangements, stay together and so effect the coming into being of compound bodies" (Kirk et al., p. 426).

This atomic clustering theory coexisted with the more continuum view of Diogenes of Appolonia. Since Thales and Anaximander had suggested water and fire, it was left to Diogenes to posit air as the fundamental element of cosmogony: ". . . the whole was in motion, and became rare in some places and dense in others; where the dense ran together centripetally it made the earth, and so the rest by the same method, while the lightest parts took the upper position and produced the sun" (Plutarch, second century A.D. in Kirk et al., p. 445). Diogenes may also have been the first to suggest that the Universe contains dark matter, an extrapolation from the fall of a large meteorite in Aegospotami (467 B.C.).

Plato and Aristotle, founders of the Academy and the Peripatetic schools, were next to dominate philosophy for a hundred years, fall into decline, and then be revived by the Romans and preserved by the Arabs until in the thirteenth century they became supreme authorities for four hundred years of scholastic church commentary. Alive, they stimulated science, but as dead authorities they were stultifying. Both of them abandoned the atomistic view that the Universe was infinite. Plato, in his "Timaeus" proposed that the Universe is analogous to a single unique living being (in his world of ideal forms), created by a good and ideal god as a model of himself. He made it spherical because it had no need of protuberances for sight, hearing, or motion. The Moon, Sun, and Planets were the first living gods, created to move around the Earth in ideal circles so they could define and preserve Time. Aristotle attempted to give somewhat more physical explanations. For example, in his "On the Heavens" the Earth is spherical because each of its parts has weight until it reaches the center. When Earth formed, all its parts sought the center, their natural place of rest. This convergence produces an object similar on all sides: a sphere. Here is one of the first physical arguments using local symmetry. Aristotle then goes on to cite two observations in support of a spherical Earth: the convex boundary of lunar eclipses and the changing positions of the stars as seen from different countries. He mentions the result of mathematicians who used this second method to determine the Earth's circumference; they found about twice the modern value.

Aristotle's arguments for a finite world and a unique Earth were less satisfactory. First he claims there are two simple types of motion, straight and circular. All others are compounds of these. Simple bodies should follow simple motions. Circular motion, being complete, is prior to rectilinear motion and therefore simple prior bodies like planets and stars should follow it. Next he says that only finite bodies can move in a circle, and since we see the heaven revolving in a circle it must be finite. Finally, the world is unique. For if there were more than one, simple bodies like earth and fire would tend to move toward or away from the center of their world. But since these bodies have the same natures wherever they are, they would all have to move to the same center, implying there could only be one world around this center.

During the temporary three hundred year decline of the Academic school following the death of Aristotle in 322 B.C., atomism resurged. It was carried to new heights by the Epicurians, culminating in its great exposition by the Roman poet Lucretius in the middle of the first century B.C.

Epicurus argued for an infinite Universe filled with an infinite number of bodies (atoms). "For if the void were infinite but the bodies finite, the bodies would not remain anywhere but would be travelling scattered all over the infinite void, for lack of the bodies which support and marshal them by buffeting. And if the void were finite, the infinite bodies would not have anywhere to be" (Long & Sedley, 1987, p. 44). Moreover, an infinite number of atoms could produce an infinite number of worlds where a world contained an earth, celestial bodies, and all the observable phenomena. This was necessary because if, as in the Epicurian philosophy, structure results by chance and not from teleology, the probability that our actual world forms will increase dramatically if an infinite number of worlds are possible. ". . . we must suppose that the worlds and every limited compound which bears a close resemblance to the things we see, has come into being from the infinite: all these things, the larger and the smaller alike, have been separated off from it as a result of individual entanglements. And all disintegrate again, some faster some slower, and through differing kinds of courses" (Long & Sedley, p. 57).

Rival Stoic philosophers who believed in a fiery continuum and a finite world surrounded by an infinite void scoffed at the atheistic Epicurian view. Indeed, Cicero, in his dialogs of the first century B.C., which translated the Greek views into Latin, has his Stoic spokesman say:

> Does it not deserve amazement on my part that there should be anyone who can persuade himself that certain solid and invisible bodies travel through the force of their own weight and that by an accidental combination of those bodies a world of the utmost splendour and beauty is created? I do not see why the person who supposes this can happen does not also believe it possible that if countless exemplars of the twenty-one letters, in gold or any other material you like, were thrown into a container then shaken out onto the ground, they might form a readable copy of the *Annals* of Ennius. I'm not sure that luck could manage this even to the extent of a single line! (Long & Sedley, p. 328)

This may have been one of the earliest probability arguments about the formation of world structure. Related arguments dominate much of the present discussion on the subject, as we shall see.

But it was Lucretius, writing about the same time as Cicero, who came closest to some ideas of modern cosmogony. The atoms of the Universe are constantly in motion. Those whose shapes interweave and stick together form dense aggregates like stone or iron. The less dense atoms recoil and rebound (we would call it elastic scattering) over great distances to provide the thin air; some atoms do not join at all but wander everlastingly through the void. To create collisions that produce structure, Lucretius introduces the idea of swerve:

> On this topic, another thing I want you to know is this. When bodies are being borne by their own weight straight down through the void, at quite uncertain times and places they veer a little from their course, just enough to be called a change of motion. If they did not have this tendency to swerve, everything would be falling downward like raindrops through the depths of the void, and collisions and impacts among the primary bodies would not have arisen, with the result that nature would never have created anything. (Long & Sedley, p. 49)

He almost seems to be portraying gravitational deflection. But then Lucretius gets muddled up by the idea that everything must move at the same speed through the unresisting void, independent of its weight. Swerve must occur by just a minimum amount, so it cannot be seen. And so the idea dissolves into an ad hoc hypothesis to save the scheme.

Nevertheless, his scheme gave rise to a statistical argument often used today: "For so many primary particles have for an infinity of time past been propelled in manifold ways by impacts and by their own weight, and have habitually travelled, combined in all possible ways, and tried out everything that their union could create, that it is not surprising if they have also fallen into arrangements, and arrived at patterns of motion, like those repeatedly enacted by this present world" (Long & Sedley, p. 59). This is his rejoinder to Cicero's Stoic.

Lucretius' most remarkable reason for believing that the Universe was indeed infinite was to consider the difference this would make to the large-scale distribution of matter.

> Besides, if the totality of room in the whole universe were enclosed by a fixed frontier on all sides, and were finite, by now the whole stock of matter would through its solid weight have accumulated from everywhere all the way to the bottom, and nothing could happen beneath the sky's canopy, nor indeed could the sky or sunlight exist at all, since all matter would be lying in a heap, having been sinking since infinite time past. But as it is the primary bodies are clearly never allowed to come to rest, because there is no absolute bottom at which they might be able to accumulate and take up residence. At all times all things are going on in constant motion everywhere, and underneath there is a supply of particles of matter which have been travelling from infinity. (Long & Sedley, p. 45)

At last in this brief sketch of nearly two dozen cosmogonies we have found one that contains the seeds both of a rational explanation and of a result that we can put into modern terms. But this seed fell on infertile soil, for it was to take more than seventeen hundred years before Isaac Newton added the motive force of gravitation.

2

First Qualitative Physics:

The Newton–Bentley Exchange

> The aim of argument, or of discussion,
> should not be victory, but progress.
> Joseph Joubert, Pensées

Isaac Newton needs no introduction.

Richard Bentley was one of England's leading theologians, with strong scientific interests and very worldly ambitions. Eventually he became Master of Trinity College, Cambridge, reigning for forty-two contentious years. Tyrannical and overbearing, Bentley tried to reform the College (as well as the University Press) and spent much of the College's income on new buildings, including a small observatory. To balance the College accounts he reduced its payments to less active Fellows, while increasing his own stipend. After ten years of this, some of the Fellows rebelled and appealed to the Bishop of Ely and Queen Anne, the ultimate College authorities, to eject Bentley from the mastership. Various ruses enabled Bentley to put off the trial for another four years. Finally the Bishop condemned Bentley in a public court. But before he could formally deprive Bentley of his mastership, the Bishop caught a chill and died. Queen Anne died the next day. Bentley now put his theological talents to work to convince his opponents that he had won "victory" by divine intervention. So he retained the mastership and raised his salary still higher. Some years later, another attempt to expel him by a fresh Bishop also failed, and he remained Master until dying in 1742 at the age of eighty. During the crucial period of these collegiate upheavals, about 1708–1713, Bentley had Newton's firm support; simultaneously he was seeing the second edition of Newton's *Principia* through the University Press.

As a young man Bentley was asked, possibly through Newton's maneuvering, to give the first Robert Boyle lectures. Although now mainly known for his result that the pressure of a perfect gas is linearly proportional to its density at constant temperature, Boyle also left an endowment for lectures in defense of religion. Earlier, Bentley had studied much of the *Principia*, having obtained a list of preliminary readings in mathematics and astronomy from Newton. By late 1692, he had a few questions to ask Newton as he prepared the final manuscript of his eight Boyle Lectures.

These lectures were supposed to confute atheists and show that Natural Science still required a role for the Creator. In modern terms, the role that emerged was to provide initial conditions, boundary conditions, and divine interventions to account for those aspects of the Universe that could not be understood with the physics of Newton's day. Among other questions, Bentley asked why some matter in the Universe formed the luminous Sun and other matter became the dark planets, whether the motions of the planets could arise from natural causes, whether the lower densities of the outer planets were caused by their distance from the Sun, and what produced the inclination of the Earth's axis. The implication of these questions was that Newton's physics could offer no complete explanation, and so Newton, who was basically theistic, agreed that these phenomena were evidence for the deity.

In answering the first question, Newton (1692) made his famous prescient comments about the distribution of matter in the Universe:

> As to your first query, it seems to me that if the matter of our sun and planets and all the matter in the universe were evenly scattered throughout all the heavens, and every particle had an innate gravity toward all the rest, and the whole space throughout which this matter was scattered was but finite, the matter on the outside of the space would, by its gravity, tend toward all the matter on the inside, and by consequence, fall down into the middle of the whole space and there compose one great spherical mass. But if the matter was evenly disposed throughout an infinite space, it could never convene into one mass; but some of it would convene into one mass and some into another, so as to make an infinite number of great masses, scattered at great distances from one to another throughout all that infinite space. And thus might the sun and fixed stars be formed, supposing the matter were of a lucid nature. But how the matter should divide itself into two sorts, and that part of it which is fit to compose a shining body should fall down into one mass and make a sun and the rest which is fit to compose an opaque body should coalesce, not into one great body, like the shining matter, but into many little ones; or if the sun at first were an opaque body like the planets, or the planets lucid bodies like the sun, how he alone would be changed into a shining body whilst all they continue opaque, or all they be changed into opaque ones whilst he remains unchanged, I do not think explicable by mere natural causes, but am forced to ascribe it to the counsel and contrivance of a voluntary Agent.

Bentley wrote back questioning whether matter in a uniform distribution would convene into clusters because by symmetry there would be no net force, either on the central particle in a finite system or on all particles in an infinite system. And so, five weeks after his first letter, Newton replied in January 1693:

> The reason why matter evenly scattered through a finite space would convene in the midst you conceive the same with me, but that there should be a central particle so accurately placed in the middle as to be always equally attracted on all sides, and thereby continue without motion, seems to me a supposition as fully as hard as to make the sharpest needle stand upright on its point upon a looking glass. For if the very mathematical center of the central particle be not accurately in the very mathematical center of the attractive power of the whole mass, the particle will not be attracted equally on both sides. And much harder it is to suppose all the particles in an infinite space should be so accurately poised one among another as to stand still in a perfect equilibrium. For I reckon this as hard as to make, not one needle only, but an infinite number of them (so many as there are particles in an infinite space) stand accurately poised upon their points. Yet I grant it possible, at least by a divine power; and if they were once to be placed, I agree with you that they would continue in that posture without motion forever, unless put into new motion by the same power. When, therefore, I said

that matter evenly spread through all space would convene by its gravity into one or more great masses, I understand it of matter not resting in an accurate poise.

...a mathematician will tell you that if a body stood *in equilibrio* between any two equal and contrary attracting infinite forces, and if to either of these forces you add any new finite attracting force, that new force, howsoever little, will destroy their equilibrium and put the body into the same motion into which it would put it were those two contrary equal forces but finite or even none at all; so that in this case the two equal infinities, by the addition of a finite to either of them, become unequal in our ways of reckoning; and after these ways we must reckon, if from the considerations of infinities we would always draw true conclusions.

Thus Newton recognized, qualitatively, the main ingredients of galaxy clustering (although he had the formation of the stars and solar system in mind): a grainy gravitational field produced by discrete objects, a multiplicity of cluster centers in an infinite (we would now say unbounded) universe, the impotence of the mean gravitational field in a symmetric distribution, and the finite force produced by perturbing an equilibrium state. Strangely, it would take nearly 300 years to quantify these insights and compare their results with the observed galaxy distribution. The difficulty was understanding nonlinear many-body gravitational clustering, whose delightful subtleties are all the more surprising for being based on a simple inverse square interaction.

Newton himself might have been able to begin the quantitative analysis. Indeed, perhaps he did. In an intriguing essay, Harrison (1986) suggests that Newton estimated the timescale $(G\rho)^{-\frac{1}{2}} \sim 10^8$ years for gravitational instability, but it conflicted so strongly with the biblical timescale since the flood ($\sim$5,000 years), and with his own estimate in the *Principia* for the cooling of the Earth from a globe of red-hot iron ($\sim$50,000 years), that he thought it wiser not to publish. This could be behind his allusion at the end of the first letter to Bentley: "There is yet another argument for a Deity, which I take to be a very strong one; but till the principles on which it is grounded are better received, I think it more advisable to let it sleep." However, Westfall (1980) thinks this remark may refer to the development of history as foretold in the prophecies. We'll never know unless someone rediscovers old documents or finds a hidden message in Newton's writings.

Newton's insights had their precursors. We saw earlier that Lucretius recognized the essentially different nature of clustering in finite and infinite universes. Although most of the medieval church scholastics who succeeded him were mainly interested in reconciling the natural world with a literal interpretation of the Bible, some of the ancient Greek knowledge and independence of mind survived (Dreyer, 1905; Duhem, 1985) at a few scattered monasteries from Bremen in Germany to Iona in the Inner Hebrides. Nicholas of Cusa (translation 1954) realized in the fifteenth century that if the universe did not have boundaries, it would have neither a fixed pole nor a center. Every point would be an equivalent center. This suggests Newton's view that many centers of clustering grow in an infinite universe. How much Newton knew of the earlier work is unclear, although in the *Scholium* of the Principia he used a straightforward example similar to one of Cusa's for discussing absolute motion. Thomas Diggs, one of England's leading mathematicians a century before Newton, had strongly promoted the idea of an infinite universe of stars surrounding a Copernican solar system, an idea also associated in a more mystical context with Giordano Bruno about the same time. But these were just pictures of the world until Newton sketched a dynamical explanation.

Fifty years after Newton, the idea of an infinite universe filled with structure whose primary cause was gravitation began to consolidate. Thomas Wright (1750) proposed that the distribution of light in the band of the Milky Way showed that the Sun lies in a great disk of stars at considerable distance from its center. The idea impressed Immanual Kant (1755) who noted: "It was reserved for an Englishman, Mr Wright of Durham, to make a happy step with a remark which does not seem to have been used by himself for any very important purpose, and the useful application of which he has not sufficiently observed. He regarded the Fixed Stars not as a mere swarm scattered without order and without design, but found a systematic contribution in the whole universe and a universal relation of these stars to the ground-plan of the regions of space which they occupy." If there was one Milky Way, Kant reasoned, why could there not be many? He even identified them as the elliptical nebulae that the French philosopher Maupertuis thought were enormous rotating single stars. Then Kant went on to suggest that even the distribution of many Milky Ways may be structured as part of a grand hierarchy of systems:

> The theory which we have expounded opens up to me a view into the infinite field of creation . . . . If the grandeur of a planetary world in which the Earth, as a grain of sand, is scarcely perceived, fills the understanding with wonder; with what astonishment are we transported when we behold the infinite multitude of worlds and systems which fill the extension of the Milky Way. But how is this astonishment increased, when we become aware of the fact that all these immense orders of star-worlds again form but one of a number whose termination we do not know, and which perhaps like the former, is a system inconceivably vast – and yet again but one member in a new combination of numbers! We see the first members of a progressive relationship of worlds and systems; and the first part of this infinite progression enables us already to reorganize what must be conjectured of the whole. There is no end but an abyss of a real immensity, in presence of which all capability of human conception sinks exhausted, although it is supported by the aid of the science of number.

This happy interplay among observations of the nebulae, Kant's essentially correct but unsupported interpretation of them as galaxies, and his speculative extrapolation to even larger scales of clustering was held together by the attraction of universal gravity. Kant also recognized that centrifugal forces were necessary for equilibrium: "The attraction which is the cause of the systematic constitution among the fixed stars of the Milky Way acts also at the distance even of those worlds, so that it would draw them out of their positions and bury the world in an inevitably impending chaos, unless the regularly distributed forces of rotation formed a counterpoise or equilibrium with attraction, and mutually produced in combination that connection which is the foundation of the systematic constitution."

At this stage Kant goes astray. Guided too strongly by analogies with the solar system, he suggests that all these Milky Ways and higher systems move around a common center despite his agreement with Cusa's view that an infinite universe has no center. To reconcile these views, he proposes a remarkably modern cosmogony:

> Let us now proceed to trace out the construction of the Universal System of Nature from the mechanical laws of matter striving to form it. In the infinite space of the scattered elementary forms of matter there must have been some one place where this primitive material had been most densely accumulated so as through the process of formation that was going on predominantly there, to have procured for the whole Universe a mass which might serve as its fulcrum. It indeed holds true that in an infinite space no point can properly have the privilege to be called the center; but by means of a certain ratio, which is founded upon the essential degrees of the density of primitive matter, according to which at its creation it is accumulated more densely in a certain place and increases in its dispersion with the distance from it, such a point may have the privilege of being called the center; and it really becomes this through the formation of the central mass by the strongest attraction prevailing in it. To this point all the rest of the elementary matter engaged in particular formations is attracted; and thereby, so far as the evolution of nature may extend it makes in the infinite sphere of creation the whole universe into only one single system.
>
> ... the creation, or rather the development of nature, first begins at this center and, constantly advancing, it gradually becomes extended into all the remoter regions, in order to fill up infinite space in the progress of eternity with worlds and systems .... Every finite period, whose duration has a proportion to the greatness of the work to be accomplished, will always bring a finite sphere to its development from this center; while the remaining infinite part will still be in conflict with the confusion and chaos, and will be further from the state of completed formation the farther its distance is away from the sphere of the already developed part of nature.

If only Kant had taken the next bold step and realized that there could be many equivalent centers of developing structure, he would have been the very model of a modern astrophysicist. He could have reconciled the geometric property that all points in an infinite universe are equivalent to the center (i.e., none are geometrically distinguished) with the structural property that any point is the center of an equivalent structure. To do this, however, he would have had to take the short but subtle step to the idea of statistical homogeneity. This idea supposes that by sampling a large number of well-separated regions, that is, regions without common structure, one obtains average statistical properties (such as density, correlation functions, distribution functions) that are independent of the actual sample. Any one or several regions on any scale may have peculiar configurations of galaxies, but these configurations will have different positions, orientations, and forms from region to region and so give common averages over samples with many regions. In this sense, all points would be statistically equivalent with respect to large-scale structure. The initial conditions needed to produce Kant's form of large-scale structure could have been present at many places, remote from each other, instead of at just one as Kant thought. Had Kant known of Newton's unpublished first letter to Bentley and combined Newton's insight into multiple clustering with his own extrapolation of Wright's Milky Way structure, he just might have realized all the essential aspects of our modern understanding of galaxy clustering.

As it happened, Kant's views survived for over a century. The questions he addressed had excited discussion in fashionable French salons ever since Fontenelle's (1686) *Conversation on the Plurality of Worlds* first popularized the new astronomy. Now a much later but timely English edition (1769) reinforced this interest:

> I confess it Madam; all this immense mass of matter, which composes the universe, is in perpetual motion, no part of it excepted; and since every part is moved, you may be sure that changes must happen sooner or later; but still in times proportional to the effect. The ancients were merry gentlemen to imagine that the celestial bodies were in their own nature unchangeable, because they observed no alteration in them; but they did not live long enough to confirm their opinion by their own experience; they were boys in comparison of us.

Perhaps Fontenelle was anticipating that he would himself live to be a hundred. Three decades later, Lambert (1800) thought he had identified Kant's center of motion of the Universe: It was the Orion nebula!

If ever astronomy needed further observations to get off the wrong track, this was the time. Newton and Kant had posed questions of the motion and distribution of matter in terms which, for the first time, could be interpreted quantitatively. There was no chance in the late eighteenth century of measuring motions at great distances, but Herschel was beginning to chart the nebulae and provide the next great clue.

3

Glimpses of Structure

For now we see through a glass, darkly.
1 Corinthians 13:12

Stubbornness, stamina, boldness, and luck enabled William Herschel to connect our Universe with Newton's and Kant's speculations. Leaving Hanover in 1757 after the French occupation, he settled in England as an itinerant teacher, copier, and composer of music, becoming organist of the Octagon Chapel at Bath in 1766. But his real interest from childhood was astronomy. He privately built a succession of larger and larger reflecting telescopes and systematically swept the heavens. His sister, Caroline, emigrating in 1772, helped with these nightly observations, to the eventual destruction of her own singing career. In 1781, Herschel had the great luck to find Uranus, the first planet discovered since the days of the ancients, although he originally thought it was just a comet. Fame followed quickly, and fortune soon after when George III granted him a pension for life. He supplemented this by building small telescopes for sale (until his wealthy marriage in 1788) and became a full-time astronomer. Career paths, like the subject itself, have changed considerably since then.

For twenty years, starting in 1783, Herschel searched for nebulae with his 20-foot telescope and its $18\frac{7}{10}$ inch speculum mirror. Messier's catalog, available in 1781, had inspired him first to try to resolve known nebulae with his superior telescope, and then to discover more. Originally, he believed all nebulae were star clusters and claimed to have resolved M31, the Andromeda galaxy, into stars. Nearly three decades later after he had discarded this belief, he also dropped the claim (Hoskin, 1963). This was to be the first of several times that Andromeda would mislead astronomers.

Although the nature of the nebulae was unknown, their projected positions at least were straightforward. Messier (1781) noticed that their distribution on the sky was irregular; 13 of the 103 in his catalog are in the Virgo constellation. As it happens, all 13 are galaxies and so Messier was first to see a cluster of galaxies. The Virgo cluster with thousands of members is one of the largest and most varied of those nearby.

Soon after, in his more extensive catalog, Herschel (1784) found the Coma Cluster with its "many capital nebulae" and noticed other inhomogeneities and voids:

In my late observations on nebulae I soon found, that I generally detected them in certain directions rather than in others: that the spaces preceding them were generally quite deprived of their stars, so as often to afford many fields without a single star in it; that the nebulae generally appeared some time after among stars of a certain considerable size, and but seldom among very small stars, that when I came to one nebula, I generally found several more in the neighborhood; that afterwards a considerable time passed before I came to another parcel; and these events being often repeated in different altitudes of my instrument, and some of them at a considerable

17

distance from each other, it occurred to me that the intermediate spaces between the sweeps might also contain nebulae; and finding this to hold good more than once, I ventured to give notice to my assistant at the clock, "to prepare, since I expected in a few minutes to come at a stratum of the nebulae, finding myself already" (as I then figuratively expressed it) "on nebulous ground." In this I succeeded immediately; so that I now can venture to point out several not far distant places, where I shall soon carry my telescope, in expectation of meeting with many nebulae. But how far these circumstances of vacant places preceding and following the nebulous strata, and their being as it were contained in a bed of stars, sparingly scattered between them, may hold good in more distant portions of the heavens, and which I have not yet been able to visit in any regular manner, I ought by no means to hazard a conjecture. The subject is new and we must attend to observations and be guided by them, before we form general opinions.

Part of this patchiness, we know now, is caused by interstellar obscuration and part is intrinsic.

Seven months later, Herschel (1785) was ready to announce his conjectures and general opinions:

By continuing to observe the heavens with my last constructed, and since that time much improved instrument, I am now enabled to bring more confirmation to several parts that were before but weakly supported, and also to offer a few still further extended hints, such as they present themselves to my present view. But first let me mention that, if we would hope to make any progress in an investigation of this delicate nature, we ought to avoid two opposite extremes, of which I can hardly say which is the most dangerous. If we indulge a fanciful imagination and build worlds of our own, we must not wonder at our going wide from the path of truth and nature; but these will vanish like the Cartesian vortices, that soon gave way when better theories were offered. On the other hand, if we add observation to observation, without attempting to draw not only certain conclusions, but also conjectural views from them, we offend against the very end for which only observations ought to be made. I will endeavour to keep a proper medium; but if I should deviate from that, I would wish not to fall into the latter error.

That the milky way is a most extensive stratum of stars of various sizes admits no longer of the least doubt; and that our sun is actually one of the heavenly bodies belonging to it is as evident. I have now viewed and gaged this shining zone in almost every direction, and find it composed of stars whose number, by the account of these gages, constantly increases and decreases in proportion to its apparent brightness to the naked eye. But in order to develop the ideas of the universe, that have been suggested by my late observations, it will be best to take the subject from a point of view at a considerable distance both of space and of time.

Theoretical view

Let us then suppose numberless stars of various sizes, scattered over an infinite portion of space in such a manner as to be almost equally distributed throughout the whole. The laws of attraction, which no doubt extend to the remotest regions of the fixed stars, will operate in such a manner as most probably to produce the following remarkable effects.

Formation of Nebulae

Form I. In the first place, since we have supposed the stars to be of various sizes, it will frequently happen that a star, being considerably larger than its neighboring ones, will attract them more than they will be attracted by others that are immediately around them; by which means they will be, in time, as it were, condensed about a center; or, in other words, form themselves into a cluster of stars of almost a globular figure, more or less regularly so, according to the size and original distance of the surrounding stars. The perturbations of these mutual attractions must undoubtedly be very intricate, as we may easily comprehend by considering what Sir Isaac Newton says in the first book of his Principia, in the 38th and following problems; but in order to apply this great author's reasoning of bodies moving in ellipses to such as there are here, for a while, supposed to have no other motion than what their mutual gravity has imparted to them, we must suppose the conjugate axes of these ellipses indefinitely diminished, where the ellipses will become straight lines.

Form II. The next case, which will also happen almost as frequently as the former, is where a few stars, though not superior in size to the rest, may chance to be rather nearer each other than the surrounding ones; for here also will be formed a prevailing attraction in the combined center of gravity of them all, which will occasion the neighboring stars to draw together; not indeed so as to form a regular globular figure, but however in such a manner as to be condensed towards the common center of gravity of the whole irregular cluster. And this construction admits of the utmost variety of shapes, according to the number and situation of the stars which first gave rise to the condensation of the rest.

Form III. From the composition and repeated conjunction of both the foregoing forms, a third may be derived, when many large stars, or combined small ones, are situated in long extended, regular, or crooked rows, hooks, or branches; for they will also draw the surrounding ones, so as to produce figures of condensed stars coarsely similar to the former which gave rise to these condensations.

Form IV. We may likewise admit of still more extensive combinations; when, at the same time that a cluster of stars is forming in one part of space, there may be another collecting in a different, but perhaps not far distant quarter, which may occasion a mutual approach towards their common center of gravity.

V. In the last place, as a natural consequence of the former cases, there will be formed great cavities or vacancies by the retreat of the stars towards the various centers which attract them; so that upon the whole there is evidently a field of the greatest variety for the mutual combined attractions of the heavenly bodies to exert themselves in.

In a paper whose abstract is its title, Herschel (1811) later illustrated some shapes of the nebulae that had led him to these conclusions. Figure 3.1 shows Herschel's drawings based on the visual appearance of these nebulae through his telescopes. Many of our modern classifications are here, apart from the spirals, which could not be resolved; with hindsight we can look for hints of their structure in these sketches. It was this wide range of patterns that prompted Herschel and many subsequent astronomers to propose that we were seeing sequential evolution.

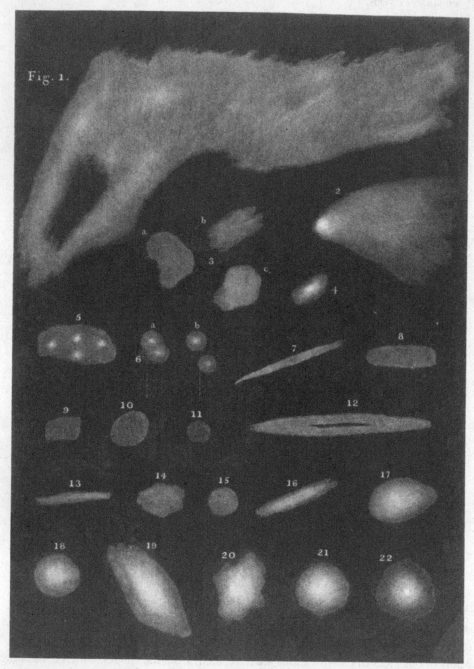

Fig. 3.1. W. Herschel's (1811) sketches of some of the nebulae in his catalog.

Fig. 3.1. (*Continued*)

So there it is. Herschel anticipated many qualitative details of gravitational clustering. And if he thought of stars rather than galaxies as the fundamental units of clustering, that was only because stars were the largest well-established objects in the astronomy of his day. Not that he hesitated to extrapolate. For he went on to contend that some groups of nebulae form by fragmentation as well. In later years, as Herschel's telescopes and observations improved, he became more confused. Planetary nebulae, reflection nebulae, globular clusters, supernova remnants, and regions of star formation all had to compete with galaxies for a limited scope of interpretation. Despite the unknown relative distances and natures of different nebulae, their distribution and Newton's *Principia* had led Herschel to the central ideas of gravitational clustering.

Following Herschel came a period of consolidation. His son, John, compiled a catalog of southern nebula in 1847. Together with his 1833 catalog of the northern hemisphere he had nearly 4,000 objects. These provided grist for the statistical mills of the next four decades, until, in 1888, Dreyer's *New General Catalog* with 7,840 objects incorporated and superseded Herschel's work. Dreyer supplemented the NGC with his two Index Catalogs of 1895 and 1908, adding another 5,386 objects and closing one epoch of discovery.

The first great debate to issue from these catalogs was whether nebulae were part of the Milky Way, or formed a separate local system, or were very distant. It was even conceivable – although proponents of "Occam's Razor" might have scoffed – that there were different types of nebulae and all three possibilities were true. John Herschel (1847) began the debate by devising an equal area (isographic) projection of the celestial sphere onto a disk in a plane and counting nebulae in areas of 3° polar distance and 15′ right ascension. His statistical distribution confirmed the general inhomogeneity and showed the Virgo Cluster with its southern extension, as well as several superclusters (e.g., Pisces). "The general conclusion which may be drawn from this survey, however, is that the nebulous system is distinct from the sidereal, though involving, and perhaps, to a certain extent, intermixed with the later." A rather ambiguous, but essentially correct, conclusion. Herschel continues on to discuss how the nebulae divide "into two chief strata, separated (apparently) from each other by the galaxy." The following year, Nichol (1848) wrote about "superb groups of galaxies separated from each other by gulfs so awful, that they surpass the distance that divide star from star . . . . Amid this system of clusters, floats the galaxy whose glories more nearly surround us." The idea that some of the nebulae might be rather special was becoming stronger. Astronomers were beginning to call them "galaxies."

Little could be decided, however, until distinctions between nebulae were better understood. Lord Rosse's (1850) mammoth 53-foot telescope with its 6-foot speculum mirror resolved some of the nebulae into spirals – a new category and a major clue. Another major discovery that differentiated nebulae was William Huggins's (1864) measurement of their spectra. Some such as Orion, the Crab, and the "planetary nebulae" were mainly gaseous; others remained mysterious. It would take some time for the meaning of Huggins's spectra to become clear. Meanwhile Alexander von Humboldt (1860), intrepid explorer of the earth and the heavens, could forecast:

> If these nebulous spots be elliptical or spherical sidereal [stellar] groups, their very conglomeration calls to mind the idea of a mysterious play of gravitational forces by which they are governed. If they be vapory masses, having one or more nebulous nuclei, the various degrees of their condensation suggest the possibility of a process of gradual star formation from inglobate matter. No other cosmical structure . . . is, in like degree, adapted to excite the imagination, not merely as a symbolic image of the infinitude of space, but because the investigation of the different conditions of *existing things*, and of their presumed connection of sequences, promises to afford us an insight into the laws of *genetic development*.

Cleveland Abbe (1867), a twenty-nine year old American astronomer who would later turn to meteorology, took the next step. He divided Herschel's 1864 "General Catalog of Nebulae and Clusters of Stars" with about 5,079 entries (a few being redundant) into ordinary clusters, and globular clusters, and resolved and unresolved nebulae. He concluded that the planetary and annular nebulae – primarily gaseous according to Huggins's spectra – had the same distribution as the clusters and were part of the Milky Way. The other resolved and unresolved nebulae, counted in cells of 0.5° in right ascension and 10° in declination, were outside the Milky Way. At first he thought their comparative absence around the Milky Way might be an illusion caused by the glare of starlight. But since it remained true at large distances from the Milky Way, and increasingly powerful telescopes failed to find many new nebulae near the Milky Way, he concluded the effect was real. His explanation was that the plane of the Milky Way "cuts nearly at right angle the axis of a prolate ellipsoid, within whose surface all the visible nebulae are uniformly distributed."

Two years later, R. A. Proctor (1869) joined the debate, concluding from the same data that the nebulae were part of the Milky Way. How could he do this? From Abbe's tables of counts he constructed an isographic (equal area) projection (initially apparently without realizing that J. Herschel had developed this earlier). The resulting maps of counts – not actual positions – looked similar to Herschel's, despite Abbe's larger and more differentiated data set. This similarity led Proctor to the dangerous conclusion "that no extension of telescope observation can appreciably affect our views respecting the distribution of the nebulae." Proctor saw the same zone of avoidance of the nebulae, but he decided that its coincidence with the Milky Way was too unlikely to be accidental. Therefore "the nebular and stellar systems are parts of a single scheme." He considers the possibility "that there is some peculiarity in the galactic stratum preventing us from looking so far out into space along its length than elsewhere." We now recognize this as obscuration by interstellar dust. Proctor dismissed this possibility because he could see both nebulae and stars intermixed in the Magellanic Clouds. He does admit, however, that a few nebulae, perhaps the spirals, might be external galaxies. In correspondence with the seventy-seven year old John Herschel, Proctor elicited Herschel's vision of a hierarchical universe

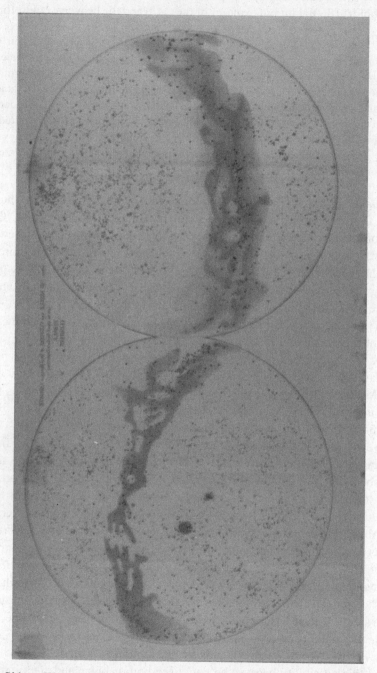

Fig. 3.2. Sidney Waters's (1873) equal area map of the positions of nebulae in Herschel's catalog.

where each level contains miniatures of itself (Hoskin, 1987). Unlike his father, John Herschel seldom speculated, but this speculation was a forerunner of the idea of "self-similar" clustering.

It was left to an amateur astronomer, Sidney Waters (1873), to produce the first accurate equal area projection of the actual positions in Herschel's catalog. Figure 3.2 shows this map, which the Royal Astronomical Society published as a three-color lithograph. Objects are again divided into star clusters, resolvable nebulae, and unresolved nebulae. Waters sees the same patterns as his predecessors, though perhaps more clearly, and concludes like Proctor that they "surely prove beyond question that not only are the clusters, which are peculiar to the Milky Way, related to the nebulae, which seem to form a distinct scheme, but that the two schemes are probably subordinate parts of our sidereal system."

And so the debate wandered inconclusively back and forth for another half century. Apparent ambiguities in the evidence were, not for the last time, often bypassed in favor of firm opinions. Collecting some of these views, the English astronomer Agnes Clerke (1890) popularized them in an even more intimidating fashion:

> The question whether nebulae are external galaxies hardly any longer needs discussion. It has been answered by the progress of discovery. No competent thinker, with the whole of the available evidence before him, can now, it is safe to say, maintain any single nebula to be a star system of coordinate rank with the Milky Way. A practical certainty has been attained that the entire contents, stellar and nebular, of the sphere belong to one mighty aggregation, and stand in ordered mutual relations within the limits of one all-embracing scheme – all-embracing, that is to say, so far as our capacities of knowledge extend. With the infinite possibilities beyond, science has no concern.

Without new physical understanding of the nebulae, astronomers could do little except refine the catalogs, reclassify the objects, remap the distributions, and rediscuss their qualitative impressions. Occasionally a new cluster was noticed, superclusters were found including Perseus–Pisces (Stratonoff, 1900) and the Local Supercluster (Easton, 1904; Reynolds, 1920, 1923, 1924; see also de Vaucouleurs, 1989), spiral nebulae were identified and mapped (Hardcastle, 1914; Hinks, 1914; Reynolds, 1920, 1923, 1924), and the dependence of the distribution on galactic latitude and longitude discussed (Sanford, 1917; Lundmark, 1920). But the general sleepiness of the subject might be characterized by Barnard's (1906) observations of groups of galaxies, which he wrote up, placed in an envelope addressed to the journal, and then forgot to put in the mail for nine years.

4
Number Counts and Distributions

I could be bounded in a nutshell and count
myself a king of infinite space.
Shakespeare, Hamlet

If, like Hamlet, you count yourself king of
an infinite space, I do not challenge your
sovereignty. I only invite attention to certain
disquieting rumours which have arisen
as to the state of Your Majesty's Nutshell.
Eddington

One dominant stroke transformed thousands of years of increasingly refined speculation on the structure of our Universe into fact. Hubble (1925a,b,c, 1926, 1929a,b)
clinched the extragalactic nature of the "white nebulae" by discovering their Cepheid
variable stars. This vastly expanded the known distance scale.

Cepheids are unusually bright stars that pulsate with regular periods ranging from
about 10 to 30 days. (The first one was found in the constellation Cepheus in the Milky
Way.) Their crucial property is the relation between a Cepheid's period and its peak
intrinsic luminosity, recognized in 1908 (Leavitt, 1912; Hertzsprung, 1913). Brighter
Cepheids have longer periods. From the observed periods of Cepheids in nebulae,
Hubble could obtain their intrinsic luminosity and thus find the distance from their
observed apparent luminosity. The main uncertainty was in calibrating their period–
luminosity relation from the independently known distances of Cepheids in our
Galaxy. Early calibrations turned out to be wrong, mainly because there are different types of Cepheids, which give somewhat different period–luminosity relations.
(Occam's Razor fails again.) These later corrections were substantial (a factor of
about two) but not qualitatively significant (Baade, 1963).

Different types of Cepheids illustrate the caution needed when applying Occam's
Razor to astronomy. It failed at least three times for Herschel (standard candle
luminosities for stars, all nebulae essentially the same objects, Uranus originally
thought more likely to be an ordinary common comet rather than a new unexpected
planet), once for Hubble (all Cepheids the same) and would be destined to mislead
on many later occasions. The Universe is far more various and heterodox than we are
apt to imagine. Moreover, Occam's Razor, naively applied, is just an empty slogan.
He said (see Russell, 1945) "It is in vain to do with more what can be done with
fewer." This means that scientific explanations should not invoke more hypotheses
than necessary. As logic it is very sensible, but it ignores the empirical question of
whether fewer hypotheses will in fact do. Nor does it address the problem that some
hypotheses are stronger than others and their implications may only partially overlap.

Controversy over whether nebulae were outside our Galaxy hinged on whether new hypotheses were necessary. Ultimately the answer is empirical. Before Hubble found Cepheids, evidence pointed both ways. There were Slipher's (1913, 1915) spectra showing Doppler shifts in the nebular absorption lines, which could be interpreted as radial velocities much larger than any in our Galaxy if no new physical hypothesis for the Doppler shift was necessary. In opposition, van Maanen claimed to observe an angular rotation of spiral nebulae. If these nebulae were at extragalactic distances, their outer arms would be moving faster than the speed of light. Many astronomers were convinced by this evidence, which turned out to be spurious (see Hetherington, 1988, for a detailed account). However, the discoveries by Ritchey and Curtis (see Shapley, 1917) of novae in the nebulae seemed to favor the extragalactic interpretation, provided the novae had the same absolute (intrinsic) brightness as those in our own Galaxy, i.e., provided no new hypothesis was necessary. But they did not. Some were supernovae. If these extra-bright novae in the nebulae were identified with ordinary novae in our Galaxy, the nebulae would seem closer. (Shapley suggested that ordinary novae were stars running into gas clouds.) A third aspect of this controversy swirled around whether individual stars had been resolved in the outer parts of the M31 (Andromeda) and M33 nebulae. They had, but the images looked soft; so perhaps they were not stars after all but something else (Shapley, 1919; Lundmark, 1921). Was a new hypothesis necessary here? The confused situation was summarized by Shapley (1919) and Curtis (1919) as a prelude to their famous quasi-debate at the National Academy of Sciences in 1920 (see Shapley, 1921; Curtis, 1921). Curtis correctly concluded that the nebulae were extragalactic, but he failed to convince many astronomers. Shapley gave apparently better arguments for the wrong conclusion. Öpik (1922) argued that the Andromeda nebula was extragalactic, based on a relation between its observed rotational velocity and its absolute magnitude. So the scene had been set for Hubble's discovery.

Once the nebulae were clearly known to be cosmological, the question of their distribution quickened and took on new life.

Hubble, this time, was first off the mark. By 1926 he realized that the Mt. Wilson photographic plates taken earlier were unsuitable for galaxy counting. Their telescopes, emulsions, exposure times, development, zenith distances, atmospheric seeing, and positions on the sky were too varied and irregular. So he began a major program using the 60″ and 100″ telescopes to provide a large and uniform survey for galaxies brighter than about 20th magnitude. Five years later he had preliminary results based on 900 plates with 20,000 galaxies (Hubble, 1931). The "zone of avoidance" found by the Herschels – a band of sky along the Milky Way where interstellar obscuration reduces the apparent number of galaxies – was delineated more precisely than ever before. Its width varies irregularly, usually from 10 to 40 degrees as a function of galactic longitude. In the direction of our galactic center there is partial obscuration to latitudes of $\pm 40°$, but much less toward our galactic anticenter. Outside the zone of avoidance the distribution appeared roughly uniform with occasional large clusters. Correlations of number counts with exposure times supported the view that the galaxies were uniformly distributed in depth, and Hubble calculated a mean density $\bar{\rho}_0$ of about 10^{-30} g cm^{-3} for luminous matter in space. Through a fortuitous approximate cancellation of errors in the absolute magnitude and estimated average mass of a galaxy, he came close to the modern value for $\bar{\rho}_0$.

With so many galaxies in his homogeneous sample, Hubble could calculate their distribution function $f(N)$, the probability for finding N galaxies in a given size area on the sky. He looked only at large areas, with more than about 25 galaxies, and in this regime found that $f(N)$ had a lognormal distribution. The distribution function would become an important but long neglected clue to understanding the dynamical clustering of galaxies; we shall return to it often in this book.

Hubble had competition. Harlow Shapley, who did not get along well with him, had previously left the Mt. Wilson Observatory staff and started his own galaxy distribution program at Harvard. This was based on an extensive survey of about 100,000 galaxies brighter than 18.2^m taken with a 24-inch refractor at the Harvard Boyden Station. The photographic plates, some dating from the beginning of the century, were a much more varied lot than Hubble's, but Shapley (1933) thought he could remove the effects of different emulsions, exposure times, seeing, etc. by comparing the limiting magnitudes of stars on each plate to obtain a normalized galaxy count. While agreeing on the zone of avoidance, he found galaxies outside this zone much less homogeneous than Hubble had claimed. According to Shapley (1931) ". . . the irregularities in apparent distribution are real and indicate actual groupings of external galaxies. The irregularities are obviously too pronounced to be attributed to chance; they are rather a demonstration of evolutionary tendencies in the metagalactic system." Soon Hubble (1936a,b) would agree, but echoes of this controversy would be heard throughout astronomy on larger and larger length scales until the late 1960s.

Meanwhile, under Hubble's instigation, the Lick Observatory was making a third great survey of the galaxy distribution. Mayall (1934) combined 184 plates of his own taken from 1923 to 1933 with 56 taken by Keeler and Perrine around the turn of the century and 249 by Curtis around 1917, all on the Crossley 36.5-inch reflector. He collected about 15,000 galaxies and concluded, like Hubble, that they were consistent with a uniform spatial distribution and had a lognormal $f(N)$. The same year Hubble (1934) published the detailed analysis of his complete number counts of about 44,000 galaxies brighter than 20th magnitude on 1,288 plates covering 2% of the three fourths of the sky north of $-30°$ declination. He concluded: "There are as yet no indications of a super-system of nebulae analogous to the system of stars. Hence, for the first time, the region now observable with existing telescopes may possibly be a fair sample of the universe as a whole." He was, of course, aware of the large groups with hundreds of galaxies, but thought them exceptional, containing only about 1% of all galaxies in his sample:

> When groups are included, the clustering tendency becomes more appreciable and significant, especially when considered in connection with the systematic advance in average type from the dense clusters to the random nebula.

On the grand scale, however, the tendency to cluster averages out. The counts with large reflectors conform rather closely with the theory of sampling for a homogeneous population. Statistically uniform distribution of nebulae appears to be a general characteristic of the observable region as a whole.

Hubble was confused. He evidently failed to distinguish between random (Poisson) clustering, in which there were no correlations, and a statistically uniform distribution, which, nevertheless, could be correlated. In the latter case, the correlations appear on average to be the same when viewed from any point in the system. This confusion increased when Shapley (1934a,b) found that the ratios of galaxy numbers in the northern to southern parts of his Harvard survey were large for bright galaxies but nearly unity for faint ($m \gtrsim 17.6$) galaxies. He concluded that this implied a large-scale gradient in the "metagalactic system" – a supersystem of galaxies. It contradicted Hubble's view, but only out to scales dominated by 18th magnitude galaxies. So perhaps the difference in their views was only a matter of scale? Shapley's results might be thought a prelude to modern superclustering of galaxies, but the plates later proved too heterogeneous to give accurate conclusions.

Bok (1934) attempted to clarify the situation. He simply and perceptively emphasized how very different $f(N)$ was from a random Poisson or normal (for large average N) distribution on all available scales. He used the Shapley–Ames catalog for galaxies brighter than 13th magnitude, the Harvard survey, which went to 18.2 magnitude, and then Hubble's own survey. Figure 4.1 shows Bok's graph for the Hubble survey. Of course, this is nothing more than Hubble's own lognormal $f(N)$ distribution, but it definitely shows we are not dealing with "random nebulae" even at the faint magnitudes and large scales that Hubble thought represented a "fair sample."

It is not clear how Hubble reacted to this. But the following year he changed his view somewhat, without referring to Bok. In the Silliman Lectures, later published as *The Realm of the Nebulae*, Hubble (1936a) strongly emphasizes (for the first time?) the difference between a Gaussian normal distribution for N, the number of galaxies per sample, and such a distribution with log N as its variable. He seems to regard the transformation from N to log N as a well-known mathematical device to remove the asymmetry of the curve, an asymmetry "presumably associated with the tendency of nebulae to cluster." Then he continues with a statement that might have been Newton's reply to a more modern Bentley:

> It is clear that the groups and clusters are not superimposed on a random (statistically uniform) distribution of isolated nebulae, but that the relation is organic. Condensations in the general field may have produced the clusters, or the evaporation of clusters may have populated the general field. Equations describing the observed distribution can doubtless be formulated on either assumption, and, when solved, should contribute significantly to speculations on nebular evolution.

Finding the equations Hubble suggested would take nearly half a century; we will meet them in Sections 15.2 and 27.2.

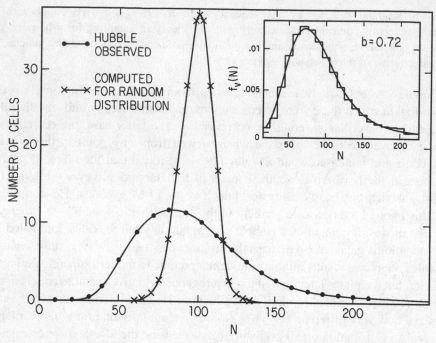

Fig. 4.1. The unnormalized frequency distribution of Hubble's galaxy counts compared to a random Gaussian distribution, after Bok (1934). The inset shows a recent fit of Bok's points to the distribution of Equation 27.24.

Still, Hubble continued to think there should be some large scale on which clustering is random, even if he had not yet discovered it. "Samples which, on the average, are large compared with a single cluster, should tend to conform with the theory of random sampling. The frequency-distributions of N per sample should approximate normal error curves. Smaller samples should give unsymmetrical frequency-distributions of N, with an excess of thinly populated fields." This appears essentially correct, but again its discovery was to take half a century (see Section 15.4 and Chapter 33).

In a subsequent and rather strange paper, mainly trying to show that galaxy redshifts are not velocity shifts and the Universe is not expanding significantly, Hubble (1936b) also elaborated his view of the distribution function:

While the large-scale distribution appears to be essentially uniform, the small-scale distribution is very appreciably influenced by the well-known tendency toward clustering. [Here, again, the idea of a uniformly clustered distribution had not occurred to Hubble. He did not distinguish between a uniformly correlated and a uniformly uncorrelated state, only between a uniform state and an irregular state.] The phenomena might be roughly represented by an originally uniform distribution from which the nebulae have tended to gather around various points until now they are found in all stages from random scattering, through groups of various sizes, up to the occasional great clusters.[10] The tendency appears to operate on a relatively modest scale − at least no clusters are known with more than a few hundred members; hence irregularities tend to average out among samples that are large compared to a single cluster.[11] In small samples irregularities are conspicuous, and, as an empirical fact, they lead to a random distribution of log N rather than of N.

The two footnotes in this paragraph are significant. The second one states "[11]Samples in the regions of great clusters were rejected in the present surveys; and, for this reason, irregularities tend to average out for samples that are large compared to single groups of nebulae." Thus Hubble had confused the "uniformity" implied by finding the same average values of log N and its dispersion in different regions of space with the much stricter requirement that the distribution be Poisson (i.e., "random" in his terminology). It is a confusion between a description in terms of "clusters" and a description in terms of "clustering"; we shall consider this problem further in Chapter 6.

The earlier footnote refers to dynamics in the spirit of Herschel, but it presents an opposite view from Newton's: "[10]This representation is purely formal and has no genetic implications. For purposes of speculation, the reverse development seems preferable, namely, that nebulae were formed in great clusters whose gradual evaporation has populated the general field." Like Herschel, Hubble imagined the observed distribution might be formed either by clustering or by fragmentation and dispersal.

Genetic implications? Did Hubble harken back to von Humboldt here (see Chapter 3), or had he something else in mind? The latter possibility arose in a conversation with Allan Sandage (Cambridge, England, August 1986) where we speculated on what originally inspired Hubble to fit a lognormal curve to his data. Hubble's statement (above) that the relation between clusters and isolated nebulae is "organic" might be a clue. At Cal Tech he was a sociable man about campus with many friends in other departments. A chance meeting with a microbiologist or bacteriologist might have led to a casual suggestion that Hubble try fitting his strange-looking skew distribution data with a lognormal function. Counting the sizes of bacterial colonies growing on agar in petri dishes was fashionable then, and their distribution was often lognormal. So perhaps Hubble tried it, and to his amazement it worked. He did not test it on smaller scales, where it would have failed, nor did he seem to think much more about the physics behind it.

Subsequent analyses confirmed the nonrandom distribution of galaxies. Mowbray (1938) essentially repeated Bok's work for different samples, and Shapley (1934a,b) continued to defend his metagalaxy. Katz and Mulders (1942), following Zwicky's suggestion, used a somewhat different technique to show that the galaxies brighter than 12.7 magnitude in the Shapley–Ames Catalog were also nonrandom. They divided the catalog into thirty-six strips of $10°$ width in galactic longitude, took different groupings of these strips, computed the dispersions of the galaxy number counts among these different groups, and found that this dispersion was significantly greater than a Gaussian distribution would have given.

Between 1943 and 1950, the effects of World War II intervened and very little was published on galaxy clustering, or on astronomy in general. What few extragalactic observations were made mostly concerned individual galaxies. But the discoveries of the decade 1925–1935 had fundamentally changed our picture of infinite space, and the theoreticians began to stir.

5

Seeds of Grand Creation

One thought fills immensity.

Blake

The clustering of galaxies became a challenge that
devoured Lemaître's research in cosmology. Time and
again Shapley demanded that the theory of the
expanding universe account for concentrations of
nebulae he was charting close to the Milky Way.
Lemaître wanted foremost to satisfy the demand. Yet
to the end of his life the solution eluded him.

Deprit (1983)

Two of the three main ingredients for understanding the universe during the first half
of the twentieth century were observational: its immense size and its expansion.
The third was Einstein's general theory of relativity. It related the force of gravity
to the structure of spacetime. Two years after his definitive account of the theory,
Einstein (1917) applied it to cosmology. His first model, introducing the cosmological constant, was static – matter without motion. Shortly afterward deSitter (1917a,b)
discovered an expanding but empty solution of Einstein's equations – motion without
matter. Then Friedmann (1922) found the intermediate solutions with both expansion
and matter, which Lemaître (1927) independently rediscovered. Eddington (1930,
1931a) was about to publish them independently yet again when Lemaître, who had
formerly been his student, gently reminded him that they were already known. So
Eddington publicized these solutions more widely and also showed that Einstein's
static universe would become unstable if condensations formed within it.

A small fraction of cosmological thought during this period strayed from the
homogeneous models to the nature and origin of structure in the universe. Charlier
(1908, 1922; see also Smart, 1922) suggested a relation between the structure and
the size of the universe. If the universe is infinite and has an average uniform density
ρ_0, then by Newton's law the total radial gravitational force on a typical particle of
mass m is

$$F = Gm \int_{r_0}^{\infty} \frac{\rho_0}{r^2} 4\pi r^2 \, dr, \tag{5.1}$$

which is infinite. (The net vectorially averaged force, however, would be zero for
strict uniformity.) Newton himself recognized this problem (see Chapter 2) and it was
emphasized by Seeliger (1895). If we substitute luminosity, which also decreases
as the inverse square of the distance, for gravity, we obtain the well-known Olbers'
paradox (Harrison 1965, 1987; Landsberg, 1972). Luminous sources of finite extent

saturate the solid angle if we can see to large enough distances. They then absorb as much as they emit, giving a uniform finite sky brightness, provided there has been enough time to reach equilibrium. Day and night are equally bright. The actual departure from equilibrium resolves Olbers' paradox. But in Newton's theory, gravity is not absorbed and it propagates infinitely rapidly; so it leads to an infinite total force in this model.

Charlier's attempt to resolve this difficulty was a hierarchical distribution of the matter. He anticipated the modern fractal description of clustering (see Chapter 12). Suppose in the simplest version of this model that the lowest level, L_0, of the hierarchy contains single objects of mass M_0. The next level, L_1, has spherical clusters each with N_1 of the basic objects. In turn, N_2 of these clusters are themselves spherically clustered to form the L_2 level and so on. The total mass at the L_2 level is $M_2 = N_2 M_1 = N_2 N_1 M_0$, and generally $M_i = N_i M_{i-1}$. If the radius of each level is R_i and we consider the gravitational force on a single basic object at the surface of an L_1 cluster, which in turn is at the surface of an L_2 cluster, which in turn ... we obtain

$$F \propto \frac{M_1}{R_1^2} + \frac{M_2}{R_2^2} + \frac{M_3}{R_3^2} + \cdots + \frac{M_i}{R_i^2} + \cdots. \tag{5.2}$$

This series converges if

$$\frac{R_i}{R_{i-1}} > N_i^{\frac{1}{2}}. \tag{5.3}$$

Notice that for the density, $\rho_i = M_i / R_i^3$, the above inequality implies $\rho_i < N_i^{-\frac{1}{2}} \rho_{i-1}$, so that the average density decreases at each level of the hierarchy. Therefore $\rho_i \to 0$ on very large scales in this model, even though there is a finite gravitational force caused by the correlations of matter on smaller scales. Using the observed numerical values of his day Charlier concluded that his inequality (5.3) was satisfied up to at least L_2 (as it is with modern values) and this was a satisfactory description of our universe. Kant's view of the Universe was now quantified.

The unsatisfactory features of this approach are its artificiality and lack of any dynamical explanation and (we now know) its instability. If such an arrangement of matter were imposed by fiat upon the Universe, the mutual gravitational forces among clusters would destroy it. A truly stationary self-consistent solution, or an origin for evolving structure, needed to be found.

Jeans proposed an origin; in fact he proposed two. Like Hubble and Shapley (and later Hoyle and Ryle), Jeans and Eddington were personal antagonists whose rivalry stimulated astronomy. Whereas Eddington asked what a perturbation can do for the Universe, Jeans asked what the Universe can do for a perturbation. A static universe, Jeans (1902a,b) found, causes gravitationally unstable linear density perturbations to grow exponentially fast with a timescale $1/\sqrt{G\rho}$. His criterion for instability, which we will see more of in Chapter 16, is essentially that the region be large enough that its gravitational potential energy, $\sim GM^2/R$, exceeds its kinetic energy, $Mv^2/2$, or approximately

$$R > R_J \approx \frac{v}{(G\rho)^{\frac{1}{2}}} \tag{5.4}$$

in terms of the average density ρ and velocity dispersion $v^2 \approx kT/m_p$, where k is Boltzmann's constant, T is the temperature, and m_p is the mass of an individual particle. Since $(G\rho)^{-\frac{1}{2}}$ is the characteristic gravitational response time of a region, R_J is roughly the distance an average particle can travel during this time.

This seemed to solve the problem of the origin of galaxies and nearly three decades later Jeans (1929) wrote:

> We have found that, as Newton first conjectured, a chaotic mass of gas of approximately uniform density and very great extent would be dynamically unstable; nuclei would tend to form in it, around which the whole of the matter would ultimately condense. We have obtained a formula which enables us to calculate the average distance apart at which these nuclei would form in a medium of given density, and this determines the average mass which would ultimately condense round each.
>
> If all the matter of those parts of the universe which are accessible to our observation, a sphere of about 140 million light years radius [here Jeans used Hubble's estimate of the useful range of the 100-inch Mount Wilson telescope], were spread out uniformly, it would form of gas of density 10^{-31} [g cm^{-3}] or thereabouts. We have calculated that the gravitational instability would cause such a medium to break up into detached bodies whose distance apart would be of the same order as the observed distance between spiral nebulae; the mass of each such body would accordingly be about equal to the mass of the average spiral nebula. We may conjecture, although it is improbable that we shall ever be able to prove, that the spiral nebulae were formed in this way. Any currents in the primeval chaotic medium would persist as rotations of the nebulae, and as these would be rotating with different speeds, they might be expected to shew all the various types of configurations ... actually observed.

Jeans's description is reminiscent of the myths of the Marquesas Islands, mentioned in Chapter 1.

Unfortunately Jeans's description also contains an element of myth because he did not properly take the expansion of the Universe into account. Remarkably, it would be nearly half a century after Jeans's original work, and two decades after the Universe was known to expand, before self-consistent solutions of gravitational instability in an expanding universe were calculated: first by Lifshitz (1946) for the relativistic case and then by Bonnor (1957) for the simpler Newtonian approximation. Equation (5.4) still gives the basic instability criterion, but in the expanding universe the instability grows much more slowly. A small relative density enhancement, $\delta\rho/\rho$, typically grows algebraically in proportion to the radius of the universe, rather than exponentially with a timescale $1/\sqrt{G\rho}$ as in the static case. This is because the universe expands on about the same $1/\sqrt{G\rho}$ timescale that the instability tries to grow. The decreasing density almost quenches the instability, leaving only a residual slow algebraic growth of perturbations. Consequently galaxies do not have enough time to form from simple $\sqrt{N}$ fluctuations (especially when the effects of the cosmic black body radiation background discovered in 1965 are considered – see Partridge, 1995). The galaxy formation problem remains essentially unsolved.

Jeans (1929) hedged his bet, however, in a very imaginative way. Difficulties in understanding the spiral structure of galaxies in the 1920s led him to propose that:

> Each failure to explain the spiral arms makes it more and more difficult to resist a suspicion that the spiral nebulae are the seat of types of forces entirely unknown to us, forces which may possibly express novel and unsuspected metric properties of space. The type of conjecture which presents itself, somewhat insistently, is that the centers of the nebulae are of the nature of "singular points," at which matter is poured into our universe from some other, and entirely extraneous spatial dimension, so that, to a denizen of our universe, they appear as points at which matter is being continually created.

Although we now have a good understanding of the galactic instabilities that produce spiral arms, new evidence of violent activity from quasars and active galactic nuclei has convinced a small minority of astronomers (see Arp, 1987) that Jeans might have been right. If so, it provides a completely different view of the origin of galaxies. Instead of condensing, they would be ejected from singularities in the structure of space-time.

A variety of other ideas, mainly explored by Eddington, Lemaître, and Milne, competed with gravitational instability to explain galaxy formation and clustering in the 1930s. In fact Eddington (1931b) did not approve of gravity as an explanation of clustering, and in his Presidential address to the Mathematical Association remarked: "It is probable that the spiral nebulae are so distant that they are very little affected by mutual gravitation and exhibit the inflation effect in its pure form. It has been known for some years that they are scattering apart rather rapidly, and we accept their measured rate of recession as a determination of the rate of expansion of the world." Although Eddington was properly more cautious than Jeans and more aware of the effects of expansion, he (like Gamow and Teller, 1939) missed two major points which accentuate the gravitational clustering of galaxies. First, when the galaxies are closer together at earlier times, the graininess of the gravitational field causes galaxy positions to become correlated during just one or two expansion timescales (this was not fully realized until four decades later). These correlations resulting from the interactions of nearest neighbor galaxies lead to the formation of small groups. Second, as the universe continues to expand each group begins to act approximately like a single particle with its large total mass. The groups themselves become correlated over larger scales and more massive clusters build up. The clustering is rescaled to larger and larger distances whose limit is determined by the time available for structure to form. Thus a hierarchical distribution can build up, although it differs significantly in detail from the form Charlier suggested.

In the same Presidential address, Eddington's comments on entropy stimulated Lemaître's concept of the Primeval Atom. Eddington had described the need for increasing entropy and disorganization in order to provide a measure of time. Lemaître (1931) connected disorganization with fragmentation and pursued this notion back to the origin of the Universe:

> ... the present state of quantum theory suggests a beginning of the world very different from the present order of Nature. Thermodynamical principles from the point of view of quantum theory may be stated as follows: (1) Energy of constant total amount is distributed in discrete quanta. (2) The number of distinct quanta is ever increasing. If we go back in the course of time we must find fewer and fewer quanta, until we find all the energy of the universe packed in a few or even a unique quantum.
>
> Now, in atomic processes, the notions of space and time are no more than statistical notions; they fade out when applied to individual phenomena involving but a small number of quanta. If the world has begun with a single quantum, the notions of space and time would altogether fail to have any meaning at the beginning; they would only begin to have a sensible meaning when the original quantum had been divided into a sufficient number of quanta. If this suggestion is correct, the beginning of the world happened a little before the beginning of space and time.

The critical difference with gravitational instability was that in Lemaître's (1950, p. 77) cosmology:

> The world has proceeded from the condense to the diffuse. The increase of entropy which characterizes the direction of evolution is the progressive fragmentation of the energy which existed at the origin in a single unit. The atom-world was broken into fragments, each fragment into still smaller pieces. To simplify the matter, supposing that this fragmentation occurred in equal pieces, two hundred and sixty generations would have been needed to reach the present pulverization of matter into our poor little atoms, almost too small to be broken again.
>
> The evolution of the world can be compared to a display of fireworks that has just ended: some few red wisps, ashes and smoke. Standing on a well-chilled cinder, we see the slow fading of the suns, and we try to recall the vanished brilliance of the origin of the worlds.

As fossil evidence for these primordial fireworks, Lemaître often cited the cosmic rays, but we now believe these high-energy particles and photons have a much later origin in supernovae, pulsars, active stars, galactic nuclei, and quasars.

Lemaître's grand picture of creation was firmly attached to his mathematical model of a universe with nonzero cosmological constant. This term in the Einstein–Friedmann equations essentially acts as a repulsive force whose strength increases with distance. It leads to three stages in the expansion of the Universe. First, there is a period of rapid expansion, which Lemaître identified with the break up of the primeval universe-atom. Second, there ensues a period of very slow expansion when the cosmological repulsion is just slightly greater than the overall gravitational attraction. Third, there is a final stage in which we now exist when the cosmological repulsion becomes much greater than gravity and the Universe again expands very rapidly.

Successive fragmentations during the first stage result in particles whose velocities are not exactly equal to the average expansion. These peculiar velocities enhance any density fluctuations that may be present and produce clouds. Then when the expansion slows (Lemaître, 1961):

> One of the effects of the fluctuations of density from place to place is to divide space into attractive regions and repellent regions. The attractive regions will exert a selection effect on the gaseous clouds that fill space and have a distribution of velocities up to a high limit. The clouds that have a velocity not too different from that of an attractive region will remain in or near it and form an assembly from which a galaxy later evolves. The important point is that these protogalaxies do not arise, as in other theories, from some effect due to numerous collisions, but only from a selection effect. Collisions would, of course, reduce the velocities, while a selection produces protogalaxies with the same velocities as the attractive regions. These velocities arise from fluctuations of density in the swarm of the swiftly moving clouds and may therefore be expected to be fairly large.

Repetition of this same process with individual galaxies instead of clouds would produce the clusters of galaxies. Therefore an individual cluster's growth cannot be considered in isolation from its neighboring clusters, and galaxies would be exchanged between clusters and the field.

Gravitational dynamics play a major role in this clustering. Overdense regions gravitationally attract galaxies from the field and can bind them if their initial total energy relative to the region is (by selection) low. As a cluster forms, it can also expel members by many-body interactions, leaving the remaining cluster more tightly bound. Galaxies with relatively high velocities will go right through the cluster unless they happen to be captured gravitationally by a rare close encounter.

Although Lemaître began to consider a mathematical formulation, and even started a numerical simulation of clustering, he did not make much progress. Interest in Lemaître-type universes diminished as it became the fashion to set the cosmological constant to zero, although there was a brief revival in the late 1960s when it appeared, temporarily, that quasar redshifts strongly clustered around $z \approx 1.95$. This could be identified with the stagnation era in the Lemaître model, and it led to a detailed linear perturbation analysis (Brecher and Silk, 1969) of the growth of galaxies and its relation to the stability of the model. This quasar clustering turned out, however, to be an artefact of observational selection. Some inflationary models of the very early Universe require a nonzero value of Λ. Recent observations of the recession velocities of high redshift galaxies, whose actual distances are found from a type of supernova which may be a good standard candle, suggest a nonzero Λ. Its value is still very uncertain.

Quite different reasons had led Milne (1935) to a nongravitational view of galaxy formation and clustering. He developed an alternative cosmology based on "kinematic relativity." His universe was flat and expanded everywhere and always at the speed of light from an initial singularity. Its smoothed out identical appearance to any observer moving with this expansion (which Milne called the Cosmological Principle) meant that such a fundamental observer would see the faster moving objects proportionally farther away. A linear redshift–distance relation thus arose solely from the kinematics of the expansion (as it would in any statistically homogeneous model). Small variations in the velocities of particles that emerged from the "preexperimental singularity" would produce regions where the faster moving particles would catch up with more distant slower moving ones and produce clumps. "No 'process of condensation' has 'caused' the nuclear agglomerations; the word 'condensation' suggests inward motions, but the relative motions have been invariably outward." Definite predictions followed for the spatial density distribution in the congested regions. They do not agree with more recent observations. Although Milne elaborated his theory with great ingenuity from simple premises, it declined rapidly in interest as the dominant role of local gravity became more clearly and widely recognized.

In its most extreme form, local gravity could cause the capture of galaxies by tidal dissipation of their orbital energy. Holmberg (1940) showed this might produce the observed clustering if the Universe were static, but not if it expands. So good was the agreement of his theory with the observed probability of finding binary and small multiple systems that he thought it strong evidence that the Universe really was static. It was a case of too many free parameters chasing too few observations. Nonetheless, several important results came out of Holmberg's analysis: He made the first reasonable estimate of the observed multiplicity function – the relative number of groups having N galaxies. He also plotted the probability for finding N galaxies in a given size area of the sky centered on a galaxy. Holmberg's technique differed from Hubble's counts, which were centred at arbitrary positions. The results of these galaxy-centered cells also had a non-Poisson distribution. This differed from Hubble's counts centered at arbitrary positions and also from a Poisson distribution. It was a forerunner of the correlation and fractal statistics. He made one of the first quantitative comparisons between galaxy type and the presence of companions. This was to search for evidence of tidal interaction, and it revealed the influence of environment on galaxies. To determine the tidal effects of close galaxy interactions, Holmberg (1941) constructed an "analog computer" in which light substituted for gravity. The mass elements of each interacting galaxy were represented by thirty-seven light bulbs, specially constructed to radiate in a plane with constant, equal, isotropic luminosities. Their inverse square law illumination at any position was detected by a photocell to provide the net effective gravitational force. The bulb at this position was then moved in response to this force, and the process repeated to mimic the system's gravitational evolution. Crude but pioneering results showed that the orbits lost enough energy for the galaxies to merge and, in merging, to produce streams of stars and spiral arms.

With hindsight it is easy to see where each of these theories was incomplete or went astray. Their great achievement was to begin to quantify ideas going back to the ancient Greeks and struggling to emerge through a cloudy filter of observations. The next stage, after the greatly reduced astronomical activity of the 1940s, was to sharpen the statistical tools for analyzing galaxy positions.

6

Clusters versus Correlations

What are they among so many?
John 6:9

How should the irregular distribution of galaxies be described statistically? Are clusters the basic unit of structure among the galaxies, or is this unit an individual galaxy itself? Two new themes and the start of a theory emerged during the 1950s and early 1960s to answer these questions. One theme built rigorous multiparameter statistical models of clusters to compare with the catalogs. The other approach looked at basic measures of clustering, mainly the two-particle correlation function, without presupposing the existence of any specific cluster form. The theory successfully began Lemaître's program to calculate kinetic gravitational clustering in an infinite system of discrete objects – the problem whose root, we have seen, goes back to Bentley and Newton. All these developments were being stimulated by the new Lick Catalog of galaxy counts. More than a million galaxies were having their positions and magnitudes measured. Although this would supercede the catalogs of the Herschels, Dreyer, Hubble, and Shapley, its refined statistics would reveal new problems.

Neyman and Scott (1952, 1959) gambled on the idea that clusters dominate the distribution. Their model generally supposed all galaxies to be in clusters, which could, however, overlap. The centers of these clusters were distributed quasi-uniformly at random throughout space. This means that any two nonoverlapping volume elements of a given size have an equal chance of containing N cluster centers, regardless of where the volumes are. The number of cluster centers in these volume elements are independent, but the number of galaxies may be correlated, since parts of different clusters may overlap.

Within each cluster, galaxies are positioned at random subject to a radially dependent number density, which has the same form for all clusters. Thus each cluster has the same scale, unrelated to the total number of galaxies in it. This hypothesis simplified the mathematics considerably but turned out to be completely unrealistic. The total number of galaxies in each cluster was chosen randomly from a single given distribution. Finally, the motions of the galaxies and clusters were ignored.

Specifying detailed forms for these distributions then defined a particular model in terms of parameters such as the average values of the distance between cluster centers, the number of galaxies in a cluster, the size of a cluster, and the ranges of these values. Projecting models onto the sky and further assuming a luminosity function for the galaxies together with a model of interstellar obscuration (usually none for simplicity) gave the average number density and spatial moments of its fluctuating projected distribution. Comparison with observation requires the intervention of additional models that describe how photographic plates record the galaxies and how

41

people, or machines, count them. Neither of these are straightforward, particularly when the emulsions, observing conditions, observers, and measurers are as varied as they were in the Lick Survey.

There are so many opportunities for oversimplification that any eventual agreement between these types of models and the observations becomes ambiguous. Proliferating parameters inhibit progress. Neyman and Scott (1959) realized this when they remarked "However it must be clear that the theory outlined, or any theory of this kind, somewhat comparable in spirit to the Ptolemean attempts to use sequences of epicycles in order to present the apparent motions of planets, cannot be expected to answer the more important question, *why* are the galaxies distributed as they actually are? The answer to this *"why,"* even if this word is taken in (quite appropriately) quotes, may be forthcoming from studies of a novel kind, combining probabalistic and statistical considerations with those of dynamics." Although the connection of this pure cluster description with the observations was too tenuous to be convincing, and it had no connection at all with dynamical theory, it did raise the discussion to a much higher level of rigor, detail, and perception. It set the tone, if not the content, for nearly everything that followed.

The first to follow was the idea that the two-particle correlation function might provide a nearly model-independent measure of clustering. Clusters were perhaps too rigid a description of the galaxy distribution. Even though it might formally be possible to decompose any distribution into clusters, their properties would be so varied that any a priori attempt to guess them would fail. Clustering statistics, however, embody fewer preconceptions. One could hope to observe them fairly directly and eventually relate them to a dynamical theory. The two-point correlation function had recently been appropriated from turbulence theory to describe brightness fluctuations in the Milky Way (Chandrasekhar and Münch, 1952), and so it was natural to use this for describing fluctuations in the galaxy density distribution. An initial attempt (Limber, 1953, 1954) faltered on a technicality – failure to distinguish completely between a smoothed out density and a discrete point density (Neyman and Scott, 1955) – but this was soon clarified (Layzer, 1956; Limber, 1957; Neyman, 1962). The two-point correlation function remains one of the most useful galaxy clustering statistics to this day.

What is it? In a system of point galaxies (or in a smoother compressible fluid) we can define the local number (or mass) density in a volume of size V. Generally this may be regarded, after suitable normalization, as the probability for finding a galaxy in V. There will be some average probability (or density) for the entire system, but the local probability (density) usually fluctuates from place to place. The presence of galaxies in one volume may alter the probability for finding galaxies in another volume, relative to the average probability. Spatial correlation functions describe how these probabilities change with position in the system. If the system is statistically homogeneous (like cottage cheese), the correlations depend only on the relative positions of volume elements, and not on their absolute position in the system. Correlations may grow or decay with time as the structure of clustering evolves. The homogeneous two-particle correlation function is the simplest one. Given a galaxy in a volume element V_1 at any position $\mathbf{r}_1$, it is the probability for finding a galaxy in another volume element V_2 a distance $|\mathbf{r}_1 - \mathbf{r}_2|$ away after the uniform average probability for finding a galaxy in any volume element V has been subtracted off. This idea is not new.

Long ago, in the theory of liquids, it became clear (e.g., Kirkwood, 1935) that the presence of one molecule in a region increased the probability of finding another nearby. To express this, consider a volume of space containing N galaxies (or molecules or, generally, particles) labelled 1, 2, 3, ..., N. Denote the probability that galaxy 1 is in a specified small region $d\mathbf{r}_1$ and galaxy 2 in $d\mathbf{r}_2$ and so on for n of the N galaxies by $P^{(n)}(\mathbf{r}_1, \mathbf{r}_2, \ldots, \mathbf{r}_n) \, d\mathbf{r}_1 \, d\mathbf{r}_2 \ldots d\mathbf{r}_n$, whatever the configuration of the remaining $N - n$ galaxies. Since all n galaxies must be somewhere in the total volume V, the normalization for each $P^{(n)}$ is

$$\int \cdots \int_V P^{(n)}(\mathbf{r}_1, \ldots, \mathbf{r}_n) \, d\mathbf{r}_1 \ldots d\mathbf{r}_n = 1. \tag{6.1}$$

Next, suppose we wish to designate the probability that there are n galaxies in the volume elements $d\mathbf{r}_1 \ldots d\mathbf{r}_n$ without specifying which galaxy is in which volume element. Then any of the N galaxies could be in $d\mathbf{r}_1$, any of $N - 1$ in $d\mathbf{r}_2$, and any of $N - n + 1$ in $d\mathbf{r}_n$, giving a total of

$$N(N - 1) \ldots (N - n + 1) = \frac{N!}{(N - n)!} \tag{6.2}$$

possibilities and the general distribution function

$$\rho^{(n)} = \frac{N!}{(N - n)!} \, P^{(n)}, \tag{6.3}$$

whose normalization is

$$\int \cdots \int_V \rho^{(n)}(\mathbf{r}_1, \ldots, \mathbf{r}_n) \, d\mathbf{r}_1 \ldots d\mathbf{r}_n = \frac{N!}{(N - n)!} \tag{6.4}$$

from (6.1).

For example, $\rho^{(1)}$ is just the probability that some one galaxy is in $d\mathbf{r}_1$ at $\mathbf{r}_1$. If this is constant over the entire volume V, then

$$\frac{1}{V} \int_V \rho^{(1)}(\mathbf{r}_1) \, d\mathbf{r}_1 = \frac{N}{V} = \rho^{(1)} = \rho \tag{6.5}$$

is just the constant number density (or mass density if all particles have the same mass) as in a perfect incompressible fluid. Similarly $\rho^{(2)}(\mathbf{r}_1, \mathbf{r}_2) \, d\mathbf{r}_1 \, d\mathbf{r}_2$ is the probability that any one galaxy is observed in $d\mathbf{r}_1$ and any other in $d\mathbf{r}_2$. If the distribution is statistically homogeneous on the scales of $r_{12} = |\mathbf{r}_1 - \mathbf{r}_2|$, then it does not matter where these volume elements are and $\rho^{(2)}$ can depend only on their separation as $\rho^{(2)}(r_{12})$. This reduces the number of variables in $\rho^{(2)}$ from six to one, a drastic simplification, which may, however, apply only over a restricted range of scales. It implies that any point can be considered the center of the system – as Nicholas de Cusa pointed out earlier.

To define the correlation function, we note that if the distribution is completely uncorrelated (analogous to a perfect gas), the probabilities of finding galaxies in different volume elements are independent and therefore multiply:

$$\rho^{(n)}(\mathbf{r}_1, \ldots, \mathbf{r}_n) = \rho^{(1)}(\mathbf{r}_1) \ldots \rho^{(1)}(\mathbf{r}_n). \tag{6.6}$$

But any correlations will introduce modifications, which we can represent by writing more generally

$$\rho^{(n)}(\mathbf{r}_1, \ldots \mathbf{r}_n) = \rho^{(1)}(\mathbf{r}_1) \ldots \rho^{(1)}(\mathbf{r}_n) \left[1 + \xi^{(n)}(\mathbf{r}_1, \ldots, \mathbf{r}_n)\right]. \tag{6.7}$$

If the general correlation function $\xi^{(n)}$ is positive, then any n galaxies will be more clustered than a Poisson (uncorrelated) distribution; for $\xi^{(n)} < 0$ they are less clustered. Obviously $\xi^{(1)} = 0$, and $\xi^{(2)}$ is the two-point correlation function.

If the distribution is statistically homogeneous on scales r_{12}, then $\xi^{(2)}$ will also depend only on $r = r_{12}$. When there is a galaxy at the origin $\mathbf{r}_1 = 0$ then $\rho(\mathbf{r}_1)\,d\mathbf{r}_1 = 1$ and the conditional probability for a galaxy to be at $\mathbf{r}_2$ in $d\mathbf{r}_2$ is, from (6.7),

$$P(r \mid N_1 = 1)\,dr = 4\pi r^2 \rho \left[1 + \xi^{(2)}(r)\right] dr. \tag{6.8}$$

Therefore $\xi^{(2)}(r)$ represents the excess probability, over the random Poisson probability, that there is a galaxy in a small volume element at r, given that there is a galaxy at $r = 0$. With statistical homogeneity, any point could be chosen as the origin, which could therefore be on any galaxy. Thus this describes clustering, without need for any well-defined clusters to exist.

Simple relations between the form of $\xi^{(2)}(r)$ and local density fluctuations provide a way to measure it observationally. Integrating (6.8) over a small volume around the origin shows that $\xi^{(2)}(\mathbf{r})$ must have a term equal to the Dirac delta function, $\delta(\mathbf{r})$, since there must be one point galaxy in this volume, however small. Any other terms in $\xi^{(2)}$ must be of order ρ or greater, since they vanish when the volume becomes so small that it can contain only 1 or 0 galaxies. Therefore we may write

$$\rho \xi^{(2)}(\mathbf{r}) = \rho \xi(\mathbf{r}) + \delta(\mathbf{r}). \tag{6.9}$$

Often $\xi(r)$ alone is called the two-particle correlation function, but the singular term is important because it describes the Poisson component of the fluctuations. To see this, suppose the number density ρ in different volumes is not exactly constant but fluctuates around its average taken over the entire system (or over a much larger volume) so that $\rho = \bar{\rho} + \Delta\rho$. The average correlation for the fluctuations of all pairs of volumes with given separation, $\mathbf{r}$ is

$$\langle \Delta \rho_1 \Delta \rho_2 \rangle = \langle (\rho_1 - \bar{\rho})(\rho_2 - \bar{\rho}) \rangle = \langle \rho_1 \rho_2 \rangle - \bar{\rho}^2 = \bar{\rho}^2 \xi(\mathbf{r}) + \bar{\rho} \delta(\mathbf{r}). \qquad (6.10)$$

The last equality follows from $\langle \rho_1 d\mathbf{r}_1 \rho_2 d\mathbf{r}_2 \rangle = \rho^{(2)} = \bar{\rho} \, d\mathbf{r}_1 \, \bar{\rho} \, d\mathbf{r}_2 [1 + \xi^{(2)}(r_{12})]$ and (6.9). Integrating over finite volumes V_1 and V_2 whose centers are a distance r apart and letting $\mathbf{r} \to 0$ gives $\langle (\Delta N)^2 \rangle \to \bar{N}$, which is the Poisson contribution for the fluctuations of N galaxies in a single finite volume. Thus, in principle, it is possible to find $\xi(r)$ by examining the correlations of fluctuations over different distances. In practice the problem is rather more subtle and will be examined, along with other properties of the correlation functions, in Chapters 14 and 16–20. Here I just introduce these ideas as background for our brief historical sketch.

During the early development of correlation functions in the 1950s, most astronomers regarded them just as a convenient alternative description of the observations. Gamow (1954) was the exception, and he tried to relate the fluctuations and correlations in a general way to his theory of galaxy formation from primordial turbulence. His was probably the first claim that quantitative details of the observed galaxy distribution (Rubin, 1954) supported a specific physical theory of cosmogony.

Comparison of the two-point correlation function and related theoretical statistics with observations began in earnest with the new Lick galaxy survey (Shane and Wirtanen, 1954; Shane, 1956; Shane, Wirtanen, and Steinlin, 1959). In the tradition of Mayall's (1934) earlier Lick catalog, described in Chapter 4, the new survey took a number of overlapping and duplicate photographic plates. All the sky visible from Lick was covered by $6° \times 6°$ plates, which generally overlapped by $1°$, and galaxies brighter than about $18^m.4$ were identified and counted in $10' \times 10'$ squares. Unfortunately the survey was not designed for precision galaxy counting; observing conditions and human measuring techniques changed from area to area. Counts in duplicated regions showed significant differences. These depended on the emulsion used; polar and zenith distances of the exposure; the seeing conditions; the apparent magnitude, type, and surface brightness of a galaxy; density of galaxies in the field; position of a galaxy on the plate; identity of the measurer; and state of the measurer. This last effect was important since a measurer, not being a machine in those days, could easily become tired or distracted. Shane is reported to have counted galaxies while talking on the telephone to prove he could do science and administration simultaneously. Attempts to correct for these effects by using the overlapping regions and averaging counts over larger areas had mixed results.

So vast was the database, however, that all these uncertainties (Scott, 1962) could not hide the existence of real clusters and the basic form of correlations. Lacking computers to analyze this database, the determination of correlations was rather crude, but even so they clearly did not fit the models dominated by clusters (Neyman and Scott, 1959). Despite a large injection of parameters, the cluster models did not survive.

The correlation functions, in their original form (Limber, 1957; Rubin, 1954), did not survive either. This was because the labor involved in a high-resolution analysis of the counts was too great to obtain $\xi(r)$, or its analog $W(\theta)$ for counts in cells projected on the sky, (14.36), accurately by direct computation. Nelson Limber once mentioned to me that his calculation was done by writing the galaxy counts for cells in contiguous strips of the sky on long strips of paper, laying these out on the floors of the large Yerkes Observatory offices, shifting the strips along the floor by a distance Δr, and reading off the shifted counts to compute the "lag correlation" with a mechanical adding machine. To ease the work, which was compounded by the need to convolve the observed counts with models for interstellar absorption and the galaxy luminosity distribution, the form of $\xi(r)$ was assumed to be either exponential or Gaussian. Neither of these assumptions was correct. The discovery in 1969 that $\xi(r)$ is essentially a power law would begin the modern age of our understanding.

Attempts to understand the physics of clustering in the 1950s also had to contend with a fundamental doubt. The earlier disagreements over whether clusters are bound or disrupting persisted. These were stimulated by discordant velocities in small groups, and by dynamical mass estimates that greatly exceeded the luminous estimates in large clusters. Disintegration of the clusters was one possibility (Ambartsumian, 1958); large amounts of underluminous matter was the other. Currently dark matter is the generally accepted explanation, based on the extended flat rotation curves of galaxies, improved dynamical mass estimates for clusters, the agreement of gravitational clustering statistics with observations, and the hope that exotic forms of dark matter will be discovered to close the universe. But the case is not completely closed.

While the debate over how to model the galaxy distribution in terms of clusters versus correlations developed, Ulam (1954) began the first numerical computer experiments to simulate this process. He started with mass points placed at random on the integer coordinates of a very long line and, in other experiments, on a plane. These masses interacted through Newtonian gravity, represented by second-order difference equations, in a nonexpanding universe. A digital computer calculated the motions of the particles. They clustered. Numerical instabilities in these early computations, however, made their detailed results uncertain. Dynamical insight from simulations would have to wait for much more powerful computers to extend Holmberg's and Ulam's pioneering work. Meanwhile, van Albada (1960, 1961) found the first important analytical description of discrete clustering dynamics.

Unlike the earlier calculations of Jeans, Lifshitz, and Bonnor for the density of a smooth gas acting like a fluid, van Albada analyzed the gravitational instability in a system of point particles. To describe the system completely requires the N-particle distribution functions $P^{(n)}$, as in (6.1), or equivalently, all the N-particle correlation functions as in (6.7). For a simpler analysis, van Albada examined how only the one-particle distribution function $P^{(1)}$ evolves if it is not affected by the local graininess of the gravitational field. This neglected all the higher order distributions and correlations, particularly those produced by near neighbors. Ignoring the correlations of near neighbors significantly underestimated small-scale nonlinear relaxation but made the problem tractable. This was a major achievement in itself.

With this "collisionless" approximation, van Albada derived kinetic equations for the evolution of velocity "moments" of the one-particle position and velocity distribution function $F(\mathbf{r}, \mathbf{v}) \, d\mathbf{r} \, d\mathbf{v}$ in the expanding universe. These moments are integrals over the distribution weighted by powers of the radial and tangential velocity. An infinite number of moments are necessary to describe the distribution, but van Albada assumed that all moments with velocity powers of three or greater vanish. This implies there is no flow of kinetic energy and that the velocity distribution at low energies is symmetrical. This assumption was needed to simplify the mathematics, but unfortunately it led to a very unrealistic description of the physics. The net effect of all these simplifications was to solve a problem only slightly more fundamental than the fluid problem. Nevertheless it gave the first results that directly related the particle density perturbations to their velocity distribution and that described spherical accretion in some detail. Large forming clusters would typically acquire a dense central core with many galaxies and a more sparsely populated halo attracting galaxies from the uniform field.

Van Albada's calculation quantified and emphasized the importance of uniform initial conditions for galaxy clustering. He even speculated that this uniformity resulted from cosmic radiation pressure – discovered five years later. But all the observations, from Herschel's survey to the great Lick catalog, had emphasized the nonuniformity of the galaxy distribution. Hubble had hoped that homogeneity would appear as the spatial scale of observations increased. As the 1960s began, opinion was mixed.

7

The Expanding Search for Homogeneity

A paradox?
A paradox!
A most ingenious paradox!
We've quips and quibbles heard in flocks,
But none to beat this paradox!

Gilbert and Sullivan

Over all scales on which astronomers had looked, from planets to clusters of galaxies, the distribution was nonuniform. Yet, the intuition of great observers like Hubble suggested there would be a scale without structure. Where was it?

Perhaps it was nowhere; perhaps it was everywhere. Every scale might be clustered, but in the same manner, so there was no distinction between scales. This self-similar picture is more subtle than the idea that the Universe suddenly becomes uniform at a particular very large scale. Carpenter (1938) suggested that it followed from his earlier (Carpenter, 1931) discovery of a single relation between the number of galaxies in a cluster and the cluster radius. He refined this relation in 1938 to the form $N \propto R_c^{1.5}$. Although he considered this to be an upper bound of the $N - R_c$ distribution, it also provides a good fit through his data. Since it applied over three orders of magnitude in N, he concluded in the later paper "that there is no essential distinction between the large, rich clusters and the smaller groups, the total populations and other characteristics varying gradually from one extreme to the other."

Without realizing it, Carpenter had found the first evidence for the form of the two-point correlation function $\xi(r)$ and, indirectly, for the fractal nature of the galaxy distribution. From Equation (6.8), the excess number of galaxies (above the average number in a Poisson distribution) around a galaxy is $N_c = 4\pi \bar{n} \int r^2 \xi(r) \, dr$. Over small scales where $\xi \gg 1$ we expect $N_c \approx N$, the number of galaxies in the cluster. Therefore if $\xi(r)$ is a power law, as Carpenter might have expected in his self-similar picture, then $N_c \propto r^{1.5}$ implies $\xi \propto r^{-1.5}$.

This is not far from the typical modern value $\xi \propto r^{-1.8}$. Furthermore, if the number density of galaxies were uniform, we would expect $N \propto r^3$. In a more general fractal distribution $N \sim r^{3-d}$ (see Chapter 12); so Carpenter essentially found the fractal dimension $d = 1.5$. Actually the situation is not quite so simple because the form of $\xi(r)$ and the fractal dimension both turn out to depend upon scale (see Chapters 12, 14, and 20).

Zwicky (1937, 1938) came to a similar conclusion that nearly all galaxies were clustered on all scales, rather than most galaxies being part of a fairly uniform field distribution with a sprinkling of dense clusters. He also described the concentration of elliptical galaxies in clusters, suggesting this was caused by mass segregation.

And he attempted to calculate the spatial distribution function of galaxies, assuming a Boltzmann energy distribution and no statistical evolution of the system of galaxies. Although this missed the essential gravitational physics of the problem, like the attempts of Lemaître and Milne, and although Zwicky's program could not be calculated, it helped open a significant question: Could statistical mechanics describe the galaxy distribution?

Thus by the late 1950s there was a consensus that galaxies did not have a Poisson distribution, even though the description of their clustering was unclear. Disputes centered over whether the essential feature was self-similar hierarchical clustering, or a few large clusters in a uniform field, or the specific distributions of cluster forms and separations that Neyman and Scott proposed, or fits to ad hoc forms of the two-point correlation function. Relations among some of these descriptions were beginning to emerge along with a rudimentary understanding of their underlying physics.

At this stage, the scale expanded. Why not repeat the galaxy clustering analyses with the clusters themselves, considering each cluster as a point mass? Are the clusters clustered? And so the great superclustering controversy began.

It was abetted by the construction of two different cluster catalogs, each defining clusters differently and subject to different selection effects. The discussion almost mimicked the earlier controversy between Hubble and Shapley, with their two different galaxy catalogs. Previously the discovery of clusters, like the earliest discoveries of galaxies, had been a rather haphazard affair. As usual, the most prominent and often least representative ones were noticed first. To systematize this process, Abell (1958) examined 879 pairs of photographs in the National Geographic Society–Palomar Observatory Sky Survey made with the 48-inch Palomar Schmidt Telescope. It covered the sky from $-27°$ to the north celestial pole, dividing it into $6°6$ square fields with $0°6$ overlaps along common edges. Abell was one of the main observers. One photograph of each field was most sensitive to red light, the other to blue.

Abell took the view that very rich clusters are superimposed on a background of field galaxies whose surface density varies. To qualify for his catalog, each cluster had to meet four criteria:

1. Richness. The number of galaxies with magnitude $m \leq m_3 + 2$ is at least fifty, where m_3 is the magnitude of the third brightest member of the cluster. Choosing the third brightest reduces the chance of confusing a nearby field galaxy with the cluster but introduces some uncertainty until the redshifts are measured.

2. Compactness. The cluster members must be within a given linear radial distance of its center. Abell chose 1.5 Mpc (adjusted to a Hubble constant of $H_0 = 100$ km s^{-1} Mpc^{-1} where one megaparsec $= 3.1 \times 10^{24}$ cm), claiming the results were insensitive to the exact distance since this scale is generally larger than the main concentration of galaxies.

3. Distance. So that clusters do not extend over much more than one $6.°6$ plate, they must have a redshift greater than 6,000 km s^{-1}. Thus Virgo is eliminated but Coma squeaks in. The upper limit is 60,000 km s^{-1}, set by requiring galaxies brighter than $m_3 + 2$ to be easily visible above the $m = 20$ plate limit. For $H_0 = 100$ km s^{-1} Mpc^{-1} this range of distances is 60–600 Mpc.

4. Galactic latitude. Regions of low galactic latitude where large numbers of stars and obscuration from interstellar dust in our own galaxy interfered with the identification of clusters were excluded.

With these criteria, Abell found 2,712 clusters of which 1,682 met more stringent conditions for a more statistically homogeneous sample. Until galaxy redshifts are measured, however, it is uncertain whether any individual cluster is an impostor produced by the chance superposition of two less rich clusters along the line of sight. Moreover, despite choosing m_3 for the richness criterion, the catalog may still be incomplete because bright field galaxies in the line of sight increase the threshold definition of a rich cluster. Subsequent analyses (e.g., Lucey, 1983) have shown that these effects are not negligible.

By counting the number of clusters in cells of equal area, Abell (1958) found they were clumped much more strongly than a Poisson distribution. The clumping could not be dominated by galactic obscuration because it persisted around the polar caps ($|b| \geq 60°$) and because the visibility of the more distant clusters (distance groups 5 and 6) did not depend on their distance. More distant clusters are often visible in regions with few closer clusters, and vice versa. Moreover, counting with cells of different sizes gives the scale on which the cluster distribution departs most from Poisson. Abell found this maximum clumping scale varies roughly inversely as the distance, suggesting a linear scale of about 45 Mpc ($H_0 = 100$) for superclustering.

Zwicky (1957) now reached the opposite conclusion: Clustering stops with clusters; larger scales are uniform. His catalog (Zwicky, Herzog, and Wild, 1961) selects clusters by using the eye to estimate regions of the Palomar Sky Survey where the surface number density of galaxies is twice the nearby density. Subtracting the average density of nearby field galaxies from the density within the contour around twice that average gives the number of galaxies in that cluster. Distant faint clusters are less likely to be included in this catalog. Zwicky's (1967) clusters showed much less statistical tendency to cluster on larger scales, leading him to deny the existence of superclustering. But he did realize that many clusters were loosely grouped into larger aggregates. Whether these were superclusters was a largely semantic distinction. Indeed the whole controversy was obscured by the uncertain physical nature and dynamics of these clusters and superclusters. Their rather arbitrary observational definitions had no known relation to any underlying physical properties.

To try to clarify the situation from a more rigorous point of view, Kiang (1967) generalized the Neyman and Scott galaxy clustering models and applied them to the clustering of Abell's clusters. The index of clumpiness (Equation 8.5) and the two-dimensional correlation function, $w(\theta)$, for clusters (see Equation 14.36) did not fit these models any better than the galaxies had. Models with superclusters of a given size did not fit the observations on more than one scale. Kiang thus suggested the hypothesis of indefinite clustering: "The indications are that galaxies are clustered on *all* scales, that there is no preferred size of clusters, and that the clusters have little physical individuality .... The hypothesis of indefinite clustering reinstates, in a sense, the *galaxy* as the ultimate building brick of the Universe. But there is no going back to the view of no clustering. Galaxies are certainly clustered, but we have not succeeded in finding any preferred sizes of clusters." Evidence began to accumulate for Carpenter's earlier view.

If Abell's clusters are themselves clustered, this should be detectable unambiguously in their two-point correlation function $\xi(r)$, independent of any a priori model. And if they are clustered on all scales, then $\xi(r)$ should be a scale-free power law. With this motivation, along with some earlier ideas from a phase transition approach (Saslaw, 1968) to gravitational clustering, Kiang and Saslaw (1969) analyzed the three-dimensional $\xi(r)$ for Abell's clusters. Using rather crude magnitude estimates of distance, they found clear evidence for correlations over scales of 50–100 Mpc (with $H_0 = 100$). However, their spatial smoothing over 50 Mpc cubes, necessitated by the inhomogeneity of the cluster distribution, made it impossible to find the functional form of $\xi(r)$ accurately.

Even as the controversy over superclustering and homogeneity continued during the next several years (de Vaucouleurs, 1970, 1971), the seeds of its resolution had already been planted. Discovery of the 2.7 degree cosmic microwave background (Penzias and Wilson, 1965) was quickly followed by tight limits on its angular variation (cf. Partridge, 1995). Although the connection of this radiation with the early universe remained controversial for many years, nearly all pockets of resistance vanished after the definitive *COBE (Cosmic Background Explorer)* satellite observations (Mather et al., 1990) showed it had a very accurate blackbody spectrum in the measured range. This relic of the thermal equilibrium between matter and radiation in the early Universe is now known to be homogeneous at the level of $\Delta T/T \approx 10^{-5}$ over angular scales of $7'$ (Bowyer & Leinert, 1990). Similar upper limits to inhomogeneity apply over scales of about $3'$ to $10°$. These largest homogeneous scales are greater than the distances over which parts of the Universe can interact before matter and radiation decouple at $z \approx 10^3$. Smaller amplitude inhomogeneities are, however, found over a range of spatial scales. For detailed discussions of the cosmic microwave background and its implications, see Partridge (1995) and more recent reviews of this rapidly developing subject.

Discovery of the blackbody spectrum and essentially uniform surface brightness of the cosmic microwave background with very small amplitude fluctuations strongly constrains the nature, size, and development of any inhomogeneities. Regions of varying density distort the microwave background both by gravitational interaction and by their different levels of ionization and radiation. The almost complete uniformity on the largest scale raises two questions: How could regions of the Universe that were outside each other's horizons (i.e., not causally connected) reach almost the same state? and what do the small radiation inhomogeneities imply for galaxy formation and clustering? Current answers to these questions involve inflationary models of the early universe and their initial spectrum of perturbations. Their results are still controversial and their exploration continues. They seem more relevant to galaxy formation than to subsequent clustering.

Direct evidence for statistical homogeneity in the distribution of matter came from the first accurate measurement of the galaxy two-point correlation function. Totsuji and Kihara (1969) solved this long-standing problem. The abstract of their paper reads: "The correlation function for the spatial distribution of galaxies in the universe is determined to be $(r_0/r)^{1.8}$, r being the distance between galaxies. The characteristic length r_0 is 4.7 Mpc. This determination is based on the distribution of galaxies brighter than the apparent magnitude 19 counted by Shane and Wirtanen (1967). The reason why the correlation function has the form of the inverse power of r is that the universe is in a state of 'neutral' stability."

Deep physical insight into the gravitational many-body problem – usually a good way to short-circuit complicated mathematical formalism – led Totsuji and Kihara to their conclusion. Previous guesses at an exponential or Gaussian form for the correlation function had been based on analogies with a stable inhomogeneous fluid where $\xi(r)$ decreases exponentially (Rubin, 1954) or on the cluster models where a Gaussian form (Limber, 1953, 1954, 1957; Neyman, 1962) was believed to have an a priori plausibility and simplicity. However, a statistically homogeneous system of gravitating objects in an unbounded expanding universe is not in stable equilibrium but is in a state more analogous to a phase transition (Saslaw, 1968, 1969). During this type of phase transition, at a critical point where forces acting on the objects are almost evenly balanced, any small perturbations grow relatively slowly with time, as Lifshitz (1946) and Bonnor (1957) had found. Under these conditions $\xi(r) \sim r^{-\gamma}$ over a substantial range of distances where it is scale free and self-similar. This provided a theoretical structure for the earlier discoveries of Carpenter (1938) and Kiang (1967). A number of views were beginning to gell. Galaxy clustering could be considered to be a phase transition from a Poisson distribution to a correlated distribution, slowly developing on larger and larger scales as the universe expands.

With this insight, to which Kihara told me he was led by his earlier work in molecular and plasma physics, he and Totsuji reworked the Neyman–Scott correlation analysis. They obtained a relation between the numbers of galaxies, N_1 and N_2, in any two regions of the sky separated by angular distance θ_{12}, the overall average $\langle N \rangle$, and a function $J(\gamma)$ of geometric integrals (see also Chapters 14 and 20):

$$\frac{\langle \{N_1 - \langle N \rangle\}\{N_2 - \langle N \rangle\} \rangle}{\langle \{N - \langle N \rangle\}^2 \rangle - \langle N \rangle} = J(\gamma). \tag{7.1}$$

The numerator is just the average of correlated fluctuations around the mean taken over all pairs of areas separated by θ_{12}; the denominator is for normalization. (The

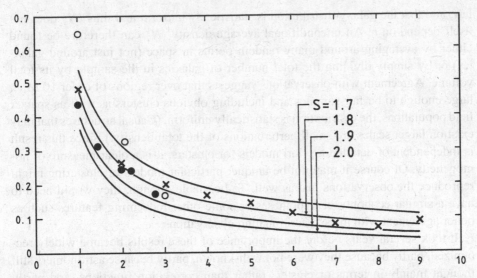

Fig. 7.1. Comparison between the observed left-hand side of Equation (7.1) (circles and crosses) and the theoretical right-hand side (solid lines) for a statistically homogeneous distribution of galaxies with a power-law two-point correlation function having exponent S (now usually denoted as γ). (From Totsuji and Kihara 1969.) The y-axis is the value of Equation (7.1); the x-axis is angular separation on the sky.

$\langle N \rangle$ term was the one Limber had initially left out; it comes from $\bar{\rho}\delta(\mathbf{r})$ in 6.10.) The right-hand side is a function just of γ, arising from angular integrals over the cones in which projected numbers are counted. Totsuji and Kihara also showed that if the three-dimensional two-point correlation function is a power law with exponent γ, the two-dimensional correlations observed on the sky will also have this form with $\gamma' = \gamma - 1$, that is, $w(\theta) = \xi(\theta) \sim \theta^{-(\gamma-1)}$.

The left-hand side of (7.1) was measured directly from the Lick catalog and compared with the theoretical calculation of $J(\gamma)$, both being functions of the separation $(x^2 + y^2)^{\frac{1}{2}}$ on the sky. Figure 7.1 from Totsuji and Kihara's paper (they used S instead of γ) shows the results with the left-hand side of (7.1) as the ordinate and the solid lines representing $J(\gamma)$ for different values of γ. The x-axis is angular separation. The solid circles are the observed values that Totsuji and Kihara found using $1° \times 1°$ squares; the open circles and crosses are values that Neyman, Scott, and Shane (1953, 1956) had previously determined for $1° \times 1°$ and $10' \times 10'$ squares respectively. Evidently there is good agreement for $\gamma = 1.8 \pm 0.1$, which also agrees with more modern values. To determine the scale r_0, Totsuji and Kihara had to assume a luminosity function for galaxies as well as an average galactic extinction. Employing the Shane–Wirtanen (1967) Gaussian luminosity function, their measurement of average surface density fluctuations $\langle (N - \langle N \rangle)^2 \rangle$ then gave $r_0 = 4.7$ Mpc, also in agreement with modern values.

The two fundamental hypotheses behind these results are that gravitational clustering resembles a neutral critical point of a phase transition, so that $\xi(r)$ is a power law, and that the galaxy distribution is statistically uniform and thus $\langle N \rangle$ does not itself depend on r. An unconditional average density $\langle N \rangle$ can therefore be found either by averaging around many random points in space (not just around galaxies) or by simply dividing the total number of galaxies in the sample by its total volume. Agreement with observations suggests that over regions of order 10 Mpc, large enough to be fair samples and including obvious clusters as well as sparser field populations, the distribution is statistically uniform. (Causal structures that may exist on larger scales are small perturbations of the total density.) Since this result is independent of detailed a priori models for clusters, it is a clear measure of homogeneity. Of course it may not be unique; particular models of clustering might reproduce the observations just as well. To be viable, though, they would need to have a similar economy of hypotheses or some other redeeming features such as better agreement with more discriminating observations.

It took several years before the importance of these results became widely recognized, partly because they were not well-known, partly because astronomers still thought mainly in terms of clusters rather than correlation functions, and partly because the applicability of ideas from thermodynamics and phase transitions was unfamiliar. Nonetheless, these results ushered in our modern understanding of galaxy clustering, and they provide a convenient conclusion to my historical sketch. Galaxy clustering was soon to change from a minor byway on the periphery of astronomy into a major industry.

PART II

Descriptions of Clustering

Quick and capacious computers, increasing realization of the importance of large-scale structure, and the first glimpses into related many-body physics all combined to change our understanding of galaxy clustering in the early 1970s. So with our historical perspective concluded (though never complete) we now change our approach and describe selected ways to characterize the galaxy distribution. With the large, automatically analyzed catalogs now available, there is no lack of positional data. Successful new observational techniques are also providing many galaxy redshifts, which are being refined into peculiar velocities relative to the general expansion. Nor is there any lack of statistical techniques for analyzing the data. Dozens of quantitative descriptions, many based on analogies in subjects ranging from archaeology to zoology, have been proposed. The main problem is to select those which give most insight into the physical causes of the structure we see. In the next chapters, I sketch several examples, their strengths and weaknesses, and some of their accomplishments. It helps provide a perspective for the two descriptions that will dominate subsequent chapters: correlation functions and distribution functions.

8

Patterns and Illusions

It's clouds' illusions I recall,
I really don't know clouds. . .
J. Mitchell

The human eye and the human mind respond strongly to contrast and continuity. Often this biases our views of constellations of stars or of clouds in the sky. Escape from this bias, or at least its mitigation, is one of the goals of an objective statistical description. Despite increasingly useful statistical descriptions, there is a tendency to look at maps and draw far-reaching conclusions about patterns in the Universe just from a visual impression. This can lead to problems.

Figure 8.1, from Barrow and Bhavsar (1987), illustrates one of the problems. Here the galaxy counts in equal area cells from the Lick Survey by Shane and Wirtanen (1967, described earlier in Chapter 6) are represented photographically. Brighter areas have higher counts. Panels a-d have different codings for this "grey scale." In Figure 8.1a the brightness, B, of a cell is proportional to $\log (\log N)$, where N is the number of counts in the cell. This very low contrast for large N gives an impression of thick filaments and large voids. Figure 8.1b has $B \propto \log N$. Here the voids have filled in, the filaments appear thinner, and the whole has a rather frothy or turbulent aspect. Figure 8.1c shows the same counts with $B \propto N$. In Fig. 8.1d the exponential dependence $B \propto e^{N/N_o}$ may, the authors suggest, combine with the approximate logarithmic response of the eye to give the most realistic (linear) visual impression of the number distribution. The contrast of the filaments is here diminished and replaced by a large-scale clumpiness. Along this sequence, the voids also become less noticeable. By strongly overemphasizing the cells with small N in Figure 8.1e, we clearly see the cell boundaries. This illustrates the inhomogeneities in the data and in the counting procedures, mentioned in Chapter 6 and examined by Smith, Frenk, Maddox, and Plionis (1991). A smaller sample of the southern sky (Figure 8.1f for $B \propto N$ and Figure 8.1g for $B \propto \log N$) shows a distribution with shallower gradients and little filamentary structure.

Manipulation of image contrast can clearly be used to overemphasize unimportant (and even spurious) structure or to deemphasize important structure. Therefore to compare theories of galaxy clustering visually with these types of observations – using modern catalogs and some three-dimensional information – requires comparisons for a very wide range of contrast functions. Selected data displayed with arbitrary contrast are bound to give a misleading impression; yet these have often provided the motivation for significant understanding! This is just one of the paradoxes of how we progress.

Continuity, too, can foster illusion. Though related to contrast, this seems a more subtle effect. The eye and brain have a visual inertia that tends to extrapolate and

57

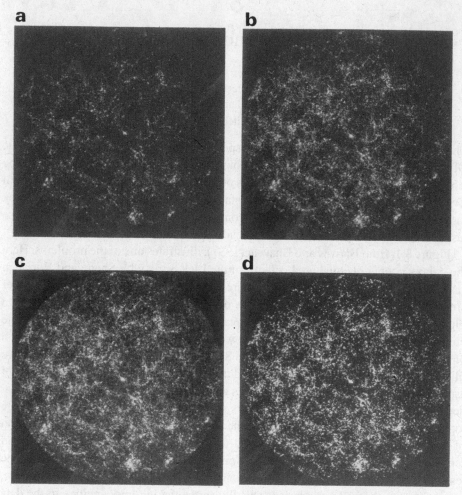

Fig. 8.1. Counts of galaxies in cells from the Lick Survey in the northern hemisphere (panels a–e) for different grey scales: The whiteness is proportional to (a) log (log N), (b) log N, (c) N, (d) e^{N/N_0}. In (e) the sensitivity to underdense regions is enhanced. Panels (f) and (g) are for the southern sky with grey scales proportional to N and log N respectively. (From Barrow and Bhavsar, 1987.)

connect close images. Figure 8.2 (from Barrow and Bhavsar, 1987, which discusses further cases) exemplifies this tendency. It is constructed by placing points in the complex plane at position $Z = re^{i\theta}$ according to the prescription

$$Z(N) = \sin(n\pi/N)e^{i\theta}, \tag{8.1}$$

$$\theta = \frac{2m\pi}{N} + (-1)^n \frac{\pi}{2N}, \tag{8.2}$$

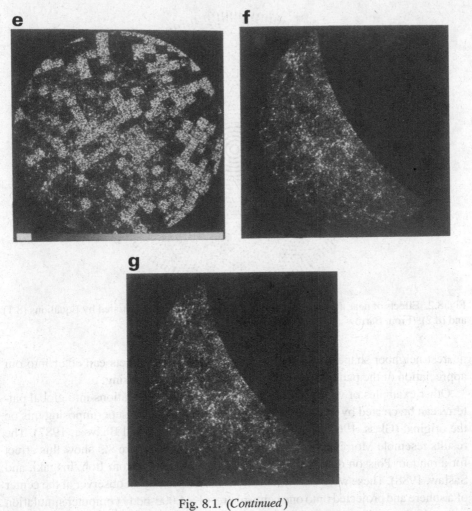

Fig. 8.1. (*Continued*)

where $n = 1, 2, \ldots, N/2$ and $m = 1, \ldots, N$. This places N (= 96 in Fig. 8.2) points on each circle along radii from the center with the points of alternate circles aligned radially. Increasing the value of N makes the pattern denser and change the perception of its structure. The relative spacings of points on a circle and between circles change from the center where a point's nearest neighbor is on the same circle, to the intermediate region where a point's nearest neighbor is on the adjacent circle, to the outermost region where a point's nearest neighbor is on the same radial line. These nearest neighbor spacings give rise (at least in most eyes) to more global patterns of concentric circles near the center, to "petals" in the intermediate region, and to radial lines near the edge. Active "op art" images, which seem to flit from one pattern to another, are often based on the use of different shapes for the elemental components of the picture in order to create ambiguity about the position of the

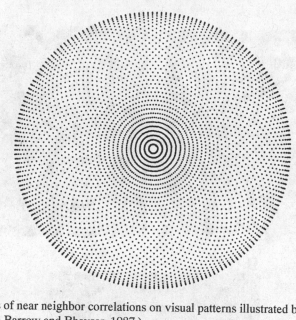

Fig. 8.2. Effects of near neighbor correlations on visual patterns illustrated by Equations (8.1) and (8.2). (From Barrow and Bhavsar, 1987.)

nearest neighbor shape. It is easy to imagine how these effects can enter into our appreciation of the patterns of constellations or galaxy clustering.

Other examples of visual inertia extrapolating local correlations into global patterns can be created by rotating or shifting a pattern and then superimposing this on the original (Glass, 1969; Glass and Perez, 1973; Barrow and Bhavsar, 1987). The results resemble Moiré fringes. The three top panels of Figure 8.3 show this effect for a random Poisson distribution of about 2,000 particles (from Itoh, Inagaki, and Saslaw, 1988). These were the initial conditions, as seen by an observer at the center of a sphere and projected onto one hemisphere, of a 4,000-body computer simulation of gravitational clustering, which will be discussed in Chapters 15 and 31. At the moment we are interested in it just as a pattern. Although Figure 8.3a is Poisson, it clearly contains large empty regions (genuine voids), larger underdense regions, and some overdense regions. Figure 8.3b shows this distribution rotated by 2 degrees and added to the original in Figure 8.3a, for a total of about 4,000 points. There are now strong correlations between nearest neighbor points, and the eye tends to continue these into a global circular pattern, aided by the similarity of the pair correlation for all pairs. Rotating the pattern further through a total of 25 degrees, in Figure 8.3c, diminishes the circular pattern as the pair separation becomes too large for the eye and brain to follow. The position where this occurs varies from person to person, and it is fun to try it out simply by photocopying these patterns onto a transparency and shifting the overlay on top of the original.

A non-Poisson pattern gives a somewhat different impression, as in Figures 8.3d–f. Figure 8.3d shows a projection of about 2,000 particles from the same *N*-body

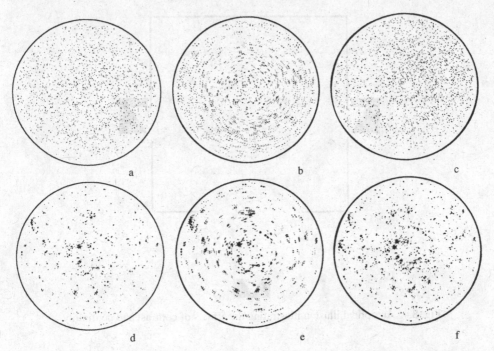

Fig. 8.3. (a) A Poisson pattern of about 2,000 points, (b) the pattern of (a) rotated by 2 degrees and added to the original, (c) the pattern of (a) rotated by 25 degrees and added to the original, (d) a gravitationally clustered pattern of about 2,000 points, (e) the pattern of (d) rotated by 2 degrees and added to the original, (f) the pattern of (d) rotated by 25 degrees and added to the original.

simulation as Figure 8.3a for a universe that is just closed (that is, the average density just equals the critical closure density: $\rho/\rho_c \equiv \Omega_0 = 1$) after it has expanded by a factor of 15.6 from its initial radius. The diagram is plotted in comoving coordinates, which are rescaled for this expansion. Gravitational attraction of the galaxies has caused the distribution to become much more clumped and clustered than the initial Poisson state. As a result, voids are much more prominent. There is also greater filamentary structure, which defines the boundaries of the voids. Rotating this distribution by 2 degrees and superimposing it on the original, as in Figure 8.3e, shows more prominent rings than in Figure 8.3b. This is mainly because the contrast is now enhanced by the clustering, even though the continuity remains similar. Rotating by 25 degrees, in Figures 8.3f, shows that even with greater contrast, the visual inertia is essentially lost.

The precise effects that contrast and continuity have on visual impressions for different types of patterns are not well understood. Nor have the interactions of many other aspects of patterns such as texture, hue, shape, boundaries, curvature, distraction, and dimensionality been thoroughly explored. Figure 8.4 shows a well-known optical illusion in which shapes and boundaries combine to give the impression of a white triangle in front of the broken square. How much might such effects influence

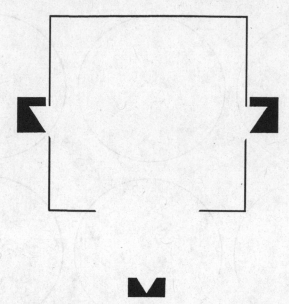

Fig. 8.4. An optical illusion illustrating the effects of contrast and continuity.

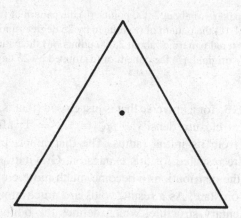

Fig. 8.5. An optical illusion illustrating the effects of distraction.

our visual impression of voids or cell structure in the galaxy distribution? Another common example of the influence of shapes on our visual judgement is shown in Figure 8.5 where the convergence of lines toward the top of the triangle makes the position of the dot seem closer to the top, whereas it is actually in the center. A slightly more subtle illusion, combining the effects of continuity with the distraction of a background pattern, in Figure 8.6, makes the sides of the triangle appear to bend inward, whereas they are actually straight. Since all these are known to be standard illusions, we can approach them with some skepticism, but how do we extend that

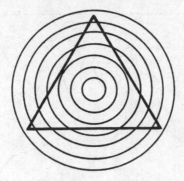

Fig. 8.6. An optical illusion illustrating the effects of continuity and distraction.

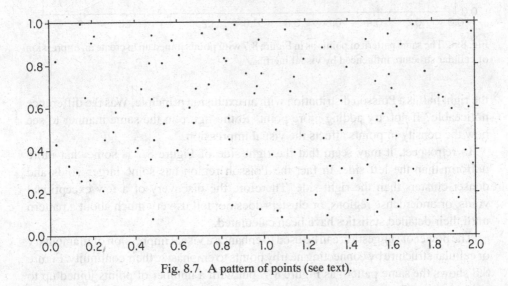

Fig. 8.7. A pattern of points (see text).

skepticism to situations where any illusions are more complex, unintentional, and unexpected?

A related problem is that some patterns may be difficult to detect. Have a look at Figure 8.7 and try to decide whether it is a random Poisson pattern or not. Try it before reading further. In this figure the way the left half is produced is simply to divide the first hundred five-digit random numbers in Abramowitz and Stegun (1964, Table 16.11) by 10^5, using the successive numbers in each row as the (x, y) coordinates for each point and plotting them on a scale of 0 to 1, giving a Poisson distribution. The right half is produced by dividing its total area into 25 equal squares, renormalizing the scale of the axes in each small square to be between 0 and 1, and plotting the same sequence of random points as on the left side with the additional constraint that only two points of the sequence are permitted in each small square. The first two points are plotted in square (1,1), the next two in (1,2), and so on. Thus

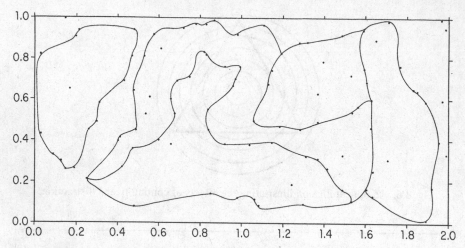

Fig. 8.8. The same pattern of points as in Figure 8.7 with points joined up to create an impression of cellular structure influenced by visual inertia.

the right half is a Poisson distribution with an exclusion principle. Was the difference noticeable? If not, try adding more points to the figure in the same manner to see how the density of points affects the visual impression.

In retrospect, it may seem that the right side of Figure 8.7 is somewhat more uniform than the left side. In fact the Poisson region has some larger voids and denser clusters than the right side. Therefore the discovery of a few exceptional voids, or underdense regions, or clusters does not tell us very much about a pattern until their detailed statistics have been calculated.

The power of suggestion can be used to enhance a visual impression of filamentary or cellular structure by connecting nearby points to emphasize their continuity. Figure 8.8 shows the same pattern as Figure 8.7, but with a number of points joined up to guide the eye. The pattern may now look rather different. Other procedures can create different appearances: Try drawing in the rough density contours on Figure 8.7, or connecting apparent filaments, or outlining the underdense regions to emphasize clustering. Alternatively, draw small circles (or hexagons that cover the total area) around each point to produce an appearance of greater homogeneity and uniformity. Another example is to estimate the ratio of the diameters of the upper and lower loops in this numeral: 8. Now turn the book upside down and see if your estimate changes. These tendencies to visualize the situation differently by guiding the eye in different ways may be exacerbated in situations where points or loops are correlated and do not have a Poisson distribution. Believing is seeing, as every good magician (illusionist) knows.

A complete escape from illusion is impossible. Even the more objective descriptions of patterns selectively emphasize certain of their properties and introduce subtle biases. Often we can at least recognize and partially compensate for these biases, sometimes by using a variety of descriptions, each providing its own insight.

Many simple statistical descriptions of clustering have been proposed to quantify patterns independently of the human eye. The rest of this chapter just mentions some representative examples. Several of these, which have been applied more extensively to the galaxy distribution, are discussed in the next few Chapters. Then we will further discuss those statistics which are not merely descriptive, but which have also been related to the physics of clustering.

These statistics fall into two general categories, depending on whether they deal with smoothed or unsmoothed data. Unsmoothed data are the detailed positions and velocities of each individual galaxy in a sample. Smoothed data are local averages of the individual galaxy data over given subsamples. These subsamples are often small spatial regions of the entire sample, but they could also be luminosity ranges, velocity ranges, ranges of relative surrounding density, nearest neighbor properties, etc. Moreover, the averages could be weighted in various ways, as in Figure 8.1. Some statistics, such as correlation functions, can be applied to both smoothed and unsmoothed data, possibly with different results. Statistics of smoothed data are usually less directly (though not necessarily less) informative since their behavior must also be explored as a function of the type and degree of smoothing. The form of smoothing is often called the "window function"; it can act like a filter.

Examples of descriptions using smoothed data are density correlations, density contours, multipole analyses, and topology genus. The first step in obtaining all these is to divide the total volume (or area) of the sample into many smaller, contiguous, mutually exclusive volume elements. The number of galaxies, N_i, in each volume element, V_i, then gives the average local number density, $n_i = N_i/V_i$. Generally this local density will vary with position: $n_i = n(\mathbf{x})$. Multiplying the densities in all volume elements separated by distance $\mathbf{r}$ and averaging over the entire sample (indicated by angular brackets) gives the two-point spatial density correlation function

$$\xi(\mathbf{r}) = \frac{\langle n(\mathbf{x})n(\mathbf{x}+\mathbf{r})\rangle}{\langle n\rangle^2} - 1. \tag{8.3}$$

Similarly one can define a three-point correlation function

$$\xi_3(\mathbf{r}, \mathbf{s}, |\mathbf{r}-\mathbf{s}|) = \frac{\langle [n(\mathbf{x}) - \langle n\rangle][n(\mathbf{x}+\mathbf{r}) - \langle n\rangle][n(\mathbf{x}+\mathbf{s}) - \langle n\rangle]\rangle}{\langle n\rangle^3} \tag{8.4}$$

along with analogous higher order n-point correlation functions. Because these correlations would be zero for a Poisson distribution, their nonzero values characterize the coherence of the distribution on different scales and levels. If the two-point correlation is isotropic and homogeneous, then at any value of $\mathbf{x}$ it is a function of just the relative separation $|\mathbf{x} - \mathbf{r}|$ rather than of all six coordinates $(x_1, x_2, x_3, r_1, r_2, r_3)$ of the two positions, effecting great simplifications. This assumption was made in analyses of the early galaxy and cluster catalogs (see Chapters 6 and 7).

Density contours are produced by lines joining contiguous volumes whose densities are within a given range. The resulting topographic maps having isodensity contours of different density levels give a pictorial representation of clustering. This can be converted into various useful statistics. Examples are the fractional area or volume covered by isodensity contours in a particular relative range, contour gradients versus peak heights, the distribution and correlation of peaks of different heights, and the topological characteristics of the contours.

Multipole analysis provides a description of the field of density contours. It represents the density field as a series of orthogonal spherical harmonic functions whose amplitudes quantify their contribution to the total projected density on the sky. The lowest order, or dipole, harmonic is usually related to the position of the density peak. This statistic can also be applied to fluctuations of the microwave background to help compare their amplitudes and positions with those of the galaxy distribution.

Topology can be quantified by the genus of a surface enclosing a volume. This is essentially the number of "holes" that cannot be deformed away in the surface: none for a sphere, one for a torus, and two for a jug with two handles. It is related to the integrated curvature over the surface. For the galaxy distribution, we can find a constant-density surface that divides the total volume of the distribution into two parts of equal volumes, or into another ratio of volumes, and determine the genus of such a surface as a function of the fractional volume it covers. This is then related to the mean curvature of the surface per unit volume as a function of the isodensity contour level. Different distributions have different values of these quantities, depending on how isolated or connected their clusters and voids are.

We turn next to examples of unsmoothed statistics for positions and velocities of individual galaxies. An unsmoothed version of the two-point correlation function, for example, measures the number of pairs of galaxy positions separated by a given distance between r and $r + dr$ and compares this number with the number expected in a Poisson distribution of the same average density. If the galaxy distribution is statistically homogeneous and the average number density n is independent of the volume (smoothing scale) used to determine the average, then the smoothed and unsmoothed two-point correlations would essentially agree (except possibly for boundary effects). Their measure of agreement can be used to check these conditions.

The fluctuation spectrum is closely related to the two-point correlation function and is basically its Fourier transform. This statistic measures the amplitude of contributions of different frequencies to the irregularities of the galaxy distribution. It is often used to provide a simple formulation of linear initial conditions for clustering, such as a power law, which may be related to more basic physical hypotheses. Sometimes it is used to describe the observed or evolved distribution, although generally it is less suitable for nonlinear conditions when high-order Fourier amplitudes interact strongly.

Percolation statistics provide a quite different description. Here the aim is to measure the "connectivity" of the distribution. As an example, start with any galaxy and find its nearest neighbor within a distance s. Then find that galaxy's nearest neighbor within s and so on until you alight on a galaxy whose nearest neighbor is farther than s. Repeat this with the other galaxies as starting points until finding the largest number of galaxies in such a "percolation group." Then change the value of s to determine the size of the largest percolation group as a function of s. This function differs for different distributions and, along with other percolation statistics, is a way to characterize clustering.

A somewhat similar statistic, which emphasizes filaments, is the minimal spanning tree. Here the idea is to connect n galaxies by the $N - 1$ straight lines having the shortest total length. This forms a unique treelike structure with no closed circuits. Many lines (branches) will connect only a small number of galaxies, and these can be "pruned" (i.e., removed) to reveal the main branches more clearly. Lines whose end galaxies are separated by too great distances can be "cut" or "separated" if these long-range connections are thought to be unphysical. The form of the filaments remaining after different degrees of pruning and cutting, their resemblance to observations, and their robustness to alterations of data all combine to provide an impression of the importance of linear structure.

Voronoi polygons (also known as Dirichlet tesselations) extend the idea of linear connections to two-dimensional areas of influence. Around each galaxy in the projected distribution draw a polygon containing the region of the plane (or a volume in space for a three-dimensional distribution) whose points are closer to that galaxy than to any other galaxy. These cells tile the plane and the number and distribution of their vertices and boundary lengths represent the "packing" of the galaxies.

Fractal descriptions are also useful for characterizing galaxy packing. The fractal dimension describes how the number of areas needed to cover a volume depends on the size of the areas. The volume occupied by N galaxies depends on how they are packed. If galaxies were smoothly and uniformly distributed in space, the number of galaxies, N_c, around any arbitrary galaxy would increase with distance just as r^3, since the average conditional number density, $\bar{n}$, would be independent of r. In fact, N_c increases more slowly than r^3 and the difference is a measure of the fractal dimension of the distribution, which, in turn, is related to the two-point correlation function. Other aspects of fractals may also be important, although observations show that galaxy clustering does not have a simple fractal form.

Repetitive cellular or wall-like structure in the galaxy distribution might be detected more readily by transect statistics. These involve looking along many lines of sight to see how the galaxy density around each line varies with distance and whether this variation has a periodic component. It can also be applied to the two-dimensional projection of galaxy positions, in which case line transects may start from different origins. Usually, properties of the (often noisy) data must be very pronounced if they are to register clearly with these one-dimensional probes.

Various means and moments of the galaxy distribution have been considered, particularly in the 1950s and 1960s when there was just enough computing power available to decide whether or not the distribution was Poisson. By dividing the projected distribution into N cells of a given area and counting the number of galaxies, n_1, in each cell, one can form the mean number $\bar{n} = \Sigma\, n_i/N$ and the variance $\sigma^2 = \Sigma\,(n_i - \bar{n})^2/(n-1)$. These would be equal for a Poisson distribution; hence the quantity

$$I_c = \frac{\sigma^2}{\bar{n}} - 1 \qquad\qquad (8.5)$$

known as the "index of clumpiness" measures departures from Poisson. If $I_c > 0$ the galaxies are clustered, if $I_c < 0$ they are anticlustered with more regular spacing, as in a lattice. Other means, variances, and higher order moments can be calculated for the distance between galaxies and random points in space or between galaxies and their nearest neighbor galaxies. Different versions include distances between a galaxy and the galaxy closest to a previously determined random point or between a galaxy closest to a random point and its nearest galaxy in a particular direction (e.g., in the upper half-plane). But all these means and moments just provide subsets of the information contained in the distribution function, so usually it is better to work directly with the distribution function itself.

Distribution functions are formed by dividing two- or three-dimensional maps of galaxy positions (or velocities) into cells according to a given prescription and counting how many galaxies are in each cell. Cells may have a variety of shapes and sizes; their locations may be contiguous and exclusive, or random and overlapping, or centered on galaxies, or weighted in some other way. At our present state of understanding, a simple random or tesselated arrangement of cells, all of the same size and regular shape, is useful, although eventually more complicated forms will yield further information. The number of cells that contain N galaxies is the distribution function. When normalized by the total number of cells, it becomes the probability $f(N)$, also often referred to simply as the distribution function. We shall see that distribution functions contain much more information about clustering than the low order correlation functions and most other statistical measures apart from the detailed positions themselves. Moreover, distribution functions can often be related to the basic physical processes of clustering, especially when gravity dominates. Consequently, they are an especially useful description for the positions and velocities of galaxies.

Groups of galaxies found by any of these statistics may or may not be gravitationally bound. Though the probability of bound clusters increases with density contrast, they cannot be distinguished by statistics alone. We need detailed positions, masses, and velocities to eliminate nearby galaxies on chance hyperbolic orbits, galaxies ejected from a cluster, and galaxies merely superposed in a telescope's gaze.

9

Percolation

And pictures in our eyes to get
was all our propagation.
Donne

Percolation describes the shape and connectivity of clustering in a quantitative way. It is related to topological and fractal patterns of a distribution. Although these descriptions have not yet been derived from fundamental dynamical theories, they are useful for characterizing the evolution of N-body experiments and for discriminating between different distributions. They are also related to basic properties of phase transitions.

Among the many applications of percolation descriptions are the spread of forest fires, the flow of liquids (particularly oil) through cracks and pores, the shapes and linkage of polymer molecules, atomic spin networks and magnetic domains, and galaxy clustering. Most of these applications occur on a lattice where the distance between interacting neighbors is fixed. Lattice models simplify the analysis greatly, while often retaining some of the essential properties of the system. Of course galaxies are not confined to a lattice, and so we will need a more general approach. However, a simple lattice defines the basic ideas and can be linked to a more continuous distribution.

Starting in one dimension, imagine a straight line divided into segments of equal length and suppose that a galaxy can be found in any segment with probability p. In the simplest case p depends neither on position along the line nor on the presence of neighboring galaxies. Now move each galaxy to the right-hand end of its segment. This sets up a lattice; each lattice site is occupied with probability p as long as we allow at most one galaxy in each segment. Thus each lattice site is either empty or contains one galaxy. These are represented by the open and filled circles respectively, in Figure 9.1.

Nearest neighbor sites that are all occupied form clusters separated by empty sites. To keep the terminology simple, even a single occupied site surrounded by unoccupied sites is usually called a cluster of one galaxy and the number of galaxies in a cluster is denoted by S. Any given lattice will contain N_S clusters of S galaxies per cluster. The density of clusters having S galaxies is N_S divided by the total number of lattice sites, or n_S. This is the usual terminology in percolation theory (see Stauffer, 1979, 1985 and Grimmett, 1989 for much more detailed reviews of the subject), where n_S is regarded as an average over many microscopic realizations of the same macroscopic type of lattice. Lattice types are characterized by features such as their dimensionality, shape (e.g., linear, triangular, square, various forms of cubic), interactions among sites, and, most importantly, their probability, p, that a lattice site is occupied. Certain properties, notably those related to spatial scaling and renormalization near a

69

Fig. 9.1. A linear lattice with open (empty circles) and occupied (filled circles) sites. The sites of a linear lattice need not be in a straight line.

critical probability, p_c, at which the lattice percolates, are essentially independent of the lattice types. All lattices having these properties are then said to belong to the some "universality class," even though the properties are not actually universal.

With these notions we can define percolation more exactly. It requires there to be a cluster of occupied sites that connects one edge of a lattice with its opposite edge in the limit as the size of the lattice becomes infinite. This is the infinite percolating network. The probability that a lattice of given type, averaged over many detailed realizations, will percolate is the fraction, P_∞, of occupied sites that belong to this infinite percolating network. It is zero below p_c and positive above p_c. The reason for taking an infinite limit is that P_∞ acquires a unique value, independent of the number of lattice sites. Finite lattices may percolate in the sense that a finite occupied cluster reaches from one edge to its opposite, but the value of p at which this occurs depends strongly on the lattice size rather than on just the more fundamental properties of its type.

9.1   Cluster Sizes and Percolation Conditions

Percolation qualitatively changes the structure on a lattice and is a useful model of a phase transition. The one-dimensional lattice of Figure 9.1 provides the easiest calculation of cluster sizes and p_c at this transition. Since the probability of occupying each lattice site is independent, the probability of occupying s contiguous sites is p^s. A cluster must be bound by two empty sites, with probability $(1 - p)^2$, and so the probability for finding a cluster of s occupied sites is

$$n_s = p^s (1 - p)^2 . \tag{9.1}$$

For an infinitely long one-dimensional lattice to percolate, all the sites must be occupied, which implies that $\lim_{s \to \infty} p^s = 1$ or $p_c = 1$. Therefore it is not possible to have $p > p_c$ in this case.

As p approaches p_c, the average size of a cluster clearly increases. This average can be defined in several ways, the most straightforward being

$$\bar{s} = \frac{\sum_{s=1}^{\infty} s n_s}{\sum_{s=1}^{\infty} n_s} = \frac{p}{p(1 - p)} = \frac{1}{1 - p} . \tag{9.2}$$

The second equality follows from (9.1) and elementary results for the sums of geometric series or alternatively from the physical interpretation of the sums. The numerator of (9.2) is the total number of lattice sites in clusters of any size normalized by the total number of all lattice sites; hence it is just the probability p that a lattice site is occupied. The denominator is the probability for finding a cluster of any size, and to define the end of a cluster requires a pair of neighboring sites, one of which is occupied with probability p and the other empty with probability $(1 - p)$. Thus $\bar{s}$ diverges as $(1 - p)^{-1}$ for $p \to 1$ from below (i.e., $p \to 1_-$).

More usually, the average cluster size is defined using the second moment of the size distribution:

$$S = \frac{\sum_{s=1}^{\infty} s^2 n_s}{\sum_{s=1}^{\infty} s n_s} = \frac{1 + p}{1 - p} \quad (p < p_c). \tag{9.3}$$

The difference between $\bar{s}$ and S is that S measures the average size of a cluster belonging to a randomly selected lattice site that is known, a priori, to be part of a finite cluster, while $\bar{s}$ is the average size of a cluster if we select the clusters rather than the sites at random. This is because if n_s is the number of clusters of size s, then $s n_s$ is the number of sites belonging to clusters of size n_s, and $\bar{s}$ is its average value. On the other hand, $s n_s / \sum_s s n_s$ is the probability for finding exactly s occupied sites in a cluster of size s among all the occupied sites $\sum_s s n_s$. Therefore S is the average value of the number of occupied sites around a randomly chosen occupied site. To calculate the sums in Equation (9.3), note that sums over $s^n p^s$ are simply related to the nth derivative with respect to p of the sum over p^s.

Either definition of cluster size shows that it diverges as $(1 - p)^{-1}$ as the linear lattice approaches percolation. We can also see this from the point of view of the pair connectivity, $c(r)$, which is the probability that two occupied lattice sites separated by distance r belong to the same cluster. Since this requires all the r sites between them to be occupied,

$$c(r) = p^r. \tag{9.4}$$

Setting $p \equiv e^{-1/L}$ gives

$$c(r) = e^{-r/L}, \tag{9.5}$$

where L is the connectivity scale length

$$L = -\frac{1}{\ln p} \approx \frac{1}{(1 - p) + \frac{1}{2}(1 - p)^2 + \cdots}. \tag{9.6}$$

Thus for $p \to 1_-$ the connectivity scale diverges similarly to $\bar{s}$ and S.

These results are early hints that physically significant quantities that depend on cluster sizes and correlations may diverge and undergo phase transitions at critical states of many-body systems. Alas, although properties of one-dimensional systems, such as the examples just given, can often be solved exactly, they seldom correspond to systems in the real world. So we move along to two dimensions. Here the results become richer.

Under what conditions do two-dimensional lattices percolate? As expected, it depends on the lattice structure, since different structures permit different types and numbers of paths. Somewhat surprisingly, it does not seem to depend on the type of lattice all that much, although only a small number of simple lattices have been explored thoroughly. Any set of polygons that cover (i.e., "tile") a plane (or other two-dimensional surface) can define a lattice. The tiling does not have to be regular or periodic. Other prescriptions that do not even cover the two-dimensional space completely can also define a lattice or network. Examples are Cayley trees (known to physicists as Bethe lattices) where each site (except those at the edge if the network is finite) is connected by a line segment to *n* other sites on the surface. Whether a lattice site is effectively occupied may be determined directly by the presence of an object on it, or by a rule concerning the distribution of objects between sites, or by whether a bond between neighboring lattice sites exists, or by longer range properties of nonneighboring sites, or by other prescriptions including combinations of these. Clearly the game can be made as complicated as the players are willing.

We shall keep the game simple and examine triangular lattices since they have some useful analytic properties. Occasionally square lattices will provide an illustrative comparison. These lattices are shown in Figure 9.2. In Figure 9.2a the dots, which are intersections of triangles, some of which are drawn, form the sites of the triangular lattice. They may or may not be occupied. If the sites were, instead, taken to be the centers of the triangles they would form a "honeycomb" lattice. The Voronoi polygons, whose sides are an equal distance from neighboring dots, are regular hexagonal cells. In Figure 9.2b they are squares and again represent the territory belonging to each lattice site. Bond percolation is defined by imagining that each site is occupied and connected to its nearest neighbors by either an open bond or a closed bond. Percolation theory had its origins in a bond model for the formation of molecular polymer chains (Flory, 1941). We shall, however, just consider site percolation.

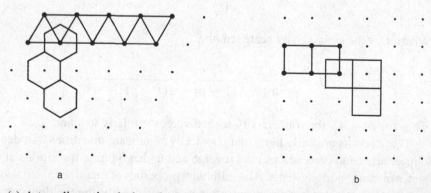

a   b

Fig. 9.2. (a) A two-dimensional triangular lattice with examples of triangular bonds between lattice sites, and hexagonal cells containing a lattice site. (b) A two-dimensional square lattice with examples of square bonds between sites, and square cells containing a site.

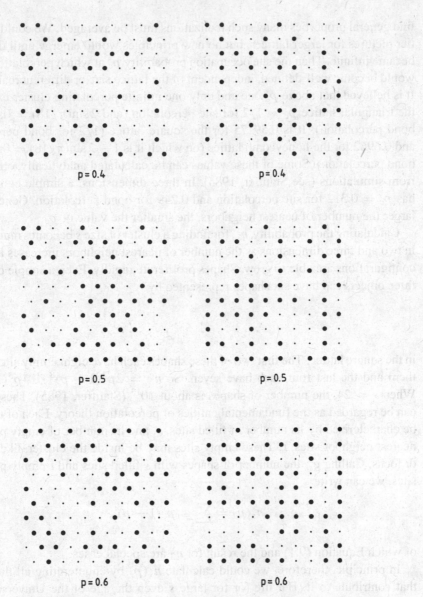

Fig. 9.3. Comparison of percolation on triangular and square lattices for different random Poisson site occupancy probabilities, *p*.

Figure 9.3 shows examples of clusters on these triangular and square lattices with random independent site occupation probabilities of 0.4, 0.5, and 0.6. At $p = 0.4$ neither the square nor triangular lattice percolates from one side to its opposite, although the square lattice misses by just one point. For $p = 0.5$, the square lattice again misses by one point, but the triangular lattice percolates across both sides now. For $p = 0.6$ both lattices percolate. Of course these are just single realizations and to find general properties many such realizations must be averaged. We could get prettier pictures for larger lattices, but no new principles would emerge until the lattice became infinite. Then the site occupation probability p_c at which percolation occurs would become well-defined, independent of the lattice size or different realizations. It is believed that above p_c one and only one infinite percolating cluster exists. For the triangular lattice $p_c = 1/2$ for site percolation (and $2\sin(\pi/18) = 0.3473$ for bond percolation). It is 0.59275 for the square lattice ($1/2$ for bond percolation) and 0.6962 for the honeycomb lattice (on which it is $1 - 2\sin(\pi/18) = 0.6527$ for bond percolation). Some of these values can be calculated analytically, others only from simulations (see Stauffer, 1985). In three dimensions, a simple cubic lattice has $p_c = 0.312$ for site percolation and 0.249 for bond percolation. Generally the larger the number of nearest neighbors, the smaller the value of p_c.

Calculating the probability, n_s, for finding a cluster of size s becomes much harder in two and three dimensions as the number of nearest neighbors increases and more configurations are able to grow. Shapes proliferate rapidly. Even a simple cluster of three objects can have six shapes represented by

$$\ldots, \;\vdots\; , \;.\hspace{-0.2em}.\hspace{-0.2em}.\; , \;\because\; , \;\cdot\hspace{-0.2em}\cdot\; , \;\ldots\; ,$$

in the square lattice. The first two of these shapes each have eight empty sites around them and the last four each have seven. So $n_3 = 2p^3(1 - p)^8 + 4p^3(1 - p)^7$. When $s = 24$, the number of shapes is about 10^{13} (Stauffer, 1985). These shapes can be regarded as the fundamental entities of percolation theory. Each of them can be characterized by its number of filled sites, s, and its number of empty perimeter nearest neighbor sites, t. These empty sites may be inside the cluster, like bubbles or lochs. Calling g_{st} the number of shapes with s filled sites and t empty perimeter sites, we can write

$$n_s(p) = \sum_t g_{st} p^s (1 - p)^t, \tag{9.7}$$

of which Equation (9.1) and the result for n_3 are special cases.

In principle, therefore, we could calculate $n_s(p)$ by enumerating all the shapes that contribute to it. But life (or for large s even the age of the Universe) is too short and much of percolation theory is concerned with just estimating the result, particularly near p_c. Knowing n_s, we can find the percolation probability, p_∞, that an occupied lattice site belongs to the infinite percolating network. To do this we note that a lattice site has only three options: It can be empty with probability $1 - p$, it can belong to a finite cluster of size s with probability sn_s, or it can belong to the infinite percolating cluster with probability pP_∞. The sum of these probabilities, including all finite clusters, must be unity. Rearranging this sum gives

$$\sum_s sn_s = p(1 - P_\infty), \tag{9.8}$$

which relates the first moment of n_s to P_∞. Once this is written down its interpretation becomes obvious: The probability that a lattice site belongs to a finite cluster equals the probability it is occupied but does not belong to the infinite cluster.

Three main techniques have been developed to search for n_s: summing over shapes in Equation (9.7), running computer experiments and Monte Carlo simulations of lattice configurations, and using the renormalization group. The last of these provides special insights, which we discuss briefly.

9.2 Scaling and Renormalization

At the percolation probability p_c the system undergoes a qualitative transition. For $p < p_c$, clusters and correlations do not extend throughout large systems whereas for $p > p_c$ they do. Even the relatively small lattices of Figure 9.3 begin to illustrate this behavior. Intuitively this suggests that if we were to look at systems with $p < p_c$ on increasing scales, that is, with coarser and coarser graining, we would eventually reach a scale on which the distribution looks nearly Poisson because the small correlated clusters are no longer resolved. However, for $p > p_c$ there would be correlated clusters on all scales and this would persist as the resolution became coarser. Just at the transition, for $p = p_c$, the lattice should be self-similar on all scales, small and large. This is a model for a dynamical system at a phase transition when, roughly speaking, the kinetic energy and the interparticle potential energy are in balance on all scales. The system does not know whether to become more correlated or less correlated. So it quavers and evolves uncharacteristically slowly. The resulting large, slow-motion fluctuations in liquids at their critical point, for example, give rise to macroscopic phenomena such as critical opalescence. We shall eventually examine whether this has an analogue for gravitational galaxy clustering.

Triangular lattices illustrate this scaling and renormalization simply and accurately, even giving the correct value for p_c and for the consequent divergence of the connectivity length L. The circles in Figure 9.4 are sites of a scaled superlattice,

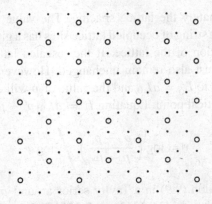

Fig. 9.4. Renormalization of a triangular lattice (dots) to a superlattice (circles). The superlattice can itself be renormalized similarly and so on.

selected so that every original lattice site belongs to one and only one superlattice site. Each superlattice site thus represents three original lattice sites. The superlattice is also triangular. Each supertriangle has three times the area of the original triangles; thus the average linear dimension of the superlattice is $a = \sqrt{3}$ times that of the original lattice. This processes can be repeated on larger and larger scales.

Next we need a rule for determining whether a superlattice site is occupied. Different rules emphasize different aspects of structure. A majority rule, by which a supersite is occupied if most of its associated sites are occupied, represents density fluctuations. A percolation rule, by which a supersite is occupied if its associated filled sites percolate, emphasizes connectivity. These rules are equivalent for a triangular lattice, but in general many variants are possible.

Adopting this rule for triangular lattices shows that the supersite will be occupied with probability p' either if all three sites are occupied, which has probability p^3, or if any two sites are occupied. The latter case has a probability $p^2(1 - p)$ for each of its three possible configurations. Therefore

$$p' = p^3 + 3p^2(1 - p). \tag{9.9}$$

If each scaling, or renormalization, is to result in a configuration similar to the previous scaling, then $p' = p$. This is the criterion for the system to have a critical point where L is so large that clusters have the same effective correlation on all scales. Equation (9.9) has three such solutions: 0, 1/2, 1. These are known as "fixed points" of the renormalization. The first and third are for lattices that are either empty or completely filled, and these are indeed similar on all scales in a very uninteresting way. The $p_c = 1/2$ solution is a half-filled triangular lattice, which (if it were infinite) would just percolate.

Near p_c we can now determine how the connectivity L behaves. Analogously to the one-dimensional case of Equation (9.6), suppose that L has the form

$$L = L_0 |p - p_c|^{-1/y}, \tag{9.10}$$

where L_0 is approximately the lattice spacing. The value of L is unchanged by renormalization since a string of occupied lattice sites has a given length independent of the scale or resolution of the lattice. If the rescaled lattice is to be similar to the original, then y must also remain unchanged. However, the lattice spacing is increased by a factor a to $L_0' = aL_0$ and the value of p will change to p' because it is not exactly at the critical point. Equating $L' = aL_0 |p' - p_c|^{-1/y}$ to L gives

$$y = \log\left(\frac{p' - p_c}{p - p_c}\right) \bigg/ \log a. \tag{9.11}$$

Expanding p' in Equation (9.9) in a Taylor series around $p_c = 1/2$ and recalling $a = \sqrt{3}$ gives $1/y = \log\sqrt{3}/\log(3/2) = 1.355$. This is often compared with an exact result of 4/3 for two-dimensional lattices as an example of the usefulness of renormalization ideas. For more complicated two-dimensional lattices, or lattices in a higher number of dimensions, renormalization techniques still agree reasonably well with results from the expansion of Equation (9.7) or Monte Carlo simulations, although the agreement is not quite so exact.

Renormalization theory is particularly useful for finding critical exponents, analogous to $1/y$, for the cluster distribution n_s and its various moments of the form

$$\sum_s s^k n_s(p) \propto |p - p_c|^{\alpha_k}, \qquad (9.12)$$

where α_k is the exponent of the leading nonanalytic term as $p \to p_c$. It may have different values, depending on whether p approaches p_c from above or below. The essential assumption is a scale-free power-law form for $n_s(p_c) \propto s^{-\tau}$ so that at p_c there is no characteristic cluster size apart from the infinite cluster. A set of relations among the α_k are derived from scaling arguments, which also give some of the α_k directly. These arguments may be made in either physical space or in Fourier space, and sometimes they use the dimensionality of the space itself as an expansion parameter. We will not discuss them in this brief sketch, but details are available from the reviews mentioned earlier.

9.3   Relation to the Galaxy Distribution

Galaxies do not sit comfortably at lattice sites, so one could be forgiven for thinking that percolation is not a helpful description of our universe. Yet it has several virtues.

Early analyses (Zeldovich, Einasto, and Shandarin, 1982) dispensed with the lattice altogether. Instead, they quantified the lengths of galaxy chains as a function of the maximum distance between neighbors. Pick a neighborhood size r, and start with any galaxy in a two- or three-dimensional catalog. All galaxies within a distance r of this galaxy are its "friends." Next, move around all of these friends and repeat the procedure. Keep repeating it until the resulting group cannot find any more such friends of friends of friends ... and loses its influence on the overall structure. This gives the maximum length, L_{max}, of that group. Then repeat the procedure starting with another galaxy. Having done this using all galaxies as initial points, one gets a distribution of L_{max} for a given r. Now repeat the entire procedure for different values of r.

The resulting distribution functions for $L_{max}(r)$ and for the total number $N_{max}(r)$ can be used in various ways. First, the distributions of L_{max} and N_{max} can be compared with each other to describe the curvature of filaments. Second, the distributions of L_{max} and N_{max} at each r can be used directly to characterize the connectivity of clustering. Third, L_{max} and N_{max} can be averaged over their distributions to obtain $\bar{L}_{max}(r)$ and $\bar{N}_{max}(r)$. Fourth, these average functions of r can be compared with similar results for N-body simulations to judge the relevance of various theoretical assumptions to the observed distribution.

Most analyses of this kind so far have just used $\bar{L}_{max}(r)$. For small r, we expect $\bar{L}_{max}(r)$ to be small and the distribution to be dominated by little lumps of close friends or tightly knit filaments. As r increases, more and more galaxies will join until a connected cluster spans the whole catalog, which then percolates. Conditions for this type of percolation differ from those discussed earlier for lattices. Nevertheless they describe some features of the clustering.

Precisely what features of clustering they describe is not well understood (see Dekel and West, 1985). One basic problem is that $\bar{L}_{max}(r)$ also depends on the average number density $\bar{n}$ of the catalog. This density dependence is related to the detailed topology of the distribution in a way that is difficult to quantify, especially a priori, since it is just what percolation properties are supposed to measure. Another basic problem is that percolation depends strongly on the properties of a low-density population between clusters. This intercluster population is difficult to detect observationally because $L_{max}(r)$ depends strongly on the magnitude limit and the distribution of obscuration in a catalog. Because of this sensitivity to intercluster galaxies, percolation may not be sensitive to major differences between theoretical models. Major competing models with galaxy formation caused by fragmentation of a large gas cloud ("top down") or by clustering of small protogalaxies ("bottom up") could easily adjust their intercluster populations to give similar percolation scales. In other words, nearest neighbor percolation may be too sensitive to the less important aspects of theoretical models. With these criticisms, the subject fell into temporary disfavor.

Interest perked up again with the realization that smoothing the data enables percolation to discriminate more usefully among models (Einasto and Saar, 1987). The simplest way to smooth is to divide the catalog into a cubic (or square if projected) grid. Each cell can be characterized by its number density. Alternatively, for a continuous number density distribution run a unit hat (or Gaussian) window function with a side (or dispersion) equal to the cell size through the system. Each place that a galaxy appears within the hat (or Gaussian) it is weighted by a positive unit value (or the value of the Gaussian at that position). This continuous density can then be averaged over a cell, and all cells whose average is above a given threshold value can be considered as occupied, whereas the others are empty. Adjusting the cell size, analogous to the neighborhood distance r, and the density threshold changes the sensitivity of the statistic to intercluster galaxies.

Once empty and filled cells have been determined, those with common sides can be connected to form clusters of cells. Studying the distribution of cluster sizes as a function of cell size and threshold value shows when a region percolates, and gives the volume filling factor of the largest cluster. Results are descriptive rather than physical. For example, large underdense regions may occur because galaxies do not form in them (biased galaxy formation), or having formed in them galaxies are disrupted by bursts of star formation, or they were regions of exceptional galaxy merging, or nearby clusters have acquired the galaxies from those underdense regions. Applications of cell percolation to small three-dimensional slices of the Universe with measured redshifts for galaxies brighter than a certain apparent magnitude (Börner and Mo, 1989a; Pellegrini et al., 1990; de Lapparent, Geller, and Huchra, 1991) confirmed that the distribution was not Poisson, but the departures from Poisson were difficult to interpret physically. The results depended on luminosity and morphological type (Mo and Börner, 1990) and the regions were too small and too peculiar to be fair samples of the Universe in this regard. On a larger scale, cell percolation confirms that the Abell cluster density is also non-Poisson with structure over scales $\gtrsim$100 Mpc. (Börner and Mo, 1989b).

Applications to computer simulations of galaxy clustering are even easier since all the three-dimensional information is available. Only the realism of the simulation is in doubt. Therefore cell percolation statistics can help constrain the free parameters of models so they match the observations (e.g., Gramann, 1990, for the case of a particular model containing cold dark matter).

An alternative approach, possibly more informative in the long run, is to understand the percolation properties of basic physical models of galaxy clustering in terms of the lattice and renormalization ideas of Sections 9.1 and 9.2. It is easy, and even useful, to turn a cell distribution into a simple lattice distribution in two dimensions. Working in three dimensions is not much harder. In both cases, results for gravity can be compared with those of other physical systems. Some aspects of percolation can also be related to other statistical measures of clustering such as the minimal spanning trees of the next chapter.

10

Minimal Spanning Trees

What! Will the line stretch out
to the crack of doom?

Shakespeare

Trees are the earth's endless effort
to speak to the listening heaven.

Tagore

Visual impressions of filamentary structure in the distribution of galaxies are easy to find, as Figure 8.1 and Chapter 8 with its caveats showed. Percolation statistics provide an objective basis for their existence. Minimal spanning trees characterize filaments in an even more refined but less physical way.

A set of points, each of which may represent a galaxy, can be connected by line segments in various ways. Each figure of this sort is called a graph, and the points are its vertices. The connecting segments, which may be straight or curved, are called edges. If every vertex is connected to every other vertex by some sequence of edges (i.e., a route) the graph is said to be "connected." It may contain circuits, which are closed routes that return to their initial vertex. But if it does not have any circuits, it is a tree. A collection of trees, which need not be connected, forms a forest.

A spanning tree connects all the vertices in the set being considered. Each edge of the spanning tree can be characterized by a property such as its length, or its relative or absolute angle, or the ratio of its length to the length of its neighboring edges, or some weighted combination of properties. The spanning tree with the smallest total value of the chosen property, for all the edges, is called the minimal spanning tree. Usually the minimal spanning tree refers simply to the total length of the edges. Minimal spanning trees based on other properties have not been much studied for the galaxy distribution, but they may turn out to be even more useful.

Finding the minimal spanning tree based on edge lengths is easy. Start with any galaxy and locate its nearest neighbor. Connect them with an edge to form a tree T_1. Next, compute the distances of all galaxies not in T_1 to their closest galaxy in T_1. Select the external galaxy nearest to a galaxy in T_1 and add it to T_1 to form a larger tree T_2. Then add the next closest one to form T_3 and so on $N - 1$ times until all N galaxies in the set are spanned. This is the minimal spanning tree and, if it is not obvious, one can prove formally that it is unique (e.g., Ore, 1963). There is also a clear generalization for minimal spanning trees based on properties other than geometric length.

Figure 10.1 shows a two-dimensional example of a set of galaxies and its constructed minimal spanning tree. Many vertices can be connected to more than one edge, creating "foliage." Often this foliage is itself useful for distinguishing distributions, but if one is mainly interested in the backbone filaments, it is easy to

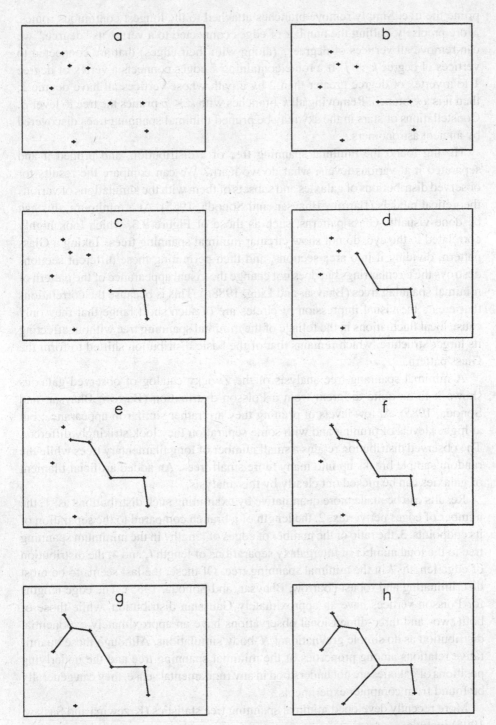

Fig. 10.1. A two-dimensional set of galaxies (a) and its minimal spanning tree (b).

prune the tree. Simply remove branches attached to the longest continuous routes. More precisely, calling the number of edges connected to a vertex its "degree," we can remove all vertices of degree j (along with their edges) that are connected to vertices of degree $k > j$. If a route containing k edges connects a vertex of degree 1 to a vertex of degree greater than 2 by a path whose vertices all have degree 2, then it is a k-branch. Removing all k-branches with $k \leq p$ prunes the tree to level p. Constellations of stars in the sky may be pruned minimal spanning trees discovered by ancient astronomers.

Having found the minimal spanning tree of a distribution, and pruned it and separated it at various levels, what do we learn? We can compare the results for observed distributions of galaxies and subsets of them with the simulations of various theoretical models (Barrow, Bhavsar, and Sonoda, 1985). At a minimum, this can be done visually. Glass patterns, such as those of Figure 8.3, which look highly correlated to the eye, do not show circular minimal spanning trees. Taking a Glass pattern, dividing it into area sections, and then permuting these different sections destroys the circular rings but does not change the visual appearance of the pattern of minimal spanning trees (Bhavsar and Ling, 1988a). This is because the correlations that create the visual impression of circles are of such short range that they only cause local fluctuations in the foliage of the minimal spanning tree without affecting its larger structure, which remains that of the basic distribution shifted to form the Glass pattern.

A minimal spanning tree analysis of the Zwicky catalog of observed galaxies shows it to be quite different from a Poisson distribution (Barrow, Bhavsar, and Sonoda, 1985). At low levels of pruning they are rather similar in appearance, but at higher levels of pruning and with some separation they look strikingly different. The observed distribution retains a small number of long filamentary trees while the random sample breaks up into many more small trees. An added artificial filament of galaxies can be picked out clearly by this analysis.

Results can be made more quantitative by examining such distributions as: 1. the number of edges per vertex, 2. the length of a branch compared to the separation of its endpoints, 3. the ratio of the number of edges of length l in the minimum spanning tree to the total number of intergalaxy separations of length l, and 4. the distribution of edge lengths l in the minimal spanning tree. Of these, the last seems to be most discriminating and robust (Barrow, Bhavsar, and Sonoda, 1985). The edge lengths for Poisson vertices have an approximately Gaussian distribution, while those of both two- and three-dimensional observations have an approximately exponential distribution as do simple gravitational N-body simulations. Although these quantitative relations among properties of the minimal spanning tree and the underlying positions of galaxies are not understood in any fundamental sense, they can generally be found from computer experiments.

More recently developed minimal spanning tree statistics (Krzewina and Saslaw, 1996) include:

1. The "edge-angle" test, which finds the frequency distribution of all nonobtuse angles between adjacent edges. This measures the overall linearity of a minimal spanning tree and can provide information about the level of relaxation in clusters of galaxies.
2. The frequency distribution, $P(M)$, for the number M, of nodes in the branches of a minimal spanning tree. This frequency appears to follow an approximate $P(M) \propto M^{-3}$ power law for a range of examples including the Poisson, observed galaxy, N-body, and gravitational quasi-equilibrium distributions described in later chapters.
3. The velocity–edgelength correlation. This gives a relation between peculiar velocities of galaxies in a simulation, or in a catalog of observed radial peculiar velocities, and the local density. Chapter 34 discusses it further.

In addition to all these statistics, it is sometimes useful to have a procedure for extracting the overall shapes of minimal spanning tree branches without the distracting jitter that arises from small fluctuations in the positions of individual nodes. An example is the use of minimal spanning trees to mark the boundaries of underdense regions. The preceding reference provides an "unwrinkling" algorithm to lessen this jaggedness in branches.

Another application is a direct test of whether observed galaxy filaments are real rather than just a visual illusion (Bhavsar and Ling, 1988b). Given, say, a cubic volume of galaxies with apparent filamentary structure, one can divide it into subcubes and shuffle them randomly. If the filaments are real, a minimal spanning tree analysis of the original cube should show them much more prominently than in the shuffled, reassembled cube. Reshuffling subcubes homogenizes real extended filaments. The differences will depend on the size of the subcubes relative to the typical length of the minimal spanning tree for a given level of pruning and separating. Filaments around the Virgo and Coma clusters analyzed this way are definitely real. Whether they are typical is another question.

To see whether filaments are typical, we can use the minimal spanning tree to estimate the Hausdorff–Besicovitch dimension, D, of a point distribution. This will be related to its fractal properties (which we take up in Chapter 12). Here I just describe the basic idea.

Consider a two-dimensional minimum spanning tree made of straight edges. Around every edge, draw a disk using the edge as the disk's diameter. Many of these disks will overlap, and they will have various sizes, all less than the diameter, ϵ, of the longest edge in the tree. Taken together, these disks provide an estimate of the minimum total area occupied by the points. If the points are closely clustered so that most of the disks overlap, they will occupy a much smaller area than if they are widely separated. If $\epsilon_i \leq \epsilon$ is the diameter of the ith disk, we can raise ϵ_i to some power $\beta \geq 0$ and sum over all the disks (even if they overlap) to get an estimate of the "β-dimensional area" of the point set. Taking the limit as the number of points becomes infinite and $\epsilon \to 0$ gives the Hausdorff–Besicovitch β-dimensional outer measure of the area A:

$$H^\beta(A) \equiv \lim_{\epsilon \to 0} \ \inf \sum_i \epsilon_i^\beta. \tag{10.1}$$

The inf (infimum) means that the limit is taken for that set of disks, of all possible sets, which gives the minimum value of the sum. The best estimate of this set is given by the minimum spanning tree. We need to have an infinite number of points so that $H^\beta(A)$ is nonzero, since taking the infimum for any finite number of points as $\epsilon \to 0$ would give $H^\beta(A) = 0$. Fortunately, we can use a finite set to estimate the result by seeing how it behaves in these limits.

There is a critical value of $\beta = D_H(A)$, known as the Hausdorff–Besicovitch dimension $D_H(A)$, where $H^\beta(A)$ changes from being infinite for $\beta < D_H(A)$ to being zero for $\beta > D_H(A)$. The minimal spanning tree estimate of $D_H(A)$ starts by selecting a random subsample containing N_R galaxies, finding the lengths l_i of the $N_R - 1$ edges in its minimal spanning tree, and calculating

$$H^\beta(A) \equiv \sum_{i=1}^{N_R-1} l_i^\beta \, (N_R - 1) = K(\beta) \, (N_R - 1)^{1-\beta/q(\beta)}. \tag{10.2}$$

The last expression is just a general functional form used to express the result. As $N_R \to \infty$, evidently $H^\beta \to \infty$ if $\beta < q(\beta)$, but $H^\beta \to 0$ if $\beta > q(\beta)$. Hence the value of β for which $q(\beta) = \beta$ is the estimate of $D_H(A)$. This is just the fixed point of $q(\beta)$ and can usually be found numerically.

The results of this approach (Martínez and Jones, 1990; Martínez et al., 1990) show that for a Poisson distribution $D_H \approx 3$ whereas for the observed spatial galaxy distribution in the CfA survey $D_H \approx 2$. Roughly speaking, this indicates that these observations are more akin to sheets than to either a random Poisson distribution or to filaments. Summarizing everything by D_H leaves out many details of the structure, but it is sufficiently refined to show that the distribution in a region dominated by several rich clusters cannot be represented as a simple fractal, as we shall see in Chapter 12. But first we examine a third description of clustering, which, like the previous two, is closely related to the connectivity of empty and occupied regions: its topology.

11

Topology

> With a name like yours, you
> might be any shape, almost.
> Lewis Carroll

Ideas of connectivity join with those of shape to describe the topology of the galaxy distribution. This addresses the question much discussed in the 1950s (see Chapter 7) of whether clusters are condensations in an otherwise uniform sea of galaxies or whether clusters are just the edges of voids and underdense regions. The question resurfaced in the 1980s when astronomers noticed fairly large three-dimensional volumes containing relatively few galaxies (Tifft & Gregory, 1976; Kirshner et al., 1981; de Lapparent et al.,1986; Kauffmann & Fairall, 1991). Consequently, much high-energy speculation arose over the origin of voids and cellular structure in the early universe. The main question was: Are clusters or voids the fundamental entities of the galaxy distribution? The answer is: both or neither.

It all depends on how you look at the distribution. If galaxies are the fundamental entities, then clusters and voids are just derivative configurations. If clusters and voids are fundamental, imposed by conditions in the early universe, then galaxies are just derivative markers. If dark matter dominates the universe, the situation becomes even more murky. In any case, topology helps quantify the conditions where relatively underdense or overdense regions dominate. It is most useful, so far, on scales at least several times that of the two-point correlation function ($\sim 5\,h^{-1}$ Mpc where h is the Hubble constant in units of 100 km s^{-1} Mpc^{-1}). On these scales, perturbations in the overall density are linear (i.e., $\delta\rho/\rho \lesssim 1$). The main measure that topologists have exploited to distinguish among distributions is the genus of their isodensity surfaces at different density levels.

To understand this, we first need to recall some simple properties of surfaces (see, e.g., Kreyszig, 1959). A surface is orientable if the positive normal direction at any point on it can be uniquely continued over the whole surface. In other words, over any closed curve in the surface one can continuously displace the positive normal from an arbitrary starting point and when it returns to its starting point it will have its original direction. Not all surfaces are like this: A Möbius strip (formed by taking a rectangle, twisting two ends by 180° relative to each other and then joining them) is a counterexample. Its returning normal will point opposite to its initial direction. We shall assume that isodensity surfaces are orientable, but it would be amusing to find one in the Universe that is not. All closed orientable surfaces, whatever their shape, can be continuously deformed at any point by a one-to-one invertible mapping (a homeomorphism) into one of an infinite set of "normal shapes." These normal shapes are spherical surfaces with handles on them, as in Figure 11.1. The number of handles, g, is called the genus of the surface. A sphere has $g = 0$; a torus has $g = 1$.

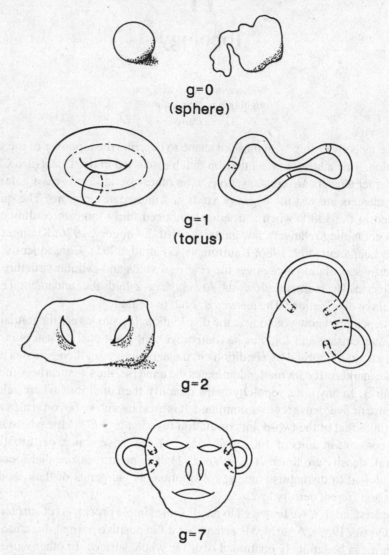

Fig. 11.1. Examples of normal shapes with different numbers of handles.

The Gauss–Bonnet theorem shows that if a surface, s, is closed, orientable, and continuous enough to be represented by a function having derivatives up to at least third order, then its genus, g, is related to the surface integral of its Gaussian curvature, K, by

$$\int\int_{s} K\, da = 4\pi(1 - g),\qquad(11.1)$$

where

$$K \equiv \frac{1}{r_1 r_2}\qquad(11.2)$$

with r_1 and r_2 the principal radii of curvature of the surface at any given point on it. These principal radii of curvature are essentially the inverse of the extreme values, over all directions, of the curvature of the normal sections of the surface. (Curvature is $|\ddot{\mathbf{x}}(s)|$, where $\mathbf{x}(s)$ parameterizes a curve by its arclength, s, and the dot derivative is with respect to s.) Normal sections are curves where the surface intersects a plane passing through both its tangent and its normal at a given point. These principal directions are usually orthogonal, except at points where the surface is flat (umbilic points). Positive Gaussian curvature implies that both r_1 and r_2 point in the same direction, and negative Gaussian curvature implies that they point in opposite directions. Although the Gauss–Bonnet theorem appears to relate a global property, g, of the whole surface to a local property, K, the value of K must be known everywhere on the surface and thus requires global information. So the genus is a concise representation of this global information. For a sphere, $K = R^{-2}$ and $dA = R^2 \sin\theta \, d\theta \, d\phi$; hence Equation (11.1) gives $g = 0$, consistent with a sphere not having any handles. A torus has as much positive Gaussian curvature as negative curvature, and so for it $g = 1$. Any sphere with more than one handle has a net negative Gaussian curvature.

Why are these notions useful for representing the galaxy distribution? In a three-dimensional distribution, the redefined genus $G_s \equiv g - 1$ is the number of holes minus the number of isolated regions (e.g., Melott et al.,1989). A hole is a region around which a closed curve cannot be shrunk to a point, and an isolated region may have either high or low density. Isodensity surfaces at density levels much greater than average will generally surround the rich clusters. If there are many such clusters, isolated like islands in the lower density sea, the high-density contours have a negative value of G_s. Alternatively, if there are many low-density regions in a high-density sea, then the low-level density contours will give negative G_s. Thus G_s as a function of density contour level, and also of smoothing length scale, describes the relative importance of holes and clusters. It can distinguish among bubble, sheet, filamentary, and random phase Gaussian distributions.

Finding isodensity surfaces and measuring their genus for the galaxies starts by smoothing the point galaxy distribution. One needs to choose a length scale and a window function for this process. Results depend significantly on these choices, since they determine the information retained from the original distribution. Therefore variations of these choices can help characterize the original distribution. Usually the length scale is greater than the intergalaxy separation and the window has a Gaussian form (to avoid the high-frequency noise introduced by sharper windows such as a top hat).

Having manufactured a smooth-varying density, $\rho(\mathbf{r})$, throughout the observed catalog or simulation, one can find the average density, $\bar{\rho}$, and the local density contrast

$$\delta(\mathbf{r}) = \frac{\rho(\mathbf{r}) - \bar{\rho}}{\bar{\rho}}. \tag{11.3}$$

Averaging the variance of the density fluctuations over the catalog gives $\langle \delta(\mathbf{r})^2 \rangle$. Thus at any position

$$v(\mathbf{r}) = \frac{\delta(r)}{\langle \delta(\mathbf{r})^2 \rangle^{1/2}} \tag{11.4}$$

describes the relative departure of the density from the average.

Selecting a particular value of $v(\mathbf{r})$ gives a threshold density $v_t(\mathbf{r})$. Contiguous three-dimensional regions with $v \geq v_t$ will have a surface that can, in principle, be deformed into a sphere with g handles to determine the genus. This may be done for different values of v_t to determine $g(v_t)$. Similarly, one can look at the complementary surface formed by $v < v_t$ and compare the properties of all these surfaces for various values of v_t.

The simplest galaxy distribution whose genus can be calculated analytically has a linear density fluctuation spectrum with Fourier components that have uncorrelated phases. These random phases imply that the density distribution is Gaussian when smoothed appropriately. A power spectrum (i.e., a spectrum of the intensity of each Fourier wavelength) is all that is needed to describe it. This case also provides a useful comparison for more complicated, possibly more realistic, distributions.

Gaussian fields have been studied extensively and a rather complicated analysis (see, for example, Adler, 1981; Hamilton, Gott, and Weinberg, 1986) shows that their average genus per unit volume, V, can be expressed as

$$g_s = \frac{G_s}{V} = \frac{1}{4\pi^2} \left(\frac{\langle k^2 \rangle}{3} \right)^{3/2} (1 - v^2) e^{-v^2/2} . \tag{11.5}$$

Here $\langle k^2 \rangle$ is the mean squared wavenumber of the density fluctuation power spectrum, $P(k)$, that is,

$$\langle k^2 \rangle = \frac{\int k^2 P(k) d^3 k}{\int P(k) d^3 k} . \tag{11.6}$$

For example, if the power spectrum has a simple power-law form $P(k) \propto k^n$, then smoothing it by convolution with a Gaussian window $e^{-r^2 \gamma^2}$ underemphasizes the short wavelengths (since they tend to average out within the window) and gives the smoothed spectrum $P_s(k) \propto k^n e^{-k^2 \gamma^2/2}$. Substituting into Equations (11.6) and (11.5) shows that

$$g_s = \frac{1}{4\pi^2 \lambda^3} \left(\frac{3+n}{3} \right)^{3/2} (1 - v^2) e^{-v^2/2} . \tag{11.7}$$

Notice that details of the power spectrum affect only the amplitude of g_s, and the amplitude involves just the second moment of $P(k)$; thus much information about the power spectrum is lost. But the dependence on v remains unchanged and is a robust indicator of Gaussian character. Since $g_s(v)$ depends on v^2, contours whose over density or under density has the same absolute value will, on average, have the same genus per unit volume. Similar results apply to the isodensity contours of projected two-dimensional distributions (Melott et al., 1989).

Systems that have this Gaussian topology can be characterized as spongelike. Neither isolated clusters nor isolated voids dominate. Rather, the clusters and the voids each form connected subsystems occupying about the same volume. It would be possible to go through the entire system by moving through either only the voids (underdense regions) or only the clusters (overdense regions). This is much like the topology of a primitive living sponge. All its cells form a continuous system (otherwise it would be two sponges) and the holes through which water and food flows are also connected to provide access to each cell. Sponges percolate.

Topological genus curves have been measured for many computer simulations of galaxy clustering and for observed galaxy catalogs. Results for the computer simulations depend on details of the models such as the initial spectrum of density fluctuations, the criteria for guessing which regions subsequently form galaxies, the relation between luminous galaxies and dark matter, the method and accuracy of the numerical integrations, the size and shape of the smoothing window, etc. (e.g., Park and Gott, 1991; Beaky, Scherrer, and Villumsen, 1992). Possibilities range widely, some of them jump in and out of fashion, and there is relatively little systematic analysis to help our understanding. Nevertheless, some results appear to be fairly general.

On large scales, where evolution is slow and initial linear Gaussian fluctuations remain linear, the topology also remains Gaussian and so Equation (11.5) continues to describe it. Typically these linear scales are several times the scale of the two-point correlation function. Some models containing cold dark matter, and others with an initial white noise spectrum ($P(k) = $ constant so $n = 0$) can also remain Gaussian at smaller scales in early evolutionary stages (or if the numerical integrations smooth the gravitational force substantially on these small scales). Other models, such as some with hot dark matter in the form of massive neutrinos, or with galaxies placed deliberately on the surface of bubbles, develop asymmetric genus curves. Their peak is shifted toward $v > 0$ (Weinberg, Gott, and Melott, 1987). Topology may depend strongly on the smoothing scale. For example, galaxy evolution seeded by very massive compact objects ($\sim 10^{11} M_\odot$) can evolve to look Gaussian on large scales and cellular on small scales (Beaky et al., 1992). Usually, gravitational evolution correlates the phases of the density field, reducing the amplitude of $g_s(v)$, and increasing its asymmetry (Park and Gott, 1991). This tends to produce more isolated cluster structure on small scales.

Observed galaxy distributions also show Gaussian topology on large scales with a tendency toward isolated clusters on small scales (Gott et al., 1989; Park et al., 1992). Comparison with simple models of bubble or cellular structure shows that these do not describe the observed distribution well. The evidence, so far, is only preliminary because it depends on comparing idealized models with partial samples having unknown selection effects, but it points in the direction of Gaussian initial conditions followed by gravitational clustering.

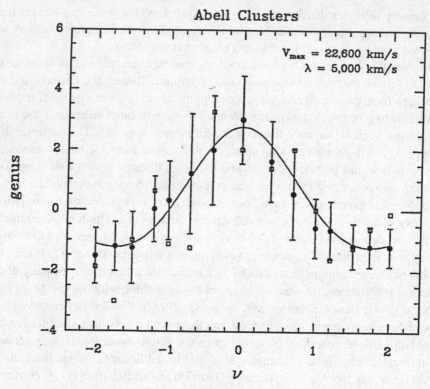

Fig. 11.2. The genus curve for the Abell clusters of galaxies. The solid line is the best fitting Gaussian curve of Equation (11.7). (From Gott et al., 1989.)

Figure 11.2 from Gott et al. (1989) shows an example of this comparison for Abell's catalog of galaxy clusters discussed in Chapter 7. The most distant members of this sample have a velocity of nearly 23,000 km s^{-1} and the density smoothing lengthscale is 5,000 km s^{-1}. The solid line is the best-fit Gaussian result of Equation (11.7). It shows the sponge topology at the average $\nu = 0$ density contour, the isolated superclusters at high ν, and the isolated underdense regions at low ν. There do not appear to be significantly more or fewer regions where the density departs substantially from the average than are expected in the Gaussian distribution.

12

Fractals

All, all of a piece throughout.
Dryden

Fractals help characterize the scaling properties of point distributions. Originally astronomers hoped that the galaxy distribution would have simple scaling properties following from the scale-free form of the gravitational potential. It was not to be. Nevertheless, fractals provide useful insights that can be related to correlation and distribution functions.

The main fractal measure of a set of points is its "dimension." A continuous set of points, such as a line, plane, or spherical ball has an integral topological dimension whose definition goes back at least to Euclid. It refers to the number of independent coordinates needed to locate a point in this set. When Cantor (1883) found his middle-thirds set, however, it became necessary to generalize the concept of dimension. Cantor's set has an uncountably infinite number of points, all disconnected from each other so that there is no continuous interval in the set, even though each point has another which is arbitrarily close. It is simple to construct (Figure 12.1). Start with the unit interval [0, 1] and successively delete the innermost third of each remaining line segment. Cantor's set, C, contains all the points common to the sequence of subsets $C_0, C_1, C_2, C_3 \ldots$. This set is self-similar since expanding the scale of C_{n+1} by a factor 3 gives, for each of the parts of C_{n+1} that contains two lines, a shape identical with C_n. The Cantor set has analogous constructions in two and three spatial dimensions with many fascinating properties. Its more general definition need not be self-similar.

What is the dimension of Cantor's set? In the limit C_n as n becomes infinite the maximum length of C_n is $(\frac{2}{3})^n \to 0$ and all the points are disconnected. A set of disconnected points each of zero topological dimension has a total topological dimension zero. Yet this does not seem quite adequate to describe an uncountably infinite number of points each with an arbitrarily close neighbor.

Hausdorff (1919) extended the idea of dimension, as mentioned in Chapter 10. A close variant of this Hausdorff dimension, introduced by Kolmogorov (1958), which is easier to determine empirically, is the box-counting or capacity dimension d_c. Suppose we have a bounded set of points in a space R^d of d topological dimensions. There will usually be many ways in which we can enclose all these points in a set of d-dimensional cubes of length ϵ and volume ϵ^d. But one configuration has the minimum number, $N(\epsilon)$, of cubes for any given size. The behavior of $N(\epsilon)$ as $\epsilon \to 0$ determines d_c. For example, when ϵ is small we can enclose the points of a line of length L by $N(\epsilon) \approx \text{int}(L/\epsilon)$ squares (two-dimensional cubes) and the points of a disk with radius r by $N(\epsilon) \approx \text{int}(\pi r^2/\epsilon^2)$ squares. By defining

$$d_c = \lim_{\epsilon \to 0} [\ln N(\epsilon)/\ln(\epsilon^{-1})] \tag{12.1}$$

91

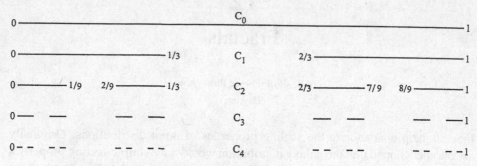

Fig. 12.1. Successive approximations to Cantor's middle-thirds set.

we obtain $d_c = 1$ for a line and $d_c = 2$ for a disk, the same as their topological dimensions because their points are continuous. But this also works for discontinuous sets of points such as Cantor's middle-thirds set. Each subset C_n has 2^n closed intervals each of length 3^{-n}; so it requires $N(\epsilon) = 2^n$ lines of length $\epsilon = 3^{-n}$ to cover C_n, and also to cover C (since $C_{n+1} \subset C_n$ and $C = \cap_{n=0}^{\infty} C_n$). As $n \to \infty$, $\epsilon \to 0$ and

$$N(\epsilon) \sim 2^n = e^{n \ln 2} = e^{(n \ln 3) \ln 2/\ln 3}$$

$$= (\epsilon^{-1})^{\ln 2/\ln 3}. \tag{12.2}$$

Applying (12.1) gives $d_c = \ln 2/\ln 3 = 0.631$.

Many other examples are illustrated in books on fractals (e.g., Feder, 1988). The Koch curve, for instance, formed by replacing the middle third of each side of an equilateral triangle by the two sides of an outward pointing equilateral triangle, and repeating this process an infinite number of times gives a set with $d_c = \ln 4/\ln 3 = 1.262$. This is very close to the observed d_c for the coastline of Britain in Richardson's (1961) famous analysis made by walking dividers with decreasing separation ϵ over maps of various scales. For much of its length, the coast looks fairly self-similar on these different scales.

The Hausdorff dimension, discussed in Chapter 10, is a bit more general than d_c since it allows the covering volumes to have any shape. This is less practical to compute and may give a different value for the dimension. However, it removes the problem that ϵ is a continuous variable in Equation (12.1) but is discrete in self-similar distributions such as C_n. In the limit $n \to \infty$, discreteness is often of little consequence for d_c, although there may be highly pathological cases.

Other definitions of dimension also are useful. Early analyses of galaxy clustering, when people hoped the distribution might be self-similar, led to counts of galaxies in volumes of increasing size. If there are N_0 galaxies in a sphere of radius r_0, and the number distribution scales so that there are $N_n = \alpha^n N_0$ within a radius that scales as $r_n = \beta^n r_0$, then eliminating n from the last two equations and applying the result to a continuum of levels gives

$$N(r) = Br^D,$$ (12.3)

where

$$D = \ln \alpha / \ln \beta$$ (12.4)

is the "fractal dimension" and $B = N_0/r_0^D$ is the amplitude. Thus the average density within a sphere,

$$\bar{n} = \frac{N(r)}{V(r)} = \frac{3B}{4\pi} r^{-(3-D)},$$ (12.5)

depends on the size of the sphere. In a continuous uniform distribution $D = 3$ and $\bar{n} = $ constant, but this is not so in general. For the galaxies, Carpenter found $D \approx 1.5$ (see Chapter 7), the first indication of large-scale correlations. Section 14.1 will show how the dependence of $\bar{n}$ on scale can affect simple computations of the two-point correlation function. By now, the reader may have realized that D is essentially the capacity dimension, d_c, for a self-similar distribution in the limit of large scale.

The "information dimension," d_i, of a set is a slight generalization of d_c with a different physical interpretation. It considers the probability p_i of finding a point in the ith cell of size ϵ^d. The information (Brillouin, 1963) contained in this distribution is

$$I(\epsilon) = - \sum_{i=1}^{N(\epsilon)} p_i \log p_i$$ (12.6)

summed over all $N(\epsilon)$ cells. If there were an equal a priori probability $p_i = \epsilon^2/\pi R^2$ for finding particles in squares of area ϵ^2 over a disk of radius R, then there would be $N(\epsilon) \approx \pi R^2/\epsilon^2$ of these cells and $I(\epsilon) \approx \log(\pi R^2/\epsilon^2)$ in this case. Its information dimension defined generally by

$$d_I = \lim_{\epsilon \to 0} I(\epsilon)/\log(\epsilon^{-1})$$ (12.7)

would be 2, equal to its topological dimension and its capacity dimension. If the values of p_i are equal, $I(\epsilon)$ will be a maximum; so the more the p_i depart from equal values (which are the most likely ones if nothing is known about the distribution a priori) the more "information" the distribution contains. Since $\log N(\epsilon) \geq I(\epsilon)$, the information dimension $d_I \leq d_c$.

These dimensions, and some more general ones we shall meet shortly, have all been used to describe the "fractal" nature of a distribution. It is neither a unique nor a complete description. The same dimensions may characterize many different distributions, having quite different properties and visual appearance. All are, to some degree "fractal" in the sense of having a range of scales over which the distribution is self-similar when D or d_c or d_I is not an integer. Mandelbrot (1982) offered a more precise definition of a fractal as a set whose dimension (such as d_c, D, or d_I) is strictly greater than its topological dimension d. Such a set fills more space, so to speak, than just the sum of the topological dimensions of its individual members. Moreover, it has this property over a range of length scales and for all length scales in a simple fractal, which can be characterized by a single value of its dimension. A fractal set is like the stories within stories within stories of the tales of Shahrazad in the Arabian Nights, except that in a simple fractal every story is the same.

Can the galaxy distribution be a simple fractal? Obviously not. It has a smallest length scale, roughly equal to the average intergalaxy spacing, at which the distribution ends. So it ceases to be self-similar on smaller scales. Limits to the largest fractal scales are less obvious. If the galaxy distribution becomes Poisson on large scales, then it ceases to be a fractal because $\bar{n}$ becomes independent of scale and $D = 3 = d$ from (12.5). This would be the natural state if the initial distribution were not correlated and there has not been enough time for correlations to grow significantly. For example, if galaxies had peculiar velocities of $\sim 10^3$ km s^{-1} relative to the average cosmic expansion (a generous estimate), then over the age of the Universe ($\sim 10^{10}$ years) they could have moved about 10 megaparsecs. Thus we would not expect to see significant correlations on scales greater than several tens of megaparsecs.

Observations on these large scales are presently ambiguous. We know (see Chapters 7 and 20) that on these scales the two-point correlation function of galaxies is negligible. But this is only a partial description, since clusters of galaxies are themselves correlated over tens of megaparsecs and their distribution is not Poisson. Furthermore, some structures appear to exist on scales of ~ 100 Mpc, but whether this is by initial cause or by subsequent chance is still unclear. It seems that these giant structures involve only a very slight perturbation of the total mass density, but the existence of dark matter makes that also unclear. If there were fractal structure over $\sim 10^2$ Mpc it would probably be a remnant of the singularity of the early universe, but this remains a matter of much controversy.

Observations on smaller scales are more clearcut, though they have their own uncertainties. Insofar as $\xi_2(r) \propto r^{-\gamma} \gg 1$ and therefore $N(r) \propto \bar{n} \int \xi(r) r^2 \, dr \propto r^{3-\gamma}$, then Equation (12.3) gives an observed "correlation fractal dimension" $D = 3 - \gamma \approx 1.2$. However, this requires that $\bar{n}$ be independent of r, which is not generally the case for most of the observed samples of galaxies (Coleman & Pietronero, 1992). Much of the variation in $\bar{n}$ is caused by intrinsic clustering, which extends over much of the sample, and by boundary effects, particularly in small samples. The conclusions may be statistically significant for a given sample, even though the sample is not representative. Large two-dimensional samples are homogeneous over scales $\gtrsim 20$ Mpc, but this may be partly a projection effect. We will not have a definitive answer until large redshift surveys (in which the peculiar velocities are either negligible or understood) become available.

Meanwhile, there are strong indications that a simple fractal model does not agree with observations on small scales (Martínez and Jones, 1990; Martínez et al., 1990) even if $\xi(r)$ is a power law. The reason is that the Hausdorff dimension, at least as measured approximately by a minimum spanning tree analysis (Martínez and Jones, 1990), is $d_H \approx 2.1$, which differs from $D \approx 1.2$. The exact value of d_H is affected by the method used to measure it (Valdarnini, Borgani, and Provenzale, 1992), but its difference from D means the distribution cannot be described by a single fractal dimension.

Multifractals make a more complicated fractal description possible. Suppose that at every point the distribution is fractal, but its dimension changes from place to place. We could then look for correlations of these dimensions and play the game over again at a higher level. The most straightforward property of these dimensions is their distribution of scaling indices. To explore this more precisely, we need to generalize the idea of a fractal dimension.

To begin, we start with the notion of the measure $\mu(A)$ of a bounded set, A, in a space R^d. Measure is a real number $0 \le \mu < \infty$ associated with the set, and is zero for the null set and additive for the union of two nonintersecting sets. One example is just the fraction $\mu_i = N_i/N_{tot}$ of galaxies in the ith cell of a distribution with a total number N_{tot} galaxies so that $\sum_i \mu_i = 1$. We can suppose each cell to be a d-dimensional cube of volume δ^d. From the μ_i we can form another measure

$$M_d(q, \delta) = \sum_{i=1}^{N(\delta)} \mu_i^q \delta^a = N(q, \delta)\delta^a \xrightarrow[\delta \to 0]{} \begin{cases} 0, & a > \tau(q), \\ \infty, & a < \tau(q), \end{cases} \quad (12.8)$$

where $N(\delta)$ is the number of cells containing at least one galaxy. Thus $N(q, \delta)$ is the number of occupied cells, weighted by q such that for large positive q the sparsely occupied cells contribute very little, whereas for large negative q they dominate. Sometimes $N(q, \delta)$, or a variant, is called the partition function. It characterizes the moments of the distribution and their singularities as $\delta \to 0$. In this limit, $M_d(q, \delta)$ either vanishes or becomes infinite, depending on the value of a (which can be negative).

For the particular value of $a = \tau(q)$, the measure $M_d(q, \delta)$ is finite and nonzero. Then (12.8) shows that

$$N(q, \delta) = \sum_{i=1}^{N(\delta)} \mu_i^q \propto \delta^{-\tau(q)}, \quad (12.9)$$

where

$$\tau(q) = -\lim_{\delta \to 0} \frac{\ln N(q, \delta)}{\ln \delta}. \quad (12.10)$$

This resembles (12.1) and indeed for $q = 0$ we see that $N(0, \delta)$ is just $N(\delta)$, the number of cells needed to enclose the galaxies; hence $\tau(0) = d_c$ is the capacity dimension. When $q = 1$, normalization of the μ_i gives $N(1, \delta) = 1$ and $\tau(1) = 0$.

The quantity

$$D_q = \tau(q)/(1-q) \tag{12.11}$$

is sometimes called the Rényi (1970) dimension and is used because $D_q = d$ if the set has constant density in a space of d topological dimensions, where $\mu_i = \delta^d$, $N(\delta) = \delta^{-d}$, and $\sum_{i=1} \mu_i^q = \sum_{N(\delta)} \delta^{qd} = \delta^{(q-1)d}$. In this case

$$D_1 = \lim_{q \to 1} D_q = \lim_{\delta \to 0} \frac{\sum_i \mu_i \ln \mu_i}{\ln \delta} \tag{12.12}$$

using

$$\ln \left(\sum_i \mu_i^q \right) = \ln \left[\sum_i (\mu_i \exp\{(q-1)\ln \mu_i\}) \right]$$

$$\to \ln \left\{ 1 + (q-1) \sum_i \mu_i \ln \mu_i \right\} \approx (q-1) \sum_i \mu_i \ln \mu_i. \tag{12.13}$$

Therefore D_1 is just the information dimension (12.7). The information dimension, d_I, is also related to the derivative of $\tau(q)$:

$$\alpha(q) \equiv \frac{d\tau(q)}{dq} = -\lim_{\delta \to 0} \frac{\sum_i \mu_i^q \ln \mu_i}{\left(\sum_i \mu_i^q \right) \ln \delta}. \tag{12.14}$$

Hence $d_I = -(d\tau/dq)_{q=1}$. We will soon show that D_2 is related to the exponent, γ, of a power-law two-particle correlation function at small scales by $D_2 = 3 - \gamma$. Often D_2 is called the correlation dimension.

Thus the sequence of exponents $\tau(q)$ for which the measure $M_d(q, \delta)$ remains finite for different "moments," q, characterizes any distributed set of points. If $\tau(q) = \tau(0)$ (or equivalently $D(q) = D(0)$) for all q, then the set can be characterized by a single dimension and would be a simple fractal. The change of τ with q characterizes the scaling properties of regions of different density. In other words, each region is dominated by a particular fractal dimension $\tau(q)$, which depends on the local density and which may fluctuate strongly over short distances. The entire set is then a combination of the fractal dimensions of each region.

The result that D_q and $\alpha(q)$ both give the information dimension d_I suggests that there are many ways to characterize a multifractal distribution. One which is often used, instead of $\tau(q)$, is its Legendre transform (cf. Equation 23.26)

$$f(\alpha) = \alpha q - \tau(q) \tag{12.15}$$

along with (12.14) for $\alpha(q)$. A multifractal set is the union of fractal subsets, each with fractal dimension $f(\alpha)$ in which the measure μ_α scales as δ^α. This describes how singular the measure becomes as $\delta \to 0$. Then $f(\alpha)$ is the fractal dimension of the set of singularities having Lipschitz–Hölder exponents α, which supports the multifractal distribution.

Multifractals have been used to describe catalogs of the galaxy distribution (e.g., Martínez et al., 1990; Coleman and Pietronero, 1992) as well as computer experiments (e.g., Itoh, 1990; Valdarnini et al., 1992). There are some difficulties in applying the basic ideas sketched here to a finite distribution of points (even disregarding problems of selection and bias in the observations). Although several different numerical estimators for the Hausdorff and Renyi dimensions usually agree reasonably well, they may differ in special cases, or for some ranges of q, so their use requires some caution. Typically $\tau(q)$ increases steeply, nearly linearly with q for negative or small positive q, and then less steeply nearly linearly for large q. The main results from the observational analyses are the need for a multifractal description and a value of $D_2 \approx 1.2$ in the limited regime where $\xi(r) \propto r^{-1.8}$. Computer simulations mainly agree with this property and show further that $D_q \approx 1$ for $q > 0$ whereas $D_q \gtrsim 3$ for $q \ll 0$ for nonlinear gravitational clustering on small enough scales that the memory of the initial conditions has been substantially lost. Values of $q \ll 0$ emphasize the unevolved underdense regions where the galaxies are much more homogeneous than in the overdense regions, so this is a quite reasonable result. Simulations show that the fractal dimensions retain some information about the initial conditions, especially on large unrelaxed scales, and about the selection and bias of samples. It is still a challenge to extract similar information directly from the observational catalogs where these effects are embedded in noise and uncertainty. Section 28.12 mentions the multifractal found indirectly from the observed distribution function.

Apart from merely describing a distribution, what virtues do multifractals have to compensate for their loss of simplicity? Where is the physics in fractals? Here the hope – so far only a hope – is that the multifractal description of galaxies may represent a strange attractor that is the nonlinear outcome of the dynamical equations of gravitational galaxy clustering. There are many definitions of a "strange attractor." Loosely speaking, a strange attractor is a fractal set of points in an n-dimensional position–momentum phase space that describes where a dynamical system tends to spend most of its time. The fractal distribution in phase space may be quite different from the distribution in physical space and represents the limits to chaotic motion (see, e.g., Drazin, 1992 for further discussion). It is not yet known how to relate these two fractal sets for galaxy clustering.

Here we consider a more modest dynamical connection (Itoh, 1990). It relates the fractal dimension

$$D'_q = \frac{1}{q-1} \frac{d \ln N(q, \delta)}{d \ln \delta} \tag{12.16}$$

to the distribution function $f(N)$ for finding N galaxies in a given volume. Note that this definition differs from the more usual version in (12.10) and (12.11), which is not so useful for a finite, discrete distribution of galaxies. As $\delta \to 0$ in a finite discrete system, cells almost always contain either one or zero galaxies so that the weighted number of occupied cells $N(q, \delta) \to N_0(1/N_0)^q$, where N_0 is the total number of galaxies in a large volume V_0. This limit is a constant, independent of the detailed galaxy distribution, and in it D_q decreases for $q \geq 0$ as q increases, whereas D_q' may increase.

To relate the D_q' to the distribution function, consider the measure μ_i in (12.9) to be the probability for finding a galaxy in the ith cubical cell of size δ in a total volume V_0. Then

$$N(0, \delta) = \sum_{i=1}^{N(\delta)} [\mu_i(\delta)]^0 = N(\delta) = \frac{V_0}{\delta^3} [1 - f(0)], \qquad (12.17)$$

which is just the total number of occupied cells since $f(0)$ is the probability that a cell is empty. Thus (12.16) gives

$$D_0' = 3 - \frac{d \log[1 - f(0)]}{d \log \delta}. \qquad (12.18)$$

For $q \neq 0$, we note that counting the i successive cells in the volume is equivalent to counting cells containing N galaxies for all values of N, so that

$$N(q, \delta) = \sum_{i=1}^{N(\delta)} [\mu_i(\delta)]^q = \frac{V_0}{\delta^3} \sum_{N=1}^{\infty} \left(\frac{N}{N_{\text{tot}}} \right)^q f(N) = \frac{V_0}{N_{\text{tot}}^q} \frac{\mu_q}{\delta^3}, \qquad (12.19)$$

where

$$\mu_q = \sum_{N=1}^{\infty} N^q f(N) \qquad (12.20)$$

is the qth order moment of $f(N)$. Therefore for $q \neq 0$

$$D_q' = \frac{1}{q - 1} \left(-3 + \frac{d \ln \mu_q}{d \ln \delta} \right). \qquad (12.21)$$

The special case of the correlation dimension D_2' (or D_2) is related to the two-particle correlation function ξ_2 through the integral of (6.10) (see also Equation 14.17) in the form

$$\mu_2 = \overline{N^2} = \bar{N}^2 + \bar{N} + \bar{N}^2 \int \xi(r) \, d^3\delta, \qquad (12.22)$$

with $\bar{N} = \bar{n}\delta^3$. For large scales, where the first term on the right-hand side dominates, $D_2' \approx 3$. Small scales with $\xi(r) \propto (\delta_x^2 + \delta_y^2 + \delta_z^2)^{-\gamma/2}$ dominant have $D_2' \approx 3 - \gamma$ (compare 12.5). So not only do the D_q' differ for different values of q, but they may also depend upon scale.

Although fractal dimensions for the galaxy distribution are difficult to measure accurately since $\delta \to 0$ is not a meaningful observational limit, they provide a useful description for particular values of δ, and as a function of δ for $\delta \neq 0$. We can also use our physical understanding of the correlation and distribution functions to give some physical insight into the fractal dimensions from (12.21). So far, fractals per se have not provided substantial new physical understanding or predictions on their own; whether they can is a considerable challenge.

13

Bound Clusters

Let there be spaces in your togetherness.

Gibran

Gravitationally bound groups and clusters of galaxies form an important subset of the galaxy distribution. The mass of galaxies and any dark matter in a cluster prevents its disruption by perturbations of nearby galaxies, by tidal forces of other clusters, and by the cosmic expansion. Prevention is not absolute; over very long times ejection, capture, and merging of galaxies modify the cluster. This chapter summarizes some observed properties of bound clusters and the physical processes that cause them to evolve. We concentrate on fundamental questions which are always under discussion, rather than on reviewing the latest results in this rapidly changing subject.

13.1 Identification of Bound Clusters, the Virial Theorem, and Dark Matter

Deciding which galaxies are bound together in an apparent cluster projected on the sky is not straightforward. If we try to apply the basic criterion that the sum of the total gravitational potential energy and the kinetic energy of peculiar velocities, v_i, of the galaxies be negative, that is,

$$-\frac{G}{2} \sum_{i \neq j} \frac{M_i M_j}{|\mathbf{r}_i - \mathbf{r}_j|} + \frac{1}{2} \sum_i M_i v_i^2 < 0, \tag{13.1}$$

we immediately encounter problems. The potential energy is summed over all $i \neq j$ pairs of galaxies with masses M_i and M_j, at positions $\mathbf{r}_i$ and $\mathbf{r}_j$. The factor $1/2$ ensures that each pair contributes only once even though it is labelled in two ways. Thus to find the total energy we need to know the total masses, velocities, and three-dimensional positions of each galaxy.

Only for a few relatively nearby galaxies can we find their mass independently by observing their internal dynamics (see reviews of these methods in, e.g., Binney and Tremaine, 1987; Saslaw, 1985a). The balance between gravitational and centrifugal forces in a spiral galaxy's observed rotation curve enables us to estimate its mass to about a factor of two. Uncertainties arise from the limited extent of the rotation curve, the detailed structure of the galaxy, and the possibility of noncircular motions. For elliptical galaxies, the balance between gravitational energy and the kinetic energy of internal motions gives a mass estimate, again uncertain by a factor of two. For small, distant, irregular or interacting galaxies, these methods do not work so well. Recourse to other, usually less certain, techniques such as relations between a galaxy's mass and some combination of its luminosity, color, morphology, amount of neutral hydrogen, isophotal diameter, etc. provide alternative mass estimates.

Underlying all this is the controversial question of how massive a galaxy's dark halo may be.

Relative positions of galaxies are also uncertain. All we really know is their angular separation on the sky. We need their three-dimensional separation to apply (13.1) directly, especially for close pairs in projection. Even in the largest clusters, it is often not clear whether a galaxy is on the near or far side. Peculiar velocities help little since most large clusters also have a large velocity dispersion. There is a reasonable presumption that the brightest most massive galaxies have formed in or moved to the center of the cluster. This is more likely to occur in large clusters than in smaller ones. In groups with just a few galaxies including one very massive galaxy, the less massive galaxies often orbit around the massive one as satellites. However, their relative positions depend on the shapes and phases of their orbits so that the whole configuration may appear very unsymmetric.

Each galaxy in the cluster has a kinetic energy $M_i v_i^2/2$ that depends on its mass and its peculiar velocity $v_i = |\mathbf{v}_i - \mathbf{r}_i H|$ relative to the average cosmic expansion radial velocity, $r H$, for galaxies at radial distance r. The Hubble constant, H, whose present value is usually written as $H_0 = 100\, h$ km s^{-1} Mpc^{-1}, is the fractional rate of expansion of the radius or scale length of the Universe: $H(t) = \dot{R}(t)/R(t)$. Most astronomers think that $0.5 \lesssim h \lesssim 1$, but its exact value is still controversial. Although angular positions of galaxies are well determined, angular velocities are almost completely uncertain. Usually we can only observe the radial velocity component of $\mathbf{v}_i$. Moreover, the peculiar radial velocity is affected by the uncertainty in radial distance. This makes the kinetic energy uncertain.

The binding energy in (13.1) does not include any optically dark matter outside the galaxies but within the cluster. Herein lies another major uncertainty. Direct evidence for this more diffuse mass component comes from its X-ray emission and from the gravitational lensing of a more distant source whose visible or radio emission passes through a cluster along our line of sight. In both these cases, the amount and distribution of the dark matter depends on detailed models and cannot be determined uniquely (e.g., Fabian & Barcons, 1991, Eyles et al., 1991). A third line of evidence for diffuse dark matter comes from the result that when the velocity dispersion of a collection of galaxies in a fairly small region is measured, it often, though not always, leads to a total kinetic energy several times greater than the estimated binding energy of the visible galaxies. For these collections to be stable over a few billion years requires either massive galactic halos or large amounts of intergalactic matter. This is the famous "missing mass" or – if one believes the mass must be present – the "missing light" problem. Although the arguments are not circular, they have acquired elements of ellipticity. Ways out include the possibilities that many clusters are unstable (but this is hard to argue for the cores of rich clusters, at least in conventional cosmologies), that the galaxies have lost substantial mass, perhaps by the decay of unconventional particles (but these particles are not observed), or that the redshifts contain a relevant nonvelocity component (but no independent evidence has convinced many astronomers).

Driving the search for dark matter is the belief of many theoreticians that we live in an Einstein–Friedmann universe that has just the critical density, ρ_c, to be closed, so that $\Omega_0 = \rho/\rho_c = 1$. For some astronomers this has an elegance devoutly to be desired; for others it is just a number to be observed. Models of the universe with an initial inflationary period of very rapid expansion do not generally require $\Omega_0 = 1$, although early examples did. The amount of invisible matter in clusters suggested so far by the observations of X-ray emission, gravitational lensing, and high galaxy velocity dispersion all add up to $\Omega_0 \lesssim 0.3$. This leaves the bulk of the matter needed to close the universe still to be discovered. The more astronomers have looked for it in the forms of dust, rocks, planets, dark stars, little black holes, big black holes, dark galaxies, hot and cold intergalactic gas, massive neutrinos, other weakly interacting massive particles, cosmic strings, exotic particles ... , the more they have not found it.

With all these uncertainties one might think the identification of bound clusters to be quite hopeless. It is not. The situation is saved by the statistical nature of the problem. Even so, we must be cautious.

Instead of viewing (13.1) as requiring detailed information about the positions and velocities of each galaxy, we can regard the kinetic and potential energies as averages over the entire cluster. We still face the uncertainties of which galaxies should be considered to belong to the cluster and how to take the averages. But we can hope, as often happens, that these do not alter some basic results. For example, a very close pair of galaxies would formally make a very large contribution to the total potential energy. However, for many purposes, we can suppose the pair to be a single bound subsystem with a very large internal gravitational binding energy. It would then contribute to the binding energy of the cluster as a single galaxy of mass $M_i + M_j$. If a cluster is bound, there is a more precise relation among the averaged galaxy masses, velocities, and separations than inequality (13.1) provides. This is the virial theorem.

The virial theorem provides the most important statistical dynamical information about bound clusters. For the simplest case, when two galaxies of masses $m^{(\alpha)} \ll m^{(\beta)}$ are bound to each other in a circular orbit of radius r, the gravitational force $Gm^{(\alpha)}m^{(\beta)}/r^2$ must balance the centrifugal force $m^{(\alpha)}v^2/r$, where v is the orbital velocity of the less massive galaxy, since the more massive one can be regarded as essentially fixed. Equating these forces and multiplying through by r gives

$$2T + W = 0, \tag{13.2}$$

where T is the kinetic and W the potential energy. If the orbits are not circular, or the masses are not very unequal, a similar result applies if we take the time averages of T and W. This theorem is very general, and even holds if the galaxy masses change with time, as we shall see next, following a more extensive discussion in GPSGS.

Consider a cluster of galaxies. Start with the gravitational potential ϕ, acting on any arbitrary galaxy with mass $m^{(\alpha)}$, contributed by all the other galaxies with masses $m^{(\beta)}$ at distances $|\mathbf{x}^{(\alpha)} - \mathbf{x}^{(\beta)}|$:

$$\phi = G \sum_{\beta \neq \alpha} \frac{m^{(\beta)}}{|\mathbf{x}^{(\alpha)} - \mathbf{x}^{(\beta)}|}. \tag{13.3}$$

The equations of motion for the ith velocity component of each galaxy are then

$$\frac{d}{dt}\left(m^{(\alpha)} v_i^{(\alpha)}\right) = F_i^{(\alpha)} = m^{(\alpha)} \frac{\partial \phi}{\partial x_i^{(\alpha)}} = -Gm^{(\alpha)} \sum_{\beta \neq \alpha} \frac{m^{(\beta)}\left(x_i^{(\alpha)} - x_i^{(\beta)}\right)}{|\mathbf{x}^{(\alpha)} - \mathbf{x}^{(\beta)}|^3}. \tag{13.4}$$

The masses may be functions of time. Mass loss may even be anisotropic, in which case the momentum derivative in (13.4) contains an $\dot{m}^{(\alpha_i)} v_i^{(\alpha)}$ term. Here we shall consider the simpler case of isotropic mass loss $\dot{m}^{(\alpha)}$.

Virial theorems are essentially moments of the equation of motion, obtained by multiplying these equations by powers of $x^{(\alpha)}$ and summing over all α galaxies. They lead to chains of coupled moment equations in which those of order x^n-depend on terms involving powers of x^{n+1}. A physical approximation for the higher order moments must be introduced to break this chain and give a closed set of equations. If we had all the moment equations, it would be equivalent to solving all the equations of motion exactly. Usually, however, we can extract the most important information from the lowest moments. In the case of galaxy clusters, we next show that the lowest moment of all gives us an estimate of their mass.

Multiplying (13.4) by $x_j^{(\alpha)}$ and summing over α we get

$$\sum_{\alpha} x_j^{(\alpha)} \frac{d}{dt}\left(m^{(\alpha)} v_i^{(\alpha)}\right) = -G \sum_{\alpha} \sum_{\beta \neq \alpha} m^{(\alpha)} m^{(\beta)} \frac{x_j^{(\alpha)}\left(x_i^{(\alpha)} - x_i^{(\beta)}\right)}{|\mathbf{x}^{(\alpha)} - \mathbf{x}^{(\beta)}|^3}. \tag{13.5}$$

Interchanging the α and β indices in the right-hand side gives the same form except for a change of sign. Therefore we can symmetrize it by taking half the sum of the two forms:

$$W_{ij} \equiv -\frac{G}{2} \sum_{\alpha} \sum_{\beta \neq \alpha} m^{(\alpha)} m^{(\beta)} \frac{\left(x_i^{(\alpha)} - x_i^{(\beta)}\right)\left(x_j^{(\alpha)} - x_j^{(\beta)}\right)}{|\mathbf{x}^{(\alpha)} - \mathbf{x}^{(\beta)}|^3}. \tag{13.6}$$

This is the symmetric potential energy tensor. Contracting it by setting $i = j$ and summing over the repeated index yields the gravitational potential energy of the system. Rewriting the left-hand side we have

$$\sum_{\alpha} x_j^{(\alpha)} \frac{d}{dt}\left(m^{(\alpha)} v_i^{(\alpha)}\right) = \frac{d}{dt} \sum_{\alpha} m^{(\alpha)} x_j^{(\alpha)} v_i^{(\alpha)} - \sum_{\alpha} m^{(\alpha)} v_i^{(\alpha)} v_j^{(\alpha)}$$

$$= \frac{d}{dt} \sum_{\alpha} \frac{1}{2} m^{(\alpha)} \left\{ \left(x_j^{(\alpha)} v_i^{(\alpha)} + x_i^{(\alpha)} v_j^{(\alpha)}\right) \right.$$

$$\left. + \left(x_j^{(\alpha)} v_i^{(\alpha)} - x_i^{(\alpha)} v_j^{(\alpha)}\right) \right\} - \sum_{\alpha} m^{(\alpha)} v_i^{(\alpha)} v_j^{(\alpha)}. \tag{13.7}$$

The antisymmetric term in (13.7) is just half the angular momentum (in the representative i and j coordinates) of the system. Since it is the only antisymmetric quantity in (13.5), it must be zero. This demonstrates the conservation of total angular momentum for the entire cluster as a bonus along the way to deriving the virial theorem.

There remains the symmetric part of (13.7):

$$\frac{1}{2}\frac{d}{dt}\sum_{\alpha} m^{(\alpha)}\left(x_j^{(\alpha)} v_i + x_i^{(\alpha)} v_j^{(\alpha)}\right) - \sum_{\alpha} m^{(\alpha)} v_i^{(\alpha)} v_j^{(\alpha)}$$

$$= \frac{1}{2}\frac{d^2}{dt^2}\sum_{\alpha} m^{(\alpha)} x_i^{(\alpha)} x_j^{(\alpha)} - \frac{1}{2}\frac{d}{dt}\sum_{\alpha} \dot{m}^{(\alpha)} x_i^{(\alpha)} x_j^{(\alpha)} - \sum_{\alpha} m^{(\alpha)} v_i^{(\alpha)} v_j^{(\alpha)}.$$

$$\tag{13.8}$$

The first term on the right-hand side is easily related to the inertia tensor

$$I_{ij} = \sum_{\alpha} m^{(\alpha)} x_i^{(\alpha)} x_j^{(\alpha)}, \tag{13.9}$$

which is the second spatial moment of the mass distribution. The second term is related to the mass variation tensor

$$J_{ij} = \sum_{\alpha} \dot{m}^{(\alpha)} x_i^{(\alpha)} x_j^{(\alpha)}, \tag{13.10}$$

which is the second spatial moment of the mass change throughout the system of galaxies. The third term is twice the kinetic energy tensor

$$T_{ij} = \frac{1}{2}\sum_{\alpha} m^{(\alpha)} v_i^{(\alpha)} v_j^{(\alpha)}, \tag{13.11}$$

which is the second velocity moment of the mass and whose contracted form T_{ii} is just the kinetic energy of all the peculiar velocities in the system.

With these physical quantities, the first-order moment (13.5) of the equations of motion becomes

$$\frac{1}{2}\frac{d^2 I_{ij}}{dt^2} - \frac{1}{2}\frac{d}{dt}J_{ij} = 2T_{ij} + W_{ij}. \tag{13.12}$$

This is the tensor virial theorem for a discrete collection of gravitationally interacting objects: in our case a cluster of galaxies. A special rate of mass loss proportional to the mass, $\dot{m}^{(\alpha)} = f(t)m^{(\alpha)}(t)$, simplifies (13.12) slightly to

$$\frac{1}{2}(1 - f(t))\frac{d^2 I_{ij}}{dt^2} = 2T_{ij} + W_{ij}. \tag{13.13}$$

Notice that to calculate the left-hand side, we need another equation for the evolution of the inertia tensor. We could find this by multiplying (13.4) by x_j^2, but that would just link the result to the evolution of the third-order moment – the start of an unending chain. Not promising.

Suppose, instead, that the system is in a steady state so that $d^2 I_{ij}/dt^2 = 0$. Then $2T_{ij} + W_{ij} = 0$ and from the positions and motions of the galaxies we could estimate their masses if we had independent evidence for the steady state assumption. Hopeful.

Alternatively, we may be able to examine a cluster long enough to obtain a time average defined for any quantity A by

$$\langle A \rangle = \lim_{\tau \to \infty} \frac{1}{\tau} \int_0^\tau A(t)\, dt. \tag{13.14}$$

For the moment of inertia

$$\left\langle \frac{d^2 I}{dt^2} \right\rangle = \lim_{T \to \infty} \frac{1}{\tau} \int_0^\tau \frac{d\dot{I}}{dt}\, dt = \lim_{\tau \to \infty} \frac{1}{\tau} [\dot{I}(\tau) - \dot{I}(0)]. \tag{13.15a}$$

This time average can be zero if either the cluster is localized in position and space so that $\dot{I}(\tau)$ has an upper bound for all τ, or if the orbits are periodic so that $\dot{I}(\tau) = \dot{I}(0)$ as $\tau \to \infty$. Additionally, if $\dot{m}(t)$ does not increase as fast as t for $t \to \infty$, then

$$2\langle T_{ij} \rangle + \langle W_{ij} \rangle = 0. \tag{13.15b}$$

Contracting and summing over indices gives

$$2\langle T \rangle + \langle W \rangle = 0, \tag{13.16}$$

which is the usual version of the virial theorem that generalizes (13.2). Quite promising.

How reasonable are the assumptions leading to (13.16) for galaxy clusters? If the masses of galaxies are changing, especially on timescales short compared with the dynamical crossing time $\sim (G\rho)^{-\frac{1}{2}}$ of the cluster, then the steady state assumption cannot hold. Nor will the instantaneous averages of T or W be close to their time averages. Even for constant $m^{(\alpha)}$, if many galaxies enter or leave the cluster, the virial theorem in the form of (13.16) will not be a good approximation. Although (13.13) with $f(t) = 0$ will hold, it will be difficult to measure the rate of change of I_{ij} accurately. Moreover, orbital periodicity as $\tau \to \infty$ occurs only for some two- or three-body systems and for very specialized larger systems.

So the simple virial theorem is most likely to apply when the cluster is gravitationally bound, relaxed, and evolving on timescales much longer than the crossing time. Under these conditions we can rewrite (13.16) as

$$M = \frac{2RV^2}{G} \tag{13.17}$$

with

$$R = \frac{\left(\sum_\alpha m^{(\alpha)}\right)^2}{\sum_\alpha \sum_\beta \frac{m^{(\alpha)} m^{(\beta)}}{|\mathbf{x}^{(\alpha)} - \mathbf{x}^{(\beta)}|}} \qquad (13.18)$$

and

$$V^2 = \frac{\sum_\alpha m^{(\alpha)} v^{(\alpha)^2}}{\sum_\alpha m^{(\alpha)}}. \qquad (13.19)$$

Individual coordinates $\mathbf{x}^{(\alpha)}$ and velocities $\mathbf{v}^{(\alpha)}$ are taken with respect to the center of mass. The summations are over all pairs of galaxies and individual galaxies (here the factor two to account for counting each pair twice is retained in M rather than in R). The numerator of (13.18) expresses the approximation that the number of pairs $N(N-1) = N^2$, and this should be modified by a term proportional to $\sum m^{(\alpha)^2}$ for small N.

In this way the virial mass, M, is expressed in terms of a mass-weighted mean square velocity and a mass-weighted harmonic radius. In practice, several subtleties and uncertainties arise when applying these results to observed clusters. These are discussed in more detail elsewhere (e.g., Aarseth & Saslaw, 1972; GPSGS) and so are just mentioned here. First, already from (13.18) and (13.19) we see that we need to know the individual galaxy masses in order to weight R and V^2 if we wish to determine M. Second, we measure instantaneous values of $\mathbf{x}$ and $\mathbf{v}$ rather than their time values. Third, the galaxies we see are projected on the sky and we do not know their radial separations or their transverse velocities. These are often assumed to satisfy statistical averages over projected angles. Individual cases could depart significantly from these averages, particularly if they contain streaming motions. Fourth, selection of subsets of galaxies for determining M will generally bias the results. The observational bias is impossible to determine a priori, but becomes apparent by examining larger samples.

Fortunately, N-body computer simulations of bound clusters show that these uncertainties often act in different directions. Usually their net effect is that M is uncertain by a factor of about two for any individual cluster. More sophisticated statistical techniques are often tried, but they introduce new biases and assumptions and the overall uncertainty remains at about this level.

Observed dynamical cluster masses, M, are often one or two orders of magnitude greater than their luminosity masses. To obtain the cluster's total luminosity mass, one converts the luminosity of each cluster galaxy to an approximate mass and sums them up. This mass–luminosity relation is based on the galaxy's morphology, size, color, and internal dynamics. It is usually determined by the properties of nearby galaxies for which these relations can be found more accurately, although they are frequently uncertain by at least a factor of two. But neither the dynamical nor the luminosity mass uncertainties are large enough to explain their difference.

It could be that many clusters are not actually bound and the $d^2 I_{ij}/dt^2$ term or mass loss dominates the virial theorem. The streaming velocities would then be greater than for a bound cluster and applying (13.17) would lead to a spuriously high value of M. For this to be significant, the overall expansion or contraction of the cluster would have to be comparable with its internal peculiar velocities. If these are $\sim 10^3$ kmp s^{-1}, then the cluster's structure changes over a scale of ~ 1 Mpc in about 10^9 years. Smaller massive clusters would almost completely disrupt during the $\sim 10^{10}$ year age of the Universe, but some larger ones might survive. The smaller clusters we now observe could then either be remnants of larger ones or could have formed recently. Any generic dynamical scheme along these lines would have to be consistent with the observed galaxy distribution function (Chapters 33 and 34). This seems difficult to contrive but needs to be explored further.

Although most astronomers believe that clusters contain dark matter – in addition to the dark matter within the galaxies themselves – the question is not completely settled. A variety of small groups or clusters appear to contain members whose redshift is substantially greater than those of its neighbors (e.g., Arp, 1987). The first (and often last) reaction of most astronomers is to regard these as distant galaxies projected along the line of sight into the group. Frequently, however, luminous protrusions or disturbances appear to connect the high redshift object with a normal redshift galaxy in the group. Are these physical connections or chance associations? A priori statistics suggest that the probability of chance association is low, but are the a priori hypotheses too restrictive? Are we seeing redshifts with a nonvelocity component? Have galaxies occasionally been ejected with high velocity? Until these questions are answered definitively for individual cases, they will cast a cloud – or perhaps a ray of light – over the subject.

If the high virial velocities in clusters are indeed "seeing" dark matter, its nature and location have remained obscure. Robust models of gravitational lensing and X-ray emission are particularly promising approaches. So far, the main reward of the general search has been a better understanding of cluster properties, which I sketch next.

13.2 Some Observed Properties of Groups and Clusters

Identifying a group or cluster as a relatively overdense region surrounded by an underdense region seems a simple working definition. Attempts to incorporate criteria of gravitational binding or specific velocity ranges run into the difficulties mentioned earlier (cf. Beers et al., 1990; McGill & Couchman, 1990). Happily, for most galaxies in most clusters the question of membership has a straightforward answer. Near the boundaries of both the definition and of the clusters themselves, however, ambiguity often sets in. The physical relevance of a particular cluster definition is, moreover, not always clear except in extreme cases. The definition of Abell clusters (see Chapter 7) illustrated these points. Many of these clusters (10–20%) are identified because two less dense clusters overlap along the line of sight. Their combined density contrast and size qualifies them for Abell's catalog where they distort the two-cluster correlation function (e.g., Dekel et al., 1989) and other cluster statistics.

Small groups (reviewed, for example, in Hickson, 1997) typically contain a few galaxies within a radius of several hundred kiloparsecs and have a velocity dispersion of several hundred kilometers per second. They can have dynamical crossing times $T_c \equiv R/V$ less than 10^8 years, which is the rotation period of a large spiral galaxy. In some cases such as H31 the galaxies interact tidally and may eventually merge (Rubin et al., 1990). A simple application of the virial theorem to several of Shakbazyan's compact groups leads to mass/light ratios of 20–300 (Kodaria et al., 1990), significantly higher than for isolated galaxies. Not all groups show strong evidence for substantial dark matter, and the amount seems to be correlated with the crossing time of the group, being less for groups with $T_c \lesssim 2 \times 10^9$ years (Field & Saslaw, 1971; Klinkhamer, 1980).

Many small groups are probably not bound but occur just by chance. Computer simulations (e.g., Walke & Mamon, 1989) illustrate that chance alignments commonly produce compact groups, especially within larger clusters. Watching films of N-body simulations of galaxy clustering, especially when the frames are rotated so the resulting parallax gives a three-dimensional impression, shows that many small groups form temporarily but then dissolve as galaxies move past one another. Sometimes a few galaxies linger together, only to be disrupted by another passing group. Sometimes they are absorbed into a larger more permanent cluster. It is impossible to predict their future from visual inspection alone, even in three dimensions.

Further evidence for unbound groups is that the members of many groups do not seem to have interacted in a way that would change their physical properties (Maia & da Costa, 1990). Neither the size of the brightest galaxy, nor the morphologies of the galaxies, nor their amount of H_α emission are related significantly to group properties such as density or velocity dispersion. Apart from their lower average surface density, regions around compact groups are usually very similar to areas of field galaxies without compact groups (Rood & Williams, 1989). This also suggests that often groups are an ephemeral accumulation.

Very small, compact groups with only several galaxies may have very low M/L ratios. Two groups each with four galaxies have $M/L \approx 24$ and 4 (Rose & Graham, 1979). These values are rather uncertain since they depend on corrections for galaxy inclination, internal absorption, dynamical fluctuations, velocity errors (which depend on the position of spectrograph slit on the galaxy), etc. Marvelously tiny groups of blue compact galaxies also exist. One example (Vorontsov-Vel'yaminov et al., 1980) has six blue interacting galaxies. Their total mass is about $10^{10} M_\odot$ and $M/L \approx 0.3$, implying that hot young stars dominate. The radial velocity dispersion of this group is about 44 km s^{-1}. Its total diameter is 15 kpc, only about one third the size of our own Milky Way.

Large clusters, in contrast, have radii ranging from several hundred kiloparsecs to 2–3 megaparsecs. Typically, ~ 1 Mpc is representative of a rich cluster. The extent of the largest clusters, such as those in the Coma and Corona Borealis constellations, is rather uncertain since we do not know how these clusters blend into the background field of galaxies. There also appear to be correlated distributions of galaxies extending over tens or perhaps even hundreds of megaparsecs. Unlike clusters whose densities are much greater than their surrounding average, these correlations represent very small density perturbations and are not dynamically bound entities. We shall meet them in more detail later in Part III.

The average peculiar radial velocity within a cluster ranges from 200 or 300 km s^{-1} to about 1,500 km s^{-1}. In Coma, for example, it is $\sim 10^3$ km s^{-1}. If the velocities are isotropic the average velocity would be a factor $\sqrt{3}$ greater.

Optical luminosities for the richest clusters are very difficult to measure, but typical estimates are in the range $L \approx (1-5) \times 10^{12} L_{\odot}$. Consequently, estimates of their total mass, using the virial theorem, give M/L ratios between about a hundred and a thousand. This is considerably greater than the average M/L ratios of individual galaxies (e.g., Faber & Gallagher, 1979), producing the "missing mass" problem for clusters, described earlier. Some rich clusters such as Abell 2029 (Uson et al., 1990) have central galaxies with enormously extended dim halos, out to ~ 1 Mpc. Their mass is uncertain, but their presence may help resolve the problem.

Masses of clusters can be estimated in several ways other than by the virial theorem or by adding luminosity masses of individual galaxies. X-ray emission from hot gas in a cluster suggests that some, perhaps most, of this gas is bound to the cluster. Detailed models of the observed X-ray distribution involve assumptions about its cooling, its relation to the galaxy mass distribution and the dark matter distribution, its temperature distribution, and either hydrostatic equilibrium or the way hot gas flows through the cluster. Applying various models of this sort to the Coma cluster (Hughes, 1989) suggests a total mass within 2.5 h^{-1} Mpc of about 10^{15} $h^{-1} M_{\odot}$, consistent with virial estimates (e.g., Zhao et al., 1991). The mass of hot gas in many clusters is comparable with that of luminous galaxies, but it is not itself sufficient to close the universe. It is an indicator of the depth of the cluster's gravitational potential well. As the X-ray emitting gas cools, it may form low mass stars, which eventually accumulate as dark matter (reviewed, e.g., in Fabian et al., 1991).

Gravitational lensing is another indicator of a cluster's mass. Images of more distant extended sources passing through a cluster are distorted as the cluster's total gravity bends the light paths (e.g., Saslaw, Narasimha, and Chitre, 1985). The bending is essentially independent of frequency (apart from scattering or absorption by galaxies and intergalactic gas). Most clusters are fairly transparent to more distant optical and radio radiation. Giant arcs were the first distorted images to be discovered. Competing explanations such as cluster shocks or ionization fronts fell by the wayside when spectra of the arcs showed their high redshifts and similarity to distant galaxies. Modeling these images has now become a major astronomical industry (e.g., Lynds & Petrosian, 1989; Bergmann et al., 1990; Hammer & Rigaut, 1989; Hammer, 1991). The models are not unique, but they become increasingly constrained as the image is mapped more sensitively at high resolution. Nearly all models require the clusters to contain substantial dark matter whose distribution may or may not follow that of the galaxies.

Masses of some rich clusters can also be estimated if smaller groups orbit around them. The fact that these groups have not been tidally destroyed puts an upper limit on the mass of the rich cluster, depending on assumptions about the orbit. Small groups around the Coma cluster, for example (Hartwick, 1981), limit its mass to $M_{\text{coma}} \lesssim 6 \times 10^{16} M_{\odot}$ if the tidal limit is applied to the present position of the groups. However, if the groups are moving away from Coma and the tidal limit is applied to a closest approach at the edge of Coma, the mass is reduced to $M_{\text{coma}} \lesssim 5 \times 10^{15} M_{\odot}$, consistent with the X-ray estimate of Hughes (1989). The uncertainty in Coma's total mass seems to be at least a factor of five.

More mysterious even than clusters' masses are their magnetic fields. Galaxies often contain global magnetic fields, but some clusters such as Coma (Kim et al., 1990) and the Virgo supercluster (Vallée, 1990) have fields on much larger scales of several megaparsecs. These are difficult measurements to make as they depend on the Faraday rotation of polarized background radio sources; they suggest cluster size fields of $1–2\mu G$ (Kim et al., 1991). The fields are not constant and, at least in Coma, seem to reverse on scales of 15–40 kpc, about the size of a galaxy. Do the flows of hot gas through the cluster produce these fields, or are they related to primordial conditions that predate the clusters?

Shapes of clusters are usually not spherical. Their elongation may indicate rotation, unrelaxed streaming velocities, substructure, and obscuration. The shape of the X-ray emission from the hot gas may differ from that of the galaxies. For small groups (Haud, 1990) this is not surprising since the orbits of their galaxies are constrained by energy and momentum conservation as well as by "isolating integrals" of the motion, which remain from initial conditions and change very slowly. As a result, these orbits will not fill position or velocity space uniformly. At any given time, the group's projected shape is most likely to be asymmetric. Larger clusters such as Coma also appear asymmetric (e.g., Baier et al., 1990), partly as a result of subclustering and extension toward neighboring clusters. This effect also illustrates the ambiguity of defining an individual cluster. A sample of 107 rich Abell clusters, in which automatic Palomar Sky Survey plate scans picked out galaxies brighter than about $m_R \approx 18.5$ magnitude in the region of each cluster, showed that about half the clusters are significantly elongated (Rhee et al., 1991a). While such results may be fairly robust to different methods of measuring position angles (such as Fourier analysis, moments, principal axes, multipoles), it is not clear how robust they are to different ways of distinguishing individual clusters. Moreover, since these are projected shapes, there is usually no unique way to reconstruct the intrinsic shape of a cluster. This is particularly true if clusters are a mixture of oblate, prolate, and triaxial configurations (Plionis et al., 1991).

Subclustering in clusters affects their apparent internal dynamics. In more extreme cases it makes the definition of a cluster ambiguous. A rich cluster may just be a superposition of smaller unrelated clusters along the line of sight. Various statistical tests (e.g., Oegerle et al., 1989; West & Bothram, 1990; Rhee et al., 1991b) have been applied to rich clusters to determine whether their subclustering is unlikely to occur by chance. These tests include percolation, bimodal clumping, minimal spanning trees, multipoles, symmetry deviations, angular spacings, and Monte Carlo simulations. The statistics work best when galaxy positions are combined with their radial velocities, but these are not always available. Present results show convincing evidence for subclustering in a small fraction ($\lesssim 10\%$) of rich clusters. Its physical significance is still unclear.

Alignment statistics are also difficult to measure and their significance is hard to determine. Motivated by the idea that neighboring clusters might form within larger scale perturbations, astronomers have searched for alignments of their major axes. Evidence for neighboring cluster alignment is very weak (Ulmer et al., 1989; Lambas et al., 1990; Fong et al., 1990). The problem is not yet sufficiently well defined, either observationally or theoretically. Not only are the cluster axes affected by questions of cluster membership and subclustering, but they may also differ in the X-ray, optical, galaxy, and possible dark matter distributions. Moreover, the ways in which small groups of galaxies agglomerate and interact tidally have not been studied realistically for a wide enough range of possible conditions to draw definite conclusions. There is some evidence that in rich clusters of galaxies, the major axis of the brightest galaxy is aligned with the cluster's axis (van Kampen et al., 1990; Struble, 1990). This may result from local tides, which reflect the overall shape of the cluster, or from previous mergers.

Morphological segregation of galaxies, with ellipticals tending to concentrate at the center of rich clusters and spirals preferentially populating the outer regions, is a common feature. The degree of segregation varies greatly among clusters and it is not yet clear how to quantify it in a physically meaningful way (e.g., Schombert & West, 1990; Yepes et al., 1991). Several causes are possible: initial conditions that form more massive or lower angular momentum ellipticals near the center, orbital interactions (few-body "dynamical friction") that decrease the average distance of massive ellipticals from the center, ram-pressure stripping by intergalactic gas, which destroys spirals preferentially near the center, and merging of galaxies near the center to produce ellipticals. Whether other physical properties of galaxies, such as their infrared or ultraviolet emission or their color gradients, are strongly affected by the local environment remains an open question (e.g., Mackie et al., 1990).

Optical luminosity functions (the fraction of galaxies in any given range of luminosities) have been determined for many clusters (e.g., Oegerle & Hoessel, 1991). Over a wide range they generally have a unimodel skew form with parameters that may differ among clusters and in different wavelength bands. Clusters may also have a large component of faint, low surface brightness galaxies, which are difficult to detect. At the other end of the luminosity function, it is important to determine whether the brightest galaxies in rich clusters have their own special luminosity function or whether they are just the chance exceedences of the individual cluster luminosity functions, or both (Bhavsar & Cohen, 1990).

All these properties suggest that rich bound clusters are somehow set apart from the hoi polloi of most other galaxies that interact weakly. Yet we shall see in Chapters 33–35 that the existence of these rich clusters can also be understood as the high density tail of a universal distribution function for all galaxies.

13.3   Physical Processes in Bound Clusters

Once a cluster becomes bound, gravity determines its future. Other physical processes such as gas dynamics are important, but even these are ultimately governed by gravity. And all we really know about the dark matter in clusters is that it interacts gravitationally. The two main physical processes we consider here are collective (sometimes violent) relaxation and two-body relaxation. These determine the overall distribution of galaxies in the cluster. Other physical processes such as gas stripping, tides, and mergers are significant in the lives of individual galaxies, but these only affect their distribution marginally.

How does a cluster form? The unknown answer to this simple question determines the importance of collective relaxation. Collective relaxation occurs when the motions of individual galaxies are dominated by the gravitational forces due to large, usually temporary, collections of other galaxies in the forming cluster. In contrast, two-body relaxation occurs when motions of individual galaxies are dominated by the gravitational forces of other single galaxies that are relatively uncorrelated. Each of these mechanisms may dominate at different times or places during the formation of a particular cluster. Both may occur together when neither dominates. Stated slightly differently, we can represent the gravitational force acting on a galaxy as the sum of the mean gravitational field, averaged over the whole forming cluster, and of a fluctuating field. If the mean field changes slowly and the fluctuations are small but change relatively rapidly, caused mainly by the changing positions of a few neighboring galaxies, then two-body relaxation dominates. But if the mean field changes rapidly, or if the fluctuations involve many galaxies and are so large and change so fast that it is not sensible to define a mean field, then collective relaxation dominates.

Theories in which massive clusters form first, out of long-wavelength high-amplitude primordial density perturbations, followed by the clusters fragmenting into galaxies, may initially have a short period of collective relaxation if the regions of galaxy formation are very inhomogeneous. Then they continue to evolve by two-body relaxation, which, as we shall see, is much slower. The same general picture may apply if the clusters are dominated by nonbaryonic dark matter, rather than by the gas that forms the galaxies. However, in theories where galaxies form first out of short-wavelength high-amplitude primordial density perturbations and then clumps of galaxies gravitationally accumulate until a few massive clumps come together to form the final cluster, collective relaxation is very important initially until the galaxy distribution in the cluster becomes smooth. Then slow two-body relaxation takes over. The first group of theories are difficult to formulate very accurately because they combine many aspects of hydrodynamics, gravity, and radiative transfer, all subject to many uncertainties. The second approach, relying on just gravitational interactions among galaxies represented by point masses, is much simpler. We mainly explore this simpler approach. In either case, gravitational clustering of point masses, is guaranteed to be a major aspect of our eventual understanding of galaxy clustering.

The keynote of any relaxation is dissipation. Initial states and structures are lost; memories of earlier times evanesce. Entropy increases and the system may evolve slowly through a sequence of quasi-equilibrium states. Thus what we see now may provide little clue to what happened before. Finding and interpreting these clues then becomes a difficult task.

The most extreme form of dissipation is "violent relaxation" (Lynden-Bell, 1967). This name is sometimes given to all modes of collective relaxation, but properly it applies to one limiting case from which much may be learned. The basic idea behind violent relaxation is that the initial state is so far from equilibrium that large-scale collective modes govern its early evolution. Since the initial distributions of masses, positions, and velocities are very irregular, individual galaxies are scattered mainly by *groups* of galaxies, rather than by other individual galaxies. The groups themselves are constantly disrupting and reforming. No particular group survives violent relaxation for long, but the statistical distribution of groups lasts longer. A galaxy may belong first to one group, then to another, and finally to none at all, or to the whole relaxed cluster.

During violent relaxation, each galaxy moves in a mean field that changes quickly in time, as well as in space. Amplitudes of these changes are large. The energy of each galaxy along its orbit is not conserved. Only the global invariants, such as the energy, of the entire cluster remain constant. This allows us to bypass structural details when seeking the outcome of violent relaxation.

It is easy to estimate the timescale for this collective relaxation. Consider the case where a protocluster consists of several large irregular groups of galaxies falling together under their mutual gravitational attraction. It is far from equilibrium. The groups collide on a timescale $\approx (G\rho_0)^{-\frac{1}{2}}$, where ρ_0 is the original average density of the whole system. Having collided, the strong gravitational tides distort the groups. Chaotic motions prevail and the system oscillates for several dynamical periods $(G\rho_1)^{-\frac{1}{2}}$, with $\rho_1 \gtrsim \rho_0$. Then it settles down as energy passes from the collective modes to individual galaxy orbits (Landau damping) and these orbits become smoothly distributed over the cluster (phase mixing). This whole process occurs over several initial $(G\rho_0)^{-\frac{1}{2}}$ dynamical timescales, typically $\sim 10^9$ years at a redshift $\gtrsim 3$.

The main result of pure violent relaxation is that a galaxy's final velocity does not depend on its mass. We can understand this quite simply. The time-changing energy of the mean field acting on the galaxy, $-m\partial\psi/\partial t$, equals the change in the galaxy's total energy, $m\,d(\frac{1}{2}v^2 - \psi)/dt$, where $\psi\,(\mathbf{r},\,t)$ is the gravitational potential of the whole cluster. The result is therefore independent of m, the mass of an individual galaxy. Alternatively, the acceleration is just equal to the gradient of the potential: Since the scattering groups and collective modes have much greater mass than the scattered galaxy, the mass of the individual galaxy cancels out. Thus there is equipartition of energy per unit mass, but not of energy itself. As a corollary, this type of relaxation does not lead to mass segregation. So greater masses do not tend to clump in the center.

What will be the distribution of these velocities? We know it does not depend on mass, but can we say more? Suppose there are no constraints apart from conservation of total mass, energy, and momentum. Each rapidly changing collective mode produces a force that is completely uncorrelated with its predecessors, its successors, and its compatriots. This idealized hypothesis implies that the orbit of each violently relaxing galaxy undergoes a series of random velocity changes. We can then use the central limit theorem of probability theory to determine the resulting velocity distribution. A useful simple version of this theorem states that if $x_1, x_2, \ldots, x_n$ are mutually independent random variables with a common distribution whose expectation is zero and variance is σ^2, then as $n \to \infty$ the distribution of the normalized sums $S_n = (x_1 + \cdots + x_n)/\sqrt{n}$ tends to the normal distribution with probability density $f(x) = (2\pi\sigma_x^2)^{-\frac{1}{2}} \exp(-x^2/2\sigma_x^2)$. Thus, if x_i is a one-dimensional velocity increment, the probability of finding a galaxy whose final velocity is between x and $x + dx$ where $x = x_1 + \cdots + x_n$ is given by this normal distribution. The theorem holds under more general conditions for multidimensional random variables with different distributions. The chief additional constraint is that the variance of each distribution be much less than the sum of all the variances.

Therefore, if the velocity increments $(\Delta v_x, \Delta v_y, \Delta v_z)$ in each of the three spatial dimensions are uncorrelated with each other as well as in time and space, the final result will have a normal distribution, which is just the product of the distributions in each dimension:

$$f(v)\, dv = \left(\frac{3}{2\pi\sigma^2}\right)^{\frac{3}{2}} e^{-3v^2/2\sigma^2}\, dv_x\, dv_y\, dv_z, \qquad (13.20)$$

where $v^2 = v_x^2 + v_y^2 + v_z^2$ and $\sigma^2 = \sigma_x^2 + \sigma_y^2 + \sigma_z^2$. Since the distribution is isotropic, the velocity components in each dimension are equivalent, and $\sigma_x^2 = \sigma^2/3$ etc. If there were equipartition of kinetic energy in each dimension, we would have $\sigma^2 = \langle v^2 \rangle = 3T/m$ with m the mass of each galaxy (assumed to be the same for all galaxies here, along with setting Boltzmann's constant $k = 1$), and (13.20) would be the Maxwell–Boltzmann velocity distribution. But there is not equipartition of energy since the collectively excited velocity of a galaxy is independent of its mass. This is equivalent to supposing that the temperature T is proportional to mass, so that σ^2 is independent of m.

While this heuristic discussion describes the essential physics and main results of violent relaxation, more rigorous derivations involve subtleties (see GPSGS for example) concerning the detailed conditions for violent relaxation to prevail (it tends to damp out on the same timescale as it relaxes). There are also subtleties involving the final state of clusters that contain galaxies of different masses and the relations to initial conditions. It is clear that the ideal case of violent relaxation can describe some important aspects of rapid early collective relaxation, which leads to the fairly smooth density distribution of galaxies in the rich clusters we see today. However, the mass and luminosity segregation found in these clusters implies that other processes have also been at work.

Two-body relaxation is one of the most important of these additional processes. It can also be important for the distribution of galaxies outside clusters. Once a cluster forms by collective relaxation, its density distribution becomes fairly smooth. Then the gravitational force acting on a galaxy can be divided roughly into two components. One is the mean field of the average smoothed density of all the other galaxies. This changes relatively slowly over the scale of the cluster although it can be considerably greater in the center. The other component is the local field of individual galaxies, which fluctuates around the mean field. This accounts for the actual granularity of the cluster. The effects of the forces of individual galaxies are often examined in the approximation of an infinite system with a constant average density, so that the mean field is zero. This is not strictly consistent, but for a large, rich cluster it usually works quite well. Our treatment here follows that of GPSGS.

Because gravity is a binary interaction, occurring between pairs of galaxies, the deflection that results from all the interactions in a system is essentially the sum of all pair interactions. The dynamical effect of each interaction is to deflect the galaxy's previous orbit. It is easy to estimate this deflection for weak scattering. Consider, as in Figure 13.1, a massive galaxy M_1 deflecting a much less massive galaxy m_2, moving initially with velocity v perpendicular to the impact parameter (distance of closest approach of the undeflected orbit) b. The gravitational acceleration Gm_1/b^2 acts for an effective time $2b/v$ to produce a component of velocity

$$\Delta v \approx \frac{2Gm_1}{bv} \tag{13.21}$$

approximately perpendicular to the initial velocity. Since $\Delta v \ll v$, the effects are linear and they give a scattering angle $\psi \approx \Delta v / v$. Although this is the physical essence of the problem, an exact treatment, which follows the orbits in detail (available from most standard texts in classical mechanics), shows that, more generally,

$$\tan\frac{\psi}{2} = \frac{G(m_1 + m_2)}{v^2 b} \tag{13.22}$$

and

$$\frac{\Delta v}{v} = 2\sin\frac{\psi}{2}. \tag{13.23}$$

Unlike collective relaxation, the velocity change now depends on each galaxy's mass. Equations (13.22) and (13.23) reduce to (13.21) for $m_2 \ll m_1$ and $\psi \ll 1$. They were originally applied to stellar systems by Jeans (1913) working in analogy with

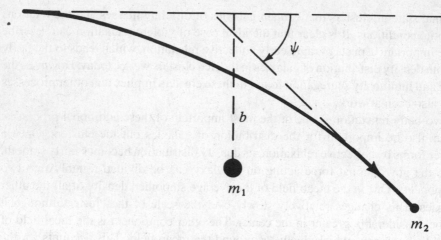

Fig. 13.1. The orbit of a deflected star.

Maxwell's kinetic theory of gases. Throughout its development, there have been many productive exchanges of techniques and results between kinetic theory – and later plasma physics – and gravitational many-body physics.

These results show that large velocity changes, and large scattering angles, occur when two galaxies are so close that their potential energy is about equal to the kinetic energy. Are these close encounters typical? Suppose N galaxies are in a cluster of radius R. Characterizing typical encounters by an impact parameter about equal to the mean interparticle separation $b \approx RN^{-\frac{1}{3}}$, an average mass $m_1 \approx M/N$, and an initial velocity of about the equilibrium velocity dispersion $v^2 \approx GM/R$ (from the virial theorem, Equation 13.17) then shows that

$$\frac{\Delta v}{v} \approx \psi \approx N^{-\frac{2}{3}} \tag{13.24}$$

for $N \gg 1$. Hence the typical neighborly interaction of galaxies in a system near equilibrium involves very little energy or momentum exchange. Of course, there will always be a few galaxies in any cluster that have low relative velocities. Usually, these are galaxies near turning points of their orbits in the averaged potential, in the outer regions where they linger longer. In these positions they are particularly susceptible to large deflections. However, the probability that they encounter a similarly slow galaxy nearby is small enough to make these large deflections rare. Nevertheless, there is one case where they are important. In inhomogeneous clusters with galaxies of different masses, the most massive members accumulate at the center and strongly alter each other's orbits. If they approach each other too closely, tidal interactions may make them merge or disrupt.

Although individual deflections are small, their cumulative effect is not. We may see this in several ways. The simplest approximation is to suppose that each small deflection in a typical encounter alters the orbit deflection angle r and velocity v

in a completely random direction, by an amount given by (13.22) and (13.23). The magnitudes of v and ψ then undergo a random walk and their root mean square values increase in proportion to the square root of the number of scatterings. The average number of scatterings of any given galaxy by surrounding galaxies having velocity between v and $v + dv$, and with an impact parameter between b and $b + db$ in time dt, is

$$2\pi b f(v,\, t)v\, dt\, db\, dv, \tag{13.25}$$

where $f(v,\, t)\, dv$ is the number of galaxies with velocity between v and $v + dv$ at time t, normalized so that its integral over all velocities is the total number density of galaxies n. In the steady state, $f(v)$ is independent of t; moreover, the fact that it depends only on the magnitude of v implies that we are taking the velocity distribution to be isotropic around any given point.

To estimate the mean square velocity change, we multiply the number of these scatterings by the square of the velocity change $(\Delta v)^2$ in each scattering described by (13.22) and (13.23) for the case $\Delta v \ll v$, $m_1 = m_2$, and $f(v,\, t) = f(v)$. Then we integrate over dt, db, and dv. This procedure overestimates the effects of the last few scatterings. But since each scattering is small, it does not seriously affect the result.

The results of each scattering add up in this linear manner for $(\Delta v)^2$ because every deflection is small and uncorrelated with any other. In a steady state, the integral over time is trivial. The integral

$$\int_0^\infty v^{-1} f(v)\, dv = n\langle v^{-1}\rangle \tag{13.26}$$

is just the average inverse velocity. The integral over impact parameter,

$$\int_{b_1}^{b_2} \frac{db}{b} = \ln\frac{b_2}{b_1}, \tag{13.27}$$

introduces a new feature of the problem. Its limits diverge at both extremes, as either b_2 becomes very large or b_1 very small, and so we must introduce a cutoff. The upper cutoff is fairly straightforward. Physically, it arises from the long-range nature of the gravitational potential and represents the largest distance over which any disturbances in the system can interact. In a finite system, this is approximately the cluster's radius R. In infinite systems, it is the distance over which disturbances can be correlated. This is closely related to the "Jeans length" at which the kinetic and potential energies of perturbations are about equal. Its plasma analog, produced by charge screening, is the Debye length.

The lower cutoff arises because very close scatterings disturb v in a violently nonlinear way. These large-angle deflections are rare and are eliminated by equating b_1 to the impact parameter for, say, $90°$ scattering. From (13.22) we then get $b_1 = 2Gm/v^2$, the distance at which the kinetic and potential energies are equal.

Thus $\ln (b_2/b_1) \approx \ln (Rv^2/2Gm)$. For a finite system close to equilibrium, the virial theorem shows that $v^2 \approx GmN/R$, that is, that the average kinetic energy of a star is about one half its potential energy in the cluster. Thus the logarithmic term is approximately $\ln (N/2)$. An alternative approach is to take the lower cutoff to be the distance at which a nearby galaxy would exert as much gravitational force as the average force of the cluster: $m/b_1^2 \approx M/R^2$. In this case $\ln (b_2/b_1) \approx 1/2 \ln N$. A third alternative would be the nearest neighbor separation: $b_1 \approx (4\pi R^3/3N)^{1/3}$. Although some interesting physical problems surround this ambiguity in b_1, its very weak effect on the result shows that it is not a great barrier to our understanding of two-body relaxation.

Putting the three integrations together gives a cumulative mean square velocity perturbation

$$\left(\frac{\Delta v}{v}\right)^2 \approx 32\pi G^2 m^2 n v^{-3} \ln (N/2)t. \qquad (13.28)$$

Apart from the ambiguity of the logarithmic factor, the main uncertainty in this formula comes from the identification of $\langle v^{-1} \rangle$ with the inverse of the velocity dispersion in the system. Thus (13.28) should be multiplied by a numerical factor, usually between 0.1 and 1.0, which depends on the velocity distribution function. For example, the more detailed discussion of Chandrasekhar (1960) shows that with a Maxwellian distribution, this factor is about 0.2 for test galaxies having speeds close to the velocity dispersion. In practice, these uncertainties are usually much less than the uncertainties in the astronomically observed values of m, n, v, and t.

Since deflections increase the expectation value of the root mean square velocity, we can easily define a a timescale τ_R, using (13.28), for this many-body relaxation effect to become significant for a typical galaxy. We just set $(\Delta v/v) = 1$, whence

$$\tau_R = \frac{v^3}{32\pi G^2 m^2 n \ln (N/2)}$$
$$= \frac{v^3 R^3}{24 G^2 m M \ln (N/2)}, \qquad (13.29)$$

where M is the total mass. In systems near equilibrium, $GM \approx Rv^2$. We can define a characteristic "crossing time"

$$\tau_c = R/v \qquad (13.30)$$

and write

$$\tau_R = \frac{0.04 N}{\ln (N/2)} \tau_c. \qquad (13.31)$$

Since for rich clusters of galaxies $N \gtrsim 10^3$, the relaxation time τ_R is much longer than the dynamical crossing time. So cumulative small deflections do not amount to much over any one orbital period, but their secular effect can dominate after many orbits, particularly for small groups of galaxies. If N is so small that $\tau_R \approx \tau_c$, it is difficult to distinguish between the mean and the fluctuating field, and (13.29) becomes invalid. In comparison to τ_R, the crossing timescale τ_c, which also governs collective relaxation, can be so short that the collective relaxation is sometimes called "violent."

Notice that (13.28) seems to imply that the mean square velocity can increase indefinitely, even though the rate of increase will slow as v^{-1}. Of course, this cannot be since the galaxy would eventually reach escape velocity and leave the cluster. We can make a rough estimate of this evaporation timescale. The escape velocity, $v_{\text{esc}} \approx (4GM/R)^{1/2}$, is typically a factor two greater than the velocity dispersion in a cluster near equilibrium. So an average galaxy evaporates from the cluster when $\Delta v \approx v$, and from (13.29) this just takes a time τ_R. Thus τ_R can also be interpreted as the timescale for a cluster to change its structure significantly by losing its members.

How many "effective encounters" does a galaxy undergo before it reaches escape velocity? This is essentially the problem of a random walk through velocity space with an absorbing barrier at the escape velocity. Figure 13.2 illustrates this random walk schematically. If the velocity increments have a Gaussian distribution with

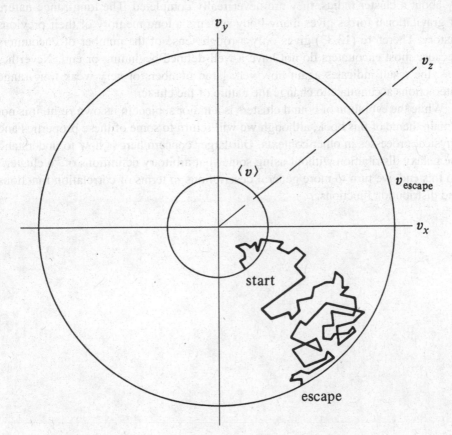

Fig. 13.2. Random walk of a galaxy in velocity space from $\langle v \rangle$ to v_{esc}.

mean value zero, so that there is no net velocity drift, and dispersion σ, then it is physically plausible (and can be shown rigorously (e.g., Cox & Miller, 1965) that the expected number I of increments necessary to reach escape velocity is given by

$$v_{\text{esc}} \approx \sigma \sqrt{I}. \tag{13.32}$$

Now in a Gaussian distribution, the velocity dispersion σ is also the most probable value of a velocity increment Δv. We can estimate it from (13.22) and (13.23) for $m_1 = m_2$ by using the volume-weighted average $1/b$ for a uniform spherical distribution: $\langle b^{-1} \rangle = \frac{3}{2} R^{-1}$. Thus $\sigma = \langle \Delta v \rangle = 6Gm/Rv$ and with $v_{\text{esc}} = 2v$ we find that a galaxy escapes after undergoing approximately

$$I = \frac{v_{\text{esc}}^2}{\sigma^2} = \frac{R^2 v^4}{9G^2 m^2} = \frac{N^2}{9} \tag{13.33}$$

encounters. Here N is again the number of galaxies in the cluster and the last equality applies for virial equilibrium.

Since most encounters are very distant ones, occurring among galaxies separated by about a cluster radius, they are never really completed. The long-range nature of gravitational forces gives many-body systems a long memory of their previous history. Therefore (13.33) gives only a rough sense of the number of encounters because most encounters do not have a well-defined beginning or end. Nevertheless, the result indicates again how very large numbers of very weak long-range interactions accumulate to change the nature of the cluster.

While the evolution of bound clusters is a major subject in its own right, it is not a main theme of this book, although we will return to some of these properties and physical processes in other contexts. Our larger concern here is how to understand the galaxy distribution without using somewhat arbitrary definitions of "a cluster." To this end, we turn to more general descriptions in terms of correlation functions and distribution functions.

14

Correlation Functions

A long pull, and a strong
pull, and a pull all together.

Dickens

14.1 Definitions

Sitting in a galaxy, it is natural to ask if nearby galaxies are closer than average. The average intergalaxy separation follows from the average number density, $\bar{n}$, so this is the first quantity to find. Even assuming we know the positions of all the galaxies in a relevant sample, which automatically accounts for their luminosity function or other selection effects, $\bar{n}$ depends on two interesting subtleties.

First, we must specify whether $\bar{n}$ is a conditional density $\bar{n}_c$, averaged over volumes known a priori to contain a galaxy at their center, or an unconditional density $\bar{n}_u$, averaged over volumes located randomly in space. If the galaxies themselves have a Poisson distribution, then $\bar{n}_c = \bar{n}_u$ for an average over sufficiently large volumes. But for a correlated distribution, these two densities will generally differ.

Second, a true number density should not depend on the size of the volumes V over which the number counts are averaged. Again, this holds for a Poisson distribution but not generally if there are correlations. So we will have to proceed a little carefully.

To illustrate the differences between conditional and unconditional distributions, consider the simple case of galaxies having a Poisson distribution with density $\bar{n}_u$ throughout space. The probability for finding N galaxies in an arbitrary volume, V, is

$$f_V(N) = \frac{\bar{N}^N e^{-\bar{N}}}{N!}, \tag{14.1}$$

where $\bar{N} = \bar{n}_u V$ and $\sum_{N=0}^{\infty} f_V(N) = 1$ (since the coefficients of $e^{-\bar{N}}$ are the expansion for $e^{\bar{N}}$). Consider the distance d, from an arbitrary point in space to the nth nearest galaxy. This distance defines a volume $V_n = \frac{4}{3}\pi d^3$ extending to the nth nearest galaxy. The probability that V_n is greater than V is

$$P(V_n > V) = P(V \text{ contains no more than } n-1 \text{ galaxies})$$

$$= \sum_{N=0}^{n-1} \frac{(\bar{n}_u V)^N}{N!} e^{-\bar{n}_u V}. \tag{14.2}$$

Thus the distribution of the distance to the nearest galaxy, $n = 1$, is an exponential in V, which is also the probability for finding a void in any region. The sum in (14.2) is essentially an incomplete gamma function,

$$P(V_n > V) = \frac{\Gamma(n, \bar{n}_u V)}{\Gamma(n)}, \tag{14.3}$$

for the nth nearest galaxy, as may be seen by integrating

$$\Gamma(n, \bar{n}_u V) = \int_{\bar{n}_u V}^{\infty} e^{-t} t^{n-1} dt \tag{14.4}$$

by parts.

Next consider the conditional probability distribution for finding N galaxies in a volume V, which is chosen to have a galaxy at its center. Suppose the Nth galaxy is at the center, and the other $N - 1$ have a Poisson distribution within V. To compare with the unconditional void probability $P(V_1 > V)$ in (14.2), we want the conditional probability that there are no galaxies in a volume $v < V$, given that there is some number $N \geq 1$ of galaxies in V. Since the probability of not finding $N - 1$ independently placed galaxies in v is $(1 - v/V)^{N-1}$,

$$
\begin{aligned}
P(\text{no galaxies in } v \mid N \geq 1) &= \frac{\sum_{N=1} \left(1 - \frac{v}{V}\right)^{N-1} f_V(N)}{\sum_{N=1} f_V(N)} \\
&= E\left[\left(1 - \frac{v}{V}\right)^{N-1} \middle| N \geq 1\right] \\
&= \frac{e^{-\bar{n}_u v} - e^{-\bar{n}_u V}}{\left(1 - e^{-\bar{n}_u V}\right)\left(1 - \frac{v}{V}\right)}
\end{aligned} \tag{14.5}
$$

using (14.1).

For large volumes $V \gg v$, which are expected to contain many galaxies so that $E(N) = \bar{N} = \bar{n}_u V \gg 1$, the a priori knowledge that V contains at least one galaxy should not matter very much. And indeed, in this limit, (14.5) reduces to $e^{-\bar{n}_u v}$, which is a distribution of the same exponential form as (14.2) for $n = 1$. When these limits do not apply, however, the knowledge that a volume contains at least one galaxy significantly changes the probability for finding an empty region within it.

To illustrate how the density could depend on volume, imagine a rather extreme case in which all the galaxies in a large volume are arranged on a linear filament with linear density λ galaxies per unit length. For spherical volumes having the filament as a diameter, the average volume density is $\bar{n} = 2\lambda r/(4\pi r^3/3) = (3/2\pi)\lambda r^{-2} \propto V^{-2/3}$. Clearly this volume dependence of $\bar{n}$ arises from the existence of coherent structure over the entire scale of the system. Only for scales larger than any structure, where galaxies fill space homogeneously, can we define an average density. If galaxies have a fractal distribution (Chapter 12), they have structure on all scales, and even though it is self-similar, the average density will still depend on the volume.

In this section, we avoid these complications, which will return to haunt us later, by supposing a homogeneous sample of galaxies to have an unconditional density $\bar{n}_u \equiv \bar{n}$ that does not depend on scale. One way to imagine such a sample is to average over an ensemble of realizations. Each realization of the system has the same average macroscopic properties such as density $\bar{n} = N/V$, where N is the total number of galaxies in the total volume V of each realization, or temperature $T = \bar{m}\langle v^2 \rangle/3$, where $\bar{m}$ is the average mass of the galaxies and $\langle v^2 \rangle$ is their peculiar velocity dispersion (setting Boltzmann's constant $k = 1$). But the detailed configurations of galaxy positions and velocities may differ among the realizations. If all possible configurations occur sufficiently randomly, that is, with an equal a priori probability, among these realizations, then their ensemble average may represent a uniform system. This is related to the ergodic hypothesis discussed in Chapter 24. Another way to imagine a homogeneous sample is to consider a single realization on a scale larger than any coherent structure in the realization. This is common in the theory of liquids, where many ideas about correlation functions arose. Intermolecular spacings in liquids are usually many orders of magnitude smaller than the size of the sample, be it a tea cup or an ocean. Samples of galaxies, and especially of clusters of galaxies, often lack this property and so run into problems.

This basic problem is known as the "fair sampling" problem. In our Universe, we must either assume that a large enough volume provides a fair sample of the averaged properties of the Universe or else that, by combining enough small samples separated by such vast distances that they do not influence each other, we can also obtain the averaged properties. A fair sample is distinct from a "representative sample," which contains the typical structure and biases found in its size region, without averaging over the microstructure of many regions.

Having obtained an average density $\bar{n} = N/V$ for a single ideal region that is homogeneous on large scales, the next step is to determine whether the positions of galaxies are correlated with each other on smaller scales. Correlations could grow naturally from gravitational clustering or remain from the initial conditions of galaxy formation. Sitting in a galaxy – any galaxy – one would find that the number of galaxies in a small volume element dV within a radius r is not generally $\bar{n}dV$. The central galaxy has perturbed its environment, or the initial distribution has not relaxed to the average density, or both. There are several ways of representing this state of affairs (e.g., Hill, 1956; Totsuji and Kihara, 1969; Peebles, 1980; Coleman and Pietronero, 1992). We had better distinguish among them to avoid the confusion sometimes found in the literature.

First, we can write the conditional probability for finding a galaxy in a volume dV_2 at $\mathbf{r}$, given that there is a galaxy in a volume dV_1, as

$$P(2 \mid 1) = \bar{n}\,(1 + \xi(\mathbf{r}))\,dV_2, \tag{14.6}$$

where $\mathbf{r} = \mathbf{r}_2 - \mathbf{r}_1$. This relation defines a two-point correlation function, $\xi(\mathbf{r})$, which allows for the influence of the galaxy in dV_1 on the probable presence of a galaxy in

dV_2. If $\xi > 0$, this probability is enhanced; if $\xi < 0$ it is decreased (anticorrelation); if $\xi = 0$ there is no effect (Poisson distribution). Since the a priori probability for finding a galaxy in dV_1 is $\bar{n}\, dV_1$, the joint probability for finding a galaxy in dV_1 and also one in dV_2 is the product:

$$P_{1,2} = \bar{n}^2\, [1 + \xi(\mathbf{r})]\, dV_1\, dV_2. \tag{14.7}$$

In general ξ will depend on the absolute positions, $\mathbf{r}_1$ and $\mathbf{r}_2$, of the two volume elements. However, if we average over all positions and all directions, then ξ will become a function only of the separation $r = |\mathbf{r}_1 - \mathbf{r}_2|$. For $\xi(r)$ to provide a good description, the system must be statistically homogeneous, so that the average over all the galaxies in a given volume will be the same as for any sufficiently large subset of galaxies in the same volume. If this is not the case, the average needs to be taken over an ensemble of independent systems.

Now we can also write $P_{1,2}$ as an average over all the galaxies in the volume of the probability of finding a galaxy at $\mathbf{r}_1 + \mathbf{r}_2$, given a galaxy at $\mathbf{r}_1$:

$$P_{1,2} = \langle n(\mathbf{r}_1) n\,(\mathbf{r}_1 + \mathbf{r}_2)\rangle\, dV_1\, dV_2, \tag{14.8}$$

where for any variable $Y(\mathbf{r})$,

$$\langle Y \rangle = \frac{1}{V} \int_V Y(\mathbf{r})\, d\mathbf{r}. \tag{14.9}$$

Comparing (14.7) and (14.8) shows that ξ can be defined consistently as

$$\xi(\mathbf{r}) = \frac{\langle n(\mathbf{r}_1) n\,(\mathbf{r}_1 + \mathbf{r}_2)\rangle}{\bar{n}^2} - 1, \tag{14.10}$$

which is the excess correlation (or anticorrelation) compared to a random Poisson distribution having n independent of r.

Integrating $\xi(\mathbf{r})$ in (14.10) over the entire volume, or $P_{1,2}$ from (14.7) over all dV_1 and dV_2, gives

$$\int_V \xi(\mathbf{r})\, d\mathbf{r} = 0, \tag{14.11}$$

provided we recognize that the same galaxy can contribute to both $n(\mathbf{r}_1)$ and $n(\mathbf{r}_1 + \mathbf{r}_2)$, so that there are N^2 possible pairs (allowing for two ways of labeling each pair). If we consider instead that each of the N galaxies has $N - 1$ pairs with other galaxies, then the integral (14.11), normalized by N^{-2}, becomes $[N\,(N - 1) - N^2]N^{-2} = -N^{-1}$. The advantage of (14.11) is that it rather naturally represents the conservation of the total number of galaxies in any statistically homogeneous volume where the positive fluctuations balance out the negative ones. The advantage of using $N(N-1)$ pairs is that it provides a more natural representation of the gravitational interaction energy. Either form can be used consistently in its appropriate context.

If the volume is not statistically homogeneous, one can still use an ensemble average density, $\bar{n}_e$. But this will not generally equal $\bar{n}$ for any particular volume in the ensemble. Therefore $\xi(r)$ will have to be redefined as an average over the entire ensemble if (14.11) is to be satisfied. Any particular volume V could have a net positive or negative correlation relative to $\bar{n}_e$.

Another related definition of the two-point correlation function follows by emphasizing the fluctuations, Δn. Writing the density in any region as $n = \bar{n} + \Delta n$, substituting into (14.10), and noting that $\langle \Delta n \rangle = 0$ leaves us with

$$\langle \Delta n(\mathbf{r}_1) \Delta n(\mathbf{r}_1 + \mathbf{r}_2) \rangle = \bar{n}^2 \xi(\mathbf{r}). \tag{14.12}$$

But something is missing. If we integrate (14.12) over $\mathbf{r}_1$ and $\mathbf{r}_2$ in a volume V containing N galaxies we get

$$\langle (\Delta N)^2 \rangle = \bar{n}^2 \int \int_V \xi(\mathbf{r}) \, d\mathbf{r}_1 \, d\mathbf{r}_2. \tag{14.13}$$

The right-hand-side will be zero if we are dealing with a single volume for which (14.11) holds since the number of galaxies in this entire volume is a given value of N, which does not fluctuate. On the other hand, if the volume is one realization in an ensemble of volumes, then N can fluctuate from volume to volume with an average value of $\bar{N}$, so that the integral on the right-hand side can be nonzero. In this case, the variance of the fluctuations in (14.13) would still be zero if $\xi(\mathbf{r}) = 0$ (i.e., if the fluctuations in the ensemble had a Poisson distribution). But this is wrong, since from (14.1) we obtain the standard result that

$$\langle (\Delta N)^2 \rangle_{\text{Poisson}} = \sum_{N=0}^{\infty} (N - \bar{N})^2 \frac{\bar{N}^N e^{-\bar{N}}}{N!}$$

$$= \sum_{N=0}^{\infty} (N^2 - \bar{N}^2) \frac{\bar{N}^N e^{-\bar{N}}}{N!} = \sum_{N=0}^{\infty} N^2 \frac{\bar{N}^N e^{-\bar{N}}}{N!} - \bar{N}^2$$

$$= e^{-\bar{N}} \sum_{N=1}^{\infty} \frac{\bar{N}^N [(N-1) + 1]}{(N-1)!} - \bar{N}^2$$

$$= e^{-\bar{N}} \left[\sum_{N=2}^{\infty} \frac{\bar{N}^N}{(N-2)!} + \sum_{N=1}^{\infty} \frac{\bar{N}^N}{(N-1)!} \right] - \bar{N}^2 = \bar{N}^2 + \bar{N} - \bar{N}^2 = \bar{N}.$$

$$\tag{14.14}$$

(In the last step let $M = N - 2$ in the first sum and $M = N - 1$ in the second and recall the normalization of 14.1.). Since (14.14) disagrees with (14.13) when $\xi = 0$, there is indeed a missing link.

This link emerges in the limit $\mathbf{r}_2 \to 0$, as mentioned in Chapter 6. A galaxy is now in both volumes dV_1 and dV_2, which coincide. If this is a small volume, it must give rise to a term of first order in $\bar{n} \Delta V$, since it can contain either zero or one galaxies as $\Delta V \to 0$. We represent this by rewriting (14.12) as

$$\langle \Delta n(\mathbf{r}_1) \Delta n(\mathbf{r}_1 + \mathbf{r}_2) \rangle = \bar{n}^2 \xi(\mathbf{r}) + \bar{n}\delta(\mathbf{r}_2 - \mathbf{r}_1), \tag{14.15}$$

where $\delta(\mathbf{r}_2 - \mathbf{r}_1)$ is the Dirac delta function whose volume integral is unity if the region includes the point $\mathbf{r}_2 = \mathbf{r}_1$ and is zero otherwise. Instead of (14.13), the integral over V is now

$$\langle (\Delta N)^2 \rangle = \bar{N} + \bar{n}^2 \int \int_V \xi(\mathbf{r}) \, dV_1 \, dV_2, \tag{14.16}$$

which contains the Poisson term, $\bar{N}$. Clearly, we must include the limit $\mathbf{r}_2 = \mathbf{r}_1$ when calculating $\langle (\Delta N)^2 \rangle$ since this mean square fluctuation in a given volume refers directly to the case when the two volume elements coincide; it is not just a matter of one definition being more convenient. If ξ depends only on the relative separation r, this simplifies to

$$\langle (\Delta N)^2 \rangle = \bar{N} + \frac{\bar{N}^2}{V} \int_V \xi(r) \, dV = \bar{N} + \bar{N}^2 \bar{\xi}. \tag{14.17}$$

Later applications of these results will be important for understanding the observations and theory of galaxy clustering.

A third related definition of the two-point correlation function is often used in the statistical mechanics of liquids and the kinetic theory of plasmas, where many of these concepts developed. We start with the basic ideas of distribution functions, whose importance will grow throughout this book.

Consider a configuration of N galaxies, each labeled by a number $1, 2, \ldots, N$. The specific distribution function $P_{(m)}(\mathbf{r}_1, \ldots, \mathbf{r}_m) \, d\mathbf{r}_1 \ldots d\mathbf{r}_m$ denotes the probability that galaxy 1 will be found in $d\mathbf{r}_1$ at $\mathbf{r}_1$, galaxy 2 in $d\mathbf{r}_2$ at $\mathbf{r}_2$, etc. For $m < N$, this probability does not depend on the positions of the remaining $N - m$ galaxies. Since each of these galaxies are found somewhere in the volume V, it is natural to adopt the normalization

$$\int \cdots \int_V P_{(m)}(\mathbf{r}_1, \ldots, \mathbf{r}_m) \, d\mathbf{r}_1 \ldots d\mathbf{r}_m = 1 \tag{14.18}$$

for any value of $m \leq N$. Often we will be more interested in the generic distribution function, $\rho_{(m)}(\mathbf{r}_1, \ldots, \mathbf{r}_m) \, d\mathbf{r}_1 \ldots d\mathbf{r}_m$, which is the probability that some galaxy is in $d\mathbf{r}_1$, another in $d\mathbf{r}_2$, and so on, independent of their labels. There are N choices for the galaxy in $d\mathbf{r}_1$, $N - 1$ for the one in $d\mathbf{r}_2, \ldots$ and $N - m + 1$ for $d\mathbf{r}_m$. The total number of these choices is the product of these factors, which is $N!/(N - m)!$. Hence

$$\rho_{(m)}(\mathbf{r}_1, \ldots, \mathbf{r}_m) = \frac{N!}{(N - m)!} P_{(m)}(\mathbf{r}_1, \ldots, \mathbf{r}_m) \tag{14.19}$$

and

$$\int_V \cdots \int \rho_{(m)}(\mathbf{r}_1, \ldots, \mathbf{r}_m) \, d\mathbf{r}_1 \ldots d\mathbf{r}_m = \frac{N!}{(N-m)!}. \qquad (14.20)$$

If $\rho_{(1)}$ is independent of position, then (14.20) immediately shows

$$\rho_{(1)} = \frac{N}{V} = \bar{n}, \qquad (14.21)$$

which is just the average (constant) number density of galaxies. Similarly, in a random distribution of galaxies defined by

$$P_{(m)} \, d\mathbf{r}_1 \ldots d\mathbf{r}_m = \frac{d\mathbf{r}_1}{V} \cdot \frac{d\mathbf{r}_2}{V} \cdots \frac{d\mathbf{r}_m}{V} \qquad (14.22)$$

or

$$P_{(m)} = V^{-m}, \qquad (14.23)$$

we obtain

$$\rho_{(m)} = \frac{N!}{N^m (N-m)!} \bar{n}^m \qquad (14.24)$$

and

$$\int_V \rho_{(2)}(\mathbf{r}_1, \mathbf{r}_2) \, d\mathbf{r}_1 \, d\mathbf{r}_2 = \int \int \rho_{(2)}(r) \, d\mathbf{r}_1 \, d\mathbf{r}_2$$

$$= V \int \rho_{(2)}(r) \, d\mathbf{r}$$

$$= N(N-1). \qquad (14.25)$$

Departures from random in these distribution functions define new correlation functions. We may write the general m-galaxy distribution function as

$$\rho_{(m)}(\mathbf{r}_1, \ldots, \mathbf{r}_m) = \rho_{(1)}(\mathbf{r}_1)\rho_{(2)}(\mathbf{r}_2) \ldots \rho_{(m)}(\mathbf{r}_m)\left[1 + \xi_{(m)}(\mathbf{r}_1, \ldots, \mathbf{r}_m)\right]. \qquad (14.26)$$

If $\xi_{(m)} = 0$ for all m, then the distributions of all the galaxies are independent and uncorrelated. Clearly $\xi_{(1)} = 0$, and $\xi_{(2)}$ is a two-galaxy correlation function, usually just denoted by ξ since this is the most often used correlation function. Analogous correlation functions could be defined using the $P_{(m)}$, but the only difference would involve multiplying these $\xi_{(m)}$ by a factor $N^m (N-m)!/N!$.

If the system has a constant density, so that (14.21) holds, then (14.26) becomes

$$\rho_{(m)}(\mathbf{r}_1, \ldots, \mathbf{r}_m) = \bar{n}^m \left[1 + \xi_{(m)}(\mathbf{r}_1, \ldots, \mathbf{r}_m)\right]. \qquad (14.27)$$

According to (14.20) and (14.21), we then have the integral property

$$\frac{1}{V^m} \int \cdots \int_V \xi_m(\mathbf{r}_1 \ldots \mathbf{r}_m) \, d\mathbf{r}_1 \ldots d\mathbf{r}_m = \frac{N!}{N^m(N-m)!} - 1 = \mathcal{O}(1/N). \quad (14.28)$$

In particular for $\xi_{(2)}$,

$$\frac{1}{V^2} \int \cdots \int_V \xi(\mathbf{r}_1, \mathbf{r}_2) \, d\mathbf{r}_1 \, d\mathbf{r}_2 = -\frac{1}{N}, \quad (14.29a)$$

which was obtained after (14.11) by excluding the case when the same galaxy contributes to the density in both volumes. Therefore this represents the situation when a specified galaxy is present at $\mathbf{r}_1$, and any other galaxy has a probability $\bar{n}(1 + \xi(r)) \, d\mathbf{r}$ of being in $d\mathbf{r}_2$ at $\mathbf{r}_2$. In this radially symmetric distribution around any specified galaxy, the average number of galaxies between r and $r + dr$ from the specified galaxy is $4\pi r^2 \bar{n}(1 + \xi(r)) \, dr$ and

$$4\pi\bar{n} \int_V (1 + \xi(r)) r^2 dr = N - 1, \quad (14.29b)$$

which just gives the other $N - 1$ galaxies in the finite volume V.

In addition to defining higher order correlation functions by (14.26), we can define them similarly by extending (14.10) to

$$\xi_m(\mathbf{r}_1, \ldots, \mathbf{r}_m) = \frac{\langle n(\mathbf{r}_1) \ldots n(\mathbf{r}_m) \rangle}{\bar{n}^m} - 1, \quad (14.30)$$

or by

$$\xi_m(\mathbf{r}_1, \ldots, \mathbf{r}_m) = \frac{\langle \Delta n(\mathbf{r}_1) \Delta n(\mathbf{r}_2) \ldots \Delta n(\mathbf{r}_m) \rangle}{\bar{n}^m} \quad (14.31)$$

as in (8.4). These various forms are readily related but not necessarily equivalent to one another. We shall develop these higher order correlations as we need them in future discussions. In general, the simplest ones, ξ_2 and ξ_3, have turned out to be the most useful theoretically and the easiest to measure accurately.

14.2   Properties

Correlation functions have a number of useful properties, some of which are summarized here as an introduction to more detailed analyses later.

From the definition (14.6), we see that the mean number of galaxies in a spherical volume of radius r around a galaxy is

$$\langle N \rangle = \frac{4}{3}\pi\bar{n}r^3 + \bar{n} \int_0^r \xi(r) \, dV$$

$$= \bar{N} + \bar{N}_c, \quad (14.32)$$

where $\bar{N}_c$ is the contribution of the correlations and $\bar{N} = \bar{n}V$ is what would be expected for a Poisson distribution or for an unconditional density. Even though the galaxies are correlated, the system remains statistically homogeneous since $\langle N \rangle$ is independent of position.

Like many other functions, the shape of $\xi(r)$ can be characterized by an amplitude and a scale length. Early analyses, referred to in Chapter 6, often parametrized $\xi(r)$ by the form $(r/r_0)^{-\gamma} e^{-\frac{r}{\lambda}}$ with the scale length λ and the amplitude r_0. Over regions smaller than about 10 Mpc where the correlations are nonlinear, $\xi(r) \gg 1$ and is very close to a pure power law, usually represented by $\xi(r) = (r/r_0)^{-\gamma}$. Gravitational many-body interactions also tend to produce this simple form of $\xi(r)$ over volumes where the galaxy distribution has relaxed. Thus within this regime the scale length $\lambda \to \infty$ and $\xi(r)$ becomes scale free. This is related to the scale-free power-law form of the two-particle gravitational potential.

Conventionally, r_0 is taken to be the value of r for which $\xi(r) = 1$. Sometimes this is called a scale length, but it is really an amplitude. If $\bar{n}$ in a sample depends on volume, particularly for a fractal distribution, then r_0 loses most of its simple physical significance. In this case, the amplitude of the correlation depends on the size of the sample. Thus $\xi(r)$ is not a useful description of a single cluster; it should be averaged over a suitable ensemble.

More general forms of correlation functions can become negative on some scale or oscillate around zero. When $\xi(r) < 0$, a positive density enhancement at $r = 0$ tends to associate with a negative one at r, or vice versa. When $\xi(r) > 0$, the associated fluctuations tend to both be positive or negative. In simple systems, the first zero of $\xi(r)$ usually indicates a typical range of either the interparticle forces or (especially for gravity) the dynamical evolution. An oscillating regime of $\xi(r)$ suggests unrelaxed evolution.

We can always Fourier transform $\xi(r)$ for a set of wavenumbers $\mathbf{k} = (2\pi/L)\mathbf{n} \neq 0$ in a normalization volume $V = L^3$ to get

$$\langle |\delta_{\mathbf{k}}|^2 \rangle = V^{-1} \int \xi(\mathbf{r})e^{i\mathbf{k}\cdot\mathbf{r}} \, d^3\mathbf{r} \tag{14.33}$$

and consequently

$$\xi(\mathbf{r}) = \frac{V}{(2\pi)^3} \int \langle |\delta_{\mathbf{k}}|^2 \rangle e^{-i\mathbf{k}\cdot\mathbf{r}} \, d^3\mathbf{k}. \tag{14.34}$$

This standard ensemble-averaged power spectrum $\langle |\delta_{\mathbf{k}}|^2 \rangle$ measures the amplitude of correlations over a distance $\sim k^{-1}$. Fourier analysis is particularly useful in the linear regime $\xi(r) \ll 1$ where the different wavenumbers are independent and evolve separately. Dynamical equations for $\mathbf{k}(t)$ are relatively simple until $\xi(r) \sim 1$ when various wavelength perturbations couple and combine in complicated ways. Then different descriptions are needed.

Guesses for the form of the power spectrum, especially as an initial condition of galaxy clustering, often assume a simple power law like

$$\langle |\delta_{\mathbf{k}}|^2 \rangle \propto k^n. \tag{14.35}$$

For $n = 0$, fluctuations on all wavelengths are equally strong and the result is white noise. For $n > 0$, short-wavelength fluctuations have higher amplitude. In these two cases, it is useful to introduce a cutoff at large k so that the integrated power spectrum converges (no "ultraviolet catastrophe"). For $n < 0$ the long-wavelength fluctuations dominate and we need a cutoff if $n < -3$ to avoid the corresponding "infrared divergence" as $k \to 0$. After theories using a simple power-law guess are found to be inadequate, more complicated postulates for the power spectrum are easily produced, though not often productive. A complete theory needs to derive $\langle |\delta_{\mathbf{k}}|^2 \rangle$ from basic physical principles. Its origin remains a cosmological mystery.

Everything discussed so far refers to the three-dimensional correlation function. This is not what we observe. The best we can do is to estimate the distance of a galaxy from its redshift, z, and, if available, from secondary distance indicators. For galaxies closer than several hundred megaparsecs (i.e., much nearer than the radius of the visible universe), the linear redshift–distance relation, $v = cz = c(\lambda_0 - \lambda)/\lambda \approx r H_0$ applies, where λ_0 is the observed wavelength of a spectral line emitted at wavelength λ and $H_0 \equiv 100 h$ km s^{-1} Mpc^{-1} is the present Hubble constant. Most observations indicate that $0.4 \lesssim h \lesssim 1$. Application of this relation does not usually give the exact distance, especially for nearby galaxies. Each galaxy has a peculiar velocity relative to the smoothed average expansion of the Universe. We shall see that these peculiar velocities are an important cosmological probe. For the present, we just note that they will distort the apparent distances of galaxies and modify $\xi(r)$ so that $\xi(r(z))$ in redshift space differs from $\xi(r)$ in real space.

Secondary distance indicators for fairly distant galaxies usually relate some easily measured integral property of a galaxy, such as its internal velocity dispersion or angular diameter at a given surface brightness, to its intrinsic luminosity in some wavelength band. Observation of its apparent luminosity then gives its distance from $L_{app} \propto L_{int} r^{-2}$. These secondary indicators are calibrated from more precise primary distance indicators, such as the period–luminosity relation of particular types of variable stars, applied to nearby galaxies. The technique works with a fair accuracy of 10–20% for a small but increasing fraction of moderately distant galaxies ($r \lesssim 100$ Mpc). Differences between the luminosity distances and the redshift distance provide an estimate of a galaxy's peculiar velocity.

Since most galaxies do not have well-determined distances, it is very useful to measure the angular correlation function of their projection on the sky. Millions of galaxies can participate. This trades less information about many galaxies for more information about fewer galaxies. As usual, these approaches are complementary.

Analogs of any of the definitions (14.6), (14.7), (14.10), and (14.15) provide suitable definitions for the angular correlation functions. Just reinterpret $\bar{n}$ as the surface density of galaxies and replace $\xi(r)$ by $W(\theta)$ to obtain the usual notation. Thus (14.7), for example, becomes

$$P_{1,2} = \bar{n}^2 [1 + W(\theta)] d\Omega_1 d\Omega_2, \qquad (14.36)$$

where $\bar{n}$ is now the average surface density of galaxies on the sky, $W(\theta)$ is the angular correlation function, and $P_{1,2}$ is the joint probability for finding a galaxy on the sky in the small solid angle $d\Omega_1$ and another in $d\Omega_2$. The spherical solid angle $d\Omega = \sin\theta\, d\theta\, d\phi$, where θ and ϕ are the polar and azimuthal angles, but $d\Omega$ could also refer to a small rectangle on the sky. Consequently, the fluctuations (14.16), for example, of counts in cells on the sky become

$$\langle (\Delta N)^2 \rangle = \bar{N} + \left(\frac{\bar{N}}{\Omega}\right)^2 \int \int_\Omega W(\theta)\, d\Omega_1\, d\Omega_2. \tag{14.37}$$

Relations between $W(\theta)$ and $\xi(r)$ are usually complicated (see Section 20.2 for observational results) and depend on details of the galaxy luminosity function. Also, structure in three dimensions tends to be smoothed out when projected onto two dimensions. In the special case when $\xi(r) \propto r^{-\gamma}$ is a pure power law on all scales, then $W(\theta) \propto \theta^{-\gamma+1}$ is also a power law whose exponent is increased by unity, although its amplitude depends on an integral of the luminosity function (Totsuji & Kihara, 1969).

There is also a simple relation (Peebles, 1980) for the amplitudes of $W(\theta)$ in statistically homogeneous catalogs observed to different magnitude limits, and therefore to different depths, D:

$$W(\theta) = D^{-1}W(\theta D). \tag{14.38}$$

From the angular correlation function analog of (14.32) we have

$$\bar{W} = \frac{\langle N \rangle - \bar{N}}{\bar{N}} = \frac{\bar{N}_c}{\bar{N}}, \tag{14.39}$$

where $\bar{W}$ is the angular correlation function averaged over an area $\Delta\Omega$. The average number of correlated galaxies, $\bar{N}_c$, in this area will be the same for all areas that have the same linear radius θD in catalogs of different depths D. However, the average number of galaxies projected into this area will increase as $\bar{N} \sim D$, which gives the angular scaling of (14.38).

Although both the angular and the three-dimensional correlation functions can be measured from galaxy counts, this does not guarantee that they represent the actual matter distribution. Dark matter may lurk among the galaxies and even dominate the total mass of the Universe. In the linear regime, $\xi(r) < 1$, the correlations of these two components are related in the simplest models by

$$\xi_{\text{gal}} = B^2 \xi_{\text{matter}}, \tag{14.40}$$

where B is the "biasing parameter" (frequently denoted by b, but we shall reserve b for another use; cosmology seems to attract many bs). This is known as "linear biasing." If $B > 1$, the galaxies are more clustered than the dark or intergalactic matter; if $B < 1$ they are less clustered. Calculating the correct physical development of B

in a model of structure formation is especially challenging. Usually it is considered as a "fudge factor," determined by forcing the model to agree with observations. When this is not sufficient, B is allowed to vary with scale, thus creating a glorified "fudge function." Other forms of biasing occur when computing $\xi(r)$ or $W(\theta)$ for subsamples of a total population, such as spirals, ellipticals, or especially luminous or massive galaxies.

It would be very useful to be able to relate the higher order correlation functions $\xi_m(\mathbf{r}_1, \mathbf{r}_2, \ldots, \mathbf{r}_m)$ defined by (14.31) to $\xi_2(\mathbf{r}_1, \mathbf{r}_2)$. Generally this cannot be done a priori. In a dynamically evolving system whose initial conditions and forces are completely known, all the ξ_m must clearly evolve consistently. Ferreting them out from the dynamical equations, however, is very difficult. Sometimes a simplifying assumption (e.g., Balian & Schaeffer, 1989) is made by supposing that all the galaxy correlation functions are scale invariant with the form

$$\xi_m(\lambda\mathbf{r}_1, \ldots, \lambda\mathbf{r}_m) = \lambda^{-\gamma(m-1)}\xi_m(\mathbf{r}_1, \ldots, \mathbf{r}_m), \tag{14.41}$$

where λ is any finite number and γ is the slope of the average two-point correlation function, assumed to be a pure power law. Its most useful implication is that the three-point correlation function has the form

$$\xi_3(r_{12}, r_{23}, r_{31}) = Q\{\xi(r_{12})\xi(r_{23}) + \xi(r_{23})\xi(r_{31}) + \xi(r_{31})\xi(r_{12})\} \tag{14.42}$$

with no term depending on all (r_{12}, r_{23}, r_{31}). This is adopted from the early theory of liquids. More generally, (14.41) implies that the volume-averaged correlations

$$\bar{\xi}_m \equiv \frac{1}{V^m} \int_V \cdots \int_V \xi_m(\mathbf{r}_1, \ldots, \mathbf{r}_m)\, d\mathbf{r}_1 \ldots d\mathbf{r}_m \tag{14.43}$$

satisfy the scaling relations

$$\bar{\xi}_m = S_m(\bar{\xi}_2)^{m-1} \tag{14.44}$$

with the S_m coefficients independent of scale. Then the game is to make various ad hoc assumptions about the algebraic dependence of the S_m on m and examine the results. Considerable observational evidence exists that (14.42) holds for ξ_3 at least approximately, but there is considerably less clear evidence that (14.44) holds for $m > 3$ in the galaxy distribution. Nevertheless, these scaling relations provide significant theoretical insight.

We have already seen how ξ_2 is related to the fractal dimension $3 - \gamma$ of the distribution in (12.21). Correlation functions have also been related to other statistics such as distribution functions, angular power spectra, and multipole moments. We will especially explore their connection to distribution functions in later sections. Now, we turn to a brief summary of observations.

14.3   Measuring $\xi(r)$ and $W(\theta)$

To measure $\xi(r)$ directly from the definition (14.10) requires us to average the discrete galaxy distribution over cells of a given size to find the local density field $n(\mathbf{r})$. Then we need three assumptions: Statistical homogeneity prevails so that $\xi(\mathbf{r}_1, \mathbf{r}_2) = \xi(|\mathbf{r}_1 - \mathbf{r}_2| = r)$, the average density $\bar{n}$ is independent of scale or of the size of the cell over which it is averaged, and the correlation $\langle n(\mathbf{r}_1)n(\mathbf{r}_1 + \mathbf{r}_2)\rangle$ is also independent of cell size.

The last assumption is often avoided, at the cost of some imprecision, by using a discrete particle estimator for $\xi(r)$. The most popular one first finds the number, $DD(r)$, of pairs of data points (galaxies) having separations in some range $r \pm \Delta r/2$. Then it sprinkles the volume, or area, of the data with a random Poisson distribution of points and gives them the same selection function as the data. The resulting artificial distribution of galaxies has a pair count $RR(r)$ and an average number density $\bar{n}_R$, compared with $\bar{n}$ for the actual galaxies. Thus the quantity

$$\xi(r)_{\text{est}} = \frac{DD(r)}{RR(r)} \frac{\bar{n}_R(\bar{n}_R - 1)}{\bar{n}(\bar{n} - 1)} - 1, \qquad (14.45)$$

which resembles (14.10), provides an estimate of $\xi(r)$. It is normalized by the ratio of the total number of random pairs to data pairs, in case $\bar{n}_R \neq \bar{n}$. In two dimensions, the analogous procedure for annuli of $\Delta\theta/2$ gives $W_{\text{est}}(\theta)$. The results are usually averaged over many realizations of the random distributions, and the dispersion provides an estimate of the statistical uncertainties (Mo et al., 1992). Figure 14.1 illustrates the observed $W(\theta)$ and is discussed further in Section 20.2.

Unfortunately, this does not solve any fundamental problems since dependence on $\Delta r/2$ replaces the dependence on cell size, statistical homogeneity is still assumed, and $\bar{n}_R$ is independent of cell size by fiat. However, it does deal with discreteness nicely, and it lends itself to generalizations of (14.45) involving DD, RR, and the cross correlation DR, with various possible weightings. Some of these have desirable statistical properties such as smaller variance (e.g., Landy and Szalay, 1993; Bernstein, 1994), although their relation to the underlying physics is more complicated. In any case (14.45) provides a simple description of any given data set that is useful for comparison with other sets.

Uncertainties in the data generate controversy at the edge of objectivity where the deepest questions form. Few astronomers now dispute the basic result of Totsuji and Kihara (1969) that on scales of about 1 to $10\,h^{-1}$ Mpc, $\xi(r)$ is approximately a power law with a slope about 2, or a little less. But answers to more detailed questions such as the extent of the power law, its dependence on galaxy morphology, $\xi(r)$ for clusters, the relation between $\xi(r)$ in real space and in velocity space, and the forms of higher order correlations are all still being debated.

Even though the data get better, they are never complete. Nor is their accuracy known with precision. For example, automatic scanning machines can measure the positions of millions of faint galaxies on photographic plates. To find where $W(\theta)$

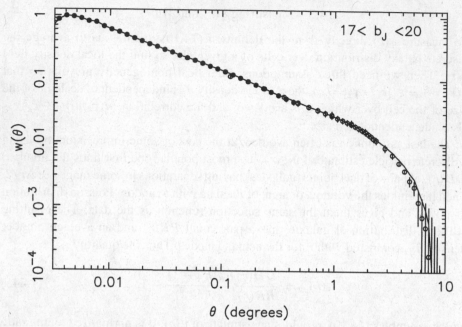

Fig. 14.1. The observed angular two-galaxy correlation function for galaxies selected from the APM galaxy survey with magnitudes in the range $17 < b_J < 20$. The line is for the full area of the survey. The open circles are means (with 1σ errors) of four separate nearly equal areas and the filled circles are means of 185 Schmidt plates. (From Maddox et al., 1996.)

breaks from a power law, we need to measure pairs across many plates to compute $DD(\theta)$. The number of galaxies in a given magnitude range on a plate depends on the absolute calibration of the zero point magnitude of that plate. If these zero point calibrations vary from plate to plate in a correlated fashion, they may introduce a spurious correlation into $\xi(r)$ for large r where $\xi(r)$ is very small. Magnitude fluctuations of only 0.1 could change the break of $\xi(r)$ from $20\,h^{-1}$ Mpc to $10\,h^{-1}$ Mpc (Fong et al., 1992) since it occurs at $\xi \approx 0.01$. Knowledge of where $\xi(r)$ breaks and how it decreases may rule out specific cosmological models of galaxy formation and clustering (e.g., Yoshii et al., 1993).

   Volume-limited samples, in which we can identify all galaxies out to a given distance, have generally been too small or unrepresentative to give definitive general results. This situation will certainly improve as it becomes possible to find small, low surface brightness galaxies, or if we learn that such galaxies share the same detailed clustering properties as their mightier companions.

   Redshift surveys of galaxies make it possible to estimate $\xi(r)$ in three dimensions directly from the observed velocities of galaxies, assuming $V_{obs} = rH_o = V_{Hubble}$. But galaxies are not always where they seem to be from their cosmological recession. Their peculiar velocity relative to the average Hubble flow adds the complication that $V_{obs} = V_{Hubble} + V_{pec}$ so that $rH_o = V_{obs} - V_{pec}$. On scales $\lesssim 5$ Mpc, gravitational clustering may induce large peculiar velocities. Galaxies in a rich cluster may have a velocity dispersion of $\sim 10^3$ km s^{-1}, making their spatial position uncertain to ~ 10 h^{-1} Mpc. On large-scales, regions of greater than average density may produce coherent bulk gravitational flows. Streaming motions may arise from initial conditions in the early universe. Although these large scale peculiar velocities are smaller than those in very dense clusters, $\xi(r)$ is also much smaller on these scales so it can still be strongly affected.

Peculiar velocity distortions of $\xi(r)$ in redshift surveys depend strongly on the details of Ω_0 and on the initial power spectrum of fluctuations that the cosmological models assume. Most of these distortions need to be investigated by N-body experiments on specific models (e.g., Gramann, Cen, and Bahcall, 1993), but somewhat more general analytic results are also available (e.g., Matsubara, 1994; Fry and Gaztañaga, 1994). They mainly show that if $\xi(r)$ is a power law in real space, then on small scales it is a somewhat less steep power law in redshift space (including the peculiar velocities). On large (linear) scales, the correlations are more alike, with perhaps a slight relative steepening in redshift space. Hierarchical and scaling properties of the higher order correlation functions in redshift space are also altered from their real space properties. These effects are less certain since the analytic models do not use velocity distributions that are consistent with their spatial distribution and the N-body simulations require very high resolution and statistical homogeneity.

Observationally, it would be ideal to have accurate distances to galaxies without having to measure their redshifts. Comparing a galaxy's secondary distance indicators with its redshift distance would give its peculiar velocity directly. Then $\xi(r)$ could be determined either by using the secondary distance indicators for those galaxies in which they are accurate – if there is a uniform sample – or by statistically connecting a larger redshift sample with a representative distribution of peculiar velocities. This is a project for the future.

Meanwhile, several general redshift surveys (e.g., Maurogordato et al., 1992; Loveday et al., 1992) of about 2,000 galaxies each and other surveys of untypical regions (e.g., Bonometto et al., 1993; Ramella et al., 1992) have been analyzed for velocity space correlations. Results of the representative surveys are generally compatible with earlier analyses of angular correlation functions. Where they differ, usually at large scales, combined uncertainties cause astronomers to dispute their significance. Eventually they will provide strong constraints on important classes of cosmological models.

Applying analogous procedures to clusters of galaxies, counting each cluster as a fundamental particle, shows that their $\xi(r)$ depends strongly on the definition of a cluster. In early analyses of $\xi(r)$ for the Abell cluster catalog, rich clusters appeared to be much more strongly correlated than individual galaxies. After a lively debate, it became clear that a large part of this effect was produced by selection and inhomogeneity in the Abell catalog. More objectively chosen catalogs of clusters show their spatial $\xi(r)$ to be comparable with that of individual galaxies (Dalton et al., 1992). The physical meaning of the cluster two-point correlation function

$\xi_{cc}(r)$, however, is very different from that of the galaxy two-point correlation. If galaxies remain the fundamental units, then $\xi_{cc}(r)$ represents correlations between regions where the very high order correlations of galaxies, say the 100-point correlation functions or greater, have a very large amplitude. This "distribution of exceedences" depends on the density distribution of the fundamental units – galaxies – and on the definition of an exceedingly dense region. We will explore it in Section 18.3.

14.4   Origin and Evolution of Correlation Functions

> Infinite shadowy isles lie silent
> before me.
>
> Dylan Thomas,   *Forest Picture*

Islands of overdensity in the early universe are widely believed to be the origin of the large-scale structure we now observe. How and when these islands arose is still mostly conjecture. Current ideas push their origins back, more and more insistently, to the extremely early universe, possibly as topological defects that occurred in particle phase transitions at enormous density.

Less fundamentally, though perhaps with greater hope of real progress, we can try to understand how the various statistics of galaxy clustering evolve as the Universe ages. Here the basic question is: "How far back in time can we extrapolate from our present observations of clustering, and what constraints does this put on the properties of earlier epochs?"

Correlation functions have been especially amenable to dynamical analysis. This is partly because we can adapt a mathematical description developed for plasma physics and kinetic theory of gases, and partly because $\xi(r, t)$ is very easy to follow in computer N-body simulations. Even so, the mathematical analysis becomes too complicated to solve analytically for general cases, and computer simulations are confined to very specific ranges of models. We will derive a simple but fundamental result here and return to a more detailed exploration in Part III.

Determining the conditions to which our result applies will help put it into perspective. Each condition here aims at simplicity. First we need to specify the fundamental unit of matter that is clustering. It seems sensible for this to be a galaxy since the correlation function refers to number counts of galaxies. At this stage we do not need to ask how the galaxies formed. For simplicity, suppose each galaxy has the same mass. Inaccurate though this may seem, it turns out to be a fairly good approximation if most of the dynamical evolution is collective and does not depend strongly on the mass of an individual galaxy. Mass segregation can always be considered at the next level of approximation. Nor will we complicate matters, at this stage, by introducing various forms of dark matter between the galaxies. All the nonuniform dark (and luminous) matter will have to be in galaxies or their halos. Multicomponent systems containing different types and amounts of galaxies, hot dark matter, cold dark matter, lukewarm dark matter, cosmic strings, black holes, decaying particles, etc. can produce a parade of effects, which may, or may not, be related to what we see.

Second, we need to specify the forces that cause clustering. These we take to be the Newtonian gravitational interactions between the galaxies acting as point particles. At the low speeds and small scales involved in galaxy clustering, local general relativity effects are unimportant. Tidal interactions and mergers enter at a more refined descriptive level. Multicomponent systems may also introduce an additional smoother gravitational potential, hydrodynamic and radiative forces (e.g., effects of protogalactic explosions), anisotropic local gravitational forces (from cosmic strings), etc., but we will again eschew these possible complications at present.

Third, we must specify the initial conditions from which clustering starts. This is our link with galaxy formation in the earlier universe. A minimalist approach would be to assume there is no information in the initial conditions and begin clustering from a Poisson distribution. This is equivalent to a white noise spectrum with power independent of frequency or scale. It implies $\Delta N \propto \sqrt{N}$ initial fluctuations on all scales. There are other possibilities. For example, a power spectrum of fluctuations $|\delta(k, t)|^2 \propto k^n$ at a given time t with $n = 1$ implies that relative mass fluctuations $(\Delta M / M)^2$ were the same for any given mass M when it entered the horizon (approximately the Hubble radius at any given time). Other models of the early universe give more complicated forms of $\delta(k, t)$. Not surprisingly the Poisson $n = 0$ case is usually the simplest to understand analytically.

Fourth, we need to specify the rate of expansion, $R(t)$, of the Universe, which is obtained from the background cosmology. At the late times and relatively small redshifts ($z \lesssim 10^2$) generally thought relevant for galaxy clustering, the classical Einstein–Friedmann models are usually deemed to apply. The observed model, determined by its value of the current density parameter $\Omega_0 = \rho / \rho_{\text{critical}}$ and Hubble constant $H_0 = \dot{R}(t) / R(t)$, is still uncertain. Since properties of galaxy clustering differ in these models, they may help decide the most likely values of H_0 and Ω_0. The simplest of these models is the Einstein–de Sitter one, which is just open with the critical density ρ_{critical}. Its spatial sections are Euclidean. Roughly speaking, in any volume its expansion energy just balances its gravitational energy. At any given time t after its singular state, it is completely characterized by its density ρ, its age t, and the gravitational constant G. Since it has no other length or time scales, it is a similarity solution determined by the only dimensionless combination of these three quantities, which is

$$Gpt^2 = \text{constant}. \tag{14.46}$$

During the matter-dominated part of its expansion, the number of particles (or galaxies) is constant in a comoving volume expanding with the Universe; so $\rho \propto R^{-3}$. Consequently,

$$R \propto t^{2/3}; \quad H = \frac{2}{3t}. \tag{14.47}$$

(In the radiation-dominated epoch long before galaxy clustering $\rho \propto R^{-4}$ and so $R \propto t^{1/2}$.)

We can determine the constant in (14.46) by a simple Newtonian argument (which gives the correct answer although it needs further justification; see Bondi, 1960). A particle at the edge of an arbitrarily located spherical volume of expanding radius $R(t)$ has a kinetic energy $\frac{1}{2}m\dot{R}^2$ with respect to the center of the sphere. Its gravitational potential energy from the mass $M = \frac{4}{3}\pi R^3 \rho$ inside the sphere, which may all be imagined to be located at its center, is GmM/R. Equating these two energies, so that the total energy is zero, gives

$$H^2 = \frac{8\pi G}{3}\rho . \tag{14.48}$$

In a universe of greater density, the gravitational energy exceeds the kinetic energy. This excess gravitational energy introduces a new characteristic scale into the model, which may be taken to be the value of R when $\dot{R} = 0$. Dimensionally we can write this as $c^2/R^2 = 8\pi G\rho/3$, where c is the speed of light. This new length scale indicates that the mass within any arbitrary sphere of radius R is just within its Schwarzschild radius: $R = 2GM/c^2$. Adding this term to (14.48) gives more generally

$$H^2 + \frac{c^2}{R^2} = \frac{8\pi G}{3}\rho \tag{14.49}$$

for a closed universe. Similarly, for an open universe with lower density in which the kinetic energy in a sphere exceeds its gravitational energy, the extra term in (14.49) is $-c^2/R^2$. In this case, the Schwarzschild radius analogy, for what is conceptually a Newtonian description, breaks down, indicating that it should not be pushed too far.

Fifth, and finally, we need to specify the scale on which to follow the evolution of $\xi(t)$. Eventually we will find its growth on all scales, but here in keeping with the spirit of examining the simplest case we look just at large scales where $\xi(r, t) < 1$. The reason is that on small scales $\xi \gg 1$ can quickly become very nonlinear. Once it becomes nonlinear, its subsequent evolution is again easy to describe. But the process of becoming nonlinear, on any scale, is very complicated because the detailed motions of intermediate numbers of particles dominate, and simple approximations break down. Another simplifying feature of large scales is that bulk gravitational forces dominate over pressure gradients.

Now we can find how the amplitude of $\xi(r, t)$ grows by noticing from (14.17) that it is related to the average density fluctuation $\langle(\Delta N)^2\rangle$. On large scales where the Poisson contribution becomes small,

$$\frac{\langle(\Delta N)^2\rangle}{\bar{N}^2} = \bar{\xi}(t). \tag{14.50}$$

For the flat Einstein–de Sitter universe, the cosmology does not introduce any length scales. Therefore if $\xi(r, t)$ has an initially scale-free power-law form, $\xi(r, t) = f(t)r^{-\gamma}$, on these scales it will retain this form as the amplitude $f(t)$ grows, as long as it remains in the linear regime. Moreover, for $\gamma < 3$, the integral $\bar{\bar{\xi}}$ is dominated by $\xi(r, t)$ on large scales, which remain linear for long times. Thus, in this regime, the time developments of $\xi(r, t)$, $\bar{\bar{\xi}}(t)$, and $\langle(\Delta N)^2\rangle$ are essentially the same.

Suppose there is a large spherical region of slightly higher than average density so that $\Delta N = N - \bar{N} > 0$ but $\Delta N/\bar{N} \ll 1$. Equivalently, since $\bar{N} \gg 1$, this means that in the spherical perturbation $\rho = \bar{\rho} + \delta\rho$ and $0 < \delta\rho/\bar{\rho} \ll 1$. By symmetry, the evolution for this region is unaffected by the universe outside it, apart from supplying an external boundary condition. This applies in Newtonian cosmology, and also in general relativity where it is an example of Birkhoff's theorem. Thus the overdense region will evolve like a closed universe, according to (14.49). The average background universe is a flat Einstein–de Sitter model evolving according to (14.48) with $\rho = \bar{\rho}$. Subtracting one from the other shows that the fractional density perturbation evolves as

$$\frac{\delta\rho}{\rho} = \frac{\rho - \bar{\rho}}{\bar{\rho}} = \frac{3c^2}{8\pi G\bar{\rho}R^2} + \frac{3}{8\pi G\bar{\rho}}[H^2(\rho) - H^2(\bar{\rho})]. \tag{14.51}$$

The "Hubble age" of the universe is $H^{-1} = R/\dot{R}$ and is the time the universe would have had to expand at its present rate to be its present size. If we compare the densities in the perturbation and the background at their same Hubble time, and neglect the small difference between $R(\rho)$ and $R(\bar{\rho})$ for $\bar{\rho} = \bar{\rho}_0 R_0^3(\bar{\rho})/R^3(\bar{\rho})$ in the denominator, then

$$\frac{\delta\rho}{\bar{\rho}} \propto R(t) \propto t^{2/3}. \tag{14.52}$$

Therefore from (14.50) and the discussion following it

$$\xi(r, t) \propto R^2(t) \propto t^{4/3} \tag{14.53}$$

on large scales in the linear regime for an Einstein–de Sitter universe with $\Omega_0 = 1$. This basic result, which also applies to a wider range of models, is that on large scales the two-point correlation function grows as the square of a density perturbation with a modest algebraic time dependence.

This is a minimalist derivation of the growth of $\xi(r, t)$ on large scales. It contains the essential physics and nothing more. Nonetheless, the result agrees with more rigorous detailed derivations (Inagaki, 1976). However, more general results require more powerful techniques (see Chapters 16 and 17). On small scales, in particular, the evolution is quite different and far more rapid. There we cannot neglect the Poisson fluctuations. When N is only a few, the relative local density perturbations $\sim N^{-1/2}$ start off almost in the nonlinear regime. Evolution now depends on detaileds orbital dynamics rather than on the large-scale fluid perturbations that led to (14.53). Moreover, the early nonlinear development is not scale free, even in a flat Einstein–de Sitter universe, because there is now a nonlinear relaxation length scale. On scales greater than the nonlinear relaxation scale, the orbits have not yet had time to interact significantly with one another. Therefore the shape of $\xi(r)$ on small scales will develop in a more complicated way.

On large scales, the shape of $\xi(r)$ is also more complicated in open and closed universes. Even if it is initially a scale-free power law, consistent with an initial power-law power spectrum, the shape generally distorts as the universe expands because the expansion is not self-similar. In the 1970s when astrophysicists first began studying $\xi(r, t)$ extensively, many thought its shape and amplitude would be a shortcut to determining Ω_0 and H_0 for our Universe. It turned out to be more subtle. On scales $\lesssim 10$ Mpc where $\xi(r)$ has evolved to nonlinear values and can be determined relatively well from observations, it is rather insensitive to both cosmology and to its initial conditions. On larger scales $\xi(r)$ is relatively unevolved and reflects its initial state more than the cosmology. But here its observational determination is less accurate. The Book of Nature was not to be read so easily. For a time, it even seemed to snap shut.

15

Distribution Functions

All is a procession,
The universe is a procession with
measured and beautiful motion.
Walt Whitman

15.1 Definitions

Although galaxy distribution functions were known to Herschel, measured by Hubble, and analyzed statistically by Neyman and Scott, a new chapter in their understanding has opened in recent years. This relates distribution functions to the gravitational clustering of point masses in an expanding Universe. Calculations of the resulting cosmological many-body problem provide new insights into observed galaxy clustering as well as into the results of computer simulations.

Why gravity? When a reporter asked Willie Sutton, a well-known American bank robber, why he robbed banks, he supposedly replied "Because that's where the money is." As money is the most obvious motivating force of banks, gravity is the most obvious motivating force of galaxy clustering. Unlike the economic parallel, studies of gravitational clustering have the advantage that the rules do not change as the system evolves and is understood better.

Still, the mutual gravitation of galaxies may not be the only significant influence on their clustering. Initial positions and velocities of galaxies when they first form as dynamical entities will help determine their subsequent distribution. This is particularly true on large scales where the distribution has not had time to relax from its initial state. On smaller relaxed scales, the nonlinear interactions of orbits will have dissipated most of the memory of the initial conditions. The nature of this relaxed state will be one of our main themes in subsequent sections and following chapters.

Apart from initial conditions, which include the rate of expansion of the Universe, dark matter outside galaxies may affect their clustering. If the dark matter has uniform density, it only affects the rate of universal expansion. By its symmetry, it does not change galaxy orbits. However, movement of a galaxy through it can produce a locally asymmetric background, which causes gravitational dynamical friction, showing the galaxy's speed. If the dark matter is generally nonuniform, it can affect galaxy orbits strongly in ways that depend on details of particular models. For example, cosmic strings or other defects left over from possible high-energy phase transitions in the early Universe may cause galaxies to form and cluster in the accretion wakes of these defects. Another possibility is that dissipative dark matter collects into large inhomogeneities before the galaxies cluster. Then the galaxies cluster around or within these inhomogeneities. Or explosions of pregalactic objects might redistribute material from which the galaxies condense. In all these cases, we would not expect the dark matter to have the same distribution as the galaxies. The difference between them is usually referred to as "bias," and it may take various forms depending on the model.

Despite much discussion, there is no clear observational evidence that any non-gravitational processes have substantially affected the overall distribution of galaxies. How do we know? The answer requires us to first understand the gravitational evolution of the distribution of point masses in the expanding universe – the cosmological many-body problem.

Then we can compare the resulting distribution with observations to see if they are consistent, or if they require further explanation. Having a standard based on known applicable gravitational physics enables us to determine whether additional physical processes are needed.

This comparison should, in principle, employ all the statistics discussed in previous chapters and any others that contain significant information. In practice, the ones that appear to be especially informative and amenable to physical understanding are the distribution functions. We have met the galaxy distribution functions before, and they will become a growing theme throughout this book. The spatial distribution function $f(N, V)$ is just the discrete probability for finding N galaxies in a given size volume V. It may also be considered as a function of V for a given N. Then it represents the continuous probability for finding N galaxies in a volume of size between V and $V + dV$. In general $f(N, V)$ will also depend on the shape of the volume. Except for pathological shapes, however (e.g., those that deliberately exclude galaxies), the shape dependence is generally small and easily calculable. The two-dimensional projection, $f(N, A)$ of $f(N, V)$, gives the probability for finding N galaxies in an area of size A placed arbitrarily on the sky (often called the counts-in-cells statistic). In the other half of phase space, the velocity distribution function, $f(v)$, describes the probability that a randomly selected galaxy has a peculiar velocity, relative to the Hubble expansion, between v and $v + dv$. We observe the peculiar radial velocity distribution $f(v_r)$. Determining peculiar radial velocities requires secondary distance indicators for galaxies, so that we can compare their observed redshift velocity with the cosmological expansion velocity at their actual distance.

These distribution functions have the same basic definitions as those of ordinary statistical mechanics. Classical distinguishable particles, which scatter off each other by isotopic short-range forces in equilibrium, satisfy Maxwell–Boltzmann statistics. Indistinguishable quantum mechanical particles with half-integral spin, which consequently obey an exclusion principle, satisfy Fermi–Dirac statistics. Indistinguishable quantum particles with integral spin and no exclusion principle satisfy Bose–Einstein statistics. Classical gravitating particles undergoing quasi-equilibrium evolution in an expanding universe satisfy a different type of statistics. One of our aims in succeeding chapters will be to understand these new statistics and some of their implications in detail. Here we summarize several of the main results as an introduction.

15.2   Theoretical Results

Amplitudes $\Delta N/\bar{N}$ for galaxy number count fluctuations on small scales are generally large and nonlinear. On large scales, their amplitudes are small. The most useful theories of gravitational clustering should therefore work in both regimes. It is possible to derive a kinetic theory based directly on the equations of motion for the galaxies and solve it for simple initial conditions in the linear regime. This BBGKY hierarchy is the topic of Chapter 17. It provides a set of coupled, nonlinear, integro-differential equations for the evolution of the correlation functions ξ_m. Volume integrals of these correlations are closely related to the $f(N, V)$. Unfortunately the solutions become mathematically intractable when the fluctuations are large.

Numerical computer experiments that simulate gravitational clustering can cover the entire range of fluctuations. We will discuss some of them soon and will return to this approach in more detail in later chapters. Simulations are valuable for checking theoretical ideas. But they are also very important for enhancing our intuition into the properties of clustering, some of which are unexpectedly subtle.

To find a useful and solvable physical description of this cosmological many-body problem, we can examine it at a level that is somewhat less detailed than either the entire BBGKY hierarchy or the numerical computations of individual orbits. This is the level of statistical thermodynamics. Its chief descriptions are the distribution functions. Here I summarize the thermodynamic approach, which is a shortcut to the statistical mechanics. The relations between them will be discussed in Chapters 22–27.

Thermodynamics is usually associated with systems in or near equilibrium. Except for the classical one-body, two-body, and restricted three-body problems, gravitating systems are almost never in equilibrium. (There are detailed discussions of these nonequilibrium systems in Saslaw (1985a).) Attractive gravitational forces always make these systems unstable, albeit sometimes over very long timescales. The detailed nature of gravitational instability is very different for finite systems such as individual clusters, than for infinite statistically homogeneous systems such as the cosmological many-body problem. In finite systems, evolution occurs mainly through the loss of high-energy particles – stars in a globular cluster or galactic nucleus; galaxies in a bound cluster. The remaining particles form a more compact subsystem, which may be a simple tight binary or a condensed core. To a zeroth-order approximation, the velocity distribution function in bound relaxed clusters may be represented by a Maxwell–Boltzmann distribution. A more accurate distribution would include an effective temperature changing with radius, a cutoff at the escape velocity, and anisotropic orbits, all depending on detailed models. The strong change in local density from its central core to its outer periphery is the most characteristic feature of a finite bound system.

By contrast, in an infinite statistically homogeneous system, the average smoothed density is independent of position and we are most interested in the local fluctuations that depart from this average. Evolution increases their amplitude and scale. Some of them eventually form bound clusters. The number and size of different fluctuations at a given time depend primarily on initial conditions and the rate of expansion of the Universe. As in finite clusters, there is a tendency for parts of an infinite system to become more compact by ejecting their high-energy members (i.e., normally those with zero or somewhat positive total energy). This process is sometimes called dynamical (or gravitational) dissipation.

Until relatively recently, the lack of rigorous equilibria for gravitating systems deterred most astrophysicists from trying to apply thermodynamic ideas. For large finite bound clusters, which evolve slowly, the pioneering work of Bonnor, Antonov, Lynden-Bell, and others (summarized in Saslaw, 1985a) showed that thermodynamics could nonetheless provide a very useful approximate description. These finite systems often relax closely to thermodynamic equilibrium states that maximize their entropy subject to internal and external constraints such as a constant total energy or angular momentum.

Infinite systems are somewhat more subtle because both particles and energy can move among fluctuations. Considering this process more deeply leads to the key assumption that makes a thermodynamic description possible:

Assumption 1
Galaxy clustering evolves in a quasi-equilibrium manner through a sequence of equilibrium states.

Several reasons suggest that this hypothesis may apply to a wide range of conditions (but not to all). If the system is statistically homogeneous, without too much high-amplitude large-scale coherence (which might be produced by cosmic strings, cosmic explosions, or strongly clustered nonequilibrium initial conditions), then its local evolution is relatively independent of any large-scale structure. This does not preclude initial homogeneous correlations with moderate power on small or large scales. The global expansion of the Universe then effectively removes the mean gravitational field, leaving only the fluctuations to be described. Local regions with fluctuations whose amplitude is high, say of order unity, will evolve initially on the local gravitational timescale $\sim (G \rho_{\text{local}})^{-1/2}$. This is shorter than the global expansion time by a factor $\sim (\bar{\rho}/\rho_{\text{local}})^{1/2}$, where $\bar{\rho}$ is the average cosmic density. Local high-density fluctuations, partly caused by particle discreteness, first tend to form small clusters, with varying degrees of gravitational binding. Later, these clusters themselves tend to cluster. Thus, after an initial period of rapid relaxation during which near neighbor interactions dominate, subsequent evolution is mainly collective.

After significant clustering occurs, and at least partially virializes, subsequent macroscopic evolution on these scales occurs on timescales longer than the global expansion time. Computer simulations confirm that this happens for a wide range of conditions (Saslaw, 1992; Sheth & Saslaw, 1996). The quasi-equilibrium hypothesis assumes that after this initial relaxation, the system can be described adequately by macroscopic thermodynamic variables, such as total internal energy, U, entropy,

S, temperature, T, pressure, P, volume, V, number of particles, N, and chemical potential, μ. Moreover, the relations among these quantities, at any given time, can be determined as if they were in an equilibrium state. However, the thermodynamic quantities themselves may change with time. This is similar to ordinary nonequilibrium thermodynamics (e.g., De Groot and Mazur, 1962) in which the macroscopic variables vary throughout space and change with time. In the atomic case, this works because the local atomic or molecular relaxation times are much shorter than the timescales for macroscopic change. In the gravitational case, the reason is different. As clusters form and become more gravitationally bound, their subsequent evolution becomes more detached from the expansion of the Universe. Thermodynamic quantities represent averages over an ensemble of volumes, which contain galaxies in many different states of clustering. These microscopic states, described by the positions and velocities of each individual particle (galaxy), change quickly on the local dynamical timescale (see Chapters 24 and 25 for details). Average properties of bound or nearly bound clusters, however, change slowly during the global expansion timescales. Spatial scales on which most galaxies are substantially clustered dominate the thermodynamics. Once these scales have relaxed into a quasi-equilibrium state, they tend to remain in it because it has the maximum entropy for given global constraints such as the average density ρ. As these constraints change, the average clustering can respond adiabatically, as described by the cosmic energy equation (see Chapter 22). The detailed conditions for relaxation to a thermodynamic quasi-equilibrium state in the first place are an aspect of kinetic theory that is only partially understood (see Chapter 25).

The character of the quasi-equilibrium state can also be described from a statistical mechanical point of view. Each entire system of N particles is represented by a single point in a $6N$-dimensional phase space with $3N$ momentum coordinates $p_{1X}, p_{1Y}, p_{1Z}, p_{2X}, \ldots, p_{NZ} = p_1, \ldots, p_N$ and $3N$ position coordinates $q_{1X}, \ldots, q_{NZ}$. The density of phase points in the region $d\mathbf{p}\,d\mathbf{q}$ at some time t is $\eta f(\mathbf{p}, \mathbf{q}; t)$, where η is the total number of phase points (systems) in the ensemble. The notation $\mathbf{p}, \mathbf{q}$ represents all $6N$ coordinates. The probability density $f(\mathbf{p}, \mathbf{q}; t)$ is the probability of finding a randomly chosen system in the ensemble to have a dynamical state in $d\mathbf{p}\,d\mathbf{q}$ around the values $\mathbf{p}, \mathbf{q}$. Its integral over all $d\mathbf{p}\,d\mathbf{q}$ is normalized to unity. The particles in each system, and therefore the phase points representing all the systems, evolve in time according to their deterministic Hamiltonian equations of motion. Along these trajectories, the total change of $f(\mathbf{p}, \mathbf{q}; t)$ with time is

$$\frac{df}{dt} = \frac{\partial f}{\partial t} + \sum_i \left(\frac{\partial f}{\partial p_i} \dot{p}_i + \frac{\partial f}{\partial q_i} \dot{q}_i \right). \tag{15.1}$$

Liouville's theorem, derived in Section 17.2, shows that $df/dt = 0$. Here we need only recognize that df/dt consists of two parts. The first, $\partial f/\partial t$, describes the change of f with time at a fixed location in phase space. The second set of terms describes how f changes as we move through phase space along the dynamical trajectory of a phase point that was at $\mathbf{p}, \mathbf{q}$ at time t.

For a system to be in rigorous thermodynamic equilibrium, its thermodynamic properties cannot change with time. However, these properties are ensemble averages of the time-averaged microscopic dynamics taken over an ensemble of systems described by $f(\mathbf{p}, \mathbf{q}; t)$. In ensembles of open systems, such as the grand canonical ensemble where individual systems (i.e., cells) are allowed to contain different numbers, N, of particles, the averages must also be taken over systems with different values of N (and generally over the different values of internal energy as well). The averages include the contributions from trajectories in systems with different values of N. All these ensemble averages that give the thermodynamic properties will depend on time if $f(\mathbf{p}, \mathbf{q}; t)$ explicitly depends on time. Thus, for rigorous equilibrium we must have $\partial f/\partial t = 0$ in (15.1). This is a necessary and sufficient condition for thermodynamic equilibrium.

The more galaxies cluster and virialize, the less rapidly $f(\mathbf{p}, \mathbf{q}; t)$ evolves in a given small coarse-grained volume $d\mathbf{p}\,d\mathbf{q}$ of phase space, and the smaller $\partial f/\partial t$ becomes relative to the second set of terms in (15.1). These gradient terms in phase space increase as the clustering becomes more inhomogeneous. This is a motivation for the quasi-equilibrium hypothesis, which is an approximation that improves as the system evolves, provided there is no strong nonequilibrium coherence that causes $\partial f/\partial t$ to remain large. Chapter 25 discusses another approach to the idea of quasi-equilibrium.

Under quasi-equilibrium conditions, a thermodynamic description follows from the equations of state. Their simplest general form is based on

Assumption 2
Galaxies interact gravitationally as point masses.

Suppose galaxies form, somehow, as bound dynamical entities and then cluster gravitationally as point particles in the expanding universe. Since scales relevant for galaxy clustering are much smaller than the universe's radius of curvature, and since the velocities are much less than that of light, their gravitational interaction is Newtonian with a potential $\phi \sim r^{-1}$. General relativity and cosmology enter only in specifying the expansion rate of the spatial metric. By this late stage of galaxy formation, some merging may still occur, but we assume its effects on the overall distribution functions are secondary. The statistics consider two merged galaxies to be a single entity. (This also removes a possible, though infrequent, divergence in the statistical mechanical partition function.)

What of the dark matter? If it is in particles or objects that dominate the gravitational field, and if they cluster very differently from the galaxies, then their gravity modifies the simple point-mass clustering of galaxies. Different models of the amount and distribution of dark matter may have detectable effects on the galaxy distribution functions. There are three exceptions: a) all the dark matter is distributed uniformly so that it does not affect the local dynamics of galaxy clustering, b) all the dark matter is in nongalactic objects, which, however, cluster like galaxies, or c) all the dark matter is in luminous galaxies or their halos. The second assumption therefore restricts the theory to these three circumstances, all plausible a priori possibilities. If we could find the distribution of galaxies and their velocities for these cases, we would at least have a useful standard to compare with observations. If the comparison is close, we can then constrain the amount and configuration of dark intergalactic matter whose gravitational effects would destroy the agreement.

To find this distribution we start with the two previous basic assumptions. These lead to the following quantitative relations, which will be derived in detail in later chapters. Thermodynamics for this infinite statistically homogeneous system of gravitating point particles is characterized by exact equations of state of the form

$$U = \frac{3}{2}NT(1 - 2b) \tag{15.2}$$

and

$$P = \frac{NT}{V}(1 - b) \tag{15.3}$$

for the total energy U and the pressure P in a volume V with a temperature given by

$$\frac{3}{2}NT = \frac{1}{2}mN\langle v^2 \rangle = K \tag{15.4}$$

and where

$$b = -\frac{W}{2K} = \frac{2\pi Gm^2\bar{n}}{3T} \int_V \xi\,(\bar{n}, T, r)\,\frac{1}{r}r^2\,dr \tag{15.5}$$

is the average ratio of the gravitational correlation energy, W, to twice the kinetic energy, K, of peculiar velocities in the ensemble. We use units in which Boltzmann's constant $k = 1$. As usual $\langle v^2 \rangle$ is the average peculiar velocity dispersion of the galaxies, whose average number density is $\bar{n}$. All galaxies are taken to have the same mass m. This simplifies the theory and is a good approximation for a wide range of masses (Itoh et al.,1993) because most of the later nonlinear relaxation is collective. The orbits of particles interacting gravitationally with collective modes are effectively perturbed by large masses and so, as Galileo pointed out, these orbits are essentially independent of the mass of the particle.

The total internal energy U includes the gravitational energy of the interacting galaxies due to correlated departures from the average smoothed cosmic density and the kinetic energy due to departures from the average cosmic expansion. Thus this is a theory of fluctuations around the average, since these average densities and velocities do not contribute directly to local clustering. (They contribute indirectly by determining the average rate of clustering, which is faster in universes with large values of $\Omega = \rho/\rho_{critical}$.) Section 25.1 shows explicitly how the expansion of the universe removes the mean gravitational field from the thermodynamic description.

It is important to realize that b differs from the usual virial ratio which may describe finite systems. For finite systems W is the total gravitational energy rather than the energy of just the correlations. The total pressure P is the force per unit area on a mathematical surface due both to the peculiar velocities of the galaxies (momentum transport) and to the correlated fluctuations in the gravitational force. Since particles (galaxies) and energy can cross the boundaries of an arbitrarily located volume V, the collection of these volumes can be regarded as a grand canonical ensemble. It is characterized by a temperature T, an average number density $\bar{n}$, and a chemical potential $\mu = \partial U/\partial N|_{S,V}$. Therefore the two-point correlation function, $\xi(\bar{n}, T, r)$, will depend on the values of T and $\bar{n}$ for the ensemble, as well as on the spatial coordinate r in a statistically homogeneous distribution.

Because gravity is a pairwise interaction whose potential $Gm_i m_j/r_{ij}$ involves only two particles, it depends directly on just the two-point correlation function and not on higher order correlations. This implies that the forms of the equations of state (15.2) and (15.3) are exact, rather than approximations to a virial expansion (e.g., Hill, 1956; see also Chapter 25). The similar forms for U and P result from the simple r^{-1} dependence of the gravitational potential. The value of b depends on the integral of ξr^{-1} over the volume V of a cell into which the system is divided. For scales beyond those where $\xi(r)$ makes no further contribution to this integral, b is independent of volume. The internal energy is extensive and the pressure is intensive on these large scales, as usual in thermodynamics. For smaller scales, b will depend on volume, but the correlation energy within a volume will generally be much greater than the correlation energy between volumes. So the thermodynamic description will remain a very good approximation, as the N-body simulations confirm (Itoh et al., 1993; Sheth & Saslaw, 1996; see Chapter 31).

To complete the thermodynamic description, we need either a third equation of state (e.g., for the entropy or the chemical potential) or, equivalently, the dependence of b on $\bar{n}$ and T in a grand canonical ensemble. We will see in Chapter 26 that the first and second laws of thermodynamics require that b depends on $\bar{n}$ and T only in the combination $\bar{n}T^{-3}$. This also follows from the scale invariance of the partition function for the gravitational potential. Another physical constraint on b follows from two limiting cases. In a noninteracting system, $b = 0$. In a gravitationally bound system that is virialized on all levels, $2K + W = 0$, and so $b = 1$. This second limit also follows from the requirement that $P \geq 0$ (in 15.3), so that increasing the temperature (peculiar velocities) for any given volume and number of galaxies also increases the pressure. Therefore

$$0 \leq b \leq 1. \tag{15.6}$$

Chapter 26 shows that the explicit form of $b(\bar{n}T^{-3})$ essentially follows from the basic assumption 2 of pairwise gravitational point interactions for galaxies:

$$b = \frac{b_0\bar{n}T^{-3}}{1 + b_0\bar{n}T^{-3}}, \tag{15.7}$$

where b_0 may depend on time but not on intensive thermodynamic quantities. It is this form which completes the basic thermodynamic description. In the original derivation (Saslaw & Hamilton, 1984) of the distribution function, (15.7) was postulated as the simplest form satisfying (15.6) in the two physical limits of a perfect gas and of complete clustering. Subsequently it was found to have a more profound physical basis, which is also related to a form of minimal clustering (for quasi-equilibrium to be maintained) and to the second law of thermodynamics, which requires the average gravitational entropy to be a nondecreasing function of b. This is related, in turn, to the fundamental statistical irreversibility of gravitational clustering in a statistically homogeneous universe. In such a universe, the maximum entropy state is highly clustered, unlike the maximum entropy state of a perfect gas.

Application of standard thermodynamic fluctuation theory to a system satisfying (15.2)–(15.7) leads to the spatial distribution function (Saslaw & Hamilton, 1984; Sheth & Saslaw, 1996; Sheth, 1995a; see Chapter 27)

$$f(N, V) = \frac{\bar{N}(1 - b)}{N!}[\bar{N}(1 - b) + Nb]^{N-1}e^{-[\bar{N}(1-b)+Nb]}, \tag{15.8}$$

for the probability of finding N galaxies in a randomly placed volume V where $\bar{N} = \bar{n}V$ is the expectation value of N. In the limit of $b = 0$, this reduces to a Poisson distribution. Moreover, on large scales where fluctuations are small it becomes Gaussian (see Section 28.1). The special case for $N = 0$,

$$f(V) = e^{-\bar{n}V(1-b)}, \tag{15.9}$$

describes the probability that a volume of size V around an arbitrary point in space is empty – the void probability. For all of these distributions, b will depend on V according to (15.5) with (14.10) for $\xi(r)$.

The dependence of b on V can be determined directly from the data using the variance of number counts in cells of volume V. Using (15.8) to calculate this variance (Section 28.4) gives

$$\langle(\Delta N)_V^2\rangle = \frac{\bar{N}}{(1 - b(V))^2}. \tag{15.10}$$

Since $\bar{n}$ is also determined directly from the data, there are no free parameters in (15.8). Computer experiments and observations test the form of (15.8) directly and the applicability of the two basic assumptions used to derive it. It seems reasonable to call (15.8) the gravitational quasi-equilibrium distribution, or GQED for short.

Except for pathologically shaped volumes, such as those that avoid galaxies based on a priori knowledge of their positions, the derivation of (15.8) will generally hold for reasonably shaped volumes. In particular, it applies to conical volumes with the apex at the observer. The number of galaxies in these cones is just the projection of the distribution along an angle of sight onto a circle on the celestial sphere. Therefore the GQED should apply to two-dimensional counts in cells on the sky (in circles, rectangles, triangles, etc.) as well as to three-dimensional counts in space (in spheres, cubes, etc.). Only the value of b will be affected somewhat by the shape through the volume integral in (15.5). This is because different shapes sample the two-point correlation differently, depending on the relative scales of different parts of the volume and of the correlations. For any shape, the variance of the counts in cells will give the self-consistent value of b simply from (15.10). Projective invariance of the form of the distribution function for a statistically homogeneous sample makes it an especially useful statistic to compare with observations. We can learn a lot without having to know distances along the line of sight. Distance information will naturally extend the two-dimensional results and may reveal interesting differences that get smoothed over in projection.

Another especially useful property of the GQED is that this fairly simple distribution applies to a very wide range of clustering. On small scales containing bound clusters, $(n - \bar{n})/\bar{n} \gg 1$ and the density fluctuations are highly nonlinear. On large scales where $(n - \bar{n})/\bar{n} \ll 1$ and an initially incoherent statistically homogeneous distribution has not had time to evolve significantly, the fluctuations are linear. Statistical thermodynamics provides a unified treatment over the whole range for these systems.

The distribution $f(N, V)$ in (15.8) can be examined as a continuous function of V for a given value of N, denoted $f_N(V)$, or as a discrete function of N for a given V, denoted $f_V(N)$. The $f_V(N)$ representation is usually referred to as counts in cells. The probability of finding the N nearest galaxies to a random point in space in a volume around that point is $f_N(V)$. Each of these representations emphasizes a different aspect of the same distribution. Often $f_N(V)$ including the special case of the void distribution $f_0(V)$ is particularly informative on small scales. It involves the slight complication that as b is also dependent on V, usually setting b equal to its value at the peak of $f_N(V)$ is a good approximation for observed distributions. Since $f_V(N)$ represents just one scale, its value of b is constant for a given volume. Figures 15.1 and 15.2 illustrate $f_N(V)$ and $f_V(N)$ for some different values of N and b.

So far we have explored only half of phase space – the configuration space half. The other half – velocity space – is equally important. The peculiar velocity distribution for the cosmological many-body problem is the analog of the Maxwell–Boltzmann velocity distribution for a perfect gas. Once again, we will discover it by taking a quasi-equilibrium shortcut through the detailed kinetic theory.

To find the velocity distribution function we make the simple additional

Assumption 3
In a given volume, the kinetic energy fluctuations are proportional to the potential energy fluctuations.

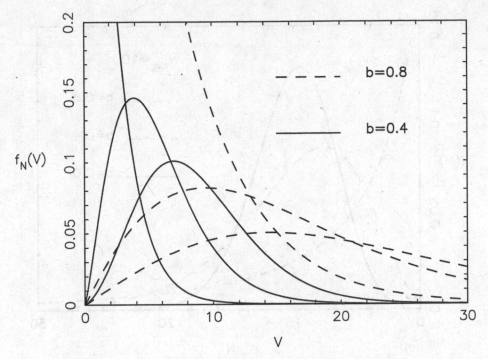

Fig. 15.1. Illustrations of $f_N(V)$ for two values of $b = 0.4$ and 0.8, each with $N = 0, 3$, and 6 (with successive peaks moving to larger V). The spatial number density is normalized to unity; so $V = \hat{N}$.

Since potential energy fluctuations are density fluctuations and these produce correlated local peculiar velocities, Assumption 3 seems quite reasonable. What may be surprising is that using the same constant of proportionality for all fluctuations is an excellent approximation. After all, the peculiar velocities generated will depend on the detailed configuration of particles (galaxies) in the fluctuation, and the theory is not meant to apply at this "microscopic" level of description (i.e., the kinetics of individual particles). The reason this approximation works so well is that the distribution of individual configurations is strongly peaked around the configuration with the most probable energy. Computer experiments indicate this (Saslaw et al.,1990), but since it has not yet been demonstrated from more fundamental principles, we give it the status of an assumption here.

The resulting velocity distribution $f(v)$ consistent with (15.8) is (Saslaw et al., 1990; see Chapters 29 and 32)

$$f(v)dv = \frac{2\alpha\beta(1-b)}{\Gamma(\alpha v^2 + 1)}[\alpha\beta(1-b) + \alpha bv^2]^{\alpha v^2 - 1}e^{-\alpha\beta(1-b) - \alpha bv^2}v\,dv \qquad (15.11)$$

in natural units with $G = M = R = 1$. Here Γ is the standard gamma function, $\beta \equiv \langle v^2 \rangle$ is a simpler symbol for the velocity dispersion, and $\alpha \equiv \beta\langle 1/r \rangle_{\text{Poisson}}$,

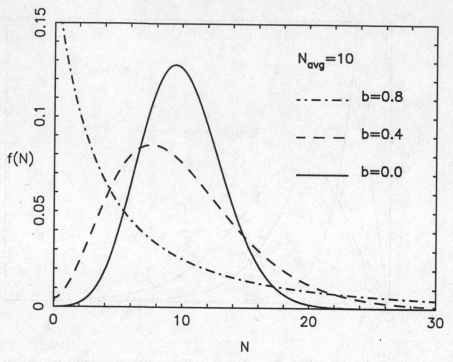

Fig. 15.2. Illustrations of $f_V(N)$ for values of $b = 0.0$, 0.4, and 0.8 and $\bar{N} = 10$.

where $\langle 1/r \rangle_{\text{Poisson}}$ is the value of r^{-1} averaged over a Poisson distribution having the average density $\bar{n}$ of the system. The quantity α can be considered to be a "form factor," which represents how the average configuration of particles in a fluctuation departs from a Poisson configuration. This velocity distribution peaks at a lower velocity and is considerably more skew than a Maxwell–Boltzmann distribution with the same variance $\langle v^2 \rangle$. This form of $f(v)dv$ represents all the galaxies, including those isolated in the general field, as well as those in the richest clusters. Figure 15.3 illustrates a range of these velocity distributions. Equations (15.8) and (15.11) solve the problem Hubble foretold (Chapter 4) for the case of the cosmological many-body problem (gravitational clustering of point masses).

It is possible to go even further and solve for the evolution of $f(N, V)$ and $f(v)$ with time. All the arguments, so far, have been thermodynamic. Time does not enter directly. So how can we determine the system's time evolution without recourse to kinetic theory? The key is that this quasi-equilibrium evolution occurs adiabatically to a very good approximation (Saslaw, 1992; Sheth & Saslaw, 1996). Heat flow caused by galaxy correlations growing across the boundary of a comoving volume element during a Hubble expansion time is small compared to the correlation energy within the volume. This approximation works best in low-Ω_0 cosmologies where the clustering timescale is longer than the Hubble timescale. But even for

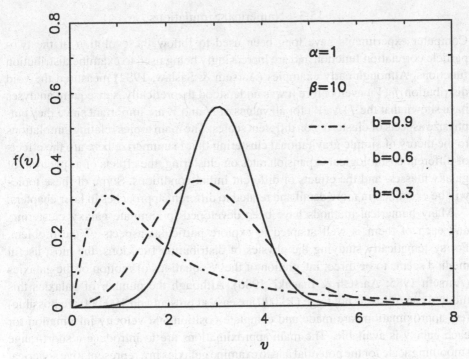

Fig. 15.3. Velocity distribution functions $f(v)$ (15.11) for $\alpha = 1$, $\beta = 10$, and $b = 0.3$, 0.6, and 0.9.

$\Omega = 1$ where these timescales are equal, numerical experiments show it remains a very good approximation. Adiabatic evolution is equivalent to the cosmic energy equation, which can be derived directly from the equation of motion (see Chapter 22), and so it need not be considered an additional assumption.

Adiabatic expansion for a system satisfying the equations of state (15.2), (15.3), and (15.7) shows that the scale $R(t)$ for any Einstein–Friedmann cosmological model is related to b by (Saslaw, 1992; see Chapter 30)

$$R = R_* \frac{b^{1/8}}{(1-b)^{7/8}}. \tag{15.12}$$

Here R_* is a constant that defines the initial values of R and b. The relevant value of b is the asymptotic value of (15.5) as $\xi(r) \to 0$. The curve $b(R)$ rises rapidly initially, levels off around $b \approx 0.8$, and increases slowly reaching $b = 1$ at $R = \infty$. Thus it is possible to trace this clustering back to high redshifts.

Equations (15.8), (15.11), and (15.12) for $f(N, V)$, $f(v)$, and $b(R(t))$ are three of the main results, so far, from the statistical thermodynamic theory of the cosmological many-body problem. Their detailed microscopic relation to kinetic theory is not yet understood. But they seem to describe our universe remarkably well.

15.3   Numerical Simulations

Computer experiments have long been used to follow the evolution of the two-particle correlation function and are increasingly being used to examine distribution functions. Although early examples (Aarseth & Saslaw, 1982) measured the void distribution $f_0(V)$ even before it was understood theoretically, subsequent analyses have shown that the $f(N, V)$ for all values of N and V are important since they emphasize aspects of clustering on different scales. The main topics relating simulations to the theory of simple gravitational clustering that I summarize here are the effects of different cosmological expansion rates on clustering, the effects of a spectrum of galaxy masses, and the effects of different initial conditions. Some of these topics will be examined in more detail and related to different approaches in later chapters.

Many numerical methods have been developed to simulate galaxy clustering, and each of them is well adapted to explore particular aspects of the problem. For systematically studying the physics of distribution functions, the most useful method seems to be direct integration of the N equations of motion of the galaxies (Aarseth 1985; Aarseth & Inagaki, 1986). Although the number of galaxies this method can follow is not large (10,000 are easy at present and 100,000 are possible) few approximations are made and complete position and velocity information for each galaxy is available. The main approximations are to introduce a short-range smoothing scale for the potential and to examine galaxies in a representative spherical volume of the universe. The smoothing scale does not expand. It represents the physical size of the galaxy and facilitates the numerical integration, but it does not have a significant effect on the distribution functions after their initial relaxation. It does affect the timescale for this initial relaxation somewhat by altering the nonlinear interactions of neighboring galaxies (Itoh et al.,1988). Moreover, the form of b is modified when the simulations are smoothed (Sheth & Saslaw, 1996). Since relatively few galaxies cross the spherical boundary (and are reflected when they do) and all the distribution function statistics exclude galaxies that are too close to the boundary, this does not affect the results significantly. Comparisons of 1,000-, 4,000-, and 10,000-body experiments under the same conditions show that while increasing the number of galaxies improves the statistics, it does not change the resulting distribution functions significantly.

Another numerical method smoothes the gravitational potential over a three-dimensional grid containing cells whose size expands with the Universe. This is well adapted for following large-scale linear and quasilinear structure, especially if the Universe contains cold dark matter not associated with the galaxies, but it minimizes the important non-linear relaxation on small scales. Distribution functions obtained with this method depend significantly on cell size and resolution (e.g., Fry et al.,1989). Moreover, this approach requires additional assumptions to define a galaxy out of the smoothed distribution. Usually a galaxy is defined as a region of peak density contrast, although there is no independent theory for choosing a particular value of the density excess. Other possibilities, such as the total mass in a volume, or the proximity of two (or more) regions of high density or mass excess, could also be chosen. The number of actual galaxies – as opposed to mass cells – in these simulations is generally comparable with those of direct N-body simulations. Presumably, with very high resolution and a suitable definition of what a galaxy is, this method will give similar results to the direct integration approach, although this problem has not been studied. Which approach best represents our actual Universe is, of course, another question. Perhaps, as often happens in this subject, both will be suitable, but on different scales. For multicomponent models with various types of dynamically important intergalactic dark matter, hybrid simulation techniques are useful.

Early results for the cosmological many-body problem (Saslaw & Hamilton, 1984; Saslaw, 1985b) showed that direct N-body simulations satisfying Assumptions 1– 3 had distribution functions that were described well by (15.8). This encouraged a more systematic exploration (Itoh et al.,1988, 1990, 1993; Inagaki et al.,1992; Saslaw et al.,1990; Suto et al.,1990; Itoh, 1990), which is still being extended. First I summarize some of the results for the spatial distribution function, then for the velocity distribution function, and then for the evolution.

Figure 15.4 (from Itoh et al.,1993) illustrates the spatial distribution function for a 10,000-body simulation with all particles (galaxies) having the same mass. They

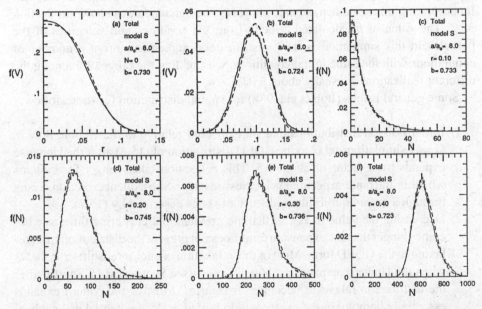

Fig. 15.4. Spatial distribution functions (from Itoh et al., 1993) for a 10,000-body simulation. Distributions $f_N(V)$ are shown for the probability of finding $N = 0$ galaxies within a volume V of radius r (the void distribution function) and for $N = 5$. Distribution functions $f_V(N)$ are shown for the probability of finding N galaxies in a volume of radius r with $r = 0.1, 0.2, 0.3,$ and 0.4. (The total radius of the simulation is normalized to unity). Histograms are the N-body simulations; dashed lines are the best fits to the distribution function of (15.8).

start with a Poisson distribution and zero velocity relative to the Hubble flow. The radius of the simulation is normalized to $R = 1$ and the results are shown here at an expansion factor $a/a_0 = 8$. This time is chosen because the distribution has achieved a two-particle correlation function whose slope and amplitude agree reasonably well with the observed values. There are two ways to portray the distribution function. By selecting a given value of N and plotting $f_N(V)$ one obtains the distribution of radii for randomly placed spheres containing N galaxies. Here the void distribution $f_0(V)$ and $f_5(V)$ are illustrated, plotted in the first two panels as a function of radius r of the volume V. For larger scales, it is more informative to locate spheres of given radius r randomly throughout the simulation (but at least a distance r from the boundary) and plot $f_V(N)$, the probability for finding N galaxies in these spheres. This is illustrated for spheres of radius 0.1, 0.2, 0.3, and 0.4 in the last four panels. Histograms represent the simulation data, and dashed lines are the theoretical distribution of (15.8). Figure 4.1 in Chapter 4 shows a Gaussian distribution for comparison.

Figure 15.4 shows that under conditions where the assumptions of the theory apply, the theoretical and experimental distribution functions agree very well. The values of b are least square fits to (15.8), and they vary by less than 3%. The fit here is not quite as good for the $f_N(V)$ as for the $f_V(N)$ because only one average value of b is fit to the $f(N, V)$, rather than taking its scale dependence into account. Incorporating this scale dependence improves the agreement, as shown in Section 31.2. The value of b calculated directly from the positions and velocities of the particles in this simulation is 0.67, also in good agreement. From a number of simulations with the same initial conditions, we find that the range of b among the different realizations is usually about ± 0.05.

Some general results (Itoh et al.,1988) for spatial distribution functions follow:

1. Initial Poisson distributions that start off cold relative to the Hubble flow in $\Omega = 1$ simulations relax to the GQED distribution of (15.8) after the Universe expands by a factor of about 1.5. This relaxation takes longer for systems whose initial random peculiar velocities are larger. Subsequent evolution occurs through a quasi-equilibrium sequence of states described by (15.7).

2. Decreasing the value of Ω_0, so that the gravitational clustering timescale becomes longer than the expansion time, does not prevent the distribution from relaxing to the GQED form. Most of this relaxation occurs at redshifts $z \gtrsim 1/\Omega_0$ when the Universe expands as though $\Omega = 1$ (see Section 30.1). Shortly after the distribution relaxes it becomes "frozen in." Most of the pattern expands essentially homologously, except within bound or deep potential wells where the distribution is relatively independent of the expansion. The pattern value of b on large scales, determined by fitting (15.8) to the distribution, increases only slowly with time when $\Omega < 1$. However, the associated physical value determined directly from W and K in (15.5) generally increases faster. This is because for low Ω_0 there are more field galaxies and they continue to lose

energy to the adiabatic expansion, decreasing their peculiar kinetic energy (see Fig. 15.7).

3. For $\Omega = 1$ models, the three-dimensional fits to (15.8) give slightly higher values of b than do two-dimensional fits to the distribution projected onto the sky. This illustrates the effects of the volume's shape – spheres versus cones. For lower Ω_0 models, the correlations are weaker so the shape is less important and the spatial and projected values of b agree very closely (see Figure. 15.6 and Section 28.5).

4. Non-Poisson initial distributions take longer to relax to the GQED, and some of them never reach it (Saslaw, 1985b; Suto et al.,1990; Itoh, 1990; Bouchet et al.,1991). This is particularly true for simulations, such as some dark matter models, that have strong initial correlations, or anticorrelations, on large scales. These do not satisfy the statistical homogeneity requirement implied by the quasi-equilibrium assumption. They cannot relax rapidly enough on large scales relative to the expansion timescale. Therefore the large-N tail of the distribution, especially over large volumes, remains non-Poisson and does not satisfy (15.8), even though (15.8) may still accurately describe the relaxed bulk of the distribution.

5. Although the theory developed so far is for systems whose components all have the same mass, (15.8) still provides a good description of two-component systems provided the mass ratio of the two components is less than about 1 to 10 (Itoh et al.,1990). The basic reason is that the relaxation leading to (15.8) is mainly collective at later times and does not depend on the masses of individual galaxies. For large mass ratios, however, the heavy galaxies tend to form the nuclei of satellite systems and if there are too many of these then the single-component theory naturally cannot describe them.

6. Systems with a continuous mass spectrum, which is more representative of observed galaxy masses, cluster rather differently. They tend to form more extended large-scale patterns in their later stages of evolution. On large scales these finite simulations become statistically inhomogeneous and (15.8) no longer applies, although it continues to represent the distribution well on smaller, more homogeneous scales.

We now turn to the velocity distribution function. Velocities of galaxies in computer simulations have also been compared with the velocity distribution predicted by (15.11) (Saslaw et al.,1990; Inagaki et al.,1992; Itoh et al.,1993; Raychaudhury & Saslaw, 1996). Remarkably, it turns out that the GQED describes velocity distributions over an even greater range of conditions than those for which it describes the density distribution. Not only does it work very well for single mass systems, but it is also accurate for systems with galaxy mass ratios much greater than 10:1. The velocity distribution depends sensitively on Ω_o, much more so than does the density distribution. It is important to compare (15.11) with observations of peculiar galaxy velocities. This stands as a clear prediction.

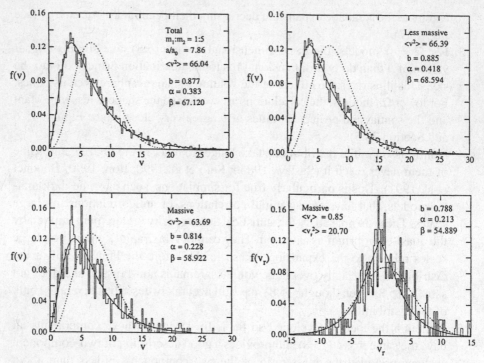

Fig. 15.5. Peculiar velocity distribution functions (from Inagaki et al.,1992) for a 4,000-body simulation described in the text. Histograms are from the simulation, dotted lines are the Maxwell–Boltzmann distribution having the same $\langle v^2 \rangle$ as the simulation, and solid lines are the best fits to the distribution function of (15.11). The lower right panel shows the peculiar radial velocities for the more massive component.

Figure 15.5 (from Inagaki et al.,1992) illustrates a typical example. This is a 4,000-body experiment with $\Omega_0 = 1$ in which 500 galaxies each have relative mass 5 compared to the other 3,500 galaxies each of mass unity. Initially both mass components have a Poisson distribution at rest relative to the Hubble expansion. They cluster rapidly and relax to the distributions (15.8) and (15.11). The figure shows velocity distributions at an expansion factor of about 7.86 when the two-particle correlation function comes into reasonable agreement with the observations. The histograms represent the experimental three-dimensional velocity distributions for all the galaxies, as well as for the two mass components separately. The lower right panel is the radial velocity distribution of the massive galaxies as seen by a representative observer at the center of the system (representing an arbitrary position in space). In each case, the dotted line is the Maxwell–Boltzmann distribution function having the same temperature (velocity dispersion) as the simulation, shown for comparison. It doesn't fit at all. The experimental distribution peaks at much lower velocities and is much more skew. The Maxwell–Boltzmann distribution cannot simultaneously represent both the field galaxies and those in rich clusters.

The solid line in Figure 15.5 is the best least squares fit of (15.11) to the experimental distribution, using b, α, and β as free parameters (although $\alpha\beta$ becomes constant as the system relaxes). One measure of agreement is the ability of (15.11) to reproduce the entire distribution. Another measure is the comparison of the fitted value of β with the value of $\langle v^2 \rangle$, which is computed directly from the peculiar velocities. For the total distribution and the less massive component, these two quantities agree within about 2%. For the more strongly clustered massive component they differ by about 8%. Comparing $\langle v^2 \rangle$ with $3\langle v_r^2 \rangle$ for the massive component measures the spatial isotropy of their velocities: These average values are 63.7 and 62.1 respectively – well within sample variations.

Figure 15.6 shows the radial peculiar velocity distribution function as it would be observed in a larger 10,000-body simulation with $\Omega_0 = 1$ and all particles having the same mass, at an expansion factor of 8 (Model 5 of Itoh et al.,1993 analyzed by Raychaudhury & Saslaw, 1996). Here the agreement between the simulations and the theoretical $f(v_r)$ obtained by integrating (15.11) over transverse velocities could hardly be better.

One of the main aspects of gravitational galaxy clustering from an initially unclustered state is that it starts growing rapidly, but after several expansion timescales the growth slows and clustering approaches an asymptotic state. This behavior appears to occur for all values of Ω_0. It is shown in Figure 15.7 (from Saslaw, 1992), which examines the evolution of single-mass 4,000-body simulations for three values of Ω_0 and an initial Poisson distribution moving with the Hubble expansion (cold). The crosses, with sampling error bars including different scales, plot the physical values of b obtained ab initio directly from (15.5) based on the positions and velocities of the particles. The open circles and filled squares plot the values of b obtained by fitting (15.8) to the spatial distribution using b as a free parameter, since $\bar{N}$ is known, for the three-dimensional and projected two-dimensional experimental distributions. These are all plotted as a function of the expansion radius R of the universe relative to its initial value R_i.

The solid line in Figure 15.7 shows the evolution predicted by (15.12) based on adiabatic clustering (R_* is determined by the initial value of b when the system has relaxed). For $\Omega_0 = 1$ the evolution differs by about 10% from the simulation. This may partly be due to effects of the softened potential used to accelerate the numerical simulations. These softening effects are more important for cases with larger Ω_0 since their clustering is stronger (Sheth & Saslaw, 1996). For lower values of Ω_0, the simulations agree with the result (15.12) for adiabatic expansion and agreement with (15.12) remains very good all the way into the highly nonlinear regime.

The value of b from the spatial pattern fit to (15.8) is less than b determined ab initio from the positions and velocities in the $\Omega_0 < 1$ cases because the spatial clustering relaxes early in the evolution and then freezes out as the expansion becomes faster than the gravitational clustering timescale. This can be understood from linear perturbation theory for the growth of density fluctuations in the system (Zhan, 1989; Zhan & Dyer, 1989; Inagaki, 1991). Chapter 30 discusses this in more detail.

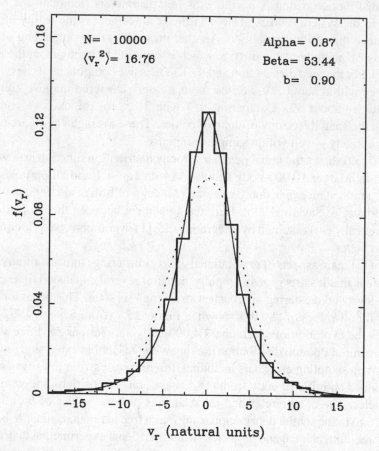

Fig. 15.6. The distribution of radial peculiar velocities for the 10,000 galaxies in the $\Omega_0 = 1$ simulation (model 5) of Itoh et al. (1993) where all galaxies are of equal mass. The dotted line is the Maxwell–Boltzmann distribution with the same velocity dispersion as the simulation. The solid line is the best fitting GQED distribution, the parameters for which are in the upper right-hand corner. (From Raychaudhury and Saslaw, 1996.)

15.4   Observed Galaxy Spatial Distribution Functions

When the gravitational statistics of (15.8) were discovered, no one knew whether they would describe the observed distribution of galaxies in our Universe. It was one of the few genuinely robust predictions for large-scale structure. Many astronomers doubted that it would be applicable since it did not employ any of the more speculative processes involving dark matter or the early universe that were then popular. So it was rather a surprise to find subsequently (Crane & Saslaw, 1986) that the projected distribution of galaxies in the Zwicky catalog agreed very well with the GQED. Its value of $b = 0.70 \pm 0.05$ even agreed with the expected value from

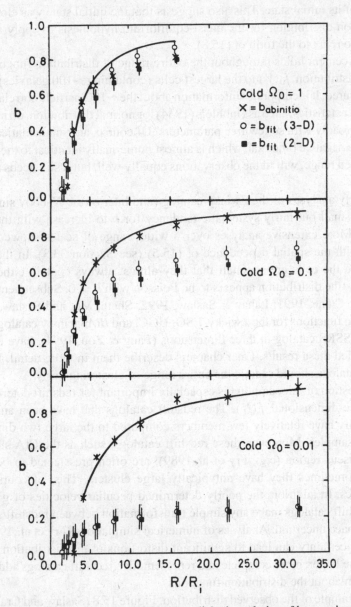

Fig. 15.7. Comparison (from Saslaw, 1992) of the adiabatic evolution of $b_{ab\,initio}$ given by (15.12) with the values (crosses) from an N-body simulation described in the text. Values of b for fits of (15.8) to counts-in-cells of three-dimensional spheres and to two-dimensional projections of three-dimensional cones are also shown.

N-body simulations. This is consistent with the galaxy distribution being primarily gravitational on scales up to at least about 10 megaparsecs. It has lost most of the memory of its initial state. This also suggests that the initial state was close enough to a Poisson distribution for the quasi-equilibrium hypothesis to apply and that it had time to relax to the form of (15.8).

The agreement holds throughout the entire range of distribution functions, from the void distribution $f_0(V)$ to the largest cells (containing $\sim$100 galaxies) that have been measured. It incorporates information about the $\sim$100-particle correlation functions. Other statistics, such as Hubble's (1934) lognormal distribution, do not seem to apply so broadly with so few free parameters. Of course, any mathematical statistic such as the negative binomial, which is almost numerically identical to the GQED in the observed range, will fit the observations equally well, but it still needs a physical basis.

Our early analyses of the Zwicky catalog (and later of the N-body simulations) showed a small but fairly systematic tendency for b to increase with the scale of the cells. More extensive analyses over a wider range of scales showed that this accords with the spatial dependence of (15.5) (see Section 33.3). In the extreme case where the cells are so small that they almost always contain either zero or one galaxy, the distribution appears to be Poisson with $b = 0$. Subsequent analyses (Saslaw & Crane, 1991; Lahav & Saslaw, 1992; Sheth, Mo, and Saslaw, 1994) of distribution functions for the Zwicky, ESO, UGC, and *IRAS* galaxy catalogs, as well as for the SSRS catalog in three dimensions (Fang & Zou, 1994), have confirmed and extended these results. Later chapters describe them in more detail. Naturally, suitable catalogs should sample the entire range of clustering fairly.

The question of fair sampling is especially important for redshift determinations of the three-dimensional $f(N)$. The redshift catalogs that have been analyzed so far generally have relatively few members compared to the large two-dimensional projected samples. Moreover, these redshift catalogs, such as the CfA slice or the Pisces–Perseus region (e.g., Fry et al.,1989), are often preselected to be unusual regions. Sometimes they have untypically large clusters. This, of course, biases the statistics. In addition, the poorly determined peculiar velocities of galaxies in many redshift catalogs make any simple transformation between redshift space and physical space uncertain. Analyses of numerical simulations (Fry et al.,1989) show that this uncertainty can lead to significant distortions of the distribution function. Even so, the larger, more complete and representative redshift catalogs add valuable information about the distribution functions.

As an example of the observed distribution, Figure 15.8 (Saslaw and Crane, 1991) shows counts in cells of different sizes from 1 to 16 square degrees. These are for the Zwicky catalog, whose galaxies are reasonably complete to a Zwicky magnitude of about 15.5. The solid lines show fits to (15.8), which apply over the whole range from isolated field galaxies to the richest cluster galaxies. They are good for a reasonable range of scales. On large scales, the smaller number of independent cells lead to greater sampling fluctuations. The value of b increases with scale as discussed after

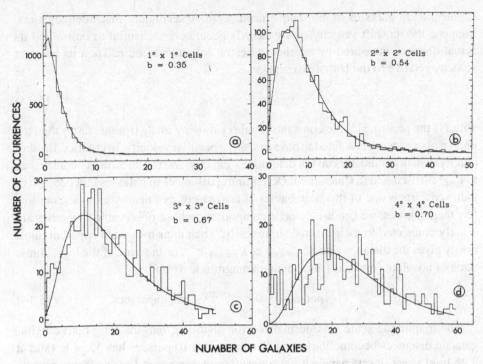

Fig. 15.8. Observed distribution functions (histograms) for counts of galaxies in cells of various sizes in the Zwicky catalog, compared with the best fits of the distribution function (15.8). (From Saslaw and Crane 1991.)

Equation (25.18) for the theory and in Sections 31.2 for numerical simulations and 33.3.3 for observations. Many insights follow from the differences between various samples and subsamples, as discussed further in Chapter 33.

15.5 Observed Galaxy Velocity Distribution Functions

Three major problems in observing the peculiar velocity distribution of galaxies are all being overcome to a reasonable and increasing degree. First, it is necessary to have a fair and representative sample. The distribution (15.11), integrated over transverse velocities (see Chapters 29 and 34), predicts the radial peculiar velocity distribution function for all galaxies. Their surroundings may range from splendid isolation in the field to jostling together in very dense groups. Observations must include all in proportion. Until the early 1990s, velocity samples were mostly biased toward either galaxies in clusters, or in small groups, or in the field. An early sufficiently unbiased sample (Mathewson, Ford, and Buckhorn, 1992) was measured to examine infall into the "Great Attractor," a suspected enormous mass inhomogeneity producing bulk flow of galaxies on scales of roughly 50 Mpc. Analysis of a subset of this sample provided the first observed $f(v)$ (Raychaudhury & Saslaw, 1996), and subsequent larger samples should give more detailed results.

Second, all galaxies in the sample need accurate secondary distance indicators. Suppose the redshift velocity of a galaxy is accurately measured to be v_z, and its actual distance measured by another indicator is known to be r. Then its peculiar velocity relative to the Hubble expansion is

$$v_{pec} = v_z - r H_0, \tag{15.13}$$

with H_0 the present Hubble constant. A galaxy moving away from us faster than the universal expansion is defined to have a positive peculiar velocity. In practice, the secondary distance indicators do not determine r exactly, and the uncertainty contributes to v_{pec}, as discussed in Chapter 34. Determining distances to galaxies independent of their redshifts is one of the main battles of astronomy. For nearby galaxies, out to a few megaparsecs, we can use galactic components whose observable properties are closely connected to their intrinsic luminosity. Then their measured apparent luminosity gives the distance since $L_{apparent} \propto L_{intrinsic} r^{-2}$, or in terms of the logarithmic scale of absolute (intrinsic) and apparent magnitude M and m,

$$r = 10^{0.2(m-M)+1} \text{ parsecs} = 10^{0.2(m-M)-5} \text{ megaparsecs.} \tag{15.14}$$

The magnitude scale was normalized for historical reasons long before extragalactic distances become important. Thus a star at 10 parsecs has $M = m$ (and at 3.26 light years or one parsec it has a *parallax* of one *second* of arc from opposite sides of the Earth's orbit around the Sun). The Sun has $M = 4.83$ in the visible band of its spectrum centred at 5,500 Å, and $m = -26.8$. The integrated visual absolute magnitude of Andromeda, our nearest large galaxy at ~ 0.7 Mpc, is about $M = -22$, just to put things in perspective. Some examples of useful distance indicators for nearby galaxies are Cepheid variables whose periods are related to their intrinsic luminosities (calibrated from similar stars in our own galaxy with other independent distance determinations), certain types of supernovae with similar maximum intrinsic luminosities, and regions of ionized gas whose intrinsic size may be similar in certain types of galaxies so that their apparent angular size indicates their distance. These techniques are discussed in most astronomy texts, and all have their uncertainties that tend to increase with distance.

At distances of tens of megaparsecs it is hard to measure individual components of galaxies, but we can relate some of their global properties to their intrinsic luminosity. For spiral galaxies, Tully and Fisher (1977, 1988) found a close correlation between intrinsic luminosity and the radio 21-cm linewidth. Some correlation is expected on simple physical grounds because the linewidth measures an average circular velocity of neutral hydrogen gas rotating around the galaxy, usually weighted by the outer regions where the rotation curve is flat. Approximate gravitational and centrifugal balance give $GM \approx r v_c^2$, to within a factor of order unity depending on details of the model and on the actual circumstances. If the weighted global mass is in turn correlated with the intrinsic luminosity of the galaxy, then v_c is related to $L_{intrinsic}$. This relation has been refined, and its dispersion decreased, by confining it

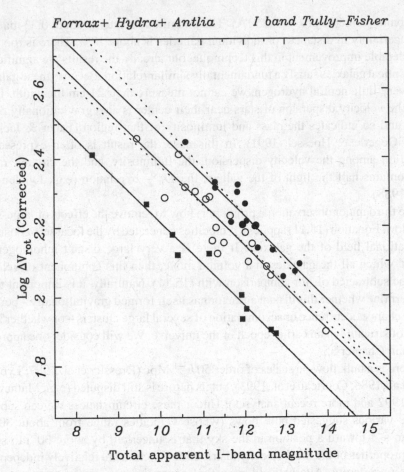

Fig. 15.9. The I-band Tully–Fisher relation for galaxies belonging to three clusters of galaxies (squares: Fornax; solid circles: Hydra I; open circles: Antlia) in the Mathewson et al. (1992) catalog. All galaxies in the same cluster are assumed to be at the same distance. The dotted lines are the best fit Tully–Fisher relations for the individual clusters. The solid lines result if all clusters are constrained to have the same slope. (From Raychaudhury and Saslaw, 1996.)

to subsets of galaxies with particular morphologies and inclinations. Measuring the luminosity at long wavelengths, and correcting for extinction effects, also tightens the correlation. Figure 15.9 shows an example (Raychaudhury & Saslaw, 1996) for thirty-seven galaxies in three clusters chosen to calibrate a subset of the Mathewson et al. (1992) sample. These give

$$\log(\Delta v) = -0.126 M_I - 0.43, \tag{15.15}$$

where Δv is the rotational velocity measured either from where the integrated neutral hydrogen 21 cm line falls to half its peak value or with optical spectroscopy, which finds the rotational velocity from the H_α linewidth. M_I is the absolute magnitude in the I band (centered at $\lambda = 8{,}250$ Å). The magnitude scatter is about 0.33 mag and the uncertainty in distances of individual galaxies is about 17%. There is room for considerable improvement in this technique, but already the results are significant.

Elliptical galaxies satisfy a fundamentally similar relation. Because these galaxies have very little neutral hydrogen, we cannot measure their 21-cm linewidth. However, the velocity dispersion of stars near their center is also gravitationally dominated and so indicates the mass and luminosity of that region (Faber & Jackson, 1976; Oegerle & Hoessel, 1991). In this case, the result is often expressed as a relation among the velocity dispersion, the luminosity, and the circular radius that contains half the light of the galaxy, the $D_n - \sigma$ relation (e.g., Lynden-Bell et al.,1988).

The third major observational problem is how to remove the effects of large-scale bulk flow. Equation (15.11) applies to velocities generated by the local self-consistent gravitational field of the galaxies. If there is a very large distant inhomogeneity toward which all the galaxies in a volume move, then this component of velocity must be subtracted off for comparison with (15.11). Naturally, it is important to try to determine whether this distant attractor has itself formed gravitationally – perhaps by the chance occurrence or accumulation of several large clusters – or whether it is a result of structure in an earlier epoch of the universe. We will consider one approach to this in Chapter 35.

There is a bulk flow on scales of order $50\,h^{-1}$ Mpc (Dressler et al.,1987; Lynden-Bell et al.,1988; Courteau et al.,1993), but its nature is still disputed (e.g., Mathewson et al.,1992 and more recent analyses). Under these circumstances we can subtract off the various suggested bulk flows (whose velocities range from about 300 to 700 km s^{-1} toward a position in the sky that is uncertain by about 30°) to see if some properties of the local velocity distribution function are relatively independent of this uncertainty. Alternatively, as a first approximation, we can correct for a bulk motion of the local rest frame by simply subtracting an average displacement velocity rH_* from each galaxy so that the average radial peculiar velocity for the sample $\langle v_r - rH_* \rangle = 0$. This determines an effective local value, H_*, for the Hubble constant in the sample. It also tends to increase the net velocity dispersion of $f(v_r)$ because it does not take the detailed flow pattern into account. Part of the uncertainty is due to the relatively small size of the samples, which usually have several hundred galaxies. Accordingly, their peculiar velocities may not represent the bulk flow fairly.

Using redshifts from a homogeneous subset of 810 galaxies with velocities $\leq 5{,}000$ km s^{-1} relative to the cosmic microwave background in the Mathewson et al. (1992) sample of $Sb - Sd$ spirals, obtaining distances from (15.15), and subtracting off a bulk motion of 600 km s^{-1} toward galactic longitude $l = 312°$ and latitude $b = 6°$ gave the first observational determination of $f(v_r)$ (Raychaudhury & Saslaw, 1996). Figure 15.10 shows the result. The solid histogram is the observed $f(v_r)$. Error bars are standard deviations for 1,000 random subsets of the data; they

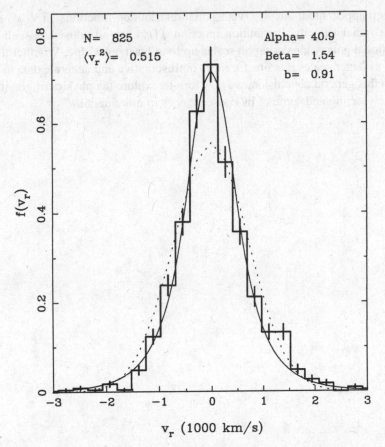

Fig. 15.10. The observed distribution function of radial peculiar velocities in a subsample of the Mathewson et al. (1992) catalog, corrected for the Dressler et al. (1987) bulk motion of 600 km s^{-1}. The dotted line is the Maxwell–Boltzmann distribution with the same velocity dispersion $\langle v_r^2 \rangle$ as the observed $f(v_r)$. The solid line is the best fitting GQED distribution (15.11) whose parameters are in the upper right. (From Raychaudhury and Saslaw, 1996.)

represent sampling uncertainties. The dotted line is the Maxwell–Boltzmann distribution with the same dispersion $\langle v_r^2 \rangle$ as the observations. The solid line is the best least squares fit of (15.11) and has the parameters listed in the upper right.

The gravitational distribution (15.11) clearly describes the observations very well, including field galaxies, rich clusters, and everything in between. This is a robust result, independent of reasonable uncertainties in the Tully–Fisher relation and the bulk flow subtraction. These uncertainties change the value of the parameters somewhat, but (15.11) remains a better fit than a Maxwellian, or even than two Maxwellians, one for the field and loose clusters and another for the dense clusters (Fang & Saslaw, 1995). The values of α, β, and b also agree reasonably well with the N-body simulations illustrated previously.

Thus it appears that the observed spatial distribution functions $f(N, V)$ and the peculiar radial velocity distribution function $f(v_r)$ are all consistent with simple gravitational galaxy clustering on scales up to at least $\sim$40 Mpc. Whether this also applies to larger scales remains for more representative and accurate data to tell.

With this general conclusion, we now turn to explore the physics of gravitational galaxy clustering and some of its consequences in more detail.

PART III

Gravity and Correlation Functions

Of all the processes that might produce correlations among galaxies in our Universe, we know only one that definitely exists. It is gravity. Inevitably and inexorably, as Newton told Bentley, gravitational instability causes the galaxies to cluster.

Of all the descriptions of galaxy clustering, the correlation functions are connected most closely to the underlying gravitational dynamics. The next several chapters develop this connection. It resembles a great fugue, starting with simple themes and variations, then combining, developing, and recombining them to obtain a grand theoretical structure whose main insights have formed over the last three decades and which still continues to expand.

16

The Growth of Correlations:

I. A Fluid Description

Simplify, simplify.

Thoreau

16.1   Introduction

Models of galaxy clustering that are meant to apply to our Universe fall into three main groups. There are models that deal primarily with very large scales by smoothing over local irregularities such as galaxies themselves and even over small groups of galaxies. The resulting coarse-grained density perturbations are usually linear, i.e., $\delta\rho/\bar{\rho} \ll 1$. The gravitational potential has comparably linear smooth perturbations and they interact to induce growing correlations. These are essentially "fluid models."

On smaller scales, if most of the inhomogeneous matter of the Universe is in galaxies or their halos, the discrete point like masses of the galaxies dominate their gravitational clustering. In these "particle models," the galaxies interact essentially as point masses and their clustering is a cosmological many-body problem.

In a third group of "hybrid models," both the point masses of galaxies and irregularly distributed dark matter may be important for clustering. These models represent a universe filled with a background of dark matter. They may also represent early clustering on small scales if galaxies form from high-amplitude local density fluctuations in a background fluid.

The relative importance of each group of models, as well as the importance of the many specific models within each group, remain centers of controversy and debate. To some extent, this debate is interwoven with models of galaxy formation from pre-existing inhomogeneities in the earlier universe. The main (scientific) reasons that the debate continues are that 1. many dissipative processes have intervened between the initial conditions and the present, dimming our view of the past, 2. observations are not yet precise enough to discriminate among many models, and 3. models do not often make unique robust predictions. We will return to some of these problems in Chapter 38. All these groups of models, however, are sufficiently basic to be interesting in their own right. After discussing the fluid criterion and reviewing relevant cosmology, we examine the linear growth of correlations, Lagrangian descriptions of density growth, and an example of nonlinear spherical contraction.

The fluid models, with which we start, apply on scales, r, where distances are large enough that the gravitational interactions between discrete galaxies of mass m are negligible compared to the interactions among large density perturbations. To develop a criterion for neglecting discrete interactions, we first compare fluctuations caused solely by discreteness with linear fluid perturbations on the same scale. The variance of discreteness fluctuations is approximately the Poisson result in (14.14) since correlations are small: $(\Delta N/\bar{N})^2 = (\delta n/\bar{n})^2 \approx \bar{N}^{-1} \approx (\bar{n}r^3)^{-1}$. For this to be small compared to density fluctuations

$$\frac{1}{\bar{n}r^3} \ll \left(\frac{\delta\rho}{\bar{\rho}}\right)^2. \tag{16.1}$$

If we imagine that the mass of each galaxy $m \to 0$ but the number of galaxies in any given volume becomes very large so that $\bar{n} \to \infty$, keeping the average mass density $\bar{\rho} = m\bar{n}$ constant, then criterion (16.1) is satisfied on any scale r. This is the usual "fluid limit" of a system; its density ρ is a continuous field. However, even a system of actual galaxies with a relatively small density $\bar{n}$ will satisfy (16.1) on a large enough scale r for a given relative density perturbation $\delta\rho/\bar{\rho}$. For example, if $\bar{n} = 1$ galaxy Mpc^{-3}, a density perturbation of 1% may be treated as a fluid on scales much larger than 10 Mpc.

Although (16.1) illustrates some essential features of discreteness, it does not quite get to the heart of the matter. In the equations of motion for gravitating correlations, the gravitational forces that are most effective in developing these correlations extend over the wavelength of the fluctuations. On scales larger than a particular wavelength, the effects of positive and negative fluctuations on that wavelength, relative to the average density, tend to cancel, leaving only the smoothed-out average density. So let us now compare the force from a discreteness fluctuation on scale r, the separation over which the correlation function is being measured, with the force from a fluid density perturbation of wavelength λ. The perturbed gravitational potential of the discreteness fluctuation, integrated over the volume, is of order $(Gm/r)(\delta n)_{\text{discrete}}r^3$. Its gradient gives the order-of-magnitude resulting force on another such fluctuation: $Gm^2\bar{n}^2(\delta n/\bar{n})^2_{\text{discrete}}r$. Similarly, the force from a fluid density perturbation is of order $Gm^2\bar{n}^2(\delta\rho/\bar{\rho})^2_{\text{fluid}}\lambda$. Comparing these, we see that the condition for the effects of discreteness to be small now becomes

$$\frac{1}{\bar{n}r^2\lambda} \ll \left(\frac{\delta\rho}{\bar{\rho}}\right)^2. \tag{16.2}$$

This same result can be derived directly from the equations of motion via the BBGKY hierarchy (Inagaki, 1976). For $\lambda = r$ it reduces to (16.1). Very long wavelength perturbations may be treated by the fluid approximation even if they have small amplitudes since their integrated gravitational force is very large compared to the force from statistical fluctuations in the numbers of discrete galaxies. A perturbation spectrum for which $(\delta\rho/\bar{\rho})^2 \sim \lambda^{-1}$ gives a criterion (16.2) that is independent of λ, so that all scales, r, greater than some limit may be treated as a fluid. To explore various fluid models, we first need to review basic properties of the expanding universe.

16.2   The Cosmological Background

A recurrent theme in the formation of large-scale structure, whatever the model, is the expansion of the Universe. Since even the largest of large-scale structures are one or two orders of magnitude smaller than the radius of curvature of the Universe, it is reasonable to approximate to the background cosmology by smoothing over all density inhomogeneities. Moreover, the high degree of homogeneity and isotropy of the observed cosmic microwave background radiation, remaining from an early hot dense epoch of the Universe, suggests that within the context of general relativity the Einstein–Friedmann models provide at least a good approximate description of the cosmological background for the purposes of galaxy clustering at redshifts of $\sim 10^3$ or less.

The Einstein–Friedmann models are the simplest most symmetric models of the Universe in general relativity. They do not explain galaxy formation or the initial conditions for galaxy clustering. At our present state of understanding these must be added by hand via specific more speculative theories of the much earlier universe. However, the Einstein–Friedmann models enable us to follow the development of subsequent clustering. Almost every text on cosmology describes these models in great detail, so here I just briefly recollect their properties relevant to galaxy clustering.

All general-relativistic cosmological models contain two major ingredients: a metric, which describes the structure of spacetime, and a dynamical equation, which specifies how that structure evolves. The Robertson–Walker metric describes the Einstein–Friedmann cosmologies and may be written in the simple form

$$ds^2 = c^2 dt^2 - R^2(t)\left[\frac{dr^2}{1 - kr^2} + r^2(d\theta^2 + \sin^2\theta\, d\phi^2)\right]. \qquad (16.3)$$

This reduces to the Minkowski metric for special relativity in spherical coordinates when $R(t) = 1$ and $k = 0$. Since the homogeneity and isotropy of space makes all points equivalent in the Einstein–Friedmann models, the origin of the coordinate system could be anywhere (recall Nicholas de Cusa). The fixed constant value of k determines whether the spatial, $t = constant$, hypersurfaces are open with constant negative curvature ($k = -1$), closed with constant positive curvature ($k = +1$), or Euclidean with no curvature ($k = 0$). The flat, $k = 0$, models contain just enough matter to almost close the Universe. Determining the value of k remains one of the main goals of cosmology.

The scale factor, $R(t)$, may change with time and therefore represents the expansion factor, often more loosely called the radius, of the Universe. Notice that $R(t)$ applies at every point. The Universe does not expand into some "preexisting space" – it is space expanding. For different observers to agree about this expansion requires the existence of a universal or cosmic time. Weyl's postulate provides for this by assuming there is a class of noninteracting "fundamental observers" whose world lines form a three-dimensional spatial set of non intersecting geodesics that are orthogonal

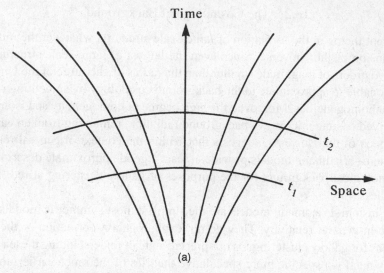

(a)

Fig. 16.1(a). World lines of fundamental observers embedded in an expanding spacetime.

to a sequence of spacelike hypersurfaces. Figure 16.1a illustrates the situation. All the observers on these geodesics agree that they are farther apart from each other on the spacelike hypersurface at t_2 than they were at t_1, and they can therefore agree that $t_2 > t_1$. Galaxies, whose complex gravitational orbits may bind them in clusters or even make them collide, clearly cannot represent these fundamental observers who are in a sense embedded in the structure of spacetime itself. Spacetimes that do not admit such a class of fundamental observers may have metrics whose coefficient of $(dt)^2$ depends on spatial coordinates. Then observers at different places would discover they also existed at different times, and they could find no time frame in common.

Einstein's field equations specify the dynamical evolution of $R(t)$. They relate the metric, and thus the curvature of spacetime, to the energy–momentum tensor of the matter (including radiation), which acts as a source for the gravitational field implied by the metric. Because the Robertson–Walker metric (16.3) is so symmetric, the six independent field equations that generally apply reduce to just two:

$$2\frac{\ddot{R}}{R} + \frac{\dot{R}^2 + kc^2}{R^2} = -\frac{8\pi G}{c^2}P + \Lambda c^2, \tag{16.4}$$

$$\frac{\dot{R}^2 + kc^2}{R^2} = \frac{8\pi G}{3c^2}\epsilon + \frac{\Lambda c^2}{3}, \tag{16.5}$$

where P is the pressure and ϵ is the energy density of the matter (including the rest mass energy) and radiation in the Universe. The cosmological constant Λ is set equal to zero in the Einstein–Friedmann models but is present in an interesting class of models explored by Lemâitre. Einstein disowned Λ after introducing it to provide a static model of the Universe consistent with astronomers' beliefs in the first two decades of this century. Had Einstein found the models with $\Lambda = 0$, later discovered by Friedmann, he could have predicted the expansion of the Universe – which he considered one of the great missed opportunities of modern science.

One of the elegant aspects of general relativity is that the field equations are constructed to contain the conservation laws. To see this in the case of (16.4) and (16.5), multiply the latter by R^2, differentiate with respect to time, and subtract $\dot{R}/R$ times (16.4) to obtain

$$\frac{d}{dt}(\epsilon R^3) + P\frac{d}{dt}(R^3) = 0. \tag{16.6}$$

Therefore in any local volume comoving with the expansion $R(t)$, the change of proper energy (i.e., in the frame of a local observer) is just the work done on its surroundings by the adiabatic change of that volume, assuming a perfect fluid. This is, of course, conservation of energy in the form of the first law of thermodynamics: $dU = -P\,dV$. If $P = 0$, then $\epsilon R^3/c^2$ is the total mass in the comoving volume and it stays constant.

To obtain Einstein–Friedmann models with $\Lambda = 0$ we must specify P and ϵ. Then we can solve (16.5) and (16.6). Although equivalent to (16.4) and (16.5) they involve just the first time derivative and so are easier to handle. The simplest models have $P = 0$. Their matter has no peculiar velocity. Technically it is called "dust." It is a good approximation to think of galaxies as dust (for this purpose!) since their peculiar velocities of $\lesssim 10^3$ km s^{-1} imply $P \lesssim 10^{-5}\epsilon$. For a dust-filled universe, (16.6) shows that in a comoving volume the mass density $\rho \sim R^{-3}$ so that the total mass $\sim \rho R^3$ is conserved. In a fixed volume, the total mass would decrease as $R^{-3}(t)$.

Next we must specify the global curvature k. Again we start with the simplest case, $k = 0$, known as the Einstein–de Sitter universe. Then (16.5) and (16.6) with $\epsilon = \rho c^2 = \rho_0 c^2 R_0^3/R^3$ integrate to

$$R = \left(\frac{8\pi G\rho_0 R_0^3}{3}\right)^{\frac{1}{3}} t^{\frac{2}{3}} \equiv R_0\left(\frac{t}{t_0}\right)^{2/3} \tag{16.7}$$

if $R = 0$ at $t = 0$, and where ρ_0 is the present density at $R = R_0$. This is familiar from the Newtonian analog (14.47, 14.48), as it should be since the $k = 0$ model is flat and, with no global scale length, is a similarity solution. Since $\dot{R}^2 \propto R^{-1}$, this model has zero expansion energy in the limit $R = \infty$. Moreover, writing (16.5) in the form $\dot{R}^2/2 - GM/R = 0$, where $M = 4\pi\rho R^3/3$, suggests that the total of the kinetic expansion energy and the gravitational potential energy is zero in a comoving volume of any size (since there is no scale length). This is a useful Newtonian mnemonic, but its general relativistic interpretation is more subtle because the gravitational energy cannot be localized completely within the volume. From (16.5) we see that at any time, and in particular at the present, the value of the Hubble constant,

$$H_0 = \left.\frac{\dot{R}}{R}\right|_{t_0} = \frac{2}{3t_0}, \tag{16.8}$$

is related to the density by

$$H_0^2 = \frac{8\pi G}{3}\rho_0 = \frac{8\pi G\rho_c}{3} \qquad (16.9)$$

for these models. Since these models are at the limit between an open and a closed universe, the value of ρ in (16.9) is usually called the critical, or closure, density, ρ_c. Its present value for $H_0 = 100\,h$ km s^{-1} Mpc^{-1} is

$$\rho_c = 2 \times 10^{-29}h^2 \text{g cm}^{-3}. \qquad (16.10)$$

Here h is a scaling factor, probably in the range $0.4 \lesssim h \lesssim 1$, which allows for the observationally uncertain value of H_0. If the Universe is closed, astronomers have not yet found most of its contents.

However, if what we see is what we get then $\rho_0 < \rho_c$, the Einstein–Friedmann models are open and $k = -1$. For any model, it is useful to define the deceleration parameter

$$q(t) = -\frac{R\ddot{R}}{\dot{R}^2} = -\frac{\ddot{R}}{RH^2} \qquad (16.11)$$

and the density parameter

$$\Omega = \frac{\rho}{\rho_c}. \qquad (16.12)$$

At the present time, $\Omega = \Omega_0 = \rho_0/\rho_c$. The field equations (16.4, 16.5) are

$$\frac{2\ddot{R}}{R} + \frac{\dot{R}^2 - c^2}{R^2} = 0 \qquad (16.13)$$

and

$$\frac{\dot{R}^2 - c^2}{R^2} - \frac{8\pi G\rho}{3} = 0. \qquad (16.14)$$

Subtracting one of these from the other, dividing by H^2, and using (16.9), (16.11), and (16.12) shows that at any time

$$q = \tfrac{1}{2}\Omega. \qquad (16.15)$$

Clearly this relation also holds for the $k = 0$ and $k = +1$ cases.

We can conveniently normalize the evolution to the present epoch, at which (16.13) and (16.14) are

$$\frac{c^2}{R_0^2} = (1 - \Omega_0)H_0^2 \qquad (16.16)$$

and

$$H_0^2 = \frac{8\pi G}{3}\rho_0 \Omega_0^{-1}. \tag{16.17}$$

Note that (16.16) and (16.17) require $0 < \Omega_0 < 1$; thus the $k = -1$ universe has less than the critical density, as expected. With these results and

$$\rho = \frac{R_0^3}{R^3}\rho_0 \tag{16.18}$$

we may rewrite the field equation (16.14) as

$$\begin{aligned}
\frac{\dot{R}^2}{c^2} &= \frac{H_0^2 \Omega_0 R_0^3}{c^2 R} + 1 \\
&= \frac{\Omega_0 c}{(1-\Omega_0)^{3/2} H_0}\frac{1}{R} + 1 \\
&= \frac{\Omega_0}{(1-\Omega_0)}\frac{R_0}{R} + 1.
\end{aligned} \tag{16.19}$$

From the third expression we see that at large redshifts z given by

$$1 + z = \frac{R_0}{R} > \frac{(1-\Omega_0)}{\Omega_0}, \tag{16.20}$$

the first term dominates. Comparing the first term in the first expression of (16.19) with (16.5) for $k = 0$ shows that, at these early times, open universes expand at the same relative rate as if they were just critical. Thus if $\Omega_0 = 0.2$, an open universe would have expanded like an Einstein–de Sitter model at $z > 3$. This will be a useful result for understanding some aspects of galaxy clustering.

When $\Omega_0 R_0/R \ll 1$ in (16.19), the universe is effectively empty and $R \approx ct$. If spacetime were flat for a special relativity model in the limit $\Omega_0 = 0$, it would also expand as $R = ct$. This is known as the Milne model. Although Milne derived it from his own theory of kinematic relativity, it also emerges formally as this special case for $k = -1$ in Einstein's theory.

The general solution of (16.19) has the parametric form in terms of an angle θ:

$$R = \frac{\alpha}{2}(\cosh\theta - 1), \tag{16.21}$$

$$ct = \frac{\alpha}{2}(\sinh\theta - \theta), \tag{16.22}$$

where α is the coefficient of R^{-1} in (16.19) and $R(t = 0) = 0$. At present

$$\cosh\theta_0 = \frac{2}{\Omega_0} - 1, \tag{16.23}$$

$$\sinh\theta_0 = 2\frac{(1-\Omega_0)^{1/2}}{\Omega_0}, \tag{16.24}$$

and

$$t_0 = \frac{\Omega_0}{2(1-\Omega_0)^{3/2}} \left\{ \frac{2(1-\Omega_0)^{1/2}}{\Omega_0} - \ln\left[\frac{2-\Omega_0+2(1-\Omega_0)^{1/2}}{\Omega_0}\right] \right\} H_0^{-1}.$$

(16.25)

As θ, or t, increases, $R(t)$ expands forever.

For the closed $k = +1$ models, a similar analysis shows that $\Omega_0 > 1$ and (16.5) becomes

$$\frac{\dot{R}^2}{c^2} = \frac{\Omega_0 c}{(\Omega_0 - 1)^{3/2} H_0} \frac{1}{R} - 1.$$

(16.26)

Writing

$$R = \frac{\beta}{2}(1 - \cos\chi) = \beta \sin^2\frac{\chi}{2},$$

(16.27)

where β is the coefficient of R^{-1} in (16.26), puts the integral of (16.26) in standard form, giving the parameterized solution

$$t = \frac{\beta}{2c}(\chi - \sin\chi)$$

(16.28)

with $R(t = 0) = 0$. This has the form of a cycloid. The universe expands, reaches a maximum size at $\chi = \pi$ when

$$R_{\max} = \beta = \frac{\Omega_0 c}{(\Omega_0 - 1)^{3/2} H_0},$$

(16.29)

and collapses to $R = 0$ when $\chi = 2\pi$ at

$$t_c = \frac{\pi\beta}{c} = \frac{\pi\Omega_0}{(\Omega_0 - 1)^{3/2} H_0}.$$

(16.30)

Comparison with the present age of such a universe,

$$t_0 = \frac{\beta}{2c}(\chi_0 - \sin\chi_0) = \frac{\Omega_0}{2(\Omega_0 - 1)^{3/2}}\left[\cos^{-1}\left(\frac{2-\Omega_0}{\Omega_0}\right) - \frac{2(\Omega_0 - 1)^{1/2}}{\Omega_0}\right]\frac{1}{H_0},$$

(16.31)

shows how much longer it will last. Denser universes with larger Ω_0 reach smaller maximum radii and collapse sooner.

Adding additional components to the universe will change the rate of expansion in (16.5) if they contribute significantly to the pressure. For example, the blackbody radiation left over from the hot dense phase of the early universe has an energy density $\epsilon_{rad} \sim R^{-4}$ and a relativistic pressure $P_{rad} = \epsilon_{rad}/3$. At present $\epsilon_{rad} \approx 10^{-13}$ erg cm^{-3} while $\rho c^2 \approx 2 \times 10^{-8} h_0^2 \Omega_0$ using (16.10). Therefore in the much earlier universe, radiation will dominate at redshifts

$$1 + z = \frac{R_0}{R} > 2 \times 10^5 h_0^2 \Omega_0. \tag{16.32}$$

At these early times, it is easy to show from (16.5) and (16.6) that $R \propto t^{1/2}$. Of course, it could be that there is an unknown component in our Universe that produces this behavior at the much later times relevant to galaxy clustering, but we have yet to find it.

Incorporating the cosmological constant Λ gives rise to new qualitative features of the expansion. For a critical value

$$\Lambda_c = \frac{4\pi G \rho_0}{c^2} = \frac{1}{R_0^2} \tag{16.33}$$

one obtains Einstein's original static model. But Eddington showed this was unstable to small perturbations, which would cause it either to expand or collapse. For $\Lambda < \Lambda_c$, the model first expands from $R = 0$ and then contracts back to $R = 0$. If Λ exceeds Λ_c just slightly, the model first expands rapidly, then enters a long quasistatic period with extremely slow expansion, and then rapidly expands again. Models with $\Lambda > \Lambda_c$ ultimately expand exponentially with time. There is also a class of models that decrease from $R = \infty$ to a nonzero minimum radius and then expand back to infinity.

Current evidence seems consistent with the view that most galaxy clustering occurred at relatively small redshifts, almost certainly at $z \lesssim 10^2$ and possibly even at $z \lesssim 10$. For these, the Einstein–Friedmann dust models are appropriate, and often the simplest $k = 0$ Einstein–de Sitter model is used as an illustrative example. If we eventually understand the clustering process well enough, it will help determine which range of these models agrees best with observations.

16.3   Linear Fluid Perturbations and Correlations

Provided we smooth a discrete distribution of galaxies over large enough scales to satisfy (16.1) or (16.2), we can use a fluid description for sufficiently large amplitude perturbations. This applies even though fluctuations may be highly nonlinear on much smaller scales. Intermediate scales, however, may combine residual effects from small-scale nonlinearities with the growth of large-scale fluid modes.

Long-wavelength modes are essentially the same gravitational fluid instabilities that Jeans analyzed for a static universe and Lifshitz and Bonnor explored in the expanding case (see Chapter 5). Since galaxy clustering involves perturbations with wavelengths much less than the horizon scale length of the universe ($\lambda \ll ct$), with peculiar velocities $v \ll c$, and with gravitational energy much less than rest mass energy $(GM/\lambda c^2 \ll 1)$, general relativity just provides $R(t)$. Apart from that, the analysis is Newtonian.

It starts with the equation of continuity for mass conservation;

$$\frac{\partial \rho}{\partial t} + \nabla \cdot (\rho \mathbf{v}) = 0, \tag{16.34}$$

Euler's force equation for momentum conservation,

$$\rho \frac{\partial \mathbf{v}}{\partial t} + \rho (\mathbf{v} \cdot \nabla) \mathbf{v} = -\nabla P + \rho \nabla \phi, \tag{16.35}$$

and Poisson's equation coupling mass and the gravitational potential,

$$\nabla^2 \phi = -4\pi G \rho. \tag{16.36}$$

The fourth relation between p, ρ, v, and ϕ is usually taken to be an equation of state $P(\rho)$. A classical gas, often used to analyze perturbations that might form galaxies when radiation pressure is negligible, assumes the adiabatic relation

$$P \propto \rho^\gamma. \tag{16.37}$$

Here γ is the usual ratio of specific heats, excluding the effects of gravity (to avoid including gravity twice). A similar relation holds for a gas of galaxies on large scales. These scales behave essentially adiabatically because during their growth very little heat (from peculiar velocities) moves into them compared with the heat they already contain. This is discussed further in Chapter 30. Meanwhile we regard (16.37) as an approximation. It is only a temporary approximation because soon we will deal with scales large enough to neglect the pressure entirely. It gives a criterion for this neglect.

The fluid variables such as density $\rho(\mathbf{r}, t)$ can be written in terms of their departures, $\rho_1(\mathbf{r}, t) \equiv \delta\rho/\rho_0$, from their average background values, $\rho_0(t)$, as

$$\rho = \rho_0(t)\,[1 + \rho_1(\mathbf{r}, t)], \tag{16.38}$$
$$P = P_0(t)\,[1 + P_1(\mathbf{r}, t)], \tag{16.39}$$
$$\phi = \phi_0(t) + \phi_1(\mathbf{r}, t), \tag{16.40}$$
$$\mathbf{v} = \mathbf{r}H(t) + \mathbf{v}_1(\mathbf{r}, t). \tag{16.41}$$

From (16.37) and (16.39) we obtain

$$P_1 = \gamma \rho_1 \tag{16.42}$$

or

$$\delta P = \left.\frac{\partial P}{\partial \rho}\right|_0 \delta\rho$$

$$= \frac{\gamma P_0}{\rho_0}\delta\rho$$

$$= c^2\delta\rho, \tag{16.43}$$

where c^2 is the nongravitational sound speed, with a value close to the velocity dispersion of the galaxies.

At this stage, the alert reader will notice a contradiction. There is no background! The system is empty. Because the infinite uniform background is symmetric, it exerts no force. So $\nabla\phi = 0$ and Poisson's equation (16.36), $\nabla \cdot (\nabla\phi) = \nabla^2\phi = 0$, implies $\rho = 0$. This is sometimes called the "Jeans swindle," after its originator. However, Jeans had the intuitively correct understanding that a uniform background could be ignored in Poisson's equation, and only local perturbations would exert forces. It took half a century, until Lifshitz solved the problem in the full context of general relativity, to verify Jeans's insight. Somewhat ironically, although the expansion of the Universe removed the relevance of Jeans's original calculation, it made his Newtonian analysis applicable to the cosmological perturbation problem. General relativity makes the analysis consistent by replacing Poisson's equation with the (perturbed) field equations (16.4, 16.5) in which ρ_0 determines the rate of background expansion and ρ_1 determines local potential gradients and forces. In the local Newtonian limit, with the correct choice of coordinates, the perturbed field equations give

$$\nabla^2\phi_1 = -4\pi G\rho_0\rho_1, \tag{16.44}$$

which is Jeans's linearized form of (16.36).

Linearizing also the equations of motion (16.35) and of mass conservation (16.34) by substituting (16.38)–(16.41) and retaining just the first-order terms gives

$$\frac{\partial \mathbf{v}_1}{\partial t} + H\left(\mathbf{v}_1 + r\frac{\partial \mathbf{v}_1}{\partial r}\right) = \nabla\phi_1 - \frac{\gamma P_0}{\rho_0}\nabla\rho_1 \tag{16.45}$$

and

$$\frac{\partial \rho_1}{\partial t} + \nabla \cdot \mathbf{v}_1 + Hr\frac{\partial \rho_1}{\partial r} = 0 \tag{16.46}$$

using (16.43) and (16.34) respectively to subtract off the zeroth-order terms. Next take the divergence of (16.45) and eliminate ϕ_1 and $\mathbf{v}_1$ by substituting (16.46), (16.44), and the vector identity

$$\nabla \cdot \left(r\frac{\partial \mathbf{v}_1}{\partial r}\right) = r\frac{\partial}{\partial r}(\nabla \cdot \mathbf{v}_1) + \nabla \cdot \mathbf{v}_1. \tag{16.47}$$

The result,

$$\left(\frac{\partial}{\partial t} + 2H + Hr\frac{\partial}{\partial r}\right)\left(\frac{\partial\rho_1}{\partial t} + Hr\frac{\partial\rho_1}{\partial r}\right) = 4\pi G\rho_0\rho_1 + \frac{\gamma P_0}{\rho_0}\nabla^2\rho_1, \quad (16.48)$$

describes the evolution of the fractional density perturbation $\rho_1(\mathbf{r}, t)$.

Part of this evolution is caused by the general expansion, and part is intrinsic to the instability. Separating these two effects becomes possible by transforming to a reference frame whose scale increases with the general expansion. Distances $\mathbf{x}$ in these comoving coordinates are just the distances in the static frame normalized continuously by the expansion scale

$$\mathbf{x} = \frac{\mathbf{r}}{R(t)}. \quad (16.49)$$

So if all the growth of size in the original static frame is due to the general expansion, then there is no growth in the comoving frame. Since the comoving frame measures only residual growth, we would expect the evolution equation (16.48) to become simpler in this frame, and it does:

$$\frac{\partial^2\rho_1(\mathbf{x}, t)}{\partial t^2} + 2\frac{\dot{R}}{R}\frac{\partial\rho_1}{\partial t} = 4\pi G\rho_0\rho_1 + \frac{\gamma P_0}{\rho_0 R^2}\nabla^2\rho_1(x, t). \quad (16.50)$$

In (16.50), ρ_1 is a function of $\mathbf{x}$ and its evolution follows from (16.48) and (16.49), bearing in mind that

$$\left(\frac{\partial}{\partial t}\right)_r = \left(\frac{\partial}{\partial t}\right)_x + \left(\frac{\partial x}{\partial t}\right)_r\left(\frac{\partial}{\partial x}\right)_t = \frac{\partial}{\partial t} - x\frac{\dot{R}}{R}\frac{\partial}{\partial x}. \quad (16.51)$$

If the Universe were static and filled with dust, then (16.50) would only have its first and third terms with $\rho_0 = $ constant. These would cause an exponentially growing instability at any point:

$$\rho_1(x, t) = \rho_1(x, t_0)e^{\frac{t}{\tau}}, \quad (16.52)$$

with the gravitational timescale

$$\tau = (4\pi G\rho_0)^{-\frac{1}{2}}. \quad (16.53)$$

This is the instability Jeans found. Expansion destroys it. The second term of (16.50) decelerates the gravitational instability. For the Einstein–Friedmann models, (16.5) shows that this deceleration has the same timescale as the instability itself. Expanding material carries momentum away from the perturbation at almost the same rate it tries to contract.

Almost, but not quite. There is a residual slow contraction. To see this, we can Fourier analyze (16.50) in the comoving frame where all wavelengths $\lambda = 2\pi R/k$

expand with the universe. Then writing

$$\rho_1(\mathbf{x}, t) = \rho_1(t)e^{i\mathbf{k}\cdot\mathbf{x}},$$ (16.54)

we have

$$\frac{d^2\rho_1(t)}{dt^2} + 2\frac{\dot{R}}{R}\frac{d\rho_1}{dt} = \left(4\pi G\rho_0 - \frac{4\pi^2\gamma P_0}{\rho_0\lambda^2}\right)\rho_1.$$ (16.55)

This explicitly shows that pressure also decreases the growth of any perturbation mode.

Consider an initial perturbation $\rho_1(t_1) > 0$ that starts growing at t_1 so $\dot{\rho}_1(t_1) > 0$. In order for this perturbation to stop growing, it must reach a maximum value at which $\dot{\rho}_1 = 0$ and $\ddot{\rho}_1 < 0$. But then the right-hand side of (16.55) must be negative, and this will never occur if

$$\lambda^2 > \frac{\pi}{G\rho_0}\frac{\gamma P_0}{\rho_0} = \frac{\pi c^2}{G\rho_0}.$$ (16.56)

Therefore, long wavelengths satisfying (16.56) keep increasing their amplitude. From (16.50) this is the same critical wavelength – the Jeans length – that would grow in a static universe, except that here it is applied in the comoving frame. At any time t, the critical wavelength in the "laboratory" frame is related to its initial value by $\lambda_J(t) = \lambda_J(t_1)R(t)/R(t_1)$. Smaller wavelengths oscillate and their amplitude decreases. Expansion does not change the basic instability criterion.

The physical meaning of the Jeans criterion becomes apparent when we rewrite (16.56) as

$$\frac{1}{2}Mc^2 < \frac{3}{8\pi^2}\frac{GM^2}{\lambda}.$$ (16.57)

Here $M = 4\pi\rho_0\lambda^3/3$ is the constant mass in a volume of comoving radius λ. Therefore when the total gravitational energy in this volume substantially exceeds its total thermal energy, the galaxies within it cluster relative to their average density. Another interpretation is that a perturbation grows gravitationally if its crossing time λ/c, for a galaxy moving with the average velocity dispersion $\approx c^2$, greatly exceeds its gravitational growth timescale $(4\pi G\rho_0)^{-1/2}$. This occurs in a cluster before it has relaxed and virialized. (In the later nonlinear stage these two timescales are comparable.) Because the mass in these density perturbations is smoothed over long wavelengths, it does not matter whether its constituents are atoms or galaxies. In the case of an atomic gas, λ/c represents the crossing time for sound waves.

Although expansion does not change the instability criterion, it dramatically alters the growth rate. We calculate this from (16.55) in the limit of long wavelength, or $P_0 = 0$, so that we can neglect the pressure term completely. It is a pure tug-of-war between local gravitational attraction and the universal expansion.

First consider the $k = 0$ Einstein–de Sitter case. Substituting (16.7) and (16.9) into (16.55) then gives

$$\ddot{\rho}_1 + \frac{4}{3t}\dot{\rho}_1 - \frac{2}{3t^2}\rho_1 = 0, \tag{16.58}$$

whose solution is

$$\rho_1(t) = At^{2/3} + Bt^{-1}, \tag{16.59}$$

where A and B are determined by initial values. Growing modes and damped modes occur. Both must often be included for a consistent solution. As time passes, only the growing mode survives, and it increases proportional to $R(t)$. Local gravitational attraction wins the tug-of-war, but this is a far cry from its exponential victory (16.52) in the static case. Slow algebraic growth of ρ_1 indicates that the universe is very close to a state of neutral stability. This is expected from the discussion following (16.7) since in any given region the kinetic expansion energy and the potential gravitational energy are almost equally balanced.

The open $k = -1$ and closed $k = +1$ models give more algebraically complicated solutions using the parametric forms (16.21, 16.22) and (16.27, 16.28) respectively. Their essential property is displayed more clearly by rewriting the field equations (16.14) and (16.26) in terms of a gravitational timescale

$$\tau_G \equiv \left(\frac{8}{3}\pi G\rho\right)^{-\frac{1}{2}} \tag{16.60}$$

and a dynamical expansion timescale

$$\tau_{\exp} = H^{-1} = \frac{R}{\dot{R}}. \tag{16.61}$$

Note that τ_G is a factor of $\sqrt{3/2}$ longer than the e-folding timescale (16.53) associated with the growth of perturbations in a static universe. In the expanding Einstein–de Sitter model, the timescale for a perturbation amplitude to double, starting at $t_0 = (6\pi G\rho_0)^{-1/2}$ (from 16.8 and 16.9), is $(4\pi G\rho_0/3)^{-1/2}$. Thus $\tau_G(\rho_0)$ is $1/\sqrt{2}$ times this doubling timescale and itself represents a characteristic timescale for gravitational perturbations to grow by a factor $(\tau_G(t_0)t_0^{-1})^{2/3} \approx 1.3$. In terms of τ_G, the field equations become

$$\frac{\tau_G^2}{\tau_{\exp}^2} = 1 \pm \frac{\tau_G^2}{R^2/c^2}. \tag{16.62}$$

The "+ sign" applies to the open $k = -1$ model and the "− sign" to the closed model.

Therefore in open models, τ_G always exceeds the prevailing expansion timescale. Clustering is weaker. In contrast, for closed models, $\tau_G < \tau_{\exp}$ and clustering is faster and stronger. The amount of clustering present at any given time depends on Ω, and observations may be able to constrain Ω_0. However, as clustering also depends significantly on the initial conditions for perturbations, and on the relative distribution of luminous galaxies and dark matter, its interpretation is not straightforward. For example, initial perturbations at high redshifts in low-Ω_0 models may lead to nearly the same present clustering as perturbations starting at smaller redshifts in high-Ω_0 models. This is particularly true if the clustering is measured by a statistic that does not characterize it adequately.

We now have the background to see how the two-point correlation function evolves in the linear fluid approximation. The Einstein–de Sitter model is simplest and gives an explicit result. From (16.59) and the definition (14.15) of ξ_2 in a statistically homogeneous system neglecting discrete particles represented by the δ-function, we have

$$
\begin{aligned}
\xi(r, t) &= \bar{n}^{-2} \langle \Delta n(\mathbf{x}) \Delta n(\mathbf{x}+\mathbf{r}) \rangle = \langle \rho_1(\mathbf{x}) \rho_1(\mathbf{x}+\mathbf{r}) \rangle \\
&= \langle [A(\mathbf{x}) t^{2/3} + B(\mathbf{x}) t^{-1}] [A(\mathbf{x}+\mathbf{r}) t^{2/3} + B(\mathbf{x}+\mathbf{r}) t^{-1}] \rangle \\
&\approx \langle A(\mathbf{x}) A(\mathbf{x}+\mathbf{r}) \rangle t^{4/3} \\
&\approx \xi(r, t_0) t^{4/3}.
\end{aligned}
\tag{16.63}
$$

The last two approximate equalities retain just the dominant term after long times. They agree with the heuristic derivation of (14.53).

From this more general derivation, it is clear that the analogous $\xi(r, t)$ for the open and closed Einstein–Friedmann models, or indeed any other model, can be calculated from the corresponding time dependence of their density perturbations. Correlations in closed models evolve relatively faster than in open models, other things such as their initial amplitudes being equal.

16.4   Other Types of Linear Fluid Analyses

Although the standard wave-mode perturbation analysis helps us understand the conditions and growth of gravitational instability, it is incomplete even within its linear context. An actual linear perturbation need not be a single mode. Its more general form is an irregular patchy inhomogeneity. This requires many wavelengths of different amplitudes for its description – a wave packet. To construct an inhomogeneity from these combined modes, we need both their power spectrum and their phase. An alternative approach, which restates this problem, yields a new set of general results for the shapes of linear fluid inhomogeneities. Instead of the Eulerian description (16.35), which focuses on a comoving volume, we use a Lagrangian description, which focuses on the elements of the fluid as they move about. It applies most directly to models of the universe filled with collisionless dark matter (e.g., massive neutrinos) moving only under its own gravity. Then each macroscopic element of the fluid is represented by a "particle" of this dark matter. As the dark matter condenses, more strongly dissipative baryonic matter may be drawn down into the resulting deep gravitational wells to form luminous galaxies. The dark matter distribution dominates galaxy clustering in this model.

Lagrangian descriptions perturb the trajectories of the "particles" (fluid elements) rather than the density in the fluid. Regions of higher density form naturally where the perturbed trajectories converge. Small linear orbital perturbations can cause many collisionless trajectories to converge in one region, forming a strong nonlinear density perturbation there. This is why a linear Lagrangian analysis may reveal structure that a linear Eulerian analysis does not.

Mass must be conserved, so that the evolving perturbed density $\rho(\mathbf{r}, t)$ in any proper volume d^3r is related to its original uniform density $\rho_0(t)$ at its initial co-moving Lagrangian position q by

$$\int_{\mathbf{r}} \rho(\mathbf{r}, t) d^3\mathbf{r} = \int_{\mathbf{q}} \rho_0 d^3\mathbf{q}. \tag{16.64}$$

Therefore the density change is determined by the coordinate transformation $\mathbf{r}(t, \mathbf{q})$ of the trajectories:

$$\rho(\mathbf{r}, t) = \rho_0 \left\| \frac{\partial q_i}{\partial r_i} \right\| = \frac{\rho_0 R^{-3}(t)}{\left| \det \left(\frac{\partial x_j}{\partial q_i} \right) \right|}. \tag{16.65}$$

The Jacobian of this transformation, indicated by the double bars, is the usual absolute value of its determinant, and $\mathbf{x}(t) = \mathbf{r}(t) R^{-1}(t)$ is the comoving position of the particle on its trajectory. If the trajectories simply expand with the universe, then $\mathbf{r}(t) = R(t)\mathbf{q}$, the Jacobian is unity, and we recover the expected result $\rho = \rho_0 R^{-3}$ for $R(t_0) = 1$.

But the trajectories are perturbed. In what is known as the Zeldovich (1970) approximation, they are assumed to have the form

$$\mathbf{x}(t) = \frac{r(t)}{R(t)} = \mathbf{q} + \eta(t)\mathbf{S}(\mathbf{q}). \tag{16.66}$$

In other words, the change, $\mathbf{x}(t) - \mathbf{q}$, in the comoving coordinate of a fluid is a separable function of time and its original position. Moreover, these functions $\eta(t)$ and $\mathbf{S}(\mathbf{q})$, are the same for all times and positions. Clearly this kinematic approximation works well only as long as the parcels follow their own ballistic trajectories and do not interact with one another.

With (16.66), we have to second order in η

$$\det \left(\frac{\partial x_j}{\partial q_i} \right) = \det \left(\delta_{ij} + \eta \frac{\partial S_j}{\partial q_i} \right) = 1 + \eta(t)\nabla_q \cdot \mathbf{S} + \mathcal{O}(\eta^2) \tag{16.67}$$

and (16.65) becomes

$$\rho_1 = \frac{\rho - \rho_0 R^{-3}}{\rho_0 R^{-3}} \equiv -\eta(t)\nabla_q \cdot \mathbf{S}(\mathbf{q}). \tag{16.68}$$

This identical time dependence for all perturbations is consistent with the pressure-free linear Eulerian analysis for all wavelengths from (16.55). Hence the solution of (16.55) for a given background cosmology determines $\eta(t)$. At early times when $\eta(t)\mathbf{S}(\mathbf{q}) \ll \mathbf{q}$, the $\mathbf{x}$ and $\mathbf{q}$ coordinates are nearly identical and the linearized equation of continuity (16.34),

$$\frac{\partial \rho_1}{\partial t} = -\nabla \cdot \mathbf{v}_1, \tag{16.69}$$

shows that $\mathbf{S}(\mathbf{q})$ is proportional to the initial peculiar velocity field.

Even without specifying $\eta(t)$ ($\propto t^{2/3}$ in the Einstein–de Sitter model, for example) and the initial velocity field, this approach gives some useful qualitative information. Imagine that initially overdense parcels of fluid are moving every which way, perhaps with some slight coordination produced by long-wavelength perturbations. As the parcels in an initially cubic volume, say, move about, the Lagrangian volume that continues to contain these same parcels will deform. The density in the volumes changes according to (16.65). If these distortions do not introduce any vorticity into an already vorticity-free velocity field (which is generally the case for linear gravitational motions as discussed later in this section), then the velocity field represented by $\mathbf{S}(\mathbf{q})$ is the gradient of a scalar velocity potential. As a result, the matrix that represents the determinant in (16.65) can be diagonalized with eigenvalues λ_1, λ_2, and λ_3 along three principal axes. Therefore (16.65) may be rewritten

$$\rho(\mathbf{r}, t) = \frac{\rho_0 R^{-3}(t)}{(1 - \eta(t)\lambda_1(\mathbf{q}))(1 - \eta(t)\lambda_2(\mathbf{q}))(1 - \eta(t)\lambda_3(\mathbf{q}))} \tag{16.70}$$

with $\mathbf{q}(\mathbf{r})$ from (16.66).

As $\eta(t)$ increases, a region along an axis with $\lambda > 0$ contributes to a high density, and one with $\lambda < 0$ contributes to a lower than average density. These are regions where fluid parcels tend to accumulate ($\lambda > 0$) or disperse ($\lambda < 0$). The largest eigenvalue represents the region (perhaps caused by initial perturbations, or by chance) where parcels accumulate so strongly they collide. The density formally becomes infinite and this description breaks down.

How the description breaks down is important. The singular eigenvectors of (16.70) are caustics (as in the theory of shock waves or of gravitational lensing) where trajectories of independently moving fluid parcels (or particles or photons) intersect. If one eigenvalue is much larger than the other two, the original cubical volume is compressed to a sheet along this corresponding principal axis. The fluid forms a flat structure, usually called a "pancake." Actually since the approximations break down before this happens, and pressure also intervenes, the structure is rather thicker. This has led some people to maintain that the proper English translation from the original Russian is not "pancake" but "muffin."

Computer simulations and analyses (e.g., Melott & Shandarin, 1990; Kofman et al., 1990) show that the Zeldovich approximation gives a good description of the locations and general shapes of the high-density regions. Details depend on the cosmology and initial density and velocity fields. Usually the simulations start with long-wavelength, low-amplitude density perturbations that look like whispy roughly cellular structure. As their orbits evolve, the original structure appears to gell. It becomes denser and more cohesive, for the most part maintaining the original linear pattern until $\rho_1 \approx 1$. Even though the density becomes somewhat nonlinear, the description (16.70) still applies because the trajectories rather than the density are being perturbed.

Several techniques have been developed to push this description further into the nonlinear regime (see Dekel, 1994). These include keeping second-order terms in the determinant (16.67), extending the relation (16.66) by a polynomial approximation, adding in gravitational forces, and introducing a mock viscosity term so that fluid elements stick together when they collide in a pancake, or muffin. None of these approaches can describe the highly nonlinear regime where fragmentation is important and the velocity flow may become rotational. Sometimes an ad hoc prescription for the formation of galaxies or clusters can be included – such as their appearance at intersections of pancakes. Whether these extensions quantitatively describe the observed galaxy distribution is still unclear.

A second approach to understanding the shapes of linear large-scale structure is also related to perturbations of trajectories. I mentioned earlier the well-known result that the linear velocity field in the expanding universe tends to be irrotational (i.e., $\nabla \times \mathbf{v}_1 = 0$). To see this, we can Fourier analyze the linear velocity field in the comoving frame where $k = 2\pi R(t)/\lambda$ into plane waves with $\mathbf{x} = \mathbf{r}/R(t)$:

$$\mathbf{v}_1(\mathbf{x}, t) = \mathbf{v_k}(t)e^{i\mathbf{k}\cdot\mathbf{x}}. \tag{16.71}$$

The mode $\mathbf{k}$ has a direction, and so the perturbed velocity will have a parallel component (of either $\mathbf{v}_1(\mathbf{x}, t)$ or $\mathbf{v_k}(t)$)

$$\mathbf{v}_\parallel = \frac{\mathbf{k}\cdot\mathbf{v}_1}{k^2}\mathbf{k}, \tag{16.72}$$

which streams along this direction, and a transverse component

$$\mathbf{v}_\perp = \frac{\mathbf{k}\times(\mathbf{v}_1\times\mathbf{k})}{k^2} \tag{16.73}$$

perpendicular to the mode: $\mathbf{v}_1 = \mathbf{v}_\parallel + \mathbf{v}_\perp$. Setting k along the $\mathbf{z}$ direction shows that the only nonzero component of $\mathbf{v}_\parallel$ is $\mathbf{v}_{\parallel z}$ and it depends only on z; so $\nabla \times \mathbf{v}_\parallel = 0$. Therefore the streaming component represents laminar irrotational flow. The transverse component represents vorticity or rotation in the flow since $\nabla \times \mathbf{v}_\perp \neq 0$.

Since $\nabla \cdot \mathbf{v}_\perp = 0$, the rotational component is the curl of a vector field. In comoving coordinates, the linearized continuity equation (16.46) together with (16.51) shows that $\mathbf{v}_\perp$ does not contribute to the growth of density perturbations.

Vorticity caused by gravitational interaction would develop according to the Euler equation (16.35). Because the gravitational force can be written as the gradient of a scalar potential, and $\nabla \times \mathbf{Grad} \equiv 0$, it does not contribute to the vorticity: $\nabla \times \mathbf{F}_1 = 0$. Similarly, the pressure force does not affect $v_\perp$. Linearizing (16.35), using the pressure–density relation (16.42) and the relation $\mathbf{v}_0 = \mathbf{r}\dot{R}/R$, and Fourier analyzing the linear equation by substituting (16.54), (16.71), and the analogous equation for ϕ_1 gives

$$\dot{\mathbf{v}}_\mathbf{k}(t) + \frac{\dot{R}}{R}\mathbf{v}_\mathbf{k}(t) + \frac{i\gamma\rho_1(t)}{R}\mathbf{k} - \frac{\phi_1(t)}{R}\mathbf{k} = 0. \tag{16.74}$$

Taking the vector product with $\mathbf{k}$,

$$\left(\dot{\mathbf{v}}_\mathbf{k}(t) + \frac{\dot{R}}{R}\mathbf{v}_\mathbf{k}(t)\right) \times \mathbf{k} = 0, \tag{16.75}$$

implies

$$v_\perp R = \text{constant}. \tag{16.76}$$

Therefore the rotational component of the velocity, and any initial vorticity, decays as $R(t)^{-1}$ due to the expansion of the universe. There is nothing mysterious about this; it is just conservation of angular momentum. In an expanding volume of radius r, the angular momentum is of order $(\rho_0 r^3)v_\perp r = \text{constant}$. Since $\rho_0 \sim r^{-3}$, the transverse velocity $v_\perp \sim r^{-1} \sim R^{-1}(t)$, for any noninteracting Fourier mode. If this vorticity decreases to zero, then Kelvin's circulation theorem (e.g., Landau & Lifshitz, 1979) implies that it remains zero as long as the trajectories of the fluid elements do not cross (no shocks or dissipation or, for particles, orbit crossings). This will generally break down in the nonlinear regime, but when it applies it connects the velocity field to the density inhomogeneities.

Continuity is the connection. For a linear, growing irrotational mode, the continuity equation may be written approximately as (Peebles, 1980)

$$\rho_1 = \frac{R}{\dot{R}f}\nabla \cdot \mathbf{v}_1, \tag{16.77}$$

where

$$f(\Omega) \equiv \frac{R\dot{A}}{\dot{R}A} \approx \Omega^{0.6} \tag{16.78}$$

with A the growing mode of (16.55) in the pressure-free case. Since $\mathbf{v}_1$ is irrotational, we may write it as the gradient of a potential, ψ, and, in principle, recover it by integrating observed radial peculiar velocities along the line of sight,

$$\psi(\mathbf{x}) = -\int_0^r v_{1r}(r', \theta, \phi) \, dr'. \tag{16.79}$$

Differentiating $\psi(\mathbf{x})$ then provides an estimate of the transverse velocities. With these velocities, one can use the Zeldovich approximation in Eulerian space,

$$\mathbf{x}(q, t) - \mathbf{q} = f^{-1}\mathbf{v}(\mathbf{q}, t), \tag{16.80}$$

or a generalization of it to obtain the density distribution $\rho_1(\mathbf{x})$ (Bertschinger and Dekel, 1989) from (16.77) directly, or by inverting its solution for $\mathbf{v}_1$, or by assuming simple models. Although this approach (reviewed in Dekel, 1994) is significantly affected by the way the data is selected and smoothed, and by observational uncertainties, it gives useful information about the density field on scales between about 30 and 100 Mpc. The main results agree with gravitational instability causing inhomogeneity on these large scales. This is not a unique solution since other processes also satisfy the equation of continuity. However, the observations seem consistent with the same time-dependent form of the solution of the continuity equation everywhere, and that, as we have seen, is an attribute of gravity.

For the third illustration, consider a single, localized spherical inhomogeneity. Sphericity is simplicity, nothing more. A region whose density is greater than average will expand more slowly than the rest of the universe; an underdense region expands more rapidly. If the local density exceeds the critical closure density ρ_c (Equation 16.9), the inhomogeneity eventually contracts. We will examine this behavior with a Lagrangian analysis (Gunn and Gott, 1972; GPSGS, 1985) analogous to the perturbations of orbits, although it originally was done in an Eulerian frame (Roxburgh and Saffman, 1965). Even though we retain the linear requirement that shells of matter do not cross, the special nature of this example enables contraction to be followed into the nonlinear regime.

Since ρ_c is the dividing line between expansion and contraction, we write the total density of the inhomogeneity as

$$\rho_c + \Delta. \tag{16.81}$$

In principle, Δ could be taken to be negative to give some indication of how a hole expands. However, holes tend to be rather more irregular than clusters. Clusters relax; holes do not. They typically involve the boundaries of several clusters and thus lack the coherence of a single cluster. Hence the spherical approximation is generally more realistic for studying the dynamics of a cluster than of a hole. Actual clusters, as well as holes, often form by mergers. Therefore this simple model, instructive as it is, only gives an approximate picture.

At the initial time, denoted by subscript "i", the average density inside a radius r_i is taken to be

$$\bar{\rho}_i = \begin{cases} \rho_{ci} + \Delta & \text{for } r_i \leq S_i, \\ \rho_{ei} + (\rho_{ci} + \Delta - \rho_{ei})S_i^3/r_i^3 & \text{for } r_i > S_i, \end{cases} \quad (16.82)$$

where S_i is the initial radius of the inhomogeneity and ρ_{ei} is the density of the universe external to the initial inhomogeneity. This distribution matches smoothly onto the background. Pressure-free matter in the inhomogeneity provides a good approximation for galaxy clustering (if not for galaxy formation) even if most of the matter is nonbaryonic. We therefore consider the background Universe to be an Einstein–Friedmann model with $\Lambda = 0$ filled with pressure-free "dust." The field equations (16.4–16.6) and definitions (16.9–16.12, 16.15, 16.19) provide the relation

$$H_i^2 = H^2[2q(1 + z_i)^3 + (1 - 2q)(1 + z_i)^2] \quad (16.83)$$

among the initial values of the Hubble ratio, its current value H, and the current value of q. Similarly, the initial external density is related to the critical density by

$$\rho_{ei} = \rho_{ci} \frac{2q(1 + z_i)}{(1 - 2q) + 2q(1 + z_i)}. \quad (16.84)$$

If the inhomogeneity starts at sufficiently high redshift that

$$\frac{1 - 2q}{2q(1 + z_i)} \ll 1, \quad (16.85)$$

then expanding (16.83) and (16.84) to first order in this ratio gives

$$\frac{\rho_{ci} - \rho_{ei}}{\rho_{ci}} = \frac{1 - 2q}{2q(1 + z_i)} \quad (16.86)$$

and

$$H_i^2 = 2qH^2(1 + z_i)^3. \quad (16.87)$$

These approximations are reasonably good for galaxy clustering.

With these preliminaries, we can follow the motion of smoothed spherical shells of galaxies, or of fluid, within and outside the inhomogeneity. Initially the radius of an arbitrary shell is r_i and its evolution may be written similarly to (16.66) in the form

$$r(r_i, t) = r_i R(r_i, t). \quad (16.88)$$

If there were no inhomogeneity, $R(r_i, t)$ would be uniform everywhere and would just be the universal scale length $R(t)$. Then (16.88) would describe comoving expansion. Indeed in the limit $r_i \gg S$, we expect $R(r_i, t) \to R(t)$ since the inhomogeneity makes a rapidly decreasing contribution to the total mass in the sphere. So at small r_i, the factor $R(r_i, t)$ measures the local departure from a uniform Hubble expansion $v = Hr$. The force acting on a spherical shell initially at r_i gives it an acceleration

$$\frac{d^2r}{dt^2} = -\frac{4\pi G \bar{\rho}_i r_i^3}{3r^2}. \tag{16.89}$$

As long as the average density $\bar{\rho}_i$ does not increase with r_i, shells initially at different r_i will not cross as the inhomogeneity contracts. This homologous contraction is analogous to the irrotational flow assumed in the earlier examples. It has a straight-forward solution by substituting (16.88) and multiplying through with $\dot{R}$ to give a first integral of (16.89). The constant of integration is settled by requiring (16.9) to be satisfied for the case $\bar{\rho}_i = \rho_{ci}$ initially with $R_i = 1$. Thus

$$\left(\frac{dR}{dt}\right)^2 = \frac{8\pi G}{3R}\bar{\rho}_i + \frac{8\pi G}{3}(\rho_{ci} - \bar{\rho}_i). \tag{16.90}$$

Note that this is a special case of (16.5) for $\rho_{\text{dust}} \propto R^{-3}$. Therefore we expect the inhomogeneity to behave similarly to the universe as a whole, but on a different timescale due to its different average density. It also illustrates the relation between the local Newtonian behavior of any small spherical region in the universe and the general relativistic expansion of the background.

Each shell expands to a maximum radius determined by setting $\dot{R} = 0$ in (16.90):

$$R_{\max} = \frac{\bar{\rho}_i}{\bar{\rho}_i - \rho_{ci}} \equiv \frac{1}{E}. \tag{16.91}$$

This last expression defines the initial fractional density excess of the inhomogeneity above the critical density. Equations (16.82), (16.88), and (16.91) show that all shells within the inhomogeneity reach a maximum radius proportional to their initial radius, all with the same constant of proportionality $\approx \bar{\rho}_i/\Delta$. Shells beyond the initial homogeneity will also reach a maximum radius and fall back onto the inhomogeneity if their initial radius is less than a critical value where $\bar{\rho}_i = \rho_{ci}$. Thus the inhomogeneity will accrete additional mass as it contracts.

Playing around with (16.90) a bit enables us to solve it exactly. First define a dynamically scaled time by

$$\tau = \left(\frac{8\pi G \bar{\rho}_i}{3}\right)^{1/2} t, \tag{16.92}$$

and substitute this, together with E from (16.91) into (16.90), which becomes

$$R\left(\frac{dR}{d\tau}\right)^2 = 1 - R(t)E. \tag{16.93}$$

Next, notice that the substitution

$$R = \frac{1}{E} \sin^2 u \qquad (16.94)$$

changes (16.93) into

$$\frac{dR}{d\tau} = E^{1/2} \cot u. \qquad (16.95)$$

We can now regard u as a parameter that relates R to τ. Hence

$$\tau = \int \frac{dR/du}{dR/d\tau} du = 2E^{-3/2} \int \sin^2 u \, du$$
$$= E^{-3/2}(u - \sin u \cos u) - E^{-3/2}\left[\sin^{-1} E^{1/2} - E^{1/2}(1 - E)^{1/2}\right]. \quad (16.96)$$

The constant of integration has been chosen so that when $\tau_i = 0$, $R_i = 1$; in other words $u_i = \sin^{-1} E^{1/2}$. Initially either the inhomogeneity may have a large finite amplitude or it may be a small perturbation for which $u_i \approx E^{1/2}$. Thus this description is intrinsically nonlinear. It may be compared with the global behavior of a closed universe for $R(t)$ in (16.26)–(16.31).

Maximum expansion occurs when $u = \pi/2$ and $R_{max} = E^{-1}$. This happens at a dynamical time

$$\tau_{max} \approx \frac{\pi}{2} E^{-3/2} \qquad (16.97)$$

for initially small perturbations when the constant term of order unity can be neglected in (16.96). After maximum expansion the inhomogeneity begins to contract and break away from its surroundings. This process is fully gravitational, unlike the formation of pancakes, which is primarily kinematic. However, it could represent a later stage in the evolution of a "spherical pancake" for which all three eigenvalues were approximately equal. Collapse is complete in this idealized illustration when $u = \pi$ and $R = 0$. Again neglecting the constant term in (16.96), collapse is complete at a dynamical time

$$\tau_c \approx \pi E^{-3/2} = 2\tau_{max}. \qquad (16.98)$$

Thus, if the initial perturbation is small, the expansion phase lasts just about as long as the collapse phase, as expected from the reversible nature of the motion. Figure 16.1b shows these results. They could also be represented on Figure 16.1a by world lines that initially diverged like those of fundamental observers but then converged toward the time axis.

The collapse time may also be written in terms of more observationally oriented quantities. Using (16.82), (16.87), (16.91), and (16.92), writing $\delta\rho \equiv \bar{\rho}_i - \rho_{ci}$, and

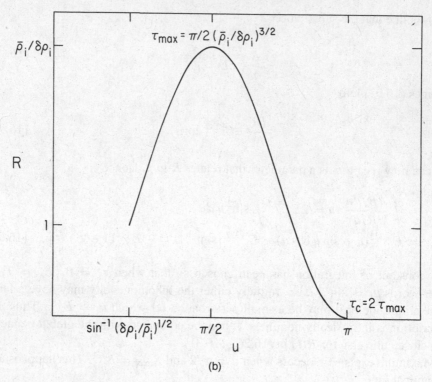

$$\bar{\rho}_i/\delta\rho_i$$

$$\tau_{max} = \pi/2\,(\bar{\rho}_i/\delta\rho_i)^{3/2}$$

$$R$$

$$1$$

$$\tau_c = 2\,\tau_{max}$$

$$\sin^{-1}(\delta\rho_i/\bar{\rho}_i)^{1/2} \qquad \pi/2 \qquad \pi$$

$$u$$

(b)

Fig. 16.1(b). The growth and collapse of a density inhomogeneity in the expanding Universe (from GPSGS).

recalling $\Omega = 2q$ gives the collapse time

$$t_c = \frac{\pi}{\Omega^{1/2}(1+z_i)^{3/2}}\left(\frac{\bar{\rho}_i}{\delta\rho}\right)^{3/2} H^{-1} \qquad (16.99)$$

in terms of the present Hubble time H^{-1}. Fractional perturbations must have formed by a redshift of approximately

$$1 + z_i = \left(\frac{\pi^2}{\Omega}\right)^{1/3}\frac{\bar{\rho}_i}{\delta\rho} \qquad (16.100)$$

in order to have collapsed by the present time.

The evolution of the single modes, pancakes, or clusters in these illustrations provides insight into the fundamental nature of gravitational instability. The full richness of this instability appears only by generalizing many of these descriptions, applying them to the interactions of different systems, and analyzing their distribution over all possible configurations. Clearly, we will need to develop some shortcuts to solve this more complicated problem. Its nature begins to emerge when we look at smaller scales where fluid smoothing is no longer adequate and the granularity of the gravitational field introduces new dimensions.

17

The Growth of Correlations:

II. A Particle Description

17.1 Introduction

Individual galaxies produce a granular gravitational field on scales too small for realistic density smoothing. From (16.2), the fluid approximation loses accuracy when $r \lesssim (\bar{\rho}/\delta p)\,(\bar{n}\lambda)^{-1/2}$. If $\bar{n} = 1$ Mpc^{-3}, then for a mode of wavelength $\lambda = 25$ Mpc, with a 1% density perturbation, granularity must be considered on scales $r \lesssim 20$ Mpc. If it has a 10% density perturbation, granularity is important on the more local scale $r \lesssim 2$ Mpc. This is only a rough guide since a realistic inhomogeneity of some well-defined shape contains many modes of different amplitudes. Moreover, the effects of granularity may be forestalled if there is dynamically important matter that behaves as a fluid between the galaxies. The last section explored the case when granularity was unimportant. Here we will see what happens when it dominates.

Under these conditions, it is the orbits of individual galaxies that determine their clustering. If the initial amplitudes of density fluctuations $\delta\rho/\bar{p} \sim \delta N/\bar{N}$ decrease with increasing scale, then inhomogeneities first grow on small scales among near neighbors. Gradually the clustering spreads to larger scales and modifies the initial fluctuations on those scales. Provided all galaxies trace the gravitational field in the same (unbiased) way, we can determine density fluctuations by counting galaxy numbers.

So our task now is to extract the correlation functions on relatively small scales from the evolution of galaxy orbits. In the linear regime, $\xi_N < 1$, this can be done rigorously for ξ_2 starting with simple initial conditions. Additional assumptions provide a good approximation for ξ_3 in the linear regime. But to explore the higher ξ_N in the linear regime, and all the ξ_N in the nonlinear regime, currently lies beyond the range of any rigorous analysis. Nevertheless, by making reasonable physical approximations, and checking them with computer simulations, we can learn a great deal about these more difficult problems.

To proceed, we need to develop a new description of this cosmological many-body problem, first based on statistical mechanics and eventually, for the nonlinear regime, on simpler statistical thermodynamics. The art of this approach is to extract enough information to be interesting and useful, but not so much that the problem becomes unsolvable. Therefore we do not analyze individual orbits directly; instead, we determine some of their statistical properties.

17.2   Liouville's Equation and Entropy

Consider all the galaxies in the Universe. We shall simplify each galaxy by representing it as a point particle of mass m, and simplify further by supposing each galaxy mass to be the same. Imagine a box containing N of these galaxies, a box small compared to the radius of curvature $R(t)$ of the Universe. Each galaxy in the box has three position and three velocity (or momentum) coordinates. All $6N$ coordinates can be represented by a single point in a $6N$-dimensional phase space. As the positions of the galaxies evolve, this phase point moves around continuously. The idea of this phase space grew out of a six-dimensional phase space where the three position and three velocity coordinates describe the motion of a single physical particle represented by a moving phase-space point. A swarm of such moving points then represents the motions of all particles in the system and its density in six-dimensional phase space represents the single particle distribution function $f(\mathbf{r}, \mathbf{v})$ of the system. Thus a system of N particles can be represented by N points in a six-dimensional space or by one point in a $6N$-dimensional space.

There is a great advantage to using the larger $6N$-dimensional phase space, as Gibbs discovered: It easily represents an ensemble of systems. In our case, each system could be a box of N galaxies. The boxes could be taken from regions of the Universe so far apart that they would not influence one another. In this Gibbs ensemble, each system would have a different internal distribution of positions and velocities (for the same number of galaxies). A swarm of points in the $6N$-dimensional phase space represents this entire ensemble. We will first examine how the evolution of this $6N$-dimensional distribution of phase space points evolves in accord with the dynamical forces governing the orbits of individual galaxies. Then we will focus on the subset of the distribution functions that describes the evolution of the two-point correlation function. Our discussion here closely follows GPSGS.

At any time, the probability density for finding a system in the ensemble within a particular range of $6N$ coordinates will be denoted by

$$f^{(N)}\left(\mathbf{x}^{(1)}, \ldots, \mathbf{x}^{(N)}, \mathbf{v}^{(1)}, \ldots, \mathbf{v}^{(N)}, t\right) d\mathbf{x}^{(1)} \ldots d\mathbf{v}^{(N)}. \qquad (17.1)$$

The value of $f^{(N)}$ is the fraction of systems in the ensemble with the desired range of velocities and positions. Thus the integral of $f^{(N)}$ over all phase space is unity:

$$\int_{-\infty}^{\infty} f^{(N)} d\mathbf{x}^{(1)} \ldots d\mathbf{v}^{(N)} = 1. \qquad (17.2)$$

The next conceptual step is to assume, following Gibbs, that the probability distribution of velocities and positions over all members of the ensemble at a given time is the same as the probability of finding a given set of coordinates in any one member of the ensemble during a long period of time. For ordinary statistical mechanics this is justified by supposing that all members of the ensemble are fairly similar and represent different microscopic realizations of systems with the same macroscopic (average) properties such as temperature and density. Then one appeals to the ergodicity of the ensemble. Some physicists find this intuitively obvious and note that it leads to many experimentally verified predictions. Others find it intuitively implausible and thus all the more remarkable for seeming to be true. They have therefore sought rigorous proofs in statistical mechanics and generated a considerable industry. In gravitational (and other explicitly Hamiltonian) systems, the situation is perhaps more straightforward: The Gibbs concept implies the exact equations of motion of the system.

To see this, we first determine how $f^{(N)}$, considered as the probability of finding a given system in the ensemble, changes with time. Initially the probability that a given system has coordinates $(\mathbf{x}_0^{(1)}, \ldots, \mathbf{v}_0^{(1)}, t_0)$ lying within a small $6N$-dimensional volume, with boundary S_0, of phase space is

$$A(t_0) = \int_{S_0} f_0^{(N)}(\mathbf{x}_0^{(1)}, \ldots, \mathbf{v}_0^{(N)}, t_0) \, d\mathbf{x}_0^{(1)} \ldots d\mathbf{v}_0^{(N)}. \tag{17.3}$$

At some later time, the coordinates $\mathbf{x}_0$, $\mathbf{v}_0$ in the system will evolve dynamically into $\mathbf{x}$ and $\mathbf{v}$, the distribution function will become $f^{(N)}(\mathbf{x}^{(1)}, \ldots, \mathbf{v}^{(N)}, t)$, and the boundary S_0 will change to S_t. The probability that the evolved system now lies within S_t is

$$A(t) = \int_{S_t} f^{(N)}(\mathbf{x}^{(1)}, \ldots, \mathbf{v}^{(N)}, t) \, d\mathbf{x}^{(1)} \ldots d\mathbf{v}^{(N)}. \tag{17.4}$$

However, since the dynamical evolution is continuous (i.e., the system does not suddenly jump into a new state), the probability that it is in the transformed region is the same as its probability of being in the original region:

$$A(t) = A(t_0) = \text{constant.} \tag{17.5}$$

This phase-space evolution is similar to the Lagrangian evolution (16.64) of density in real space. The invariant A may be regarded as the fractional volume of phase space occupied by systems having a particular evolving range of coordinates. The volume is conserved during the evolution, as shown schematically in Figure 17.1.

Consider in more detail how this volume stays the same. Its value at time $t_0 + \Delta t$ is related to its value at an arbitrary earlier time t_0 by

$$A(t_0 + \Delta t) = \int_{S_{t_0} + \Delta t} f^{(N)}(\mathbf{x}^{(1)}, \ldots, \mathbf{v}^{(N)}, t_0 + \Delta t) \, d\mathbf{x}^{(1)} \ldots d\mathbf{v}^{(N)}$$

$$= \int_{S_{t_0}} f^{(N)}(\mathbf{x}^{(1)}, \ldots, \mathbf{v}^{(N)}, t_0 + \Delta t) \frac{\partial(\mathbf{x}^{(1)}, \ldots, \mathbf{v}^{(N)})}{\partial(\mathbf{x}_0^{(1)}, \ldots, \mathbf{v}_0^{(N)})} \, d\mathbf{x}_0^{(1)} \ldots d\mathbf{v}_0^{(N)}.$$

$$\tag{17.6}$$

So far we have just changed the region of integration. We next relate the distribution

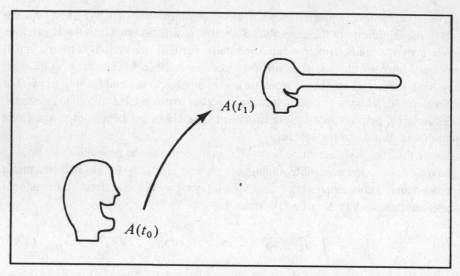

Fig. 17.1. Schematic evolution of volume in $6N$-dimensional phase space.

function at the two times by a Taylor series expansion

$$f^{(N)}\left(\mathbf{x}^{(1)}, \ldots, \mathbf{v}^{(N)}, t_0 + \Delta t\right) = f_0^{(N)}\left(\mathbf{x}_0^{(1)}, \ldots, \mathbf{v}_0^{(N)}, t_0\right) + \left(\frac{\partial f_0^{(N)}}{\partial \mathbf{x}_0^{(1)}} \cdot \frac{d\mathbf{x}_0^{(1)}}{dt} + \cdots\right.$$

$$\left. + \frac{\partial f_0^{(N)}}{\partial \mathbf{v}_0^{(N)}} \cdot \frac{d\mathbf{v}_0^{(N)}}{dt} + \frac{\partial f_0^{(N)}}{\partial t}\right) \Delta t + \mathcal{O}[(\Delta t)^2] + \cdots.$$

$$\text{(17.7)}$$

The coordinates and velocities at the two times are related by

$$\mathbf{x}^{(n)} = \mathbf{x}_0^{(n)} + \frac{d\mathbf{x}_0^{(n)}}{dt} \Delta t + \cdots \tag{17.8}$$

and

$$\mathbf{v}^{(n)} = \mathbf{v}_0^{(n)} + \frac{d\mathbf{v}_0^{(n)}}{dt} \Delta t + \cdots, \tag{17.9}$$

which give the Jacobian

$$\frac{\partial\left(\mathbf{x}^{(1)}, \ldots, \mathbf{v}^{(N)}\right)}{\partial\left(\mathbf{x}_0^{(1)}, \ldots, \mathbf{v}_0^{(N)}\right)} = 1 + \left(\frac{\partial \dot{x}_0^{(1)}}{\partial x_0^{(1)}} + \frac{\partial \dot{y}_0^{(1)}}{\partial y_0^{(1)}} + \frac{\partial \dot{z}_0^{(1)}}{\partial z_0^{(1)}} + \cdots + \frac{\partial \dot{v}_{0x}^{(N)}}{\partial v_{0x}^{(N)}}\right.$$

$$\left. + \frac{\partial \dot{v}_{0y}^{(N)}}{\partial v_{0y}^{(N)}} + \frac{\partial \dot{v}_{0z}^{(N)}}{\partial v_{0z}^{(N)}} + \right) \Delta t + \mathcal{O}[(\Delta t)^2] + \cdots$$

$$= 1 + \sum_{\alpha=1}^{N} \left(\nabla_{\mathbf{x}_0} \cdot \dot{\mathbf{x}}_0^{\alpha} + \nabla_{\mathbf{v}_0} \cdot \dot{\mathbf{v}}_0^{\alpha} \Delta t + \cdots\right), \tag{17.10}$$

using Cartesian component notation. Multiplying (17.7) and (17.10), we get

$$f^{(N)} \frac{\partial \left(\mathbf{x}^{(1)}, \ldots, \mathbf{v}^{(N)} \right)}{\partial \left(\mathbf{x}_0^{(1)}, \ldots, \mathbf{v}_0^{(N)} \right)} = f_0^{(N)} + \sum_{\alpha=1}^{N} \left[\nabla_{\mathbf{x}_0} \cdot \left(f_0^{(N)} \dot{\mathbf{x}}_0^{(\alpha)} \right) + \nabla_{\mathbf{v}_0} \cdot \left(f_0^{(N)} \dot{\mathbf{v}}_0^{(\alpha)} \right) \right] \Delta t$$

$$+ \frac{\partial f_0^{(N)}}{\partial t} \Delta t + \cdots. \tag{17.11}$$

Note that the position and velocity coordinates are treated symmetrically. Substituting (17.11) into (17.6), recalling (17.3), and taking the limit as $\Delta t \to 0$ gives

$$\frac{dA(t)}{dt} = \int_{S_{t_0}} \left(\frac{\partial f_0^{(N)}}{\partial t} + \sum_{\alpha=1}^{N} \left[\nabla_{\mathbf{x}_0} \cdot \left(f_0^{(N)} \dot{\mathbf{x}}_0^{(\alpha)} \right) + \nabla_{\mathbf{v}_0} \cdot \left(f_0^{(N)} \dot{\mathbf{v}}_0^{(\alpha)} \right) \right] \right) d\mathbf{x}_0^{(1)} \ldots d\mathbf{v}_0^{(N)}. \tag{17.12}$$

Because $dA/dt = 0$ for any volume S_{t_0}, from (17.5), the integrand of (17.12) must be zero. Moreover, since one initial time is as good as another, we may drop the subscript in the integrand to obtain

$$\frac{\partial f^{(N)}}{\partial t} + \sum_{\alpha=1}^{N} \left[\nabla_{\mathbf{x}} \cdot \left(f^{(N)} \dot{\mathbf{x}}^{(\alpha)} \right) + \nabla_{\mathbf{v}} \cdot \left(f^{(N)} \dot{\mathbf{v}}^{(\alpha)} \right) \right] = 0. \tag{17.13}$$

This is Liouville's equation. It applies more generally to any phase space representing a set of continuous first-order differential equations, not just to gravitational dynamics. Immediately, we notice its close resemblance to the equation of continuity (16.34) in ordinary space. Liouville's equation is indeed no more than a generalized equation of continuity for $6N$-dimensional phase space. It says there are neither sources nor sinks as the probability ebbs and flows throughout phase space.

Regarding $f^{(N)} (q_i, p_i)$ as a function of $6N$ coordinates and momenta, rather than velocities, and substituting Hamilton's equations $\dot{q}_i = \partial H / \partial p_i$ and $\dot{p}_i = -\partial H / \partial q_i$ for the Hamiltonian H into (17.13) gives

$$\frac{\partial f^{(N)}}{\partial t} + \left\{ f^{(N)}, H \right\} = 0, \tag{17.14}$$

where the Poisson bracket is summed over all N galaxies:

$$\left\{ f^{(N)}, H \right\} = \sum_{i=1}^{3N} \left(\frac{\partial f^{(N)}}{\partial q_i} \frac{\partial H}{\partial p_i} - \frac{\partial f^{(N)}}{\partial p_i} \frac{\partial H}{\partial q_i} \right). \tag{17.15}$$

In other words, the Jacobian (17.10) of the dynamical transformation was really just unity, as the alert reader may have noticed. Therefore density is conserved in phase space.

The result (17.14) can thus be derived directly by requiring the total time derivative of $f^{(N)}$ to be zero,

$$\frac{df^{(N)}}{dt} = \frac{\partial f^{(N)}}{\partial t} + \sum_{i=1}^{3N} \left(\frac{\partial f^{(N)}}{\partial q_i} \frac{dq_i}{dt} + \frac{\partial f^{(N)}}{\partial p_i} \frac{dp_i}{dt} \right) = 0, \qquad (17.16)$$

and substituting Hamilton's equations. This also shows that the total time derivative of any dynamical quantity is its partial time derivative plus its Poisson bracket with the Hamiltonian. If the Hamiltonian is explicitly independent of time, for example, then the total energy of the system is constant since $\{H, H\} = 0$. Physically, the reason for constant phase-space density is that all the positions and velocities are determined completely and forever by their values at any one time. Since Hamilton's equations are first order in time, they determine a unique trajectory through any point in six-dimensional phase-space for each object in a system, and thus a unique trajectory in $6N$-dimensional phase space for each point representing an entire system in the Gibbs ensemble. Phase points cannot cross the changing boundary of any initial region in phase space, and because the volume of this region is invariant as the ensemble evolves, the phase space density is constant.

Liouville's equation (17.14) is really very powerful. This is because it contains complete information about all the orbits in the system, with no assumptions or approximations. Indeed the characteristics of (17.14),

$$\frac{dt}{1} = \frac{dq_1}{\partial H/\partial p_1} = -\frac{dp_1}{\partial H/\partial q_1} = \frac{dq_2}{\partial H/\partial p_2} = -\frac{dp_2}{\partial H/\partial q_2} = \cdots = -\frac{dp_{3N}}{\partial H/\partial q_{3N}},$$

$$(17.17)$$

are the exact equations of motion for all N objects. We now have seen how the distribution function for an ensemble describes the orbits within a member system. Though powerful, this result is somewhat formal because H involves a sum or integral over the distribution function, which is not known until the orbits are all solved. Despite this difficulty, Liouville's equation is very useful, not only as a compact representation of all the orbits. It provides significant insights into the properties of complex systems. As an illustration, we consider the nature of gravitational entropy, which will be useful for understanding galaxy distribution functions.

If $f^{(N)}$ contains all the information about the system, then the entropy in $6N$-dimensional phase space should remain constant since no information is lost as the system evolves. This is in contrast with the six-dimensional Boltzmann entropy, which is constant only in equilibrium and otherwise increases. Define the $6N$-dimensional entropy $S^{(N)}$ analogously to Boltzmann's entropy, that is, proportional to the average value of the logarithm of the probability $f^{(N)}$,

$$S^{(N)} = -\int f^{(N)} \ln f^{(N)} \, dV^{(N)} \qquad (17.18)$$

(where Boltzmann's constant of proportionality $k = 1$ and the integral is over $6N$ phase space). Its derivative is

$$\frac{dS^{(N)}}{dt} = -\int \frac{\partial f^{(N)}}{\partial t} \left(\ln f^{(N)} + 1 \right) dV^{(N)}. \tag{17.19}$$

Now we develop the first term on the right-hand side using Liouville's equation (17.14) and an integration by parts assuming that $f^{(N)}$ vanishes on the boundary:

$$\int \frac{\partial f^{(N)}}{dt} \ln f^{(N)} dV^{(N)} = -\int \left\{ f^{(N)}, H \right\} \ln f^{(N)} \, dV^{(N)}$$

$$= \int f^{(N)} \left\{ \ln f^{(N)}, H \right\} dV^{(N)}$$

$$= \int \left\{ f^{(N)}, H \right\} dV^{(N)}$$

$$= -\int \frac{\partial f^{(N)}}{dt} dV^{(N)}. \tag{17.20}$$

Hence the two right-hand terms of (17.19) cancel and

$$\frac{dS^{(N)}}{dt} = 0. \tag{17.21}$$

To obtain an entropy that increases with time, we must relax the exact description of the system. There are two ways to do this. Following Gibbs, we could coarse grain the distribution function, replacing $f^{(N)}$ by its average over a finite volume element of phase space:

$$\bar{f}^{(N)} = \frac{1}{\Delta V^{(N)}} \int_{\Delta V} f^{(N)} dV^{(N)}. \tag{17.22}$$

As an initial set of points representing systems in phase space – an element of "phase fluid" – evolves, its boundary will generally become very convoluted even though its volume is constant. This is because different systems in the ensemble evolve at different rates and their representative points move off in different directions. After a while, a fixed volume of phase space that was initially filled may have some gaps in it; so $\bar{f}^{(N)} \leq f^{(N)}$. Consequently,

$$\bar{S}(N) = -\int \bar{f}^{(N)} \ln \bar{f}^{(N)} dV^{(N)} \geq S^{(N)}. \tag{17.23}$$

The temporal increase of coarse-grained entropy,

$$\bar{S}^{(N)}(t_1) \leq \bar{S}^{(N)}(t_2 > t_1), \tag{17.24}$$

is a result of the tendency of most regions of the phase fluid to become more and more filamentary as systems evolve. Normally there are only a few small regions of phase space, representing objects locked into resonances and periodic orbits, that do not participate. Eventually systems may evolve so that their phase fluid is so finely interwoven that a sort of "detailed balance" applies. As one filament moves out of the averaging volume $\Delta V^{(N)}$, another moves in to take its place, so that for any two times $\bar{f}^{(N)}(t_1) = \bar{f}^{(N)}(t_2)$. This is the sign that the system has reached equilibrium. Although it does not happen exactly for gravitating systems with $N > 3$, in some cases it may be a good approximation.

A problem with Gibbs's coarse graining is that its detailed result will clearly depend on the size of $\Delta V^{(N)}$. On the one hand, if the averaging volume is extremely fine, the orbits of individual objects will show up in the motion of the representative point, and in this limit we are back to $S^{(N)}$ rather than $\bar{S}^{(N)}$. Using very large averaging elements, on the other hand, means that the representative point will spend a long time in one element and then quickly move to another, possibly distant, element. Obviously in the extreme case, averaging over the entire phase space, nothing ever happens. Sudden jumps between large averaging elements will therefore depend on the system's evolution over a finite segment of its previous history, during which it actually moved around within the element, but where this motion was unnoticed in the averaging. In this case the description of the phase-space evolution becomes non-Markovian.

The standard statistical mechanical description of entropy depends upon there being an intermediate case. It must be possible to find a scale of averaging on which the representative points appear to move about in a random Markovian manner. The broader the range of scales to which this applies, the better the description. On these scales particles appear to diffuse through phase space, and this diffusion generates entropy.

There is another way to define entropy that avoids this "eye of the beholder" aspect (see Grad, 1958). In Section 17.3 we will find that it also provides a most useful way to extract information from a system. Basically the idea is to form reduced distribution functions $f^{(n)}$ by integrating out all the $6(N - n)$ phase-space coordinates,

$$f^{(n)} = \int \cdots \int f^{(N)} dV^{(N-n)}. \tag{17.25}$$

Thus $f^{(2)}(\mathbf{x}_1, \mathbf{v}_1, \mathbf{x}_2, \mathbf{v}_2, t) \equiv f^{(2)}(1, 2)$, for example, is the probability of finding an object with position $\mathbf{x}_1$ and velocity $\mathbf{v}_1$ in $dV^{(1)}(1) = d\mathbf{x}_1 d\mathbf{v}_1$, as well as an object with position $\mathbf{x}_2$ and velocity $\mathbf{v}_2$ in $dV^{(1)}(2) = d\mathbf{x}_2 d\mathbf{v}_2$. In this notation, the argument or subscript refers to the particle label; the superscript refers to the number of particles. Consider, for the present, just the one- and two-particle distribution functions. Since the objects must be somewhere in phase space, the simplest normalizations are

$$\int f^{(1)}(1) \, dV^{(1)}(1) = \int f^{(1)}(2) \, dV^{(1)}(2) = \int f^{(2)}(1, 2) \, dV^{(2)}(1, 2) = 1$$

$$\tag{17.26}$$

and

$$\int f^{(2)} dV^{(1)}(1) = f^{(1)}(2); \quad \int f^{(2)} dV^{(1)}(2) = f^{(1)}(1), \qquad (17.27)$$

where $dV^{(2)}(1, 2) \equiv dV^{(1)}(1) \, dV^{(1)}(2)$. Now if the objects did not interact at all, their distributions would be independent. Since independent probabilities multiply, we would have

$$f^{(2)}(1, 2) = f^{(1)}(1) f^{(1)}(2). \qquad (17.28)$$

Any interactions, however, will introduce correlations between the objects. The probability of finding two objects in a given element of phase space will no longer be the probability of finding one object times the probability of finding another object. The presence of one object may increase or decrease the probability of finding another object nearby. To allow for this, we modify (17.28) by introducing a pair correlation function $P(\mathbf{x}_1, \mathbf{v}_1, \mathbf{x}_2, \mathbf{v}_2, t)$ so that

$$f^{(2)}(1, 2) = f^{(1)}(1) f^{(1)}(2) + P(1, 2). \qquad (17.29)$$

In the limit as the interaction vanishes, $P(1, 2)$ vanishes also. If the interaction is weak we might also expect $P(1, 2) \ll f^{(1)}(1) f^{(1)}(2)$. However, this is not always the case in gravitating systems since the gravitational energy is nonlinear with mass and does not saturate. Therefore a large enough number of objects, each interacting weakly, can produce large pair correlations. From (17.29) and the normalizations (17.26), we see that, as in (14.11) with slightly different normalization,

$$\int P(1, 2) \, dV^{(2)}(1, 2) = 0, \qquad (17.30)$$

which is just a sophisticated way of saying that mass is conserved. Positive correlations in one region must be balanced by negative correlations elsewhere. The entropy of the single-particle distribution function,

$$S^{(1)}(1) = - \int f^{(1)}(1) \ln f^{(1)}(1) \, dV^{(1)}(1), \qquad (17.31)$$

is the form most similar to the standard Boltzmann entropy. It increases with time (unless there is equilibrium) because it is not a complete description of the system. But we can now also define a two-particle entropy:

$$S^{(2)}(1, 2) = - \int f^{(2)}(1, 2) \ln f^{(2)}(1, 2) \, dV^{(2)}(1, 2). \qquad (17.32)$$

If there were no interactions, substituting (17.29) and (17.27) shows that $S^{(2)}(1, 2)$ would just be the sum of the one-particle entropies. Moreover, a similar result would hold for $S^{(3)}(1, 2, 3)$ and so on up the line to $S^{(N)}$. Thus when the single-particle entropies increase with time, so do the multiparticle entropies up to $S^{(N-1)}$. The result ceases to hold for $S^{(N)}$ because $f^{(N)}$, and only $f^{(N)}$, satisfies Liouville's equation and is equivalent to complete orbital information.

How do interactions change the entropy? Looking at $S^{(2)}$, we see from (17.32) and (17.29) that it will contain the somewhat awkward factor $\ln[f^{(1)}(1)f^{(1)}(2) + P(1, 2)]$ in which both terms may be any size. We can cast this into a neater form using the Taylor series expansion for any analytic function around a value x:

$$f(x + y) = \sum_{j=0}^{m} \frac{y^j}{j!} \frac{d^j f(x)}{dx^j} + \frac{y^{m+1}}{(m+1)!} \frac{d^{m+1} f(x')}{dx'^{m+1}}\bigg|_{x'=x+ay}, \qquad (17.33)$$

where the number α is bounded by $0 \le \alpha \le 1$. The last term is called the remainder; for normal use $y \ll x$ and m is large so that the remainder will be as small as desired. Here, however, we take the opposite extreme with $m = 0$, so the result may be nearly "all remainder" for large P. This gives

$$\ln[f^{(2)}(1, 2)] = \ln[f^{(1)}(1)f^{(1)}(2) + P(1, 2)]$$

$$= \ln[f^{(1)}(1)f^{(1)}(2)] + \frac{P(1, 2)}{f^{(1)}(1)f^{(1)}(2) + \alpha P(1, 2)}. \qquad (17.34)$$

From (17.26)–(17.34) we next calculate the difference

$$S^{(2)}(1, 2) - [S^{(1)}(1) + S^{(1)}(2)]$$

$$= -\int P(1, 2)\left[1 + \frac{(1 - \alpha)P(1, 2)}{f^{(1)}(1)f^{(1)}(2) + \alpha P(1, 2)}\right] dV^{(2)}(1, 2)$$

$$= -\int \frac{(1 - \alpha)P^2(1, 2)}{f^{(1)}(1)f^{(1)}(2) + \alpha P(1, 2)} dV^{(2)}(1, 2), \qquad (17.35)$$

using sophisticated mass conservation (17.30) for the last step. With no interaction or correlation we clearly recover our previous result that the single-particle entropies add to produce the two-particle entropy. If the pair correlation is positive, then the integrand in (17.35) is also positive. If $P(1, 2)$ is negative, then

$$f^{(1)}(1)f^{(1)}(2) + \alpha P \ge f^{(1)}(1)f^{(1)}(2) + P = f^{(2)}(1, 2) \ge 0 \qquad (17.36)$$

and the integrand is still positive. Therefore in both cases

$$S^{(2)}(1, 2) \le S^{(1)}(1) + S^{(1)}(2). \qquad (17.37)$$

Correlations always decrease the two-particle entropy.

In the discussion of distribution functions, we shall see how this result provides insight into the tendency of gravitating systems to cluster. The gravitational contribution to the two-particle entropy is negative and the correlations of clustering decrease this entropy. Higher order entropies $S^{(N)}(1, 2, \ldots, N)$, which dominate very strong clustering, can have more complicated behavior. If we retain the notion that systems evolve in the direction of an entropy extreme, then we should expect infinite systems of galaxies to form tighter and tighter clumps over larger and larger scales. Naturally the expansion of the Universe modifies the nature of clustering, but the qualitative features remain. This theme also occurs in stellar dynamics. Spherical systems of stars evolve toward an extreme gravitational entropy. Clustering within the constraint of spherical symmetry leads the core of the system to become denser and denser as it ejects high-energy stars into its halo. However, finite systems (see GPSGS) behave quite differently than infinite systems because they do not have translational and rotational symmetry at every point (i.e., they are not statistically homogeneous when averaged over an ensemble of similar systems).

17.3   The BBGKY Hierarchy

Delving into Liouville's equation, we now aim to extract the evolution of the two-point correlation function. By systematically removing most of the information in $6N$-dimensional phase space, what remains becomes manageable. The initials BBGK and Y are from the names of the physicists who independently, or in different contexts, developed a rigorous way to extract knowledge from the information in many-variable partial differential equations: Born, Bogoliubov, Green, Kirkwood, and Yvon. Their basic idea was to turn one equation with a large number of variables into many equations, each with a small number of variables. These many equations cannot be the orbit equations since each orbit equation contains the positions and velocities of all the other galaxies. However, we can average over most of these orbits, represented by the high order (i.e., large N) distribution functions, by integrating over their coordinates. The resulting hierarchical set of coupled equations for the reduced distribution functions (17.25) may provide just the right amount of detail.

To develop the hierarchy, we write Liouville's equation (17.13) in component form. For simplicity, suppose all the masses are the same, so that the gravitational acceleration on any galaxy, labeled α, from all the other galaxies is

$$\frac{dv_i^{(\alpha)}}{dt} = \frac{\partial \phi}{\partial x_i^{(\alpha)}} = -Gm \sum_{\beta \neq \alpha} \frac{\left(x_i^{(\alpha)} - x_i^{(\beta)}\right)}{\left|\mathbf{x}^{(\alpha)} - \mathbf{x}^{(\beta)}\right|^3} \tag{17.38}$$

for the ith Cartesian coordinate. Thus (17.13) becomes

$$\frac{\partial f^{(N)}}{\partial t} + \sum_{\alpha} v_i^{(\alpha)} \frac{\partial f^{(N)}}{\partial x_i^{(\alpha)}} + Gm \sum_{\alpha} \sum_{\beta \neq \alpha} \frac{\left(x_i^{(\beta)} - x_i^{(\alpha)}\right)}{\left|\mathbf{x}^{(\beta)} - \mathbf{x}^{(\alpha)}\right|^3} \frac{\partial f^{(N)}}{\partial v_i^{(\alpha)}} = 0, \tag{17.39}$$

where terms are summed for Cartesian components over the repeated i index. Since the distribution function cannot distinguish among objects of identical mass, it must be symmetric in the Greek indices. Therefore the summation over all $\beta \neq \alpha$ just gives a factor $(N - 1)$. Changing the normalization of $f^{(n)}$ from (17.25) to

$$f^{(n)}(\mathbf{x}_1, \mathbf{v}_1, \ldots, \mathbf{x}_n, \mathbf{v}_n, t) \equiv f^{(n)}(1, \ldots, n)$$

$$= \frac{(N/p)^N}{(N-n)!} \int f^{(N)}(1, \ldots, N)\, d\mathbf{x}_{n+1} d\mathbf{v}_{n+1} \ldots d\mathbf{v}_N \qquad (17.40)$$

and

$$\int f^{(N)} dV^{(N)} = (p/N)^N N! \qquad (17.41)$$

will make the resulting equations appear less encumbered with coefficients. Here p denotes the average number density of objects. The factorials in these normalizations arise from the number of ways of interchanging labels among the identical objects in the symmetric distribution functions $f^{(n)}$. Thus

$$f^{(n)}(1, \ldots, n)\, d\mathbf{x}_1\, d\mathbf{v}_1 \ldots d\mathbf{x}_n\, d\mathbf{v}_n \equiv f^{(n)} dV^{(n)} \qquad (17.42)$$

gives the probability of finding any n objects simultaneously in the different volume elements $d\mathbf{x}_1, \ldots, d\mathbf{x}_n$ at positions $\mathbf{x}_1, \ldots, \mathbf{x}_n$ and in the different velocity elements $d\mathbf{v}_1, \ldots, d\mathbf{v}_n$ at velocities $\mathbf{v}_1, \ldots, \mathbf{v}_n$.

Multiply (17.39) by $(N/p)^N/(N - n)!$ and integrate over $dV^{(N-n)} = d\mathbf{x}_{n+1} \ldots d\mathbf{v}_N$ for an arbitrary value of n. The first term is just $\partial f^{(n)}/\partial t$. In the second set of terms, those terms for which $\alpha < n$ give

$$\sum_{i=1}^{n} \mathbf{v}_i \cdot \frac{\partial f^{(n)}}{\partial \mathbf{x}_i},$$

where the notation is simplified by replacing the i index of (17.39) with the vector dot product, and replacing the (α) index by the dummy summation variable i over the particles. The remaining terms for $\alpha > n$ can be integrated by parts. In each of them the integral is zero since the v_i and x_i are independent variables. The contribution at the limits

$$v_i\, f^{(N)}(\mathbf{x}_{n+1})\, \Big|_{\mathbf{x}_{n+1} \to -\infty}^{\mathbf{x}_{n+1} \to \infty}$$

will vanish if the system is finite. It also vanishes in infinite systems if they remain homogeneous on a large scale so that the value of $f^{(N)}$ at the two limits is the same, and we will take this to be the case. A special example is a spherically symmetric distribution or one that is an even function of the x_{n+1}. The reduction of the last set of terms involves a similar integration by parts. Now $f^{(n)}$ depends on all the $f^{(n')}$ with $n' > n$ since the acceleration depends on the positions of all the other particles. However, the symmetry of $f^{(N)}$ allows all these higher distribution functions to be represented by $f^{(n+1)}$.

The result of this integration of (17.39) is the BBGKY hierarchy for gravitating systems with no external force:

$$\left[\frac{\partial}{\partial t} + \sum_{i=1}^{n} \mathbf{v}_i \cdot \frac{\partial}{\partial \mathbf{x}_i} - \sum_{i \neq j}^{n} \mathcal{V}(i, j) \right] f^{(n)}(1, \dots, n)$$

$$= \sum_{i=1}^{n} \int \mathcal{V}(i, n+1) f^{(n+1)}(1, \dots, n+1) \, d\mathbf{x}_{n+1} \, d\mathbf{v}_{n+1}. \quad (17.43)$$

Here we have written

$$\mathcal{V}(i, j) \equiv \frac{\partial \phi(i, j)}{\partial \mathbf{x}_i} \cdot \frac{\partial}{\partial \mathbf{v}_i} \quad (17.44)$$

for the interaction of the (i, j) pair whose gravitational potential is

$$\phi(i, j) = -\frac{Gm}{|\mathbf{x}_i - \mathbf{x}_j|}. \quad (17.45)$$

The result (17.43) is actually a set of N equations, one for each value of n. The equation for each value of n is coupled to that for $n + 1$, until $n = N$ whereupon (17.43) becomes Liouville's equation. The case $n = 1$ would be the collisionless Boltzmann equation if the integral term involving $f^{(2)}$ were zero. Explicitly, this shows how the collisional term of the Boltzmann equation represents the combined effects of "scattering" by all the other objects when not subject to the usual simplifications of billiard ball fluids. Each equation of the hierarchy may also be viewed as the "projection" of the system's complete dynamics onto the evolution of its partial distribution functions.

Nothing has been gained or lost; the BBGKY hierarchy contains as many equations as there are galaxies. But by recasting the problem into this form we hope to find a suitable physical approximation that enables us to truncate the hierarchy at a solvable level. Originally (17.43), or its equivalent, was devised to treat problems involving the distribution of molecules in fluids and of ions in a plasma. These are systems close to equilibrium. Gravitating systems act differently, but it is still possible to solve (17.43) for the two-point correlation function $\xi_2(r, t)$ under restricted circumstances.

17.4 Gravitational Graininess Initiates Clustering

The most fundamental aspect of the solution for $\xi_2(r, t)$ is the distinction between the terms in the BBGKY hierarchy that describe the discrete nature of the gravitational field and those that describe the smoothed fluid approximation whose consequences Chapter 16 explored. The graininess terms introduce a new class of instabilities. These involve the orbits of nearest neighbor particles (galaxies). Initially they dominate on scales where the distribution is unrelaxed, i.e., far from being virialized. For an initially unclustered Poisson distribution, graininess dominates on small scales where $\delta\rho/\rho \approx N^{-1/2}$ is relatively large. For more general hierarchical clustering, graininess may be important on scales just larger than the most recently bound substructures. Its relative importance naturally depends on other perturbations that may initially be embedded in these scales.

The two-point correlation function is hiding in the BBGKY hierarchy. To find it (Fall and Saslaw, 1976) we must first relate it to the distribution functions. Again define the distribution function $f^{(n)}dV^{(n)}$ as the joint probability of finding any n galaxies in the differential position–velocity volume element $dV^{(n)}$ of phase space. The single-particle distribution function $f^{(1)}$, when averaged over a volume containing many galaxies, is the fluid density. Since we do not take that average here, it applies as a probability density for any size volume. The form of Equation (14.26) suggests defining a sequence for volume distributions in which each joint probability is related to lower order joint probabilities according to

$$f^{(1)}(1) \equiv f(1), \tag{17.46}$$

$$f^{(2)}(1, 2) \equiv f(1)f(2)[1 + g(1, 2)], \tag{17.47}$$

$$f^{(3)}(1, 2, 3) \equiv f(1)f(2)f(3)[1 + g(1, 2) + g(2, 3) + g(3, 1) + h(1, 2, 3)],$$

$$\tag{17.48}$$

etc. Equation (17.47) is formulated a bit differently from (17.29) so that $g(1, 2)$ will eventually lead directly to $\xi(r)$ in the usual notation. Normalizations are the same as in (17.40) and (17.41). So far there are no restrictions on the size of the correlation functions.

Theories should first be developed to describe the simplest conditions containing the essential physics of a problem. Assume, therefore, that the galaxy distribution is initially uniform, as also suggested by the near isotropy of the microwave background. Then the single-particle distribution function does not depend on position. When clustering starts it has an equal probability of occurring anywhere.

"Macroscopically" the system remains uniform and its early clustering is described by the two-particle and higher order correlation functions. Initially these higher order correlations are zero. This cosmological situation is opposite to the case of nonequilibrium plasmas, where correlations start large and decrease as the plasma relaxes to a uniform Maxwellian distribution. High order cosmic correlations may grow from nothing, strengthen through a linear regime, and then evolve into great density inhomogeneities. After the system becomes nonlinear it is no longer plausible to expect uniformity with $f^{(1)}(1) = f^{(1)}(2)$ (unless these distributions are averaged over an ensemble). Our goal here is to find what happens in the linear regime and how long the system takes to pass through it.

Evolution of distribution functions is described by the BBGKY hierarchy (17.43). To find the two-point correlation function with uniform initial conditions we need only consider the first two equations of the hierarchy. Substituting (17.46–17.48) into (17.43) and rearranging terms, the first two equations of the hierarchy become

$$\left[\frac{\partial}{\partial t} + \mathbf{v}_1 \cdot \frac{\partial}{\partial \mathbf{x}_1} - \frac{\partial \Phi(1)}{\partial \mathbf{x}_1} \cdot \frac{\partial}{\partial \mathbf{v}_1}\right] f(1) = \mathcal{I}(1) \tag{17.49}$$

and

$$f(1)f(2)\left[\frac{\partial}{\partial t} + \sum_{i=1}^{2}\left(\mathbf{v}_i \cdot \frac{\partial}{\partial \mathbf{x}_i} - \frac{\partial \Phi(i)}{\partial \mathbf{x}_i} \cdot \frac{\partial}{\partial \mathbf{v}_i}\right)\right] g(1,2)$$

$$= \sum_{i \neq j}^{2}[\mathcal{J}(i,j) + \mathcal{K}(i,j)]. \tag{17.50}$$

Equation (17.49) has been used in deriving (17.50), and it contains the following abbreviations:

$$\Phi(i) \equiv \int f(3)\phi(i,3)\,d\mathbf{x}_3\,d\mathbf{v}_3, \tag{17.51}$$

$$\mathcal{I}(i) \equiv \int f(3)\frac{\partial \phi(i,3)}{\partial \mathbf{x}_i} \cdot \frac{\partial}{\partial \mathbf{v}_i}[f(i)g(i,3)]\,d\mathbf{x}_3\,d\mathbf{v}_3, \tag{17.52}$$

$$\mathcal{J}(i,j) \equiv \left[f(j)\frac{\partial f(i)}{\partial \mathbf{v}_i} + f(i)f(j)\frac{\partial g(i,j)}{\partial \mathbf{v}_i} + g(i,j)f(j)\frac{\partial f(i)}{\partial \mathbf{v}_i}\right] \cdot \frac{\partial \phi(i,j)}{\partial \mathbf{x}_i}, \tag{17.53}$$

$$\mathcal{K}(i,j) \equiv f(i) \int f(3) \left\{ g(i,3)\frac{\partial f(j)}{\partial \mathbf{v}_j} + \frac{\partial}{\partial \mathbf{v}_j}[f(j)h(i,j,3)] \right.$$

$$\left. - g(i,j)\frac{\partial}{\partial \mathbf{v}_j}[f(j)g(j,3)] \right\} \cdot \frac{\partial \phi(j,3)}{\partial \mathbf{x}_j}\,d\mathbf{x}_3\,d\mathbf{v}_3. \tag{17.54}$$

There are a couple of pages of straightforward algebra in this derivation, which can provide a useful exercise.

The terms in these BBGKY equations have been separated according to their physical import. On the left-hand side of (17.49) is the familiar convective time derivative (following the motion) and the mean field gravitational force that accelerates galaxies. This much is the collisionless Boltzmann equation. On the right-hand side is the analog of a "collision integral," which couples changes in the single-body velocity–density distribution with changes in the two-body correlations. On the left-hand side of (17.50) is the complete derivative of the two-body correlation function. It is summed over $i = 1, 2$ because the changes occur at two places in phase space. The source terms on the right-hand side are of two types. Those in $\mathcal{J}$ involve the distribution functions, the correlations, and the gravitational forces at each point. They represent local inhomogeneities in the system due to the discrete nature of the mass particles and are referred to as "graininess terms." Terms in $\mathcal{K}$, on the other hand, are integrated over the entire system and represent global inhomogeneities in the correlations of particles. They are known as "self-inducing" terms since they describe a back-reaction of the averaged correlation on its growth at a particular place. The second self-inducing term also couples with the three-body correlation function.

We can see why the graininess terms do not appear in the continuum approximation. (This approximation is sometimes called the "fluid limit" but should not be confused with taking moments of the collisionless Boltzmann equation to derive the usual fluid equations.) In the continuum limit the mass of each particle $m \to 0$ and the total number of particles $N \to \infty$ in such a way that the total mass mN remains finite. (The analogous statement often used in a plasma is that the total volume $V \to \infty$ so that the density N/V remains finite.) Taking the continuum limit, we see that $\phi(i, j) \to 0$ and since the distribution functions retain finite normalized values, the graininess terms $\mathcal{J}$ vanish. However, the self-inducing terms $\mathcal{K}$ are larger by a factor N because of the integral and they remain finite in the continuum limit. They are the terms that would describe a generalization of Jeans's perturbation analysis of a gas to take correlations of the perturbations into account.

When there are no correlations to start with, $\mathcal{K}$ and all but the first term of $\mathcal{J}$ vanish. This remaining term is a remarkable one. It represents the graininess of the single-particle distribution function and causes spontaneous growth of two-particle correlations. No initial perturbations, not even infinitesimal ones, are necessary to produce instability. The seeds of galaxy clustering are contained in the discrete nature of the galaxies themselves.

Once g has begun to grow, its graininess will also promote the growth of correlations through the second and third terms of $\mathcal{J}$. However, by that time, terms in $\mathcal{K}$ will make an even larger contribution. This is because the integrals of $\mathcal{K}$ effectively extend over the entire system to contribute self-inducing terms of order N times the graininess terms. So the last two terms in $\mathcal{J}$ are not important.

By the time most galaxies become bound into clusters the higher order correlations will become important. However, in the early stages, we may neglect the self-inducing term involving h in comparison with the others since initially h grows more slowly than g. Mathematically, this is because the source terms of the third hierarchy equation vanish when there are no correlations; thus h cannot grow until g grows. Physically, this is because with a two-particle potential the triple correlations are more likely to form through the correlation of a pair of correlated galaxies with a third uncorrelated galaxy than through the simultaneous correlation of three initially uncorrelated galaxies. The third term of $\mathcal{K}$ is also small compared with the first during the early clustering since it is quadratic in g.

Thus we are left with the first term of $\mathcal{J}$ and the first term of $\mathcal{K}$. Introducing the definition

$$\Psi(i, j) \equiv \int f(3) g(i, 3) \phi(j, 3) \, d\mathbf{x}_3 \, d\mathbf{v}_3, \qquad (17.55)$$

we can write the first term of $\mathcal{K}(i, j)$ as

$$f(i) \frac{\partial f(j)}{\partial \mathbf{v}_j} \cdot \frac{\partial \Psi(i, j)}{\partial \mathbf{x}_j}. \qquad (17.56)$$

Comparing this with the first term of $\mathcal{J}(i, j)$ shows that a sufficient condition for the graininess term to dominate is

$$|\Psi(1, 2)| < |\phi(1, 2)| \tag{17.57}$$

for all $\mathbf{x}_1$ and $\mathbf{x}_2$. Section 17.5 explores this regime and shows how long it lasts.

Initially the collision integral, $\mathcal{I}$, of the Boltzmann equation (17.49) is zero, and the system is strictly collisionless. Comparing $\mathcal{I}$ with the first term of $\mathcal{K}$ shows that the system remains approximately collisionless as long as condition (17.57) holds.

Therefore, the early evolution of an initially uncorrelated system is described by the simplified equations

$$\left[\frac{\partial}{\partial t} + \mathbf{v}_1 \cdot \frac{\partial}{\partial \mathbf{x}_1} - \frac{\partial \Phi(1)}{\partial \mathbf{x}_1} \cdot \frac{\partial}{\partial \mathbf{v}_1}\right] f(1) = 0 \tag{17.58}$$

and

$$f(1)f(2)\left[\frac{\partial}{\partial t} + \sum_{i=1}^{2}\left(\mathbf{v}_i \cdot \frac{\partial}{\partial \mathbf{x}_i} - \frac{\partial \Phi(i)}{\partial \mathbf{x}_i} \cdot \frac{\partial}{\partial \mathbf{v}_i}\right)\right] g(1, 2)$$

$$= \sum_{i \neq j}^{2} f(i)\frac{\partial f(j)}{\partial \mathbf{v}_i} \cdot \frac{\partial \phi(i, j)}{\partial \mathbf{x}_i}. \tag{17.59}$$

This is a basic result. There is an analytical solution for $g(1, 2)$ that will answer our question: How long does it take correlations to become significant and what do they look like?

17.5 Growth of the Two-Galaxy Correlation Function

Even though $g(1, 2)$ is the lowest order correlation function in the BBGKY hierarchy, it is not possible to obtain rigorous general solutions of (17.59), and we must simplify still further. Fortunately, the simplifications include a particularly interesting case that provides much of the physical insight into the development of correlations. This is the case starting from Poisson initial conditions with a Maxwell–Boltzmann velocity distribution at time t_0 having mean thermal speed $\langle v \rangle \equiv (2/\beta_0)^{1/2}$. Thus there are no initial correlations. All galaxies have the same mass m for simplicity. For an Einstein–de Sitter universe, a detailed analysis (Fall and Saslaw, 1976; GPSGS, Section 23) when the graininess terms dominate gives

$$\xi(r, t) = Gm\beta(t)r^{-1}\text{erfc}[r/\lambda(t)], \tag{17.60}$$

where

$$\lambda(t) \equiv \frac{2(t - t_0)R(t)}{R_0\sqrt{\beta_0}} \tag{17.61}$$

and

$$\beta(t) = \frac{\beta_0 R^2(t)}{R_0^2} \tag{17.62}$$

in proper coordinates with $R(t) \sim t^{2/3}$ for the Einstein–de Sitter model. As an intermediate step in this derivation, the full much more complicated solution including velocities for $g(1, 2)$ is obtained. The condition (17.57) that the graininess terms dominate becomes

$$\left(\frac{t - t_0}{t_0}\right)^2 \left(\frac{t}{t_0}\right)^{2/3} \lesssim \frac{3}{2}. \tag{17.63}$$

So if clustering begins when the Hubble age is t_0, these results are valid for about one more Hubble time.

During a Hubble time, clustering can grow from nothing to dominate the distribution on scales of order λ. For $r \ll \lambda(t)$, Equation (17.60) shows that $\xi(r) \sim r^{-1}$. Correlations often have a scale-free power-law behavior at a critical point during a phase transition when the distribution changes from being locally homogeneous to containing bound inhomogeneous clusters (while remaining statistically homogeneous when averaged over an ensemble of systems). For $r \gg \lambda(t)$, the asymptotic expansion of the erfc in (17.60) shows that

$$\xi(r) \sim \pi^{1/2} Gm\beta(t)\lambda(t)r^{-2}e^{-r^2/\lambda^2}\left[1 - \frac{\lambda^2(t)}{2r^2} + \cdots\right]. \tag{17.64}$$

Thus $\lambda(t)$ is a scale length for the correlations. It characterizes the spatial extent of the "phase transition." Were there no universal expansion, (17.61) shows that λ would grow linearly with time at a rate proportional to the galaxies' mean thermal speed. In other words, a given galaxy's sphere of influence at time t would extend just as far, on average, as another initially nearby galaxy could have traveled since time t_0. Expansion, however, causes the correlation length to increase in proper coordinates. This occurs because in the early stages almost none of the galaxies are bound into clusters and so, to a greater or lesser degree, correlated galaxies, like uncorrelated galaxies, essentially separate with the general expansion.

Physical properties of clustering are also brought out by other measures of correlation. One such quantity is the correlation energy. Since $\nu(r, t) = n(t)\xi(r, t)$ represents the excess density of galaxies around a given galaxy above the density of an uncorrelated state, the excess potential energy per galaxy above that of the uncorrelated state is simply

$$U(t) = -\frac{1}{2}\int_0^\infty \frac{Gm^2}{r}\nu(r, t)\, 4\pi r^2 dr = -2\sqrt{\pi}G^2m^3\beta(t)n(t)\lambda(t). \tag{17.65}$$

The factor $1/2$ accounts as usual for the sharing of energy between pairs of galaxies, and $\nu(r, t)$ comes from (17.60). Thus correlation energy increases linearly with time.

To conserve energy in a gravitational system, the kinetic energy of thermal motion must increase to compensate the decrease of correlation potential energy. When the magnitude of the correlation energy becomes of the same order as the mean thermal energy, the correlations dominate the system's behavior. Our neglect of the coupling integral in (17.49) prevents correlations from altering the evolution of the single-particle distribution function and therefore from affecting the mean kinetic energy. However, until correlations actually dominate, it is reasonable to approximate the mean kinetic energy per galaxy by $3m/\beta(t)$. Therefore the condition that correlations *not* dominate becomes

$$\frac{4}{3}\pi^{1/2}G^2m^2\beta^{3/2}n(t)(t-t_0)R^2(t)/R_0^2 \lesssim 1. \tag{17.66}$$

Another way to understand the condition (17.66) is to rewrite it in terms of the typical initial impact parameter $b_0 = 2Gm/u_0^2$ needed to produce a large gravitational deflection between nearest neighbor galaxies. For the Einstein–de Sitter cosmology we then have

$$\frac{b_0/u_0}{t_0}\tau\left(\frac{t}{t_0}\right)^{4/3} \lesssim 9\sqrt{\pi} \tag{17.67}$$

with $\tau \equiv (t-t_0)/t_0$. So the shorter the orbital time b_0/u_0 during which a close encounter deflects a galaxy significantly, the less important are correlations. Consequently, correlations take longer to grow when the random initial velocities u_0 are larger.

Perhaps the simplest measure of clustering that can be derived from the pair correlation function is the excess number of galaxies expected in some neighborhood of a given galaxy compared with the number in the corresponding uncorrelated state. The number expected within a sphere of radius r about a given galaxy at time t is

$$N_c(r,t) = \int_0^r v(s,t)\,4\pi s^2\,ds = 2\pi Gmn(t)\beta(t)\lambda^2(t)D[r/\lambda(t)], \tag{17.68}$$

where

$$D(x) = x^2\,\mathrm{erfc}\,x - x\pi^{-1/2}\exp(-x^2) + \frac{1}{2}\,\mathrm{erf}\,x. \tag{17.69}$$

In the limits of small and large spheres

$$N_c(r,t) = \begin{cases} \pi Gmn(t)\beta(t)2r^2, & r \ll \lambda(t), \\ \pi Gmn(t)\beta(t)\lambda^2(t), & r \gg \lambda(t). \end{cases} \tag{17.70}$$

The excess number can be compared with $N_0(r,t) = 4\pi r^3 n(t)/3$, the number expected when no correlations exist. A natural radius to use for this comparison is the mean interparticle separation $\sim n^{-1/3}(t)$. From (17.70), for small λ/r, this ratio becomes

$$\frac{N_c}{N_0} = 3Gmn(t)(t - t_0)^2 \left(\frac{R(t)}{R_0}\right)^4. \tag{17.71}$$

When this ratio approaches or exceeds unity the system will begin to appear highly clustered. For the Einstein–de Sitter cosmology this condition that clustering not be important becomes

$$\tau^2 \left(\frac{t}{t_0}\right)^{8/3} \lesssim 2\pi. \tag{17.72}$$

This is very similar to the condition (17.63) that self-inducing terms be negligible in the kinetic equations.

Fluctuations are closely related to correlations, as shown by Equation (14.15). For a system to have just $\sqrt{N}$ fluctuations on a length scale r requires the correlation contribution $N_c(r, t) \ll 1$. Comparing (17.68) with (17.63) using (16.9) and the fact that $0 \leq D(x) < 1/2$, we see that the condition for an expanding gravitating system to have $\sqrt{N}$ fluctuations on all scales is also equivalent to condition (17.63). Hence this condition is a multiple measure of when gravitational clustering is important.

Finally, we will see how the rate of clustering is simply related to graininess. At any time t and radial distance r, $N_c(r, t)/N_0(r, t)$ is proportional to m and independent of $n(t)$. Thus a system with galaxies of mass αm will cluster α times faster than a system with mass m. The rate of clustering will also be greater for the more massive galaxies within a system. This was expected since ϕ in the graininess source term in (17.50) is also proportional to m and vanishes in the fluid limit.

All these results are strictly valid only for $\tau \equiv (t - t_0)/t_0 \lesssim 1$. Even in this short time clustering can become significant starting from an initially unclustered distribution of galaxies. For $\tau \gtrsim 1$ correlations should grow even faster than for $\tau \lesssim 1$ since the self-inducing terms spring into action. These source terms in the kinetic equation correspond to the correlation of galaxies with groups of already clustered galaxies, a more efficient process than the early correlation of uncorrelated pairs of galaxies. When these terms are still linear, we can use the fluid approximation of Chapter 16 to extend the analysis to larger scales. Eventually $\xi(r, t)$ becomes much greater than unity and the clustering becomes nonlinear for both the graininess terms and the fluid terms.

How are we to analyze this nonlinear era, which we know is demanded by present observations? At first it would seem tempting to add the self-inducing terms and step up a rung on the ladder of the BBGKY hierarchy. But the work is enormous and the rungs are very slippery.

At the next level the mathematics becomes so intractable that many major and uncertain approximations are necessary, even for numerical solutions. The main problem is that as gravitational clustering becomes nonlinear its translational and rotational symmetries are broken on ever-enlarging scales: Inhomogeneities dominate.

When the correlations are small $g(1, 2) = g(r, u_r, u_t, t)$, but when they are large $g(1, 2) = g(\mathbf{x}_1, \mathbf{x}_2, \mathbf{u}_1, \mathbf{u}_2, t)$. The number of necessary variables increases from four to thirteen, and that is just for the two-point function!

Another major problem with extending the BBGKY approach is that higher order correlations become important in the nonlinear regime. This can be seen from the hierarchy equation for the three-point correlations, which start growing when the $g(i, j)$ grow. Soon after clustering begins, N-body experiments (to be described in Chapters 19 and 20) tell us that clumps of increasingly many galaxies accumulate. It would therefore seem that several-point correlation functions become important very quickly. Thus the next stage, after $\xi(r, t)$ becomes of order unity, is that *all* the low level correlation functions become large. Pursuit of just two or three of them will not give a convergent description of clustering.

In the next chapter we shall examine some of the approximations that allow us to understand aspects of clustering when more rigorous analyses are no longer possible. To check the range and validity of the approximations, we can use N-body simulations. These simulations also act as experiments to examine specific examples of clustering and to compare them with observations. In more complicated models that include a fluid component of dark matter as well as individual galaxies, approximations and simulations are all the guidance we have.

18

General Correlation Properties

> Would it not be better to get *something* done even
> though one might not quite understand *what*?
> J. Synge

> Apparent confusion is a product of good order.
> Sun Tzu

Despite difficulties in solving the BBGKY kinetic equations directly, judicious approximations give valuable insights into their physical properties. As soon as the system leaves the linear regime, these approximations replace a more rigorous analysis. Computer simulations, although they require further approximations, can tell if these insights are genuine and provide a useful description. In this chapter we discuss several physical approximations; following chapters summarize simulations and observations.

18.1 Scaling

Scaling has acquired many meanings in different contexts of large-scale structure and thereby caused some confusion. One form of scaling is the simple geometric relation between angular correlation functions $W(\theta)$ in catalogs of various limiting magnitudes, given by (14.38). This is closely related to the manufacture of catalogs rather than to the underlying dynamics of clustering. We mention it further in Chapter 20. A more physical form of scaling results from noticing that the $\Omega_0 = 1$ Einstein–de Sitter cosmology contains no scale length. Nor is there any scale length in the Newtonian gravitational force between galaxies. Therefore, the argument goes, correlation functions should not contain any scale lengths either and should be power laws.

Unfortunately this argument neglects the actual dynamical development of correlations. Initially, if the positions of newly formed galaxies are not correlated in a gravitationally self-consistent way on all scales, gravitational interactions of galaxies will tend to produce self-consistent correlations. However, if discrete density fluctuations are larger on smaller scales, then these smaller scales evolve faster. This process sets up a scale length, at any time, within which the local relaxation to a quasi-equilibrium self-consistent distribution is reasonably complete, and beyond which relaxation still continues. The detailed nature of this relaxation length scale depends on the specific cosmology, initial conditions, distribution of dark matter, and definition of relaxation. The important point is that for almost all initial conditions it exists.

Even though dynamic evolution implies that there cannot be one power law for all scales, there may be several ranges of scales, each with its own approximate

216

power-law correlation. It would be reasonable to expect a power law on small scales where the clustering is highly relaxed and nonlinear. Very large unrelaxed or linear scales might have a different power law reflecting initial conditions. On intermediate partially relaxed scales, the shapes of correlation functions may be more complicated.

Formally, scale-free correlations would have the property of (14.41):

$$\xi_N (\lambda r_1, \ldots, \lambda r_N) = \lambda^{-\gamma (N-1)} \xi_N (r_1, \ldots, r_N) \qquad (18.1)$$

for the reduced N-point correlation functions (i.e., the terms connecting all N particles). For $N = 2$ the resulting functional equation clearly has a power-law solution $\xi_2 \propto r^{-\gamma}$. For $N > 2$, solutions have the form of homogeneous sums of products of lower order correlation functions, as in (14.42):

$$\xi_3 (r_{12}, r_{23}, r_{31}) = Q \{ \xi (r_{12}) \xi (r_{23}) + \xi (r_{23}) \xi (r_{31}) + \xi (r_{31}) \xi (r_{12}) \}. \qquad (18.2)$$

Here the simplifying approximation is that Q is the same for all three terms, independent of the detailed configuration of the triplet $(\mathbf{r}_1, \mathbf{r}_2, \mathbf{r}_3)$. As N increases, the number and variety of terms increase faster. They rapidly become too complex to calculate informatively or measure accurately. Many attempts have been made to impose simplifying assumptions on the forms of the coefficients of these terms and their relations to the coefficients of terms in lower order correlations. Often this ad hoc algebraic approach eventually leads to inconsistencies such as negative probability densities or disagreements with dynamical simulations. Basically these difficulties arise because the system contains a relaxation scale length.

Sometimes a simpler type of scaling is conjectured. This involves the volume integrals (14.43) of ξ_N rather than ξ_N itself. These integrals average out the detailed configurations of particles. Equation (14.44), $\bar{\xi}_N = S_N (\bar{\xi}_2)^{N-1}$, is assumed to relate the integrals of higher order correlation functions to lower order ones. The S_N are assumed to be constants, independent of scale, and are allowed to depend only on N. Various functions $S(N)$ can be assumed or derived from suggested forms of distributions and compared with simulations and observations.

A third type of dynamical scaling attempts to relate the nonlinear evolution of $\bar{\xi}$ on small scales to its linear evolution on large scales. From (16.63) we see that in the linear regime $\bar{\xi}(t) = R^2(t) \bar{\xi}_0 (t_0)$. Rewriting (14.32) in terms of comoving coordinates $x = r/R(t)$ we have

$$x_0^3 \equiv R(t)^{-3} \int_0^x (1 + \xi) \, dV = x^3 (1 + \bar{\xi}), \qquad (18.3)$$

where x_0 (which incorporates a geometric factor and the difference between conditional and unconditional densities) is an effective Lagrangian comoving radial distance containing $\langle N \rangle$ galaxies around a typical given galaxy. Since this represents the average conserved number of galaxies around a given galaxy, it is sometimes called the conserved pair integral. In the very nonlinear regime ($\bar{\xi} \gg 1$) the explicit time evolution of $\bar{\xi}$ is slow and for a fixed x_0 we see that $\bar{\xi} \propto x^{-3} \propto R^3(t)$ to a good approximation. This suggests that in general $\bar{\xi}$ will remain a function only of the scaled quantity $R^2 \bar{\xi}_0 (x_0)$. Hamilton, Kumar, Lu, and Matthews (1991) went a step further and hypothesized that the same function $\bar{\xi}(R^2 \bar{\xi}_0)$ applies to all systems with scale-free cosmologies ($\Omega_0 = 1$) and scale-free power-law initial forms of $\xi_N(x)$. This works reasonably well for several such computer simulations from which they derived an empirical fitting formula for $\xi(x)$ containing eight free parameters as coefficients and fitted powers of $R^2 \bar{\xi}_0$. In principle this formula can be inverted to obtain the initial $\bar{\xi}_0 (x_0)$ from the presently observed $\bar{\xi}(x)$, though it is difficult to find a unique solution within the uncertainties. Using (14.17), (18.3), and the relation $\bar{\xi}(R^2 \bar{\xi}_0)$, we can relate the nonlinear density contrast on small scales to the linear density contrast on large scales. The catch is that even for scale-free systems there is significant orbit crossing as the clustering becomes very nonlinear. This makes the initial Lagrangian volume containing the original galaxies more and more distorted and difficult to retrieve from the evolved distribution, although it may be possible in some average sense. Despite its lack of rigor, this scaling can still be informative for the two-point correlations even though it might not hold for higher orders.

Dynamical derivations of scaling laws are difficult. There is no rigorous proof of Equations (18.1), (18.2), or the extension of (18.3) from first principles yet. Nor, for the physical reasons mentioned earlier, is there likely to be. Nonetheless, they remain useful "working approximations" in some circumstances.

A fourth type of scaling is temporal scaling, i.e., a similarity solution. The restrictions necessary for this scaling are illustrated by an attempt (Davis and Peebles, 1977) to derive it directly from the BBGKY hierarchy. This involved four main physical and two mathematical assumptions. The physical assumptions were: 1. An $\Omega = 1$ Einstein–de Sitter universe so that the cosmology did not introduce any scale length, 2. An initial power law for the spectrum of density perturbations so that these did not introduce a scale length either, 3. Ignoring the discrete nature of the galaxies (particles) on small scales so that the interparticle separation does not appear as a scale length, and finally, 4. Assuming that the distribution of relative velocities of particle pairs has zero skewness, $\langle \Delta v^3 \rangle = 0$, so that the hierarchy of velocity moments can be truncated (but see Yano and Gouda, 1997). The mathematical assumptions were that (18.2) holds with one value of Q and without any triplet terms of the form $\xi(r_{12}) \xi(r_{23}) \xi(r_{31})$ and, second, that the momentum dependence of the full two-point correlation function is a separable function of the average and relative momenta of the two particles. These assumptions have consequences. A consequence of ignoring the discrete nature of the galaxies is that the calculation is essentially done in a fluid limit $m \rightarrow 0, n \rightarrow \infty$, but $\rho = nm =$ constant. This ignores particle pair interactions and just retains the interactions of a particle with clusters of particles. A consequence of truncating the velocity moments is that once small groups become bound they no longer evolve (e.g., by ejecting galaxies, interacting with other groups, or accreting galaxies).

With these assumptions, the BBGKY hierarchy admits similarity solutions (Davis and Peebles, 1977) in terms of scaled comoving distances: $\xi_2(x, t) = \xi_2(xt^{-\alpha})$ and $\xi_3(x_{12}, x_{23}, x_{31}, t) = \xi_3(x_{12}t^{-\alpha}, x_{23}t^{-\alpha}, x_{31}t^{-\alpha})$. For an initial density perturbation power spectrum $\langle |\delta_\mathbf{k}|^2 \rangle \propto k^n$ with a single power law, results give $\alpha = 4/(9 + 3n)$ and $\xi_2 \propto (x/t^\alpha)^{-\gamma}$ with $\gamma = (9 + 3n)/(5 + n)$. An initial value of $n = 0$ would give the observed $\gamma = 1.8$, but $n = 1$ would not be far off and $n = -1$ would still be within the observed range.

Naturally, just because a similarity solution exists for scaled variables does not mean that a system actually evolves along it. The similarity solution must also be shown to be stable, which has not been done in this case. Indeed Hansel et al. (1986) imply it is unstable because they find that the separability assumed for the momentum dependence of ξ_2 only holds for very small x, which is in the highly nonlinear region. There the assumption that ignores discrete particle interactions breaks down, as does the truncation of velocity moments (Yano & Gouda, 1997). Detailed analysis (Ruamsuwan & Fry, 1992) shows that modes generally exist that evolve away from the similarity solution. Neutral stability or slow growth of such modes is perhaps the best hope.

Despite apparent dynamical inconsistencies, this scaling approach may be a useful approximation for an interesting range of problems. Its applicability to our own Universe remains quite unknown, especially since other approaches also lead to $\gamma \approx 1.8$. (cf. GPSGS, Chapter 32.) More modern models suggest further that a single power law may be an oversimplified initial condition in any case.

The four types of dynamical scaling outlined here – involving $\xi_N(\lambda r_1, \ldots, \lambda r_N)$, $\xi_N[\xi_2^{N-1}]$, $\bar{\xi}_2[R^2\bar{\xi}_2(0)]$, and $\xi_2(xt^{-\alpha})$ – are often sought in N-body simulations and observations. Sought but seldom clearly found, as we shall see in later examples.

18.2   Real Space and Redshift Space

Scaling relations are delicate creatures and may appear differently when viewed in redshift space than in real space. In fact redshift space distorts the correlation functions, distribution functions, and most other statistics as well. Two-dimensional angular projected statistics, however, survive this distortion relatively unscathed, being affected only by slightly different limits to their outer boundaries in real and redshift space.

The cause of this distortion is that in real space distances may be known exactly, but in redshift space distances are more complicated. The observed redshift, which is usually measured very accurately, contains two components. One is the global Hubble expansion of the Universe; the other is the local peculiar velocity usually associated with clustering. Uncertainty in the Hubble constant is not serious in this regard because it just affects the scale of the global picture. However, contamination by peculiar velocities of $\sim 10^3$km s^{-1} in strongly clustered regions can alter the impression of the galaxy distribution on scales up to ~ 100 Mpc. Beyond this scale the effect becomes $\lesssim 10\%$, which may still influence sensitive statistics. The cure is

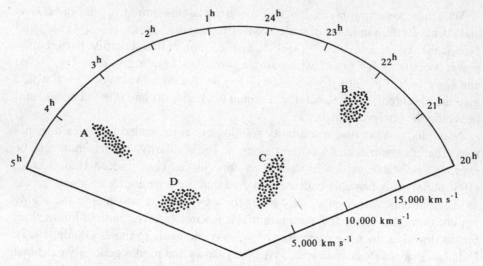

Fig. 18.1. Schematic representation of distortions in a redshift space "wedge diagram" of right ascension versus redshift velocity. A, B, and C are representative nonlinear relaxed spherical clusters; D is an unrelaxed linear cluster contracting slowly.

accurate secondary (i.e., nonvelocity) distance indicators for galaxies (see Chapter 34). Then, for a given Hubble constant, we get the peculiar velocity directly.

To see how these distortions come about, first consider a simple, bound, virialized cluster of many galaxies. In real space it looks spherical. In redshift space, portrayed by the wedge diagram of Figure 18.1, such clusters appear extended toward the observer. In these standard wedge diagrams, all galaxies within some range of declination are projected onto a single plane (a distortion in itself) of right ascension versus redshift. Here the spherical cluster covers a small range of right ascension, but its internal velocity dispersion produces the elongated distortion toward the observer, shown schematically in the diagram. Some astronomers, rather waggishly, call these "fingers of God."

Next consider a spherical cluster with a slight overdensity where galaxies are coherently moving toward the center, as in Chapter 16. The real space–redshift space distortion of this linear flow differs considerably from the previous relaxed nonlinear cluster. Now cluster galaxies beyond the center of the cluster, as seen by a distant observer, will have redshifts less than the Hubble velocity corresponding to their real distance. Their redshift is the sum of their Hubble velocity and a negative apparent peculiar velocity. So if their redshift is used to measure their distance, $r = v/H_0$, they will appear to be closer than they actually are. Similarly, galaxies on the near side of the cluster, whose peculiar velocity toward the center is positive, will seem farther away than they actually are.

In the wedge diagram, this linear peculiar velocity contamination compresses points along the line of sight, equivalent to an apparent elongation transverse to the line of sight, shown schematically in group D of Figure 18.1.

Coherent linear motions and fairly random nonlinear motions, along with departures from spherical geometry, are all present in realistic clustering. This makes retrieval of real space positions from redshift space information very difficult. Only good secondary distance indicators will really do the job.

How will these distortions affect correlation functions? Generally the answer depends on the linear, nonlinear, and geometric properties of each detailed model and must be sought by particular numerical experiments. Qualitatively one expects, and the experiments show (cf. Kaiser, 1987; Suto and Suginohara, 1991; Matsubara and Suto, 1994), that in the nonlinear regime the real space $\xi_2(r)$ has a higher amplitude and steeper slope than the velocity space $\xi_2(r)$. This is because on small scales where strong clustering dominates, the high peculiar velocities smear the distribution significantly over redshift space and decrease the resulting apparent correlation. On large scales where the clustering is linear, average peculiar velocities are generally smaller, and so are the differences between correlations in real and redshift space.

Consequently, we would generally expect the moments of ξ_2 for statistically homogeneous systems to be smaller in redshift space than in real space. Numerical experiments (Lahav et al., 1993) demonstrate this. A further related consequence is that hierarchical clustering should generally appear to hold better in redshift space than in real space. Thus Q in (18.2) and the S_N in (14.44) should be less scale dependent in redshift space than in real space. This dependence has been checked for Q in ξ_3 for a low-density model containing cold dark matter ($\Omega_0 = 0.2, \lambda_0 = 0.8, k = 1$) with numerical simulations (Matsubara and Suto, 1994). Over ~ 2 Mpc, Q typically varied by a factor of ~ 3 in real space, but it hardly changed at all in redshift space. Observations in this model could lead to the erroneous conclusion that ξ_3 was hierarchically clustered.

The moral is that patterns of clustering observed in redshift space may be quite different from the physical clustering in real space. To approximate one by the other can be deceptive. Comparing these differences for numerical simulations having complete information is easy. But retrieving detailed structure in real space from observations in redshift space without good secondary distance indicators is ambiguous.

18.3   Bias

How well do galaxies trace the dark matter? If there is substantial dark matter between the galaxies, then the statistics of bright galaxies give a biased impression of the total distribution. This possibility has been known since the "missing mass" problem was discovered (cf. Zwicky, 1957 and earlier references therein). Another well-known variant is that galaxies of different masses or luminosities or morphologies may be distributed differently – biased relative to one another or to the dark matter.

Complete characterizations of these biases require comparison of all their correlation functions ξ_N. The next best description may be to compare their distribution functions $f(N, V)$. More practically, but still less informatively, we can compare their two-particle correlation function as in (14.40):

$$\xi_{gal} = B^2 \xi_{matter} \tag{18.4}$$

or using (14.16) in the nonlinear regime

$$\left.\frac{\langle(\Delta N^2)\rangle}{\bar{N}^2}\right|_{gal} = B^2 \left.\frac{\langle(\Delta M^2)\rangle}{\bar{M}^2}\right|_{other\ matter}. \tag{18.5}$$

Similar expressions may represent biased clustering of rich clusters among themselves, relative to clustering of individual galaxies.

Theoretical arguments have not yet given compelling insight into the value of B (often denoted by b). Observational analyses of different samples, and sometimes different analyses of the same sample, give values of $0.5 \lesssim B \lesssim 2$. So the situation is unclear. A priori one might expect that if the formation of galaxies in one region inhibited their formation nearby, either because they used up all the regional gas or because rapid star formation and other violent activity blew surrounding gas away, then galaxies might be less correlated than dark matter: $B^2 < 1$. In the extreme case, if galaxies were anticorrelated relative to some other type of matter, they would have an imaginary bias: $B^2 < 0$ (which was not imagined when the original notation was suggested!). This could occur if galaxies formed from seeds whose distribution was determined by the much earlier universe. On the other hand, if most of the dark matter is in or closely around galaxies then $B \approx 1$. Finally, if galaxies are more strongly clustered than the dark matter, or if rich clusters are more strongly clustered than average galaxies or poor clusters, then their $B > 1$. All these possibilities can depend on scale, leading to a $B^2(r)$ that would implicitly incorporate information from higher order correlations as well as from any window function used to smooth the density field. Bias may also depend on time, both locally and globally.

One form of biasing arises quite naturally if clustering occurs in regions that are already denser than average. These overdense regions could result either from random fluctuations or from underlying perturbations. In the case of underlying perturbations, the average density would itself be scale dependent and could lead to a fractal distribution (see Chapter 12) modifying the amplitude of ξ_2 so that $B^2 \neq 1$. As an example (Kaiser, 1984; Bardeen et al., 1986), consider a field $\eta(\mathbf{x})$ (e.g., density) with a small-amplitude long-wavelength signal (such as a linear density perturbation) superimposed on background noise (which need not be Gaussian). The coherence length of the signal is much greater than that of the noise: $l_S \gg l_N$, and the total field $\eta = \eta_S + \eta_N$. Now suppose we are interested in regions of the field of size l between l_N and l_S. The variance, σ, of the field among these regions is essentially σ_N, the variance of the noise. Suppose further that the objects we wish to find (e.g., galaxies or rich clusters) occur where the field exceeds a certain threshold $\nu\sigma$. Generally these regions will be places where both the signal and the noise are greater than average. Therefore objects in these regions will be correlated more strongly than the underlying noise; the signal supports them.

To estimate this correlation enhancement, we require $\eta = \eta_S + \eta_N > \nu\sigma$ or $\eta_N > (\nu\sigma - \eta_S)$. If we know the noise probability $P_N(\eta > \nu\sigma)$ for finding a fluctuation greater than $\nu\sigma$ in the absence of the signal, then a first-order Taylor series expansion around this when the signal is present gives

$$P[\eta > \nu\sigma] \approx P_N[\eta_N > (\nu\sigma - \eta_S)]$$

$$\approx P_N(\eta > \nu\sigma) + \frac{dP_N(\eta > \nu\sigma)}{d\nu}\Delta\nu$$

$$\approx P_N(\eta > \nu\sigma)\left[1 - \frac{1}{\sigma}\frac{d\ln P_N(\eta > \nu\sigma)}{d\nu}\eta_S(\mathbf{x})\right] \quad (18.6)$$

since $\Delta\nu = -\eta_S/\sigma$ is the effective change in the required noise threshold produced by the presence of the signal. The second term in (18.6) reduces the relative probability of η exceeding its threshold $\nu\sigma$ and therefore acts like a density contrast $\Delta\eta$. The correlation function amplitude varies as $(\Delta\eta/\eta)^2$ and therefore from (18.5) the bias in this case is an amplification factor

$$B^2 = \left[\frac{1}{\sigma}\frac{d\ln P_N(\eta > \nu\sigma)}{d\nu}\right]^2. \quad (18.7)$$

The range of validity can be checked by requiring the second-order term of the Taylor series expansion to be relatively small and the earlier assumptions to hold. Details depend on specific forms of $P_N(\eta > \nu\sigma)$, but $\xi_S < 1$ is a fairly general restriction. Mo (1988) has examined several examples for cluster–cluster correlations, which may be amplified above the galaxy–galaxy correlation. When η_N or η_S are continuous fields, the amplification will also depend on the window functions used to smooth them.

Various ad hoc forms of bias have been used to patch up theories that are otherwise difficult to reconcile with observations. Bias does however have a legitimate use when calculated from basic physics in a way that reduces the free parameters, functions, or assumptions in a problem. Such calculations, as well as observational interpretations, demand great self-consistency. This is why bias is hard to pin down.

18.4 A Relation among the Amplitude, Range, and Slope of ξ

In the relaxed regime $r \leq R$ the two-point correlation is close to a power law: $\xi(r) = (r/r_0)^{-\gamma}$. Why should the amplitude r_0, range R, and slope γ have the values they do? Observed values for different catalogs may depend somewhat on limiting magnitude, on morphological type, on how the galaxies are selected (e.g., by optical or infrared luminosity, or by angular diameter), and on the detailed method of analysis. But the slope is almost always between 1.5 and 1.8, the amplitude is between about $3h^{-1}$ and $6h^{-1}$ Megaparsecs, and the range, less well determined, may be as great as $\sim 30 h^{-1}$ Mpc.

Naturally these quantities evolve in some dynamically self-consistent way, as N-body simulations show. Indeed, a wide range of computer experiments (see Chapters 19 and 20) for different cosmologies, initial conditions, and mixtures of hot and cold dark matter can reproduce the observed γ and r_0, reasonably well. Determining R is more difficult because it is influenced strongly by the boundaries of the simulations, which often have insufficient dynamical range. The ubiquity of these values for many models and many observations suggests a more general explanation.

After a statistically homogeneous galaxy distribution relaxes to a single power-law form for ξ $(r \leq R)$, the values of γ, r_0, and R tend to change more and more slowly with time. This implies a quasi-equilibrium evolution in which the system evolves through a sequence of equilibrium states (see Chapter 25 for more detailed discussion and conditions for quasi-equilibrium). Often such states have the property that their energy is an approximate local extremum – approximate, because gravitating systems always continue to evolve and have no rigorous equilibrium.

To explore this for an essentially unbounded, statistically homogeneous system of point particles (galaxies) in an expanding universe, consider the gravitational correlation energy

$$W = \frac{\bar{N}\bar{n}m}{2} \int_0^R \phi(r)\xi(r)4\pi r^2 \, dr \tag{18.8}$$

in a spherical volume V of radius R. Here $\phi(r) = -Gm/r$ is the gravitational potential for particles that all have mass m (including massive galactic halos), average number density $\bar{n}$, and total average number $\bar{N} = \bar{n}V$. Since gravity is a pairwise interaction, W does not depend on higher order correlation functions. Moreover, the expansion of the Universe compensates the long-range average gravitational field, leaving W to depend on just the local correlations (see Chapter 25). Substituting $\phi(r)$ and $\xi(r)$ into (18.8), we get

$$W = -2\pi G\bar{N}\bar{n}m^2 r_0^2 \frac{1}{(2-\gamma)} \left(\frac{R}{r_0}\right)^{2-\gamma} \tag{18.9}$$

for $\gamma < 2$. The case $\gamma = 2$ has an infinite logarithmic divergence at the lower limit, unless it is cut off at $r = r_* \ll R$ giving $W(\gamma = 2) \propto -\ln(R/r_*)$. Thus if $r_* \to 0$ and galaxies cluster so that W tends to a minimum value, not subject to any constraints, $\gamma \to 2$ (see also GPSGS, Section 32).

But there are constraints. In quasi-equilibrium when the total energy is an approximate extreme, the absolute rate of change, $|dU/dt|$, of the total internal energy $U = K + W = K - |W|$ is an approximate minimum for any value of U characterized by a given r_0 and R and kinetic energy K in the volume V. Chapter 22 shows that as the Universe expands adiabatically, U evolves according to the cosmic energy equation

$$\frac{dU}{dt} = -H(t)(2K + W) = -H(t)(2U + |W|), \qquad (18.10)$$

where $H = \dot{a}(t)/a(t)$ is the Hubble parameter, which describes the relative expansion rate of the Universe. (We now use $a(t)$ for the radius of the Universe, so as not to confuse it with the range, R). Evidently, to minimize $|dU/dt|$ for any given U, we must minimize $|W|$ as a function of γ for given r_0 and R. At the minimum $d|W|/d\gamma = 0$; so taking the derivative of $\ln |W|$ from (18.9) we obtain the simple relation

$$\gamma = 2 - \frac{1}{\ln \frac{R}{r_0}}. \qquad (18.11)$$

The further requirement $d^2|W|/d\gamma^2 > 0$ for a minimum is easily verified. Equation (18.11) is an approximate relation since a region of complete equilibrium would satisfy the virial theorem $2K+W = 0$, effectively detaching itself from the expansion of the Universe, and have $dU/dt = 0$. As the region approaches this equilibrium, $\dot{U}$ is minimized for a given amplitude and scale of correlations in quasi-equilibrium. The corresponding correlation energy from (18.9) and (18.11) is

$$W = -2\pi G \bar{N} \bar{n} m^2 r_0^2 \left(\frac{R}{r_0}\right)^{1/\ln(R/r_0)} \ln \frac{R}{r_0}. \qquad (18.12)$$

In the limit $R/r_0 \to \infty$, correlations extend throughout the entire system, as in a critical point phase transition. In this limit the value of $\gamma \to 2$ and $W \to -\infty$. These are the same limits as for the unconstrained case where W is simply minimized, since R is not constrained to have a finite value. However, any infinite expanding universe that is initially correlated only over a finite (usually small) scale will require an infinite time for these correlations to extend throughout the entire universe. And if the universe expands at a faster rate than the scale of correlations grows, it will never reach the limit $R/r_0 = \infty$. At best, this limit can be approached asymptotically (see Chapter 30 for more details in the analogous limit $b \to 1$).

At first sight, it may seem surprising to have even an approximate relation among the amplitude, range, and slope of ξ. Dynamically this occurs because within the relaxed range R, galaxy clustering on one scale has strongly influenced the clustering on other scales. For example, the formation of small bound clusters can only proceed by local ejection of higher energy galaxies, which may eventually cluster on larger scales $\sim R$. In turn, the larger scale clusters tidally influence those on smaller scales. All these complicated orbits are approximately summarized by (18.11).

Nonlinear clustering scales mainly determine the slope γ, whereas linear scales determine R, and the scale r_0 represents the transition between these two regimes. Starting with a given initial spectrum of fluctuations, linear perturbation theory provides the time development of r_0 and R. Thus (18.11) enables one to calculate the time development of the nonlinear slope $\gamma(t)$ from linear theory.

The relation in (18.11) may be compared with observations for samples where r_0, R, and γ are well determined, preferably in real (rather than in redshift or in projected) space. The *IRAS* survey of infrared galaxies provides an example, although improved catalogs are always under construction. Analysis (Saunders et al., 1992) of the *IRAS* survey gives $r_0 = 3.79 \pm 0.14 \, h^{-1}$ Mpc, $R \approx 30 \, h^{-1}$ Mpc, and $\gamma = 1.57 \pm 0.03$. Equation (18.11) for these values of r_0 and R gives $\gamma = 1.52$. Within the range of observational uncertainty, and recognizing that the clustering may still be growing slowly so that the requirement $d|W|/d\gamma = 0$ is not exact, we see that (18.11) agrees reasonably well with observed galaxy clustering.

19

Computer Simulations

> All computer simulations sacrifice
> information to the great God Time.

19.1   Direct Methods

For inspiring new insights into galaxy clustering, for testing our understanding of gravitational many-body physics, and for detailed comparisons with observation, nothing works better than computer experiments. But they also have trade-offs and dangers. The trade-offs are among detailed physical information, computational speed, and number of physical particles. The dangers are lack of uniqueness and a tendency to examine only a small range of models based just on different parameters rather than on different basic ideas.

A variety of numerical techniques, all compromises, have been developed for different types of problems. The simplest problem considers the evolution of a distribution of N points with the same or different masses in the background of an expanding universe. We shall call this the cosmological many-body problem. Each point mass represents a galaxy and its associated halo. This is a good approximation if we are not concerned with galaxies' tidal interactions and mergers, or with inhomogeneous dynamically important intergalactic matter. Such complications can be added using other techniques to determine their significance.

Cosmological many-body problems are usually solved by integrating all the N particles' equations of motion. This is the direct method. Since only particle–particle interactions occur, it is also called the particle–particle (PP) method. Though direct, it is not straightforward. There are N equations, each with N terms, leading to $\sim N^2$ operations. Moreover, when particles come close together, their high accelerations require short time steps. All this costs computer time and memory. For values of N large enough to represent a typical region of the Universe, sophisticated techniques can speed up the computations. Let's see how this works.

Galaxy clustering involves peculiar velocities much smaller than the speed of light and regions of space much smaller than the radius of the Universe. Therefore it is essentially a Newtonian problem in a locally flat expanding spacetime metric. A particle of mass m_i is gravitationally accelerated by all other $j \neq i$ particles so that

$$\ddot{\mathbf{r}}_i = -G \sum_{j \neq i}^{N} \frac{m_j (\mathbf{r}_i - \mathbf{r}_j)}{\left(|\mathbf{r}_i - \mathbf{r}_j|^2 + \epsilon_0^2\right)^{3/2}}. \tag{19.1}$$

The only unfamiliar part of this equation is the softening parameter ϵ_0. It is added to keep the numerical integrations well behaved as the separation r_{ij} of any two particles becomes very small. It may also be a somewhat more realistic approximation to overlapping galaxies if their internal density varies as $\rho \propto r^{-2}$ leading to a force roughly proportional to $\rho r^3 \propto r$ in this limit. This is meant to be only schematic as $r_{ij} \to 0$. The usual procedure for a given problem is to determine the importance of ϵ_0 by varying it. Then we take the results of numerical simulations seriously only on scales where these variations have negligible effects. Alternatively, for a "pure" N-body problem with $\epsilon_0 = 0$, there are useful regularization techniques that transform the variables of the equations of motion to remove or displace the singularities (cf. Stiefel and Scheifele, 1971; Bettis and Szebehely, 1972). These have been generalized to three-body and multiple close encounters (Aarseth and Zare, 1974; Mikkola and Aarseth, 1993). Regularization is particularly useful for finite systems that form close binaries, but usually (19.1) suffices for galaxy clustering. Although galaxy mergers may be important on small scales, expansion of the Universe decreases their global role.

Transformation to comoving coordinates, $\mathbf{x} = \mathbf{r}/R(t)$, incorporates this global expansion. A convenient representation of the expansion follows by differentiating (16.5) for $\epsilon = \rho c^2$ and $\Lambda = 0$ but any value of k to give

$$\ddot{R} = -\frac{GM}{R^2}. \tag{19.2}$$

From (16.89) we see that this is equivalent to the acceleration of a particle at the edge of a uniform density sphere of mass M, arbitrarily placed anywhere in the Universe. Substituting $\ddot{\mathbf{r}} = R\ddot{\mathbf{x}} + 2\dot{R}\dot{\mathbf{x}} + \mathbf{x}\ddot{R}$ and (19.2) into (19.1) and letting $\epsilon = \epsilon_0/R(t)$ gives the comoving equations of motion,

$$\ddot{\mathbf{x}}_i = -\frac{2\dot{R}}{R}\dot{\mathbf{x}}_i - \frac{G}{R^3}\left(\sum_{j \neq i} \frac{m_j(\mathbf{x}_i - \mathbf{x}_j)}{(|\mathbf{x}_i - \mathbf{x}_j|^2 + \epsilon^2)^{3/2}} - M\mathbf{x}_i \right). \tag{19.3}$$

The first term on the right is the decay of peculiar velocity caused by expansion. For numerical integration, (19.3) has the disadvantage that the R^{-3} factor complicates the higher order derivatives used in the numerical integration scheme. But the time transformation

$$d\tau = R^{-3/2} dt \tag{19.4}$$

removes this problem:

$$\mathbf{x}_i'' = -\frac{R'(\tau)}{2R}\mathbf{x}_i' - G \sum_{j \neq i} \frac{m_j(\mathbf{x}_i - \mathbf{x}_j)}{(|\mathbf{x}_i - \mathbf{x}_j|^2 + \epsilon^2)^{3/2}} + GM\mathbf{x}_i, \tag{19.5}$$

where the primes indicate derivatives with respect to τ and

$$R'' = \frac{3}{2}\frac{R'^2}{R} - GMR \tag{19.6}$$

from transforming (19.2). This also creates more elegant equations of motion in the transformed time coordinate.

Usually simulations are scaled to units in which $G = 1$ and the comoving radius is represented by a unit sphere. Particles crossing this sphere are reflected back with an equal negative comoving radial velocity. Since the sphere is supposed to represent a typical volume of the Universe, it must contain enough particles to give a reasonable statistical description. The actual number of particles needed depends on the specific physical questions asked. For rough estimates of correlation functions and distributions, 10^3 particles are about the minimum needed, 10^4 are now typical of these simulations using workstations, and 10^5 are possible. Generally, the more particles the better, especially since particles close to the boundary are not moved by typical near-neighbor forces and so the statistics exclude them.

Numerous numerical tricks and techniques have been developed to integrate (19.5) quickly and accurately (see Aarseth, 1994 for a detailed review). The main time saver is the realization that forces on any given particle from its near neighbors generally change faster than the forces from more distant particles. Therefore the distant forces do not have to be updated as often or as precisely as those nearby. Starting with initial positions and velocities, (19.5) provides the total acceleration of every particle and updates their velocities and positions after each timestep. Simultaneously $R(t)$ and $\epsilon(t)$ are also updated. Once the forces are determined at, say, four earlier times, they can be represented by a fourth-order polynomial or equivalent Taylor series expansion. This provides the basis for a fourth-order predictor–corrector integration technique. Then the computation steps along until either the physical questions are answered or the computer time runs out.

To compare these experiments with observed clustering, we need to escape from the natural units $G = R = m = 1$. All galaxies may have the same mass, m, or m may be the average mass of the galaxies. Assuming that all of the mass M in a sphere of radius R_{Mpc} is associated with the N particles, each representing a galaxy, we can write the average mass of a galaxy in the simulation as

$$m = \frac{M}{N} = 6.5 \times 10^{11} N^{-1} R_{\mathrm{Mpc}}^3 h_{75}^2 \Omega_0 \quad \text{solar masses}, \tag{19.7}$$

where $h_{75} = H_0/75 \text{ km s}^{-1} \text{ Mpc}^{-1}$. Note that for given values of N and the current cosmological expansion h_{75} and density Ω_0, we have the freedom of fixing either m or R. This shows the definite advantage of using larger N for more representative volumes. The velocities are converted into physical units by a scaling factor

$$v_{\mathrm{scale}} = \frac{v_{\mathrm{physical}}}{v_{\mathrm{natural}}} = \left[Gm \left\langle \frac{1}{r} \right\rangle \right]^{1/2} = [1.35 Gm N^{1/3} R^{-1}]^{1/2} \tag{19.8}$$

so that one natural unit of the simulation velocity equals v_{scale} units of physical velocity (e.g., km s^{-1}). The value of $\langle 1/r \rangle$ is the average inverse separation the particles would have in a uniform (Poisson) distribution. For a fixed value of m, the last two equations give

$$v_{\text{scale}} = 143 \left(\frac{N}{10^4}\right)^{-1/3} \left(\frac{R}{50 \text{ Mpc}}\right) \left(h_{75}\Omega_0^{1/2}\right) \text{ km s}^{-1}. \qquad (19.9)$$

We will see many results of direct simulations in later chapters, where they help us explore the basic many-body physics of gravitational clustering in cosmology. First I summarize other more approximate simulation techniques suitable for related problems.

19.2   Other Methods

Tree codes are closest in spirit to direct integrations. Their basic idea is to represent the forces from distant groups of particles by a multipole expansion for the group, rather than by the forces of its individual members. This multipole expansion generally changes more slowly than the positions of individual particles, so it need not be reevaluated as often (e.g., Barnes and Hut, 1986; Hernquist, 1987; McMillan and Aarseth, 1993). To put this idea into practice, the system is divided into hierarchical sets of cells: the tree. Its root is the largest cell containing all the particles. This is divided into 2^n subcells until at the nth level, which may vary from place to place, a cell has only one particle. Every particle sees a hierarchy of cells of different sizes s, at distances r, subtending angular diameters $\theta = s/r$. The gravitational force on a given particle from all the particles in a cell at any level n that has an angular diameter smaller than a specified value, $\theta < \theta_c$, is represented by a multipole expansion for that entire cell (ignoring its division into subcells) around its center of mass. Larger values of θ_c require higher order multipole expansions to preserve accuracy. Typically values of θ_c of 0.7 or 1 radian are chosen, and the monopole and quadrupole terms are included (the dipole vanishes because the cell is chosen to reside at its center of mass). Therefore the force on a given particle is the sum of single particle interactions, usually from nearby cells, and multipole expansions. As a result, the numerical integration time is reduced from the order of N^2 to the order of $N \log N$. The catch here is that these codes require considerable memory to keep track of the changing cell structure seen by each particle.

Incessant developments of computer technology have also made direct integrations and tree codes more rapid. One technique is to use thousands of processors in parallel (e.g., Warren et al., 1992). Every processor almost simultaneously contributes part of the computation for each timestep, rather than having one processor working sequentially. Another development is to design very fast chips with hard-wired programs just to compute the gravitational forces (e.g., Makino, Kokubo, and Taiji, 1993). These can work at teraflop speeds, much more nimble than software. They can also be combined into parallel machines of great power. Ten million galaxies in a radius of 100 Mpc is not too much to hope for. It will provide the most exact solution of the cosmological many-body problem in a large enough region of space to represent a fair sample of galaxy clustering.

Meanwhile, there are other techniques to explore problems where high resolution is not so essential. Large-scale linear or quasi-linear growth of clustering is one example; the role of possible intergalactic dark matter is another. For these problems, particle–mesh techniques (see Hockney and Eastwood, 1981; Efstathiou et al., 1985; Couchman, 1991) are fast and convenient. First, a grid of mesh points is constructed in two or three dimensions. Particles are laid down with some initial configuration in the mesh cells. Then a mass weighting function assigns these particles to the mesh point intersections of the cells. The gravitational potential at each intersection is the sum of the densities assigned to all other mesh points multiplied by their Green function. This potential can be calculated on a timescale of order $N \log N$, where N is the number of mesh points, using fast Fourier transforms. Differencing the potential between mesh points gives the gravitational force at each mesh point. Then interpolating these forces between the mesh points gives the accelerations of the original particles and the computation steps forward in time.

Usually the boundary conditions for particle–mesh computations are assumed to be periodic. This does not represent the forces on particles near the boundary quite correctly. Direct integrations, we recall, face the same general problem, but it is not too serious for statistics that exclude particles near the boundary. Modified boundary conditions are possible, assuming extra information, and their effects can be tested. A more important disadvantage of particle–mesh computations is their low resolution on scales of two or three cells.

A natural solution for low resolution is to calculate the forces from each particle's nearest neighbor, within say three mesh cells, directly and then add on the longer range particle-mesh forces. Unsurprisingly, this is called the particle–particle–particle–mesh (PPPM) method. As its resolution is a compromise between direct integrations and particle–mesh computations, so is its running time. Various techniques such as incorporating finer submeshes for the particle–particle part of the calculations can make them run much faster (Couchman, 1991; Bertschinger and Gelb, 1991). Here the challenge, as in computations that incorporate hydrodynamic effects (Kang et al., 1994), is to model the local physical processes accurately. In many of these cases mass points may represent parts of galaxies or elements of fluids. So although the simulations may contain many millions of mass points, they contain far fewer galaxies. They often have the more ambitious goal of trying to account for the internal structure of galaxies along with their clustering.

In addition to variations of direct integration and particle–mesh techniques, there is a third major approach to numerical computations of clustering, particularly for fluids. This is to approximate the development of clustering by a simplified physical model. The technique closest to N-body simulations in spirit is smoothed-particle hydrodynamics (e.g., Evrard, 1988; Hernquist and Katz, 1989). The fluid is replaced by a set of particles obtained from weighted averages of density, pressure, momentum, and energy over all fluid volumes. These particles then interact with the

appropriate averaged pressure gradient and gravitational forces. They can be incorporated into PPPM or tree codes. Another approximation more appropriate for a
collisionless gas, perhaps representing weakly interacting dark matter, truncates the
Euler equation in various ways. The Euler equation (16.35) for the peculiar velocity
$\mathbf{v}(t, \mathbf{x}) = \dot{\mathbf{r}} - H\mathbf{r}$ in a pressure free gas takes the form

$$\frac{\partial \mathbf{v}}{\partial t} + \frac{1}{R}(\mathbf{v} \cdot \nabla)\mathbf{v} + H\mathbf{v} = \frac{1}{R}\nabla\phi, \tag{19.10}$$

where the spatial derivatives are with respect to $\mathbf{x} = \mathbf{r}/R(t)$. Now using $R(t)$ as the
time variable and transforming to a comoving velocity

$$\mathbf{u} = \frac{d\mathbf{x}}{dR} = \frac{1}{R}\frac{\dot{\mathbf{r}}}{\dot{R}} - \frac{\mathbf{r}}{R^2} = \frac{\mathbf{v}}{R\dot{R}} \tag{19.11}$$

gives

$$\frac{\partial \mathbf{u}}{\partial R} + (\mathbf{u} \cdot \nabla)\mathbf{u} = -\frac{3}{2R}\left(\mathbf{u} - \frac{2}{3H^2R^3}\nabla\phi\right). \tag{19.12}$$

This velocity field may be written as the gradient of a velocity potential as long as
the motion is linear and streams of particles do not cross.

The various fluid approximations (discussed and compared in Sathyaprakash et al.,
1995) involve neglecting different terms of (19.12). If the entire right-hand side is
set equal to zero, we obtain the Zeldovich approximation (16.66) as the solution.
This neglects everything except inertia, which is fine in the linear regime. It can
be extended a bit into the nonlinear regime, where thick condensations (pancakes)
form, by artificially filtering out the power on all condensed scales smaller than
the one becoming nonlinear at any particular time (called the truncated Zeldovich
approximation). If instead one replaces the right-hand side of (19.12) by an artificial
viscosity, $\nu\nabla^2\mathbf{u}$, the result is the adhesion approximation. This causes the elements
of fluid to stick together in regions of large velocity gradients. For small ν it reduces
to the Zeldovich approximation elsewhere. It preserves the general nonlinear struc-
ture but does not describe it internally very well. If one just uses $\partial\mathbf{u}/\partial R = 0$ instead
of (19.12), one gets the frozen flow approximation. This keeps the velocity field
always equal to its linear comoving value, ignoring inertia as well as gravitational
forces. It is not very accurate at late times. The frozen potential approximation, in
which the initial value of $\phi = \phi_0(\mathbf{x}, t_0)$ is substituted into (19.12) forever fares
somewhat better, at least in the linear perturbation regime of a flat matter dominated
universe. All these approximations are designed to hasten the integrations at the
expense of accuracy. By comparing several statistics for the distributions resulting
from these approximations with those of N-body computations, Sathyaprakash et al.
(1995) showed that the adhesion approximation generally works best furthest into
the nonlinear regime.

Many examples of computer simulations using these techniques and others have been worked out to test ideas of galaxy formation and clustering (e.g., Bertschinger, 1998). To compare them with observations, one needs to define a galaxy in the simulations. This is easy to do for the direct N-body experiments in which all the dynamically important matter is in galaxies, each point represents a galaxy, and the points may have a range of masses to inject a greater note of realism. The next simplest examples add an inhomogeneous background of dark matter, but the galaxies still dominate the gravitational forces. When the dark matter dominates, it essentially becomes a fluid problem. If the galaxies have already formed (say from pregalactic seeds or cosmic strings) they just go along for the ride. But if the galaxies are simultaneously forming from this fluid as they cluster, it is necessary to introduce a whole new layer of physical theory – more usually assumptions – to determine which fluid elements compose galaxies.

Once the galaxies' positions are known, all the different statistics discussed earlier in this book can be compared with observations. In practice the two-point correlation function has provided the most popular, though hardly the most sensitive, comparison. Other statistics are becoming more widely used. Since the observations of ξ_2 are still the most extensive, we turn to them next in more detail.

20

Simulations and Observations
of Two-Particle Correlations

Hundreds of examples of two-point correlations have been published for simulated and observed systems. Here we will stick to the basics. First we see how point masses, each representing a galaxy, can start with a Poisson distribution and then correlate gravitationally for different initial velocity dispersions in universes with different critical densities Ω_0. Then I summarize some effects of incorporating a range of galaxy masses, and of other initial distributions, and of intergalactic dark matter.

Originally, astronomers hoped that these sorts of results could be compared directly with observations to read the value of Ω_0 off the sky. But it was not to be. Too many ambiguities and combinations of conditions gave similar results for $\xi_2(r)$. Current observations are not yet sensitive enough to distinguish among all views of the past. Here I describe just some modern examples of the observed form of ξ_2 and consider how they may differ for different types of galaxies.

20.1   Simulations

Starting a simulation from a Poisson distribution has the attractive feature of starting with minimum structural information. The initial power spectrum (14.35) has $\hat{n} = 0$ and thus there is equal power on all scales: a democratic beginning. Then we can watch how pure gravitational interaction builds up different structures. This approach also helps isolate the effects of more complex initial conditions and processes.

First, we consider a related set of simulations with $N = 4,000$, values of $\Omega_0 = 1, 0.1,$ and 0.01, and a Poisson initial spatial distribution. This does not completely specify the initial state, for we must also decide the initial velocity distribution. It could be cold, with no initial peculiar velocities relative to the Hubble flow. This is inconsistent with an initial Poisson distribution. However, after one or two initial Hubble expansion times, the gravitational forces, especially from nearest neighbors, will cause the spatial and velocity distributions to relax into a self-consistent total phase space distribution. Or, it could start warm, with an initial Gaussian velocity distribution whose dispersion is $\langle v^2 \rangle^{1/2} = \bar{r} H_i$, where $\bar{r}$ is the average initial distance between galaxies and H_i is the initial Hubble constant. Hot initial velocities with $\langle v^2 \rangle^{1/2} = 3\bar{r} H_i$ give additional insight.

In most numerical experiments, it is important to average any statistics over several simulations that are different microscopic realizations of the same macroscopic conditions. This provides a reasonable appreciation of the effects of "noise." After all, we are mainly interested in general properties of the system, not which particle has what velocity where. Moreover, some systems will occasionally form peculiar structures by chance, and we need to know how representative they are.

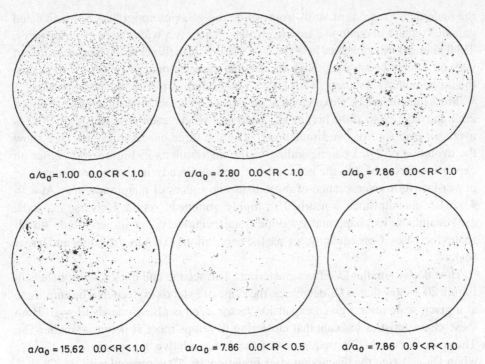

a/a₀ = 1.00 0.0 < R < 1.0   a/a₀ = 2.80 0.0 < R < 1.0   a/a₀ = 7.86 0.0 < R < 1.0

a/a₀ = 15.62 0.0 < R < 1.0   a/a₀ = 7.86 0.0 < R < 0.5   a/a₀ = 7.86 0.9 < R < 1.0

Fig. 20.1. Clustering projected onto the sky at different expansion scales $a(t)/a_0$ in an N-body simulation of 4,000 galaxies starting from a cold Poisson initial distribution for $\Omega_0 = 1$ (Itoh et al., 1988). R is the comoving radial coordinate of the expanding sphere.

Figure 20.1 (from Itoh et al., 1988) illustrates the development of clustering in an initially cold, $\Omega_0 = 1$ model. These are all projected views of about 2,000 galaxies in the same hemisphere as seen by an observer at the center. They are illustrated at representative values of the expansion scale $a(t)$ in terms of its initial value $a_0 \equiv 1$. (Earlier, $a(t)$ was often denoted by $R(t)$, but here we shall use R for the comoving radial coordinate of the simulation.) The maximum comoving radius of the simulation is always scaled to unity. The last two projections, at $a/a_0 = 7.86$, are subsets of the upper right-hand distribution for those galaxies nearby within $0 < R < 0.5$ and those more distant in $0.9 < R < 1.0$, respectively.

Rapidity of clustering is the first striking result. Within one or two expansion times, one sees clear clusters beginning to form. This occurs first on small scales where some near neighbors happen to be closer than average. Locally the gravitational timescale $\sim 1/\sqrt{G\rho}$ is shorter than the global average $\sim 1/\sqrt{G\bar{\rho}}$. If, by chance, these neighbors are moving toward each other in comoving coordinates (so they do not move apart as fast as the universe expands), they will tend to accumulate, perhaps also accreting their more distant companions. A few galaxies are ejected with positive energy, so that the others may fall deeper into their gravitational wells. Gradually small groups, then larger groups, condense into great clusters. Around these clusters are the underdense regions or voids from which the galaxies emigrated. Some isolated galaxies remain in equipoise among clusters. There is a wide range of group sizes. Tidal interactions of clusters help produce filaments. In this case the filaments are not very prominent because although there is the same amount of power on all scales, the initial number fluctuations $\delta N/N = 1/\sqrt{N}$ diminish with increasing scale.

More quantitatively, we can look at the two-point correlation functions in Figures 20.2 and 20.3 (also from Itoh et al., 1988 and where each case in Figure 20.2 is averaged over four or five simulations). The top three panels in Figure 20.2 show the evolution of $\xi_2(r)$ corresponding to the distributions in Figure 20.1. After an expansion factor of 2.8, the correlation function already has an approximate r^{-2} power-law form over a range of more than two orders of magnitude in r. At $r \approx 4 \times 10^{-3}$ its amplitude is nearly 10^4, highly nonlinear. As the Universe expands, $\xi(r)$ maintains its shape and its slope quite accurately, but its amplitude on all scales increases. Consequently, we see the correlations extending to larger and larger scales.

How does a smaller Ω_0 affect clustering? The second and third rows of panels in Figure 20.2 show that as Ω_0 decreases, the slope of ξ_2 tends to steepen. The amplitude at a given scale after a given expansion factor a/a_0 is also somewhat lower. Both these characteristics indicate that clustering develops more slowly for smaller Ω_0. These are reasonable consequences of the faster relative expansion for $z \lesssim \Omega_0^{-1}$ when $\Omega_0 < 1$ (see the discussion after Equation 16.20) compared with the $\Omega_0 = 1$ case. In other words, to reach a particular state of higher clustering in an $\Omega_0 < 1$ universe requires a greater expansion factor than in an $\Omega_0 = 1$ universe. Therefore to get a similar degree of clustering in a less dense universe, the clustering must begin at a higher redshift in these models. One might conclude that galaxy formation had to start sooner in a lower density universe. But that presumes identical initial fluctuations, so there is a loophole. For the Poisson case, these simulations show that to attain the present observed slope and amplitude of ξ_2 requires that galaxies start clustering at $z \approx 5$–10 for $\Omega_0 \approx 1$ and $z \approx 25$–35 for $\Omega_0 \approx 0.1$.

How does the initial velocity dispersion affect clustering? Figure 20.3 (which has one representative simulation for each case) shows that as the initial velocities increase, it takes longer for clustering to develop. This is natural because the ratio of gravitational interaction energy to thermal energy decreases. In fact, for the hot, low-density model, correlations do not grow much until adiabatic expansion has cooled the peculiar velocities. Even so, the approximately r^{-2} power-law spectrum eventually takes over for these cases also.

How does a more realistic range of galaxy masses affect clustering? To a first approximation, not very much. This is because the relaxation process only depends strongly on mass initially when near neighbors interact nonlinearly to form the first clusters. Once these clusters begin to interact with one another, and with remaining unclustered galaxies, the masses of individual galaxies are not so important. Galileo would have recognized this phenomenon. Nonetheless, there are noticeable residual effects of a mass spectrum. For example, in a region with most of the mass in a

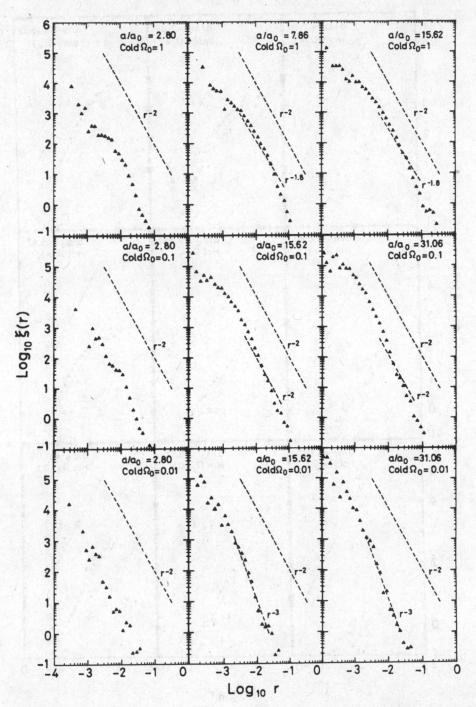

Fig. 20.2. Evolution of the two-point correlation function for the simulation in Fig. 20.1 and similar simulations for $\Omega_0 = 0.1$ and $\Omega_0 = 0.01$ (Itoh et al., 1988).

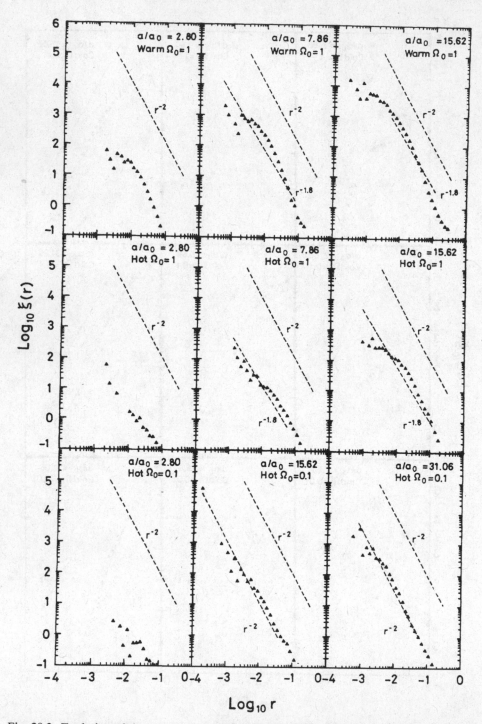

Fig. 20.3. Evolution of the two-point correlation function for simulations similar to those in Fig. 20.1 but with warm and hot initial conditions for $\Omega_0 = 1$ and $\Omega_0 = 0.1$ (Itoh et al., 1988).

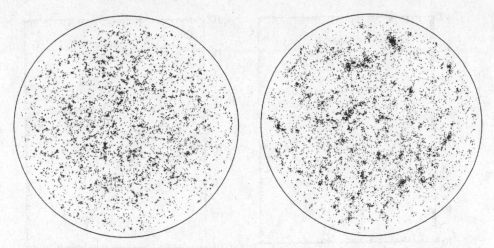

Fig. 20.4. Projected clustering in 10,000 galaxy experiments for $\Omega_0 = 1$ and a cold initial Poisson distribution. On the left all galaxies have the same mass; on the right the galaxies have the gamma function mass distribution of Equation 20.1 (Itoh et al., 1993).

few massive galaxies, and the rest in low-mass galaxies, these low-mass galaxies will tend to form satellite systems around more massive ones. The massive galaxies cluster along with their camp followers, and the whole distribution is more irregular and less homogeneous than the single-mass case.

Figure 20.4 shows a representative example from Itoh et al. (1993). It is a 10,000-body experiment, again with $\Omega_0 = 1$ and cold Poisson initial conditions. The expansion factor $a/a_0 = 8$, which is where the slope and amplitude of $\xi_2(r)$ are consistent with the values now observed. On the left, for comparison, is a single-mass model. Its projection greatly resembles the similar example in Figure 20.1, showing that an increase of N from 4,000 to 10,000 does not introduce any qualitative new features. On the right is an experiment whose only change is to incorporate a spectrum of masses

$$dN(m) \propto \left(\frac{m}{m_*}\right)^{-p} \exp\left(-\frac{m}{m_*}\right) d\left(\frac{m}{m_*}\right). \tag{20.1}$$

This is closely related to the luminosity function of galaxies (Schechter, 1976) provided their mass/luminosity ratio is constant. Parameters of (20.1) in the simulation of Figure 20.4 are $p = 1.25$, with a range $m_{\min}/m_* = 0.01$ to $m_{\max}/m_* = 10$. The least massive 20% of all these galaxies have only 1.7% of the total mass, while the most massive 20% contain 74% of the total mass.

A range of masses enhances clustering. Not only do clusters become denser and more extended, they also tend to join up into longer filaments. Galaxies of greater mass cluster more strongly, and clusters generally contain galaxies with a wide range of mass. Often more massive galaxies form the nuclei of these clusters.

The two-point correlation functions in Figure 20.5 for the same model bear out these visual impressions. Correlations of the most massive 20% grow very quickly.

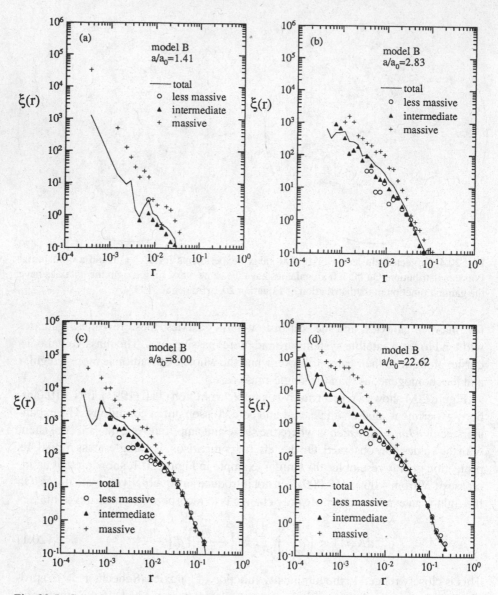

Fig. 20.5. Growth of the two-point correlation function for different mass ranges in the simulation of Fig. 20.4 (Itoh et al., 1993).

Then those of intermediate mass join in the clustering, followed by low-mass galaxies. The low-mass galaxies do not have enough mutual gravity to cluster on small scales, and so their $\xi_2(r)$ does not extend to $r \lesssim 10^{-3}$ over the time of these simulations. On large scales, galaxies of all masses have nearly the same $\xi_2(r)$ for $a/a_0 \gtrsim 3$. Many of these are field galaxies outside large clusters, so their mass is irrelevant when responding to the clusters, as Galileo would have predicted.

Throughout the whole evolution, the most massive and potentially most luminous galaxies remain the most strongly correlated.

How do non-Poisson initial conditions influence clustering? Naturally there is an infinite variety of possible initial conditions. A great deal of work has been done trying to connect these initial conditions with possible properties of the early universe. The results, so far, are highly uncertain. We can gain general insight from simple models where the initial density fluctuations have a power-law spectrum

$$|\delta_k|^2 \propto k^n, \tag{20.2}$$

where n has a typical value of $1, 0, -1$, or -2 for each different model. For $n < 0$, the initial fluctuations have more power at longer wavelengths. At small wavelengths there is hardly any power at all. Clearly this situation will not be realistic, so sometimes limits are introduced to such spectra at both small and large wavelengths, or it is otherwise modified. For $n > 0$, the fluctuations are largest on small scales and die away over distance.

Representative N-body simulations (Itoh, 1990) for galaxies with all the same mass in an $\Omega = 1$ cosmology, with $N = 4,000$, and values of $n = 1, 0, -1, -2$ illustrate the results. As the value of n decreases, fewer but larger clusters form. These larger clusters are extended into more obvious filamentary structures on greater scales. On these greater scales where the evolution is linear, remains of the initial fluctuations can clearly be seen. On smaller scales, where the evolution is relaxed, the initial conditions are mostly washed out. The division between these regimes depends upon n and the cosmic expansion.

Two-point correlation functions in Itoh's examples have a slope of about -2 over the range $\epsilon \lesssim r \lesssim 0.05$ (ϵ is the softening parameter), which covers about an order of magnitude for $n = 1$ at $a/a_0 \approx 20$ and for $n = 0$ at $a/a_0 \approx 10$. At longer scales, $\xi_2(r)$ drops more steeply, roughly as r^{-4} and r^{-3} for these two cases respectively. As a/a_0 increases, this structure remains similar with an increasing amplitude and a slope that settles down near $\xi(r) \propto r^{-2}$ in the strongly nonlinear regime. Models with $n = -1$ or $n = -2$ behave a bit differently. Here the larger initial fluctuations on larger scales cause the clustering to relax more rapidly on those scales. When these scales become nonlinear, $\xi_2(r)$ again becomes approximately proportional to r^{-2}. At the same time, the supression of initial fluctuations on small scales causes them to relax more slowly; hence their resulting amplitude is lower and their slope is less steep when n is more negative. These results are qualitatively intuitive and can be extended to more complicated initial conditions.

How does dark matter change the picture? There are two aspects: the initial amount of the dark matter and its spatial relation to the galaxies. We now have a two-component system, which is naturally more complicated. If dark matter dominates the total mass of the universe, but is hot and nearly uniformly distributed, it will not affect the local clustering significantly except by determining the global expansion $a(t)$. If the dark matter is cooler, it may itself cluster at earlier times and influence the locations of galaxy formation. Cold dark matter will cluster more strongly, usually in a complicated way. Previously fashionable "standard" cold dark matter models suggest a power spectrum for dominant dark matter with $n = -3$ at small scales and $n = 1$ at large scales and a smooth transition between these limits. The $n = 1$ large-scale limit corresponds to mass fluctuations independent of mass when each mass scale enters the horizon, which may appeal philosophically to some cosmologists. Since the galaxies are essentially test particles in dark matter dominated models, their relation to the dark matter depends on a bias parameter or function, as discussed in Chapter 18.

Representative simulations for hot (Sathyaprakash et al., 1995) and cold (Weinberg and Gunn, 1990; Park, 1990) dark matter, using particle–mesh, adhesion, or other codes described in Chapter 18, also give power-law forms for $\xi_2(r)$ with the usual exponents around -1.8 or -2 for suitable bias. The clustering of the dark matter on large scales leads to stronger extended filaments and underdense regions. In real and in redshift space a great variety of configurations are produced, some of which resemble observations quite closely. Detailed statistical comparisons with observations are not yet conclusive, even for $\xi_2(r)$, which has been the most extensively studied. This has led to other modifications such as cold dark matter with different power spectra, and hybrid models that combine both hot and cold dark matter, also adding more complicated biasing and a cosmological constant. The end of this story is not yet in sight.

All these complications to basic gravitational clustering – the value of Ω_0, the initial velocity distribution, a spectrum of masses, non-Poisson initial conditions, various forms of dark matter and bias, and perhaps others not yet explored – may contribute to the observed correlation function to which we now turn.

20.2   Observations

The early results of Totsuji and Kihara (1969), Peebles (1980) and his collaborators, and many others have generally held up quite well as more extensive and precise galaxy catalogs extended and refined the observed $\xi_2(r)$. An example is the survey (Maddox et al., 1990) using the Cambridge Automatic Plate Measuring (APM) system to examine about two million galaxy positions on 185 U.K. Schmidt plates. These cover a contiguous area of 4,300 square degrees and include images with J magnitudes $b_J \leq 20.5$. Problems such as the completeness of the sample, merging of images, galactic obscuration, and variations in the magnitude standards from plate to plate crop up at the level of a few percent. They are generally larger at fainter magnitudes but can be corrected to some extent.

Two estimators of the angular correlation function, $W(\theta) = A\theta^{1-\gamma}$, essentially those of (14.10) and (14.45), give a very good fit to a power law for $W(\theta)$. Figure 14.1 (from Maddox et al. (1996), who discuss its analysis in detail) gives a slope $1 - \gamma = -0.7 \pm 0.03$ and amplitude $A = 2.8 \times 10^{-2}$ for the range $0.01° \leq \theta \leq 1°$. There is a break with this power law at about $3°$ and $W(\theta)$ decreases faster to essentially zero at around $6°$. This is for galaxies in the magnitude range $17 \leq b_J \leq 20$. A test of the geometric scaling relation (14.38), using an $\Omega_0 = 1$ (i.e., $q_0 = 0.5$), Einstein–de Sitter cosmology and a model for the evolution of the galaxy luminosity function shows that this scaling works quite nicely for different magnitude ranges of $W(\theta)$. The corresponding slope and amplitude for this APM three-dimensional spatial correlation function (see the discussion after Equation 14.37) would be $\gamma = 1.7$ and $r_0 \approx 4.7\,h^{-1}$ Mpc (Maddox et al., 1990, 1996).

Redshift measurements for large numbers of galaxies provide more direct estimates of the three-dimensional spatial correlation functions than using the luminosity function to obtain $\xi_2(r)$ from $W(\theta)$ (Limber, 1954). These distance estimates would be even better if we had reliable secondary distance indicators that could remove the effects of peculiar velocities. Meanwhile several methods, partially incorporating distances, appear to give reasonable results.

A good illustration (Saunders, Rowan-Robinson, and Lawrence 1992) is the cross-correlation between galaxies in the *Infrared Astronomical Satellite* (*IRAS*) catalog and a subset of these with measured redshifts. The *IRAS* catalog is particularly useful because it is sampled quite uniformly over a large part of the sky with a known selection function. This statistical cross-correlation method works best if the distances to the galaxies are generally much greater than their clustering scale, so that galaxies projected onto the sky close to a galaxy with measured redshift are also likely to be close to it along the line of sight. Similarly, it works best if all the angular separations of pairs of galaxies are small. One still needs to know the selection function, which may depend on distance, for galaxies in the catalog, and this requires an estimate of their luminosity function. For about 13,000 *IRAS* sources having 60 micron fluxes greater than 0.6 Jansky, of which about 2,100 had measured redshifts, Saunders et al. (1992) found a very close fit to a power law

$$\xi_2(r) = \left(\frac{r}{r_0}\right)^{-\gamma}. \tag{20.3}$$

The parameter values are $r_0 = 3.79 \pm 0.14\,h^{-1}$ Mpc and $\gamma = 1.57 \pm 0.03$ for the range $0.1\,h^{-1} \le r \le 30\,h^{-1}$ Mpc. This range is greater, and the slope somewhat shallower, than for all the APM sample. Whether this represents a physical property of the *IRAS* sample, which is dominated by spiral galaxies (typically less strongly clustered than ellipticals), or whether it represents other selection effects is uncertain.

Redshifts have been measured for about 5,300 of the brighter *IRAS* galaxies with fluxes greater than 1.2 Jansky to estimate ξ_2 in redshift space as well as in real space (Fisher et al., 1994). Like all magnitude-limited samples, these are subject to distance-dependent selection effects, even if the galaxy sample in a given volume is complete to the absolute magnitude limit appropriate for the redshift distance of that volume. To some extent this can be compensated by giving less weight to the more distant relatively undersampled volumes. Different weighting schemes modify the results somewhat but do not appear to be critical. This sample also shows the redshift space distortions along the line of sight caused by nonlinear clustering on small scales and transverse to the line of sight caused by linear infall on large scales (see Section 18.2). In redshift space the resulting ξ_2 has a power-law form with a slope of about -1.3 and amplitude $\sim 4.5\ h^{-1}$ Mpc. The corresponding real space ξ_2 has a slope of about -1.7 and amplitude $\sim 3.8\ h^{-1}$ Mpc. These numbers have estimated uncertainties of about 0.1 to 0.2, representing one standard deviation for the effects of selection, sampling, and weighting. Thus these results are compatible with those of Tutsuji and Kihara, the APM, the earlier *IRAS* cross-correlation, and most others. They illustrate the shallower slope in redshift space found also in numerical simulations. Since this survey is restricted to fewer, brighter *IRAS* galaxies, the power-law range of $\xi_2(r)$ without dominant noise is about $1 \lesssim r \lesssim 20\ h^{-1}$ Mpc, somewhat less than in the cross-correlation analysis.

All these observations suggest greater clustering on large scales than predicted by numerical simulations of a variety of highly specific cold dark matter (CDM) models. They have therefore been used to rule out these CDM models, including the one with an initial power-law perturbation spectrum in (20.2) of $n = -3$ on small scales and $n = 1$ on large scales, which used to be called the "the standard model." However, modified models rise to take their place and be tested against $\xi_2(r)$ as well as more informative statistics.

Observational progress is also being made using $\xi_2(r)$ to distinguish different degrees of clustering for different types of galaxies. This requires large amounts of data to explore many categories with low noise. If the noise level in the data is high, then so is the noise level in the published literature. This should not detract from the pioneering spirit that impels astronomers to reach preliminary conclusions. Like the pioneering towns of the American West, a certain rowdiness prevails until the population settles down and law and order take over.

So, at present, there are pioneering discussions over whether $\xi_2(r)$ depends significantly on the luminosity or surface brightness of the galaxies and on whether it shows evolutionary effects up to moderate ($z \approx 0.7$) redshifts. Some evidence (Benoist et al., 1996) from about 3,600 galaxies with measured redshifts in the Southern Sky Redshift Survey suggests that for galaxies more luminous than about $M = -19.5$ the brighter galaxies cluster more strongly. This is based on measuring $\xi_2(r)$ directly in volume-limited samples in redshift space. Because there are no major clusters in this sample the velocity distortion may be small. Another example (Mo, McGaugh, and Bothun, 1994) used the less direct technique of cross-correlating low surface brightness galaxies with about 2,400 other galaxies having redshifts in the CfA survey and with about 2,700 *IRAS* galaxies having redshifts. Their results gave the same slope ≈ -1.7 for the cross-correlation $\xi_2(r)$, but with a lower amplitude than for the self-correlation functions. Other examples (referenced in these last two papers) have not revealed this differential clustering. This may be because the samples were too small, or too varied, or the data were too overcorrected, or the statistics were too insensitive. Or, perhaps, the effect is too small.

My prediction is that differential clustering is real and will emerge increasingly clearly with better and more extensive data and more sensitive statistics. The reason is that lower luminosity generally implies lower mass. Whatever other effects may be present, gravitational interactions are certainly present. The N-body simulations described earlier in this chapter, and related gravitational theory, show that gravity produces differential clustering dependent on mass, though it may be a secondary effect (and therefore not easy to detect unambiguously).

There are also environmental effects that might contribute to differential clustering. Less luminous galaxies may have formed preferentially in lower density regions. This could reduce their clustering on intermediate scales. Or many of them may have merged, reducing their apparent clustering on small scales. Or those in high-density regions might have been disrupted by intergalactic material or become more prone to starburst disruption. "Many a little makes a mickle" as the old proverb says, and accurate data with many statistics are needed to distinguish among all these possibilities.

Evolution of $\xi_2(r)$ with redshift is also just beginning to be understood. From a survey of 1,100 galaxies with magnitudes $17 < B < 22$ in 33 long thin volumes with $z \lesssim 0.5$ and median redshift $z = 0.16$, Cole et al. (1994) found that $\xi_2(r)$ again had a slope of -1.7 and amplitude $r_0 = 6.5 \pm 0.4 \, h^{-1}$ Mpc. There was no clear evidence for evolution of the comoving $\xi_2(r)$ over this redshift range. This is consistent with the gravitational clustering simulations described earlier, since their rate of clustering generally slows down considerably for $z \lesssim 1$. However, merging or starburst self-destruction can still dominate subpopulations of galaxies during this epoch without affecting the total $\xi_2(r)$ substantially.

Despite small uncertainties and possible different behavior of subsets of galaxies, there is a remarkable robustness in the form of $\xi_2(r)$, particularly in the well-measured nonlinear regime. This suggests that a single physical process dominates clustering, even to the extent of washing out most initial conditions on small and intermediate scales. The best candidate for this physical process is gravity.

Future observations will check this view and perhaps find more direct evidence for initial conditions on larger scales. Surveys are in hand to measure millions of redshifts. A small fraction of those will be for galaxies with improved secondary distance indicators. Clearly it is the case that "more is better," but it is not so obvious that "much more is much better." The samples will have to be chosen wisely to answer the most important questions.

PART IV

Gravity and Distribution Functions

Number counts of galaxies in cells on the sky or in space, near neighbor statistics, and distributions of peculiar velocities all enhance our understanding of how galaxies cluster. Even though they contain more information than low-order correlations, we can extend these distributions into highly nonlinear regimes where gravity dominates. General physical principles, rather than detailed orbital dynamics or models, simplify and guide these extensions.

After reviewing some basic mathematical properties of distribution functions, we examine how dynamics describes their linear evolution. As the evolution becomes more and more nonlinear, however, the dynamical arguments give out, just as they did for correlation functions. Mathematical pertubation theory becomes more intractable; its returns diminish. Nonlinearity, spreading from smaller to larger scales, destroys memories of the initial state. For many initial conditions, nonlinear evolution can lead to quasi-equilibrium. Somewhat unexpectedly, perhaps, this is amenable to a statistical thermodynamic description. Later we will compare the predicted distributions with detailed simulations and observations.

21

General Properties of Distribution Functions

It Don't Mean a Thing.
If It Ain't Got That Swing.
Duke Ellington

Here I collect some of the basic properties of distribution functions for use in later chapters. They may be found in much greater detail and rigor in most probability or statistics texts, of which Feller (1957), Kendall and Stuart (1977), and Moran (1968) are classic examples.

21.1   Discrete and Continuous Distributions

Distributions come in two basic categories: discrete and continuous. Discrete distributions occur when the outcome of an event, experiment, or observation is a member of a finite, enumerable set of possibilities. Often these possibilities are positive integers, as for counts of galaxies in cells. If many events are observed, we denote the relative probability of finding a particular outcome x_i by $f(x_i)$ and call it the distribution function. This is the terminology usually used by physicists, although it is often called the "probability mass function" by probabilists and statisticians (who call it a "probability density function" in the continuous case and refer to its integral between $-\infty$ and x as the cumulative distribution function, or sometimes just as the distribution function). This distribution function is obviously real and non negative. It covers all possible outcomes of a particular measurement, so its sum over all these possibilities is normalized to unity:

$$\sum_i f(x_i) = 1 \qquad (21.1)$$

with $0 \leq f(x_i) \leq 1$ for any x_i. If the enumerable events x_i are mutually exclusive then

$$f\left(\sum_i x_i\right) = \sum_i f(x_i). \qquad (21.2)$$

Whether or not N events in a set are mutually exclusive, they satisfy Boole's inequality

$$f\left(\sum_{i=0}^N x_i\right) \leq \sum_{i=0}^N f(x_i). \qquad (21.3)$$

If the probabilities of N events are independent, then their joint probability is

$$f(x_1 x_2 \ldots x_N) = f(x_i) f(x_2) \ldots f(x_N) \tag{21.4}$$

as in (17.28). Any subset of a set of independent events is obviously also independent. However, if all pairs of members of a set are independent, it is not generally true that all the members are themselves independent. Higher order correlations may still occur.

Conditional distribution functions are often useful, especially when selection effects are present in a sample. Suppose we are given that an event x_1, say, occurred and we ask for the conditional probability $f(x_2 \mid x_1)$ that another event x_2 also occurs. This is just the relative frequency of the two events; so

$$f(x_2 \mid x_1) = \frac{f(x_1 x_2)}{f(x_1)}. \tag{21.5}$$

Generalizing this we see that

$$f(x_3 \mid x_1 x_2) = \frac{f(x_1 x_2 x_3)}{f(x_1 x_2)} = \frac{f(x_1 x_2 x_3)}{f(x_2 \mid x_1) f(x_1)} \tag{21.6}$$

and so on. Two events are independent if either $f(x_2 \mid x_1) = f(x_2)$ or $f(x_1) = 0$.

For a set of mutually exclusive events y_j that cover the same set as the x_i, we have the normalization

$$\sum_j f(y_j \mid x_1) = 1 \tag{21.7}$$

since this is equivalent to (21.1). Also,

$$f(x) = \sum_j f(y_j) f(x \mid y_j) \tag{21.8}$$

setting $f(y_j) f(x \mid y_j) = 0$ for the case $f(y_j) = 0$. Combining (21.5) and (21.8) for $f(x) > 0$ shows that

$$f(y_j \mid x) = \frac{f(xy_j)}{f(x)} = \frac{f(y_j) f(x \mid y_j)}{\sum_j f(y_j) f(x \mid y_j)}, \tag{21.9}$$

which is known as Bayes's theorem.

Applied to conditional distributions, Bayes's theorem is quite innocuous. It becomes controversial when used by statisticians to infer whether a statistical model fits certain data. For that purpose $f(y_j)$ is the prior probability of model y_j, the conditional probability $f(x \mid y_j)$ is called the likelihood of finding the result x given the model y_j, and $f(y_j \mid x)$ is the a posteriori probability that the model y_j is correct, given that the result x was found. (If the possible outcomes are represented by a continuous set of values of some parameter y and the distributions are continuous, just replace y_j in (21.9) by y and the sum by an integral over dy.) The controversy arises in assigning meaning, exclusivity, and actual values to the prior probability $f(y_i)$. One approach is Bayes's postulate that all prior probabilities are equal if we have no knowledge otherwise. In physical problems this runs into the difficulty that models might not be specified completely; they could contain physical constraints that are not built into the prior probabilities. In practice, astronomers' belief in a model seldom depends on Bayesian formalism, although it could encourage consideration of wider ranges of a priori models.

Continuous distributions often arise when the outcome of an event can be any number. Then sums such as (21.1) are replaced by integrals:

$$\int_{-\infty}^{\infty} f(x)\, dx = 1, \tag{21.10}$$

with the continuous distribution function satisfying $0 \le f(x) \le 1$. (Statisticians sometimes call this the probability density.) The cumulative distribution function is

$$F(x) = \int_{-\infty}^{x} f(y)\, dy. \tag{21.11}$$

All the previous discrete formulae are modified in the obvious way for continuous distributions.

Neither discreteness nor continuity is immutable. A discrete distribution can be approximated by a continuous one by smoothing it over a window with a continuous weighting function. And a continuous distribution can be discretized by representing its arguments by their nearest integers. Moreover, a distribution can be continuous over one range and discrete over another.

21.2 Expectations, Moments, and Cumulants

If $z(x_i)$ is a discrete function of the x_i, then its expectation value (or average) is

$$E(z) = \sum_i z(x_i) f(x_i), \tag{21.12}$$

which is also sometimes denoted μ, $\bar{z}$, or $\langle z \rangle$ depending on the states (e.g., spatial or temporal) over which the average is taken. For continuous functions the sum is again replaced by the integral over x. Clearly

$$E(z(x) + cu(x)) = E(z) + cE(u) \tag{21.13}$$

for two functions of random variables z and u and a constant c. In general $E(zu) \ne E(z)E(u)$, however.

When $z(x) = x^r$ for any real value of r we obtain the rth moment of $f(x)$ about zero from (21.12),

$$\mu'_r = E(x^r), \tag{21.14}$$

as long as the sum is finite. This gives the moments of the distribution function in which $f(x)$ is weighted by various powers of x. Large values of r emphasize the properties of $f(x)$ at large x. This can be extended to more variables (e.g., $\mu'_{rs} = E(x^r y^s)$ for $f = f(x, y)$), such as when the distribution function depends on both position and velocity.

Moments may also be taken about any other constant value of x to obtain $E[(x - c)^r]$. In particular when c is the mean this gives the rth central moment, also known as the rth moment about the mean (written without the prime):

$$\mu_r = E[(x - \mu)^r] = E[(x - \bar{x})^r]. \tag{21.15}$$

Thus μ_1 is zero. The second central moment is the variance, $\mu_2 = \text{var}(x) \equiv \sigma_x^2$, and its positive square root is the standard deviation σ_x. It is one partial measure of typical departures from the mean, and the dimensionless form σ_x/μ is known as the coefficient of variation. Ratios of higher order moments such as the skewness

$$\alpha_3 = \mu_3 (\mu_2)^{-3/2}, \tag{21.16}$$

the kurtosis

$$\alpha_4 = \mu_4(\mu_2)^{-2}, \tag{21.17}$$

and in general

$$\alpha_r = \mu_r(\mu_2)^{-r/2} \tag{21.18}$$

are also partial characterizations of distributions. For linear functions $a + bx$ having $b > 0$, all the α_r are the same (and $\alpha_{2n+1} = -\alpha_{2n+3}$ if $b < 0$). It is important to realize that the low order α_r do not generally characterize interesting functions very well. With a little effort the reader can find functions that look quite different but have very similar mean, skewness, and kurtosis.

We shall also find factorial moments to be useful, particularly the rth descending factorial moment

$$\mu'_{[r]} = E[x!/(x - r)!]. \tag{21.19}$$

These turn up, for example, in the expansions of exponential functions. Other types of moments, such as the moments of the absolute values $|x - \mu|^r$, occur occasionally and can be introduced when needed.

All these moments are related, not unexpectedly. Sometimes it is easier to calculate one type of moment and convert it into another than to calculate the desired moment directly. Standard formulae relating low order moments around zero, μ'_r, to the moments μ_r around the mean, $\mu'_1 \equiv \mu$, are

$$\mu_2 = \mu'_2 - \mu^2, \tag{21.20}$$

$$\mu_3 = \mu'_3 - 3\mu'_2\mu + 2\mu^3, \tag{21.21}$$

$$\mu_4 = \mu'_4 - 4\mu'_3\mu + 6\mu'_2\mu^2 - 3\mu^4. \tag{21.22}$$

The inverse of these relations is

$$\mu'_2 = \mu_2 + \mu^2, \tag{21.23}$$

$$\mu'_3 = \mu_3 + 3\mu_2\mu + \mu^3, \tag{21.24}$$

$$\mu'_4 = \mu_4 + 4\mu_3\mu + 6\mu_2\mu^2 + \mu^4. \tag{21.25}$$

Factorial moments are related to the moments centered around zero by

$$\mu'_{[1]} = \mu, \tag{21.26}$$

$$\mu'_{[2]} = \mu'_2 - \mu, \tag{21.27}$$

$$\mu'_{[3]} = \mu'_3 - 3\mu'_2 + 2\mu, \tag{21.28}$$

$$\mu'_{[4]} = \mu'_4 - 6\mu'_3 + 11\mu'_2 - 6\mu \tag{21.29}$$

and

$$\mu = \mu'_{[1]}, \tag{21.30}$$

$$\mu'_2 = \mu'_{[2]} + \mu, \tag{21.31}$$

$$\mu'_3 = \mu'_{[3]} + 3\mu'_{[2]} + \mu, \tag{21.32}$$

$$\mu'_4 = \mu'_{[4]} + 6\mu'_{[3]} + 7\mu'_{[2]} + \mu. \tag{21.33}$$

All these relations follow directly from the basic definitions, although expanding the factorials in terms of Stirling numbers facilitates their derivation.

Cumulants offer another way to describe distributions. They have some advantages: The cumulant of a sum of independent random variables equals the sum of the cumulants, and the cumulants κ_r are invariant under a translation of the origin, except for κ_1. The cumulants are defined formally by the identity in the dummy variable t for moments μ'_r around zero (or any arbitrary value)

$$\exp\left\{\kappa_1 t + \frac{\kappa_2 t^2}{2!} + \cdots + \frac{\kappa_r t^r}{r!} + \cdots\right\} = 1 + \mu'_1 t + \frac{\mu'_2 t^2}{2!} + \cdots + \frac{\mu'_r t^r}{r!} + \cdots. \tag{21.44}$$

Equating coefficients of the same powers of t on both sides gives

$$\kappa_1 = \mu'_1 \equiv \mu, \tag{21.45}$$

$$\kappa_2 = \mu'_2 - \mu'^2_1, \tag{21.46}$$

$$\kappa_3 = \mu'_3 - 3\mu'_2\mu'_1 + 2\mu'^3_1 \tag{21.47}$$

and the inverse formulae

$$\mu_1' = \kappa_1, \tag{21.48}$$
$$\mu_2' = \kappa_2 + \kappa_1^2, \tag{21.49}$$
$$\mu_3' = \kappa_3 + 3\kappa_2\kappa_1 + \kappa_1^3. \tag{21.50}$$

For moments around the mean, $\kappa_1 = 0$ and

$$\kappa_2 = \mu_2, \tag{21.51}$$
$$\kappa_3 = \mu_3, \tag{21.52}$$
$$\kappa_4 = \mu_4 - 3\mu_2^2, \tag{21.53}$$

etc.

21.3   Generating and Characteristic Functions

All these results can be represented more compactly, simply, and often usefully by generating functions (originally discussed by Laplace in his *Theorie Analytique des Probabilités*). This device enables us to explore the properties of distribution functions with minimal algebraic fuss. Writing $f(N)$ for the probability that $x_i = N$ in $f(x_i)$, the generating function for the distribution of $f(N)$ is

$$g(s) = \sum_{N=0}^{\infty} f(N)s^N. \tag{21.54}$$

This works most naturally when N is a nonnegative integer, as will be the case for counts of galaxies in cells; hence $f(N) = 0$ if $N < 0$. Then $g(s = 1) = 1$ from (21.1) and the series always converges for $|s| \le 1$. Actually the convergence of (21.54) is not always critical since s is an indicator variable whose value s^N just serves to keep track of the probability $f(N)$. If $f(N)$ decreases sufficiently fast as $N \to \infty$, the series may converge in a region even if $s > 1$. When the summation has a closed form for $g(s)$, it is often easy to manipulate algebraically to determine many properties of $f(N)$.

To retrieve $f(N)$ from the generating function we just calculate the Nth derivative evaluated at $s = 0$:

$$\frac{1}{N!} \frac{d^N g(s)}{ds^N}\bigg|_{s=0} = f(N). \tag{21.55}$$

We may also think of $g(s)$ as the expectation value of a function s^x where $x = 0, 1, 2, 3 \ldots$.

A related generating function gives the tail of the distribution, that is, the probability

$$q(N) = f(N+1) + f(N+2) + \cdots \tag{21.56}$$

that $x_i > N$. This generating function is just the sum

$$Q(s) = q(0) + q(1)s + q(2)s^2 + \cdots = \frac{1 - g(s)}{1 - s}. \tag{21.57}$$

This last equality follows for $|s| < 1$ by noting that $(1 - s)Q(s) = 1 - g(s)$, and convergence is assured since $q(N) \leq 1$. Similarly the probabilities that $x_i \leq N$ are generated by

$$f(0) + (f(0) + f(1))s + (f(0) + f(1) + f(2))s^2 + \cdots = \frac{g(s)}{1-s} = 1 - Q(s). \tag{21.58}$$

From the generating functions flow the moments. First, for the expectation value,

$$E(N) = \mu = \sum_{N=1}^{\infty} N f(N) = \sum_{N=1}^{\infty} N f(N) s^{N-1} \Big|_{s=1} = g'(s) \Big|_{s=1} = g'(1). \tag{21.59}$$

The second factorial moment is

$$\mu'_{[2]} = \sum_{N=1}^{\infty} N(N-1) f(N) = g''(1), \tag{21.60}$$

and in general

$$\mu'_{[r]} = g^{(r)}(1), \tag{21.61}$$

where $g^{(r)}(1)$ is the rth derivative with s evaluated at 1. Thus (21.59) and (21.60) directly give the second moment about the origin,

$$\mu'_2 = \sum_{N=1}^{\infty} N^2 f(N) = g''(1) + g'(1). \tag{21.62}$$

From this we immediately get the variance

$$\mathrm{var}(N) = \sum_{0}^{\infty} (N - \mu)^2 f(N)$$
$$= \langle N^2 \rangle - \langle N \rangle^2 = \mu'_2 - \mu^2 = g''(1) + g'(1) - [g'(1)]^2. \tag{21.63}$$

We will see that this is particularly useful for calculating the variance of distribution functions that are modified by selection effects.

To turn (21.54) into a generating function for all the moments $\mu'_r = \sum N^r f(N)$ about the origin, we substitute $s = e^t$ and expand in powers of t to get

$$g(e^t) = \sum_{N=0}^{\infty} f(N)e^{Nt}$$

$$= \sum_{r=0}^{\infty} t^r \sum_{N=0}^{\infty} N^r f(N)/r!$$

$$= \sum_{r=0}^{\infty} \frac{\mu'_r}{r!} t^r. \tag{21.64}$$

This is the moment generating function since

$$\mu'_r = g^{(r)}(e^t)|_{t=0}. \tag{21.65}$$

The generating function for moments around the mean is

$$\sum_{r=0}^{\infty} \frac{\mu_r}{r!} t^r = \sum_{r=0}^{\infty} \frac{t^r}{r!} \sum_{N=0}^{\infty} (N-\mu)^r f(N)$$

$$= \sum_{N=0}^{\infty} f(N)e^{t(N-\mu)}$$

$$= e^{-\mu t} g(e^t). \tag{21.66}$$

Similarly, the cumulant generating function is

$$\ln g(e^t) = \sum_{N=0}^{\infty} \frac{\kappa_N}{N!} t^N. \tag{21.67}$$

The factorial moment generating function is

$$g(1+t) = \sum_{N=0}^{\infty} f(N)(1+t)^N$$

$$= \sum_{r=0}^{\infty} \frac{\mu'_{[r]}}{r!} t^r, \tag{21.68}$$

consistent with (21.62). From these generating functions, one can derive the relations in Section 21.2 among the various moments and cumulants.

Why bother with all these relations? The answer is that if the generating functions and moments are known, they can usually be inverted to give the distribution $f(N)$, and this is sometimes easier than calculating $f(N)$ directly. Even if all the moments are not known, those we have can constrain $f(N)$, at least to some extent. For example, given the factorial moments, substitute $u = 1 + t$ in (21.68) so that

$$g(u) = \sum_{N=0}^{\infty} f(N)u^N = \sum_{r=0}^{\infty} \frac{\mu'_{[r]}}{r!}(u - 1)^r. \qquad (21.69)$$

Assuming that around $t = 0$ both series are analytic in a circle around $u = 1$ enables us to identify the coefficients of all powers of u on both sides to get

$$f(N) = \sum_{r=N}^{\infty} (-1)^{r-N} \frac{\mu'_{[r]}}{N!(r - N)!}. \qquad (21.70)$$

Thus each $f(N)$ depends on all the factorial moments with $r \geq N$. This is a general feature of statistical distributions: To transform from one basic representation (e.g., $f(N)$) to another (e.g., moments) requires a complete set of the basis functions. This is of course familiar from the analogous case of eigenfunction solutions of linear differential equations. In both cases it makes sense to solve the problem directly in the most physically relevant representation. There are exceptional cases, however, in which all the moments do not uniquely determine a discrete distribution function, either because some of them are infinite, or analyticity fails, or knowledge of $g(1+t)$ around $t = 0$ does not specify it uniquely for all t. The lognormal distribution and the Cauchy distribution are well-known examples where $f(N)$ cannot be found from its moments.

For a continuous distribution, the analog of a generating function is its characteristic function. This may be any functional transform, but the most useful is the Fourier transform

$$\phi(t) = \int_{-\infty}^{\infty} e^{itu} f(u)\, du, \qquad (21.71)$$

with its inverse

$$f(u) = \frac{1}{2\pi} \int_{-\infty}^{\infty} e^{-itu} \phi(t)\, dt. \qquad (21.72)$$

Rigorous discussions of the conditions for which this transform holds are available in many texts on probability or analysis. Generally, if $\phi(t)$ has derivatives $\phi^{(r)}(t)$ up to degree r, then differentiating (21.71) with respect to t shows directly that the rth moment of $f(u)$ around zero is

$$\mu'_r = (-i)^r \phi^{(r)}(0), \qquad (21.73)$$

where $\phi^{(r)}(t)$ is evaluated at $t = 0$. So if we expand the closed form of $\phi(t)$ in powers of t, it will give the moments from

$$\phi(t) = \sum_{r=0}^{\infty} \mu_r' \frac{(it)^r}{r!} \tag{21.74}$$

provided they exist. For many purposes we can formally substitute $\theta = it$ and ignore its imaginary property. Next, we examine other uses for generating functions.

21.4   Convolutions, Combinations, and Compounding

Suppose we have two discrete random variables x and y with independent distributions $f^{(1)}$ and $f^{(2)}$ and we want to know the distribution of the sums $x + y$. To find the probability that $x + y$ is equal to some integer N, we add the probabilities that one of the distributions gives $N - M$ and the other gives M, summing over all values of M:

$$f_N^{(1)} f_0^{(2)} + f_{N-1}^{(1)} f_1^{(2)} + \cdots + f_0^{(1)} f_N^{(2)}. \tag{21.75}$$

This sum is just the coefficient of the term s^N in the product of the generating functions $g^{(1)}(s)g^{(2)}(s)$. Therefore the distribution of the sums has the generating function

$$g^{(1)+(2)}(s) = g^{(1)}(s)g^{(2)}(s), \tag{21.76}$$

which is the product of the individual generating functions. The probability distribution of $x + y$ is denoted by the convolution of the distributions of x and y,

$$\{f_N^{(1)}\} * \{f_N^{(2)}\}, \tag{21.77}$$

and is given by the sequence of sums with the form of (21.75). The operation $(*)$ of convolution is readily seen to be associative and commutative, but division $g^{(1)}(s)/g^{(2)}(s)$ does not generally give a new generating function. So the convolution operation is not a group but is called a semi group. One can, of course, have the distribution of any number of sums, or even of the sum of the same variable M times. In this last case the "Mth convolution power"

$$\{f(N)\}^{M*} \tag{21.78}$$

is generated by $[g(s)]^M$.

   Distributions of sums should not be confused with sums of distributions. Sometimes the latter are called mixtures, or combinations. They are just superimpositions of distributions in specified proportions. These may be characterised by their cumulative probabilities or by their probability density; the distributions may have different parameters or different functional forms. In terms of probability densities

$f_i(N)$, having proportions $a_i \geq 0$ where $\sum_{i=0}^{M} a_i = 1$, we may write

$$f(x_1, x_2, \ldots, x_N) = \sum_{i=0}^{M} a_i f_i(x_1, x_2, \ldots, x_N) \tag{21.79}$$

for the combined or mixed distribution.

Often a discrete distribution will depend on some parameter, say θ, as $f(N; \theta)$. If θ is a continuous variable with a distribution $p(\theta)$, then the average of $f(N; \theta)$ over all θ,

$$f(N) = \int f(N; \theta) p(\theta) \, d\theta, \tag{21.80}$$

is the result of "compounding" $f(N; \theta)$ with $p(\theta)$. Thus $f(N)$ has the generating function

$$g(s) = \sum_{N} f(N) s^N = \int \sum_{N} f(N; \theta) p(\theta) s^N \, d\theta = \int g(s, \theta) p(\theta) \, d\theta, \tag{21.81}$$

where $g(s, \theta)$ is the generating function of $f(N; \theta)$. Moments, cumulants etc. of the compound distribution are readily calculated from $g(s)$ as in Section 21.3. If $p(\theta)$ is not continuous, a sum over the discrete values of θ replaces the integrals in (21.80) and (21.81). For example, consider the sum $y_N = \sum_{i=1}^{N} x_i$ of random independent variables x_i all from the same distribution $f(x)$ having generating function $g(s)$. For any fixed value of N, the generating function of y_N is $g(s)^N$. But if N is itself a random variable with a distribution $p(N)$ and generating function $h(s)$, then the generating function of y is analogous to the last equality in (21.81),

$$\sum_{N} p(N) g(s)^N = h[g(s)]. \tag{21.82}$$

The left-hand side of this equation just adds the generating functions of y_N for each value of N times $p(N)$ so that the coefficients of s_N give the total probability of Y for all values of N. This has the form of the generating function (21.54) with s replaced by $g(s)$ and is thus "compounded."

A simple concrete example, which we will later use for many comparisons, is the Poisson distribution

$$f(N) = \frac{\lambda^N}{N!} e^{-\lambda}, \tag{21.83}$$

with $\lambda > 0$ being both the mean, $\bar{N}$, and the variance. It is a completely uncorrelated distribution for which $\xi_2 = \xi_3 = \cdots \xi_n = 0$. Its generating function is

$$g(s) = \sum_{N=0}^{\infty} \frac{1}{N!} e^{-\lambda} (\lambda s)^N = e^{\lambda(s-1)}. \tag{21.84}$$

The generating function (21.64) for moments about the origin,

$$g(e^t) = \exp[\lambda(e^t - 1)],$$
(21.85)

gives

$$\mu_1' = \lambda = \bar{N}, \quad \mu_2' = \lambda + \lambda^2, \quad \mu_3' = \lambda + 3\lambda^2 + \lambda^3,$$
$$\mu_4' = \lambda + 7\lambda^2 + 6\lambda^3 + \lambda^4.$$
(21.86)

Similarly, the generating function (21.66) for moments about the mean gives

$$\mu_2 = \lambda, \quad \mu_3 = \lambda, \quad \mu_4 = \lambda + 3\lambda^2$$
(21.87)

and the cumulant generating function (21.67) shows that $\kappa_N = \lambda$ for all N.

Another useful example is the binomial distribution, which is based on the probability, p, that an event will occur in an independent random trial. The event does or does not occur with probability p and $q = 1 - p$ respectively. If there are n trials, the probability that it occurs (success) r times and does not occur (failure) $n - r$ times is $p^r q^{n-r}$. There are $n!/r!(n-r)! = \binom{n}{r}$ possible sequences for success and so the probability that the event occurs exactly r times is the binomial distribution

$$f(r) = \binom{n}{r} p^r q^{n-r} \qquad (r = 0, 1, \ldots, n).$$
(21.88)

Each value of $f(r)$ is a term in the binomial expansion of $(q + p)^n$. Therefore the generating function of the binomial distribution is

$$g(s) = \sum_{r=0}^{\infty} \binom{n}{r} p^r q^{n-r} s^r = (q + ps)^n,$$
(21.89)

and its corresponding moment generating function is $(q + pe^t)^n$.

Now suppose there is a Poisson distribution of galaxies with mean value $\bar{N}$ in some volume of space. We select galaxies from it at random with probability p. The selected galaxies will also have a Poisson distribution, since no information apart from p has been changed, but their mean will be $p\bar{N}$. More formally, this is a binomial distribution (21.88) whose index n is a Poisson variable with mean $\bar{N}$. It represents the probability p of choosing r galaxies from a Poisson number n in a cell. From (21.89) and the discrete analog of (21.81), the generating function of the selected sample is

$$\sum_{n=0}^{\infty} g(s, n)p(n) = \sum_{n=0}^{\infty} (q + ps)^n \bar{N}^n (n!)^{-1} e^{-\bar{N}} = e^{-\bar{N}p(1-s)},$$
(21.90)

where the last equality follows from the expansion of the exponential and $q = 1 - p$. Comparing with (21.84) we see that this is indeed the generating function of a Poisson distribution with mean $p\bar{N}$.

As a more general example, consider a Poisson distribution compounded with another distribution. This could arise if there is a Poisson distribution of particles with mean $\bar{N}$ in many independent volumes and the means themselves have a distribution $p(\tilde{N})$ with generating function $g(s)$. Then the sum obtained by adding particles from different volumes has a distribution given by the generating function (21.82) with (21.84) for $h(s)$:

$$h[g(s)] = e^{-\bar{N} + \bar{N}g(s)}. \tag{21.91}$$

Under some conditions, $g(s)$ might also be a Poisson generating function with a mean different from $\bar{N}$.

By compounding two or more distributions, one can obtain many of the distributions of statistics and probability theory. These mostly provide formal mathematical representations, but sometimes they also carry interesting physical content.

21.5 Infinite Divisibility

Distribution functions with the form of (21.91) have an interesting feature related to the physical properties of statistical homogeneity and of projection from three dimensions onto two dimensions. Writing $\bar{N} = \tilde{n}V$ in (21.91) shows that if $V = V_1 + V_2$ is divided into two independent volumes then

$$h(s; V_1 + V_2) = h(s; V_1) h(s; V_2). \tag{21.92}$$

Therefore the contributions of the statistical distribution functions in each volume just add (via convolution because their generating functions multiply) to give the distribution of the sum in the total volume. The generalization to infinite divisibility is simply that for any positive integer n, the nth root of an infinitely divisible generating function $h(s)$ is also a generating function

$$h^{1/n}(s) = H(s), \tag{21.93}$$

where $H(s)$ generates another distribution. Texts on probability theory prove (theorems by Lévy and Feller) that the form on the right-hand side of (21.91) is both necessary and sufficient for infinite divisibility.

Thus we may regard an infinitely divisible distribution as the distribution of the sums of any number of smaller independent volumes (or other subsets). In particular, if the distribution of points in different spatial volumes is infinitely divisible and we add them by stacking the volumes and projecting all their points onto a two-dimensional plane, the resulting projected distribution will also have the form of (21.91). Indeed we may write

$$h(s; V) = h^n(s; V/n). \tag{21.94}$$

The requirement that the volumes be independent expresses a form of statistical homogeneity. Because all volumes should be equivalent, it does not matter where they are found throughout space. In other words, the distribution function generated by $g(s)$ in (21.91) should be essentially the same in each cell into which the volume V is divided. We will later see that this is not quite the case for the conical volumes that project the spatial galaxy clustering onto the sky, but it is often a very good approximation.

There is another useful property of infinitely divisible distributions. Substituting the expansion (21.54) for $g(s)$ into (21.91) and recalling from (21.1) that $f(0) \equiv f_0 = 1 - f_1 - f_2 - \cdots$ enables us to rewrite (21.91) with the notation $\bar{N} = \lambda$ and $f(N) \equiv f_N$ as

$$
\begin{aligned}
h(s) &= \exp\left[\lambda(f_0 + f_1 s + f_2 s^2 + \cdots - 1)\right] \\
&= e^{\lambda f_1(s-1)} e^{\lambda f_2(s^2-1)} e^{\lambda f_3(s^3-1)} \cdots
\end{aligned}
\tag{21.95}
$$

Comparing this with (21.84) shows that each factor generates a Poisson distribution with mean value $\lambda_i = \lambda f_i$ and $\sum_i \lambda_i = \lambda$. The factor involving the exponent $(s-1)$ generates the usual Poisson distribution of single particles; the factor involving $(s^2 - 1)$ represents a Poisson generating function for a series $\sum_{n=1}^{\infty} a_{2n} s^{2n}$ and so it generates a Poisson distribution of paired particles; similarly the factor involving $(s^3 - 1)$ generates a Poisson distribution of triplets; and so on. Evidently, any infinitely divisible function can also be represented as the distribution of the sums of singlets, pairs, triplets, and higher order clusters each having a Poisson distribution, and λ_i is the average probability that each type of cluster occurs.

21.6 Relation to Correlation Functions

Intuitively it should be clear that distribution functions are related to correlation functions and that each distribution function $f(N)$ must be related to all the correlation functions $\xi_2, \xi_3, \ldots$. For example, the void probability, $f(0)$, for finding an empty region depends on the negative (or anti-) correlation of all particles at the boundary of the void with all possible numbers of particles within its boundary. Similar statements apply to all the $f(N)$.

In fact these relations, familiar in probability theory (e.g., Daley & Vere-Jones, 1988), have been discussed in the astronomical literature from several related points of view (White, 1979; Balian and Schaffer, 1989; Szapudi and Szalay, 1993; Sheth, 1996a,c). The main result is that the generating function for the counts in cells of volume V may be written in terms of the volume averages, $\bar{\xi}_m$, of all the mth-order correlation functions as

$$
g(s) = \exp\left[\sum_{m=1}^{\infty} \frac{\bar{N}^m (s-1)^m}{m!} \bar{\xi}_m\right],
\tag{21.96}
$$

where

$$
\bar{N} = \bar{n}V
\tag{21.97}
$$

as usual and

$$\bar{\xi}_m \equiv \frac{1}{V^m} \int_V \xi_m\,(\mathbf{r}_1,\ldots,\mathbf{r}_m)\,d\mathbf{r}_1\ldots d\mathbf{r}_m.$$ (21.98)

The probability distribution functions for counts in cells of a given volume V follow from (21.55) as

$$f_V(N) = \frac{1}{N!}\frac{d^N}{ds^N}\exp\left[\sum_{m=1}^{\infty}\frac{\bar{n}^m V^m (s-1)^m}{m!}\bar{\xi}_m(V)\right]_{s=0}$$

$$= \frac{(-\bar{n})^N}{N!}\frac{d^N}{d\bar{n}^N}\exp\left[\sum_{m=1}^{\infty}\frac{(-\bar{n})^m V^m \bar{\xi}_m(V)}{m!}\right]_{\bar{\xi}_m=\text{constant}}.$$ (21.99)

Since the $\bar{\xi}_m$ do not depend on s or $\bar{n}$, the second equality follows when $\bar{\xi}_m$ is treated as a constant in the differentiation. These results can be derived (with rather a lot of algebra) either by relating the combinatorics of counting particles in cells to the correlation functions or by using characteristic functionals, which are a generalization of generating functions (see the previous references).

Because every $f_V(N)$ is related to all the $\bar{\xi}_m$, it should be possible to formally extract all the $f_V(N)$ from any particular $f_V(N)$ given in appropriate form as a function of the $\bar{\xi}_m$. The simplest example follows by noticing that the summation in the second equality of (21.99) is just $g(s = 0)$ in (21.96) and recalling that $g(s = 0) = f_V(N = 0)$. We then have

$$f_V(N) = \frac{(-\bar{n})^N}{N!}\frac{d^N}{d\bar{n}^N}f_V(0)\bigg|_{\bar{\xi}_m=\text{constant}}.$$ (21.100)

In this case the void distribution generates the other $f(N)$. There is nothing magical about this, when we remember that we must know all the $\bar{\xi}_m$ so that they can be kept constant when taking the derivatives: The relation (21.100) does not enable us to determine $f_V(N)$ from reduced or incomplete information about the $\bar{\xi}_m$. From the fundamental definition (14.31) of the correlation functions, we see that they are invariant under a random dilution that just decreases n equally, without bias, in the $\Delta n\,(\mathbf{r}_i)$ and in $\bar{n}$. Therefore the derivatives in (21.100) may be thought of as being taken under the constraint of random dilution in order to obtain the $f_V(N)$ from $f_V(0)$. By applying the properties of generating functions discussed earlier in this chapter to (21.96), one can obtain a plethora of formal relations among distribution functions, their various types of moments, and their associated correlation functions. These often provide useful insights into gravitational clustering, and we will develop them further as they are needed.

22

Dynamics of Distribution Functions

The universe is change.

Marcus Aurelius

22.1 Introduction

One way to determine the evolution of distribution functions follows immediately from Section 21.6. If we could extract the time development of the $\bar{\xi}_m$ from the BBGKY hierarchy of Chapter 17, then (21.96)–(21.99) would give the development of the generating function and of $f(N)$. The difficulty here we have recognized before: The BBGKY hierarchy is not solvable in any convenient way except for ξ_2 and ξ_3 in the linear regime. Nevertheless, the form of the generating function contains a clue to its development.

Expanding (21.96) in powers of the dummy variable s, we may rewrite it for small $\bar{\xi}_2$ and $\bar{\xi}_3$, neglecting the higher order correlations, as

$$g(s) = \exp\left\{-\bar{N} + \frac{1}{2}\bar{N}^2\bar{\xi}_2 - \frac{1}{6}\bar{N}^3\bar{\xi}_3 + \left(\bar{N} - \bar{N}^2\bar{\xi}_2 + \frac{1}{2}\bar{N}^3\bar{\xi}_3\right)s\right.$$
$$\left. + \frac{1}{2}(\bar{N}^2\bar{\xi}_2 - \bar{N}^3\bar{\xi}_3)s^2 + \frac{1}{6}\bar{N}^3\bar{\xi}_3 s^3\right\}. \tag{22.1}$$

Thus using (21.55), the void distribution and $f(1)$ have the form

$$f(0) = g(s = 0) = \exp\left(-\bar{N} + \frac{1}{2}\bar{N}^2\bar{\xi}_2 - \frac{1}{6}\bar{N}^3\bar{\xi}_3\right) \tag{22.2}$$

and

$$f(1) = \left(\bar{N} - \bar{N}^2\bar{\xi}_2 + \frac{1}{2}\bar{N}^3\bar{\xi}_3\right)\exp\left(-\bar{N} + \frac{1}{2}\bar{N}^2\bar{\xi}_2 - \frac{1}{6}\bar{N}^3\bar{\xi}_3\right), \tag{22.3}$$

with naturally more complicated expressions for the other $f(N)$. Notice that as the $\bar{\xi}_m \to 0$, the distribution reverts to a Poisson form. This occurs either if there are no correlations, so that all the $\xi_m = 0$, or if the spatial dependences of the ξ_m are sufficiently nonsingular and the volumes of the cells are small enough that $\bar{\xi}_m \to 0$. In this last case, different small volumes are effectively independent. As their volume tends to zero, the probability that they contain more than one point (galaxy) tends to zero even faster. If we divide a large volume V into η equal nonoverlapping cells, and $\bar{n}$ is the average number density of points, then the probability that there is just one point in one of the η cells is $\bar{n}V\eta^{-1} + o(\eta^{-1})$, where $o(\eta^{-1})$ is a function tending to zero faster than η^{-1}. The probability that the volume has more than one point is

$o(\eta^{-1})$. Therefore the probability that N points are in N of the η cells and no points are in any other cell (counting all the different possible arrangements of the points in the cells) is

$$f(N) = \binom{\eta}{N} [\bar{n}V\eta^{-1} + o(\eta^{-1})]^N [1 - \bar{n}V\eta^{-1} - o(\eta^{-1})]^{\eta-N}. \qquad (22.4)$$

As we take the limit $\eta \to \infty$, so that there are a very large number of very small cells, each with a very small probability of containing a galaxy, but with $\bar{n}$ remaining constant, the previous expression tends to

$$f(N) = \frac{\bar{N}^N}{N!} e^{-\bar{N}}, \qquad (22.5)$$

which is just the Poisson distribution (14.1) or (21.83) with $\lambda = \bar{N} = \bar{n}V$ for any size volume V. Its void probability is $f(0) = e^{-\bar{n}V}$, which is the first factor in (22.2).

As the volume of a cell increases in a correlated system, the terms of $o(\eta^{-1})$ become important and the strength of the departure from Poisson is described by the terms $\bar{N}^m \bar{\xi}_m$ in (22.1)–(22.3) and their extensions to higher $\bar{\xi}_m$. However, the variance (21.63) depends only on $\bar{\xi}_2$ as obtained most conveniently using $g(s)$ in the form (21.96):

$$\langle (\Delta N)^2 \rangle = \bar{N} + \bar{N}^2 \bar{\xi}_2, \qquad (22.6)$$

which was derived previously (14.17) from a more physical point of view. For this and the higher moments, in terms of the density contrast in cells of volume V,

$$\delta(V) = \frac{N - \bar{N}}{\bar{N}}, \qquad (22.7)$$

one similarly finds

$$\langle \delta^2 \rangle = \frac{1}{\bar{N}} + \bar{\xi}_2, \qquad (22.8)$$

$$\langle \delta^3 \rangle = \frac{1}{\bar{N}^2} + \frac{3}{\bar{N}} \bar{\xi}_2 + \bar{\xi}_3, \qquad (22.9)$$

$$\langle \delta^4 \rangle = \frac{1}{\bar{N}^3} + \frac{1}{\bar{N}^2}(3 + 7\bar{\xi}_2) + \frac{6}{\bar{N}}(\bar{\xi}_2 + \bar{\xi}_3) + 3\bar{\xi}_2^2 + \bar{\xi}_4, \qquad (22.10)$$

etc. Thus the mth moment involves the volume integrals of the correlation functions up to mth order. These relations can be inverted and either the moments or the correlation integrals can be used as a basis for describing clustering.

One way to attach these formal relations to actual dynamics is to use the equations of motion to evolve the $\langle \delta^m \rangle$ and thus approximate $g(s)$. This has been done (Bernardeau, 1992) for a smoothed fluid approximation starting with a Gaussian distribution of density fluctuations for $\delta(V) = \delta(\mathbf{r})$ defined by (22.7). This means that its Fourier components $\delta_{\mathbf{k}}$ given by

$$\delta(\mathbf{r}) = \int \delta_{\mathbf{k}} e^{i\mathbf{k}\cdot\mathbf{r}} \, d^3\mathbf{k} \tag{22.11}$$

are complex random variables whose phases are independent. Since $\delta(\mathbf{r})$ is real, $\langle \delta_{\mathbf{k}_1} \delta_{\mathbf{k}_2} \rangle = \delta_3 (\mathbf{k}_1 + \mathbf{k}_2) \, P(k_1)$ with $\delta_3(\mathbf{k})$ being the three-dimensional Dirac delta function, and initially the two-point correlation function has the form

$$\xi(r) = \int P(k) e^{i\mathbf{k}\cdot\mathbf{r}} \, d^3\mathbf{k}, \tag{22.12}$$

which is a specialized form of (14.34). Under these conditions, the mean values of the moments $\langle \delta^m \rangle$ at different separations involve only ξ_2 and not the higher ξ_m.

In this smoothed fluid approximation, applicable to a model in which galaxies are smoothed over large scales as discussed in Section 16.1 or to a model of dark matter in which galaxies are dynamically irrelevant either because they have not yet formed or because smooth dark matter dominates the dynamics, the fluid equations in Section 16.3 enable the $\langle \delta^m \rangle$ to be calculated into the weakly nonlinear regime. Like the attempts to calculate the ξ_m directly from the BBGKY hierarchy, this analysis for the $\langle \delta^m \rangle$ rapidly becomes mathematically complicated, even with additional simplifying assumptions.

We can try a more physical approach to gravitational clustering that does not require the fluid approximation and is more directly related to the nonlinear regime, especially when discrete particles (galaxies) dominate. As galaxies cluster more strongly, starting from a Poisson or near-Poisson initial state, the average ratio of gravitational correlation energy, W, to the kinetic energy, K, of peculiar velocities tends to increase. We may characterize this ratio for an average cell of volume V by the quantity

$$b = -\frac{W}{2K} = -\frac{\frac{1}{2}\bar{N}m\bar{n}}{3\bar{N}T} \int_V \phi(r)\xi_2(r) \, dV$$

$$= \frac{2\pi Gm^2\bar{n}}{3T} \int_0^R \xi_2(r)r \, dr. \tag{22.13}$$

Here $\phi(r) = -Gm/r$ is the interaction potential, the usual factor of $1/2$ counts all pairs of particles only once, the temperature is given by

$$K = \frac{3}{2}\bar{N}T = \frac{1}{2}\sum_{i=1}^{N} mv_i^2 \tag{22.14}$$

with Boltzmann's constant set equal to unity, v_i is the peculiar velocity of the ith galaxy relative to the Hubble expansion, all N galaxies are assumed to have the same average mass m for simplicity (this can be generalized), and the last expression is for a spherical volume (whose radius should not be mistaken for the scale length $R(t)$ of the metric). The symbol b does not come out of the blue. It has been used since the early twentieth century for this ratio in the kinetic and thermodynamic theories of interacting particles, which we will explore in later chapters. The value $b = 0$ describes noninteracting particles, or equivalently, approximates the condition $K \gg |W|$. The other limit, $b = 1$, describes the state when all the particles in the volume are "virialized," that is, moving as fast as they can, on average, for their total gravitational interactions. However, the ratio b should not be confused with the standard virial theorem, which refers to the total gravitational energy in a finite inhomogeneous distribution. Here b involves the correlation energy in an infinite statistically homogeneous system.

For small values of $|W|$, we expect $\bar{N}\bar{\xi}_2$ to grow from an initial Poisson state as $-W/K = 2b$. Therefore assume an expansion around $b = 0$ of the form

$$\bar{N}\bar{\xi}_2 = 2b + \alpha(2b)^2 + \mathcal{O}(2b)^3, \tag{22.15}$$

with α some constant. There is no term independent of b since as $b \to 0$, $\bar{\xi}_2 \to 0$. Let us also assume that $\bar{\xi}_3$ satisfies the hierarchical scaling (18.2) so that

$$\bar{\xi}_3 = 3Q\bar{\xi}_2^2. \tag{22.16}$$

Substituting these last two relations into (22.1) gives, to order b^2,

$$g(s) = \exp \bar{N} \left\{ -1 + b + \frac{1}{2}(\alpha - Q)(2b)^2 + \left[1 - 2b - \left(\alpha - \frac{3}{2}Q \right)(2b)^2 \right] s \right.$$
$$\left. + \frac{1}{2}[2b + (\alpha - 3Q)(2b)^2]s^2 + \frac{1}{2}Q(2b)^2 s^3 \right\}. \tag{22.17}$$

For higher orders of b, there will be additional α_m coefficients from higher order terms in (22.15) and additional Q_m terms if a scaling generalization of (22.16) continues to hold. This provides a concise and relatively simple illustration of how dynamical information can enter into the distribution functions.

We can go a step further by noting that the clustering becomes stronger and stronger as $b \to 1$. If the higher order terms in b that multiply s^0 in (22.17) also contain the factor $\alpha - Q$, then the condition $\alpha = Q$ describes the case when $f(0) \to 1$ as $b \to 1$, since $f(0) = g(s = 0)$ from (22.2). In other words, the state of maximal clustering is one where an arbitrarily large volume, placed at random in the universe, has probability unity of being empty. The clusters occupy a volume of measure zero. Later chapters will show that $\alpha = Q$ is effectively often the case for the cosmological many-body problem.

To pursue this approach further, we need the dynamical evolution of $b(t)$. This dynamics is partly contained in the cosmic energy equation, which applies directly to both linear and nonlinear evolution.

22.2   The Cosmic Energy Equation

Despite its grand-sounding title, this equation is really quite useful. We examine it here from several points of view, each providing its own insight.

Originally the cosmic energy equation was derived directly from the comoving equations of motion (Irvine, 1961; Layzer 1963; see also Zeldovich, 1965). In comoving coordinates (16.49) $\mathbf{x} = \mathbf{r}/R(t)$, the local Newtonian equation of continuity (16.34) embedded in the cosmological expansion becomes

$$\frac{\partial}{\partial t}(R^3 \rho) + \frac{\partial}{\partial \mathbf{x}} \cdot (R^2 \rho \mathbf{v}) = 0. \tag{22.18}$$

Here

$$\mathbf{v} = R\dot{\mathbf{x}} = \dot{\mathbf{r}} - H\mathbf{r} \tag{22.19}$$

is the peculiar velocity relative to the uniform Hubble expansion. This says that since mass is conserved, the rate of change of the total mass within a comoving volume is equal to the net flux of mass across the surface of that volume. Since the local peculiar velocities are much less than the velocity of light, and the scales of galaxy clustering are small compared with the scale of the Universe, the Newtonian approximation is excellent. Transforming (22.18) to the locally inertial physical coordinate r using $\partial/\partial \mathbf{x} = R(t)\partial/\partial \mathbf{r}$ and writing the total density as the sum of an average and a fluctuating component,

$$\rho(\mathbf{r}, t) = \bar{\rho}(t) + \Delta\rho(\mathbf{r}, t), \tag{22.20}$$

gives

$$\frac{\partial}{\partial t}(R^3 \Delta\rho) + \frac{\partial}{\partial \mathbf{r}} \cdot (R^3 \rho \mathbf{v}) = 0 \tag{22.21}$$

since $R^3 \bar{\rho} \propto R^3 \bar{N}\bar{m}/r^3 = \bar{N}\bar{m}/x^3$ is independent of time [as in (16.17) where $\rho = \bar{\rho}$]. Note that throughout this discussion $\mathbf{x}$ is held constant in the partial time derivative $\partial/\partial t$, so that $\mathbf{r} = \mathbf{r}(\mathbf{x}, t)$. Also, the density ρ may either be continuous or a sum of discrete particles.

Next, we proceed to the equation of motion. This follows from differentiating (22.19):

$$\frac{\partial \mathbf{v}}{\partial t} = -H\mathbf{v} + \ddot{\mathbf{r}} - \frac{\ddot{R}}{R}\mathbf{r}. \tag{22.22}$$

The last two terms on the right-hand side are the difference between the acceleration in the local inertial frame and the acceleration caused by the changing global expansion of the Universe, as may be seen from (19.2) and (19.3). This difference is produced by the gradient of fluctuations in the gravitational potential: $-\partial\phi/\partial \mathbf{r}$.

These fluctuations $\phi(\mathbf{r}, t)$ satisfy Poisson's equation for the density fluctuation:

$$\nabla^2 \phi = 4\pi G \Delta\rho(\mathbf{r}, t). \tag{22.23}$$

For a statistically homogeneous and isotropic distribution in which the average fluctuating potential is zero, this has the solution in the standard form

$$\phi(\mathbf{r}, t) = -G \int_V \frac{\Delta\rho(\mathbf{r}', t)}{|\mathbf{r} - \mathbf{r}'|} \, d^3\mathbf{r}'. \tag{22.24}$$

This integral, the standard Green function solution of (22.23), may be thought of as an inverse spatial moment of the density fluctuations. Thus it weights nearby fluctuations quite strongly. In principle, all fluctuations everywhere throughout the Universe contribute to ϕ. A statistically homogeneous distribution, however, will have a scale beyond which the distribution is so smooth that its contribution to the weighted integral is negligible. This scale is generally the correlation length, especially since ξ_2 is already averaged over many volumes. The integral can also be taken over smaller scales and then represents an average potential depending on scale. Note that (22.23) has the form of the linearized Poisson equation (16.44) in Jeans's analysis, where again only the perturbed potential produces any local force. In (22.23), however, there is no linearization and the fluctuations may have any magnitude.

The first term on the right-hand side of (22.22) is essentially a kinematic reduction of peculiar velocity, which results from expansion even if there are no accelerations. As the coordinate system expands, a particle moving at some peculiar velocity in physical coordinates will move a smaller physical distance per unit of time, so $v \propto R^{-1}$. We may see this directly by setting $\ddot{r} = \ddot{R} = 0$ and solving (22.22) to find $v \propto R^{-1}$. This also corresponds to the adiabatic decay of a freely expanding perfect gas with a ratio of specific heats $\gamma = 5/3$, since $P \propto \bar{\rho}^{-\gamma} \propto R^{-5}$ and also $P \propto \bar{\rho}v^2 \propto R^{-5}$ so $v \propto R^{-1}$. The forces from the fluctuating gravitational potential produce departures from this free expansion. To incorporate these fluctuating forces into (22.22), we substitute $-\partial\phi/\partial\mathbf{r}$ for its last two terms, use $H = \dot{R}/R$, and rearrange the terms to give the dynamical equation

$$\frac{\partial}{\partial t}(R\mathbf{v}) = -\frac{\partial}{\partial\mathbf{r}}(R\phi). \tag{22.25}$$

This immediately shows again that if there are no fluctuating forces on the right-hand side, then $v \propto R^{-1}$.

The cosmic energy equation is just a moment of (22.25). This is analogous to the virial theorem derived in Chapter 13. But instead of a spatial moment as in the virial theorem, we now take a velocity moment. Moreover, the virial theorem applies only to a relaxed finite system, but since (22.25) applies anywhere throughout an unbounded statistically homogeneous system (much larger than the fluctuation scale lengths and therefore effectively infinite), the cosmic energy equation applies to any typical volume. It follows from taking the scalar product of (22.25) with $R\mathbf{v}\rho\,dV$ and integrating over all the particles in the volume V. The total density ρ is the sum of delta functions, each of which represents the position of a point (galaxy) of mass m_i. For the left-hand side of (22.25) we find

$$\frac{1}{2}\int \frac{\partial}{\partial t}(Rv)^2\rho\,dV = \frac{1}{2}\sum_i m_i \frac{\partial}{\partial t}(Rv_i)^2 = \frac{d}{dt}\left(R^2\sum_i \frac{1}{2}m_i v_i^2\right)$$

$$= \frac{d}{dt}(R^2 K), \tag{22.26}$$

where K, as in (22.14) for any masses, is the kinetic energy of the peculiar velocities within the volume. The right-hand side of (22.25), after integrating by parts and substituting the continuity equation (22.21) and the fluctuating potential (22.24), gives

$$-\int R\rho\mathbf{v}\cdot\frac{\partial}{\partial\mathbf{r}}(R\phi)\,dV = R^2\int \phi\frac{\partial}{\partial\mathbf{r}}\cdot(\rho\mathbf{v})\,dV$$

$$= GR^2\int \frac{\partial}{\partial t}\Delta\rho(\mathbf{r},t)\,dV\int \frac{\Delta\rho'(\mathbf{r}',t)}{|\mathbf{r}-\mathbf{r}'|}\,dV'$$

$$= GR^2\int \frac{\partial}{\partial t}\Delta\rho'(\mathbf{r}',t)\,dV'\int \frac{\Delta\rho(\mathbf{r},t)}{|\mathbf{r}-\mathbf{r}'|}\,dV. \tag{22.27}$$

In the last two expressions we have interchanged the dummy variables so that, similarly to (13.6), we can write the result in terms of the gravitational correlation energy, which is symmetric in $\mathbf{r}$ and $\mathbf{r}'$:

$$W = -\frac{G}{2}\iint \frac{\Delta\rho(\mathbf{r},t)\Delta\rho(\mathbf{r}',t)}{|\mathbf{r}-\mathbf{r}'|}\,dV\,dV'$$

$$= -\frac{G}{2}\bar\rho^2\iint \frac{\xi(\mathbf{r},\mathbf{r}',t)}{|\mathbf{r}-\mathbf{r}'|}\,dV\,dV'. \tag{22.28}$$

The second expression employs (14.15) without the Poisson component since that just represents the uniform background. From $\mathbf{r} = R(t)\mathbf{x}$ and $H = \dot R/R$ we obtain

$$\frac{\partial}{\partial t}\frac{1}{|\mathbf{r}-\mathbf{r}'|} = -\frac{H}{|\mathbf{r}-\mathbf{r}'|}. \tag{22.29}$$

With these last two results, (22.27) becomes

$$-R^2 HW - R^2\frac{dW}{dt}. \tag{22.30}$$

Equating the left (22.26) and right (22.30) sides of the velocity moment of the equation of motion (22.25) gives the cosmic energy equation

$$\frac{d}{dt}(K + W) + \frac{\dot{R}}{R}(2K + W) = 0. \tag{22.31}$$

This is a basic dynamical result, applicable to both linear and nonlinear evolution. The requirement that the comoving volume be large enough that external fluctuations do not contribute enters in three ways: the vanishing of ϕ at the boundary of the integration by parts in (22.27), the vanishing of the volume integral of $\Delta\rho$ in the second equality of (22.27), and the lack of contributions to W from outside the volume in (22.28). If these conditions do not hold rigorously, then the cosmic energy equation will be an approximation, which can still be very useful. From (22.31) we immediately see that there are two cases in which the total energy $U = K + W$ is conserved: $\dot{U} = 0$. The first is when $\dot{R} = 0$ and the Universe does not expand, so there are no adiabatic energy losses. The second case is when all the particles in the system satisfy the virial theorem $2K + W = 0$. In this case they are bound together and relaxed, so they do not participate in the cosmic expansion. This answers the oft-asked question of whether bound systems like the solar system expand with the Universe: They do not. Partially bound systems, on the other hand, partially participate in the general expansion according to their value of $2K + W$. Since (22.31) is equivalent to

$$\frac{d}{dt}[R(K + W)] = -\dot{R}K \tag{22.32}$$

and the right-hand side is always negative in an expanding Universe, the total energy, U, in a comoving volume will decrease as the Universe expands. This is a consequence of adiabatic expansion modified by increased local clustering. If U is small initially, it eventually becomes more and more negative.

Guided by this detailed derivation, we can rederive the cosmic energy equation more swiftly from two other points of view. If we know the explicit time dependence of the Hamiltonian $H(\mathbf{p}, \mathbf{x}, t) = K + W$ in comoving coordinates, then Hamilton's energy equation

$$\frac{dH}{dt} = \frac{\partial H}{\partial t} \tag{22.33}$$

is the cosmic energy equation. Well, we know that the Lagrangian for the peculiar velocities $v_i = R\dot{x}_i$ is

$$L = \frac{1}{2}\sum_i m_i R^2 \dot{x}_i^2 - W, \tag{22.34}$$

giving the comoving momentum

$$\mathbf{p}_i = \frac{\partial L}{\partial \dot{x}_i} = m_i R^2 \dot{\mathbf{x}}_i, \tag{22.35}$$

and thus the comoving Hamiltonian

$$H = \sum_i \frac{p_i^2}{2R^2 m_i} + W = K + W. \tag{22.36}$$

So $K \propto R^{-2}(t)$. Transforming W in (22.28) into comoving coordinates we see that if the total mass in a comoving volume V is constant then the volume factors in the $\Delta \rho$ cancel those of the volume integral, leaving $W \propto r^{-1} \propto (Rx)^{-1} \propto R^{-1}(t)$. With these time dependences, substituting (22.36) into (22.33) immediately gives the cosmic energy equation (22.31). Again no assumptions concerning linearity are needed. However, the result does depend on the condition that the only time dependence in H is from the dependence of the peculiar velocities and comoving coordinates on $R(t)$. The correlation energy within the comoving volume V can change as the galaxies cluster and the time derivative in (22.31) describes this. But no provision is made for correlation energy moving across the boundary of V. This transfer must be small relative to the energy within V, or slow compared to the rate of change of energy in V in order for (22.31) to apply. Usually this is the case because for large volumes the relative contributions of correlations and motions at the boundary are small, whereas for small volumes they are dominated by conditions at the center through the x^{-1} weighting in W if ξ is sufficiently steep there.

The third derivation of the cosmic energy equation we have nearly done already. From the Einstein field equations, we derived the conservation of energy (16.6), which is equivalent to the first law of thermodynamics in an adiabatically expanding local volume. To apply this, we need an equation of state (Irvine, 1961) that expresses the total internal energy $U = \epsilon V$ and the pressure in the volume as a function of its kinetic and potential energy. Since the energies add we have simply

$$\epsilon V = U = K + W. \tag{22.37}$$

The pressure relation is not quite so simple. It will have a contribution from the momentum transport produced by the kinetic energy density, which for a perfect gas is the usual $2K/3V$. But there will also be a contribution from the correlated force of the gravitational interaction. In Chapter 25 we show that for the cosmological many-body problem this interaction pressure is $W/3V$, a special case of a more general result in standard kinetic theory. Thus

$$P = \frac{1}{3V}(2K + W). \tag{22.38}$$

Substituting (22.37) and (22.38) into (16.6) for a representative spherical volume immediately gives (22.31), the cosmic energy equation again. We now turn to some of its consequences.

22.3   Dynamical Implications of the Cosmic Energy Equation

Representing the relative importance of clustering by $b(t)$ from (22.13), we may eliminate K from the cosmic energy equation and express the time evolution $d/dt = \dot{R}d/dR$ in terms of the scale length to obtain

$$\frac{db}{dR} = \left(\frac{2}{R} + \frac{W'}{W}\right) b - 2\left(\frac{1}{R} + \frac{W'}{W}\right) b^2. \tag{22.39}$$

The primes denote derivatives with respect to R. We see that if b starts off small and $W \propto R^\alpha$ initially, then $b \propto R^{2+\alpha}$ for short times when the term in b^2 is negligible. This is essentially adiabatic expansion with $K \propto R^{-2}$. However, as $R \to \infty$ and $W'/W \to 0$, the value of b approaches an asymptotic limit.

The evolution equation (22.39) has the form of a simple Bernoulli differential equation whose general solution is

$$b = \frac{b_0 R^2 W}{R_0^2 W_0 + 2b_0(R^2 W - R_0^2 W_0) - 2b_0 \int_{R_0}^{R} R W dR}. \tag{22.40}$$

Here W_0 is evaluated at the initial expansion scale R_0 when $b = b_0 \neq 0$. Any departure from perfect homogeneity will start the correlation energy growing, as Newton realized (Chapter 2).

At early times, the first term in the denominator dominates and we recover the linear result $b = b_0 R^2 W / R_0^2 W_0$. After a very long time, if the system is virialized on all scales

$$\lim_{R \to \infty} W(R) = \text{constant} \tag{22.41}$$

and

$$\lim_{R \to \infty} b = 1, \tag{22.42}$$

independent of b_0. More generally, if $W(R) \propto R^\alpha$ with $\alpha \geq 0$, then

$$\lim_{R \to \infty} b = \frac{2 + \alpha}{2(1 + \alpha)} \leq 1. \tag{22.43}$$

To carry this further requires $W(R)$. The linear solution (17.65) of the BBGKY hierarchy (in which W is denoted by U), combined with (17.61), (17.62), and $\bar{n} \propto R^{-3}$, gives $W \propto t \propto R^{3/2}$ for the Einstein–de Sitter model. Thus for initial evolution starting close to a Poisson distribution, (22.40) yields $b \propto R^{7/2} \propto t^{7/3}$. This early rapid growth slows considerably in less than a Hubble expansion timescale.

Substituting $b(R)$ into the generating function (22.17) then gives the early dynamical development of the distribution functions as they begin to depart from the near-Poisson state.

Although the exact evolution depends on $W(R)$, we can use the approximation $W \propto R^\alpha$ to derive two important general properties of the system. This approximation will hold for any short interval. For longer intervals when $\alpha(R)$ evolves more slowly than R^α these results can be made more rigorous by a multiple timescale analysis. The first general result is that if we start off with an ensemble of systems having an initial distribution of values of b_0, then this distribution tends to narrow with time. The members of this ensemble might be different large volumes of space, all of a particular size and shape. Thus the evolution governed by the cosmic energy equation leads to a stable attractor for the value of $b(R)$. The second general property shows how perturbations of the cosmic energy equation directly determine the initial form into which the distribution function $f(N)$ evolves.

We start with the collapse of the initial distribution of b_0 in an ensemble of evolving volumes. Suppose the average initial value of $b(R_0)$ is $b_0 = 0$. Since there can be volumes in which the galaxies are initially correlated or anticorrelated, the integral of $\xi_2(r)$ in (22.13) for different volumes, all of the same size and shape, may fluctuate around zero. The simplest assumption is that b_0 has a Gaussian distribution,

$$f(b_0) = \frac{1}{\sqrt{2\pi}\,\sigma_0} e^{-b_0^2/2\sigma_0^2} . \tag{22.44}$$

Substituting the approximation

$$W = W_0 \left(\frac{R}{R_0}\right)^\alpha \tag{22.45}$$

into (22.40) and solving for $b_0(b)$ gives

$$f(b) = f[b_0(b)] \left|\frac{db_0}{db}\right|$$

$$= \frac{1}{\sqrt{2\pi}\,S(\alpha, R, b)\sigma(b)} e^{-b^2/2\sigma^2(b)} . \tag{22.46}$$

Here

$$S(\alpha, R, b) \equiv 1 - \frac{2(1+\alpha)}{2+\alpha} \left[1 - \left(\frac{R_0}{R}\right)^{2+\alpha}\right] b \tag{22.47}$$

and

$$\sigma(b) \equiv \sigma_0 \left(\frac{R}{R_0}\right)^{2+\alpha} S(\alpha, R, b). \tag{22.48}$$

At $R = R_0$ and $b = b_0$, with $S(\alpha, R_0, b_0) = 1$, we recover the original Gaussian (22.44), which peaks at $b_0 = 0$. However, as the Universe expands with increasing R/R_0, the distribution of b departs more and more from the original Gaussian. The peak of $f(b)$ moves to higher values until for $(R/R_0)^{2+\alpha} \gg 1$ it approaches

$$b_{peak} \rightarrow \frac{1}{\frac{2(1+\alpha)}{2+\alpha} + \frac{1}{2\sigma_0}\left(\frac{R_0}{R}\right)^{2+\alpha}}. \tag{22.49}$$

The memory embodied in σ_0 of the original distribution eventually decays and b_{peak} reaches its asymptotic value (22.43), independent of b_0. Consequently, $f(b)$ must become more and more sharply peaked as R/R_0 increases. This occurs around the value of b for which $S \rightarrow 0$ in (22.47).

As a result, the initial values of b_0 in the ensemble members are all attracted to the particular value of b at the peak of the evolving $f(b)$ distribution. They congeal to a common value of b for the entire ensemble. Notice that the S^{-1} factor that promotes this attraction does not come from the original distribution itself, but from the determinant of the transformation of $b_0 \rightarrow b$ in (22.46). Thus it essentially describes the changing phase space available to the transformed distribution. This is governed by the dynamics of the orbital motions via the cosmic energy equation. Therefore it is relatively independent of the form of the initial distribution(although some initial distributions may not have had the time or ability to relax adequately).

Next we examine how the cosmic energy equation guides the evolving form of the spatial number distribution function $f(N)$. Consider an ensemble of comoving volumes each containing $N = \bar{N} + \Delta N$ galaxies, where $\bar{N} = \bar{n}V$ is constant but ΔN may fluctuate. The temperature and potential energy may fluctuate from one volume (ensemble member) to another, and part of these fluctuations will be caused by the fluctuations in N. Without averaging over the ensemble, we may explicitly identify the fluctuations caused by different N by writing

$$K = \frac{3}{2}T(t)N, \tag{22.50}$$

$$W = -w(t)N^2, \tag{22.51}$$

$$b(t) = \frac{w(t)}{3T(t)}N \equiv \beta(t)(1 + \delta(t)), \tag{22.52}$$

with

$$N = \bar{N}(1 + \delta(t)) \tag{22.53}$$

and

$$\beta(t) \equiv \frac{\bar{N}w(t)}{3T(t)}. \tag{22.54}$$

Here T and W also contain average and fluctuating parts due to the different and evolving configurations of galaxies within the volume. Fortunately, we will not need to identify these explicitly to see how the cosmic energy equation constrains the fluctuations of N.

Substituting (22.50)–(22.54) into the cosmic energy equation, and using the expansion radius $R(t)$ rather than time directly, we obtain

$$\frac{d}{dR}[T(1-2\beta) + T(1-4\beta)\delta - 2T\beta\delta^2] = -\frac{2T}{R}[1 - \beta + (1-2\beta)\delta - \beta\delta^2]$$

(22.55)

for the cosmic energy equation including particle fluctuations. It reduces to the standard result when $\delta = 0$ and is valid for any value of δ. As the Universe evolves, it describes how any given particle fluctuation $\delta(R)$ changes. Of course to solve for $\delta(R)$ exactly, we would have to know $T(R)$ and $W(R)$, which would again require detailed kinetic theory solutions. But we can still extract some less detailed dynamical understanding from (22.55).

The first two terms on each side of (22.55) are just the average cosmic energy equation, which is satisfied for the ensemble and so they cancel out. This leaves

$$\frac{d}{dR}[T\delta - 2T\delta(2+\delta)\beta] = -\frac{2T}{R}\delta + \frac{2T}{R}\delta(2+\delta)\beta.$$

(22.56)

Initially we may regard $\beta(R)$ as a small parameter if the distribution starts close to a Poisson. Then we can solve (22.56) separately for the terms not involving β and the terms linear in β (there being no terms of higher order). The lowest order terms give

$$T\delta = T_0\delta_0 \left(\frac{R_0}{R}\right)^2.$$

(22.57)

Thus we recover the essentially adiabatic evolution for T modified by any change in $\delta(R)$. Next, solving for the terms proportional to β, we get

$$T\delta(2+\delta)\beta = T_0\delta_0 (2+\delta_0) \beta_0\frac{R_0}{R}.$$

(22.58)

Here we are not so interested in $T(R)$, $\delta(R)$, or $\beta(R)$ as in the developing form of the distribution function. This just requires $\delta_0(\delta)$, which from (22.58) is

$$\delta_0 = -1 + [1 + a^{-1}\delta(2+\delta^2)]^{1/2},$$

(22.59)

where

$$a(R) \equiv \frac{T_0\beta_0R_0}{T\beta R}$$

(22.60)

and we take the positive square root since $1 + \delta_0 = N_0/N > 0$.

Suppose the initial distribution of N is Poisson:

$$f(N_0, R = R_0) = \frac{\bar{N}^{N_0}}{N_0!} e^{-\bar{N}}. \tag{22.61}$$

Then for large enough $\bar{N}$, standard texts on probability theory show that this can be approximated by a Gaussian

$$f(N_0) \approx \frac{1}{\sqrt{2\pi\bar{N}}} e^{-(N_0 - \bar{N})^2/2\bar{N}}, \tag{22.62}$$

because the Poisson is a limit of binomial distribution that can be approximated by a Gaussian. Consequently, from (22.53)

$$f(\delta_0) = \left(\frac{\bar{N}}{2\pi}\right)^{1/2} e^{-\bar{N}\delta_0^2/2}, \tag{22.63}$$

remembering the Jacobian. At later times $f(\delta_0)$ is therefore transformed by (22.59) into

$$\begin{aligned}
f(\delta) &= f\left[\delta_0(\delta)\right] \frac{d\delta_0}{d\delta} \\
&= \left(\frac{\bar{N}}{2\pi}\right)^{1/2} \frac{(1+\delta)}{ag(a,\delta)} e^{-\bar{N}[g(a,\delta)-1]^2/2},
\end{aligned} \tag{22.64}$$

where

$$g(a, \delta) \equiv [1 + a^{-1}\delta(2 + \delta)]^{1/2}. \tag{22.65}$$

One can readily see from this how the original Gaussian is modified, especially in the limits $g(a, \delta) \approx 1$ and $g(a, \delta) \gg 1$. Moreover, the same technique can be applied to other statistically homogeneous initial distributions to get a general idea of their evolution. For a more complete discussion, however, we need a change of strategy.

23

Short Review of Basic Thermodynamics

Now the resources of those skilled in the use of
extraordinary forces are as infinite as the heavens
and earth, as inexhaustible as the flow of the
great rivers, for they end and recommence –
cyclical as are the movements of the sun and moon.

Sun Tzu

23.1 Concepts

Gravity is an extraordinary force and understanding its more profound implications for the cosmological many-body problem requires many strategies. So far, we have followed two broad avenues of insight into the instability and clustering of infinite gravitating systems: linear kinetic theory and numerical N-body simulations. Now we turn onto a third avenue: thermodynamics. Classical thermodynamics is a theory of great scope and generality. It survived the relativity and quantum mechanical revolutions of physics nearly intact. In part, this was because among all theories of physics thermodynamics has the least physical content. Its statements relate very general quantities that must be defined anew, through equations of state, for each specific application. With this view, it is natural to ask whether thermodynamics also subsumes gravitating systems.

The answer is yes, with certain caveats and qualifications. Results of gravitational thermodynamics – gravithermodynamics, or GTD for short – are often surprising and counterintuitive compared to the thermodynamics of ordinary gases. Specific heats, for example, can be negative and equilibrium is a more distant ideal. Basically, these differences are caused by the long-range, unsaturated (unshielded) nature of gravitational forces. As a result, rigorous understanding of GTD is less certain than for ordinary thermodynamics. The present situation is a bit similar to the early thermodynamic gropings of Watt, Carnot, Kelvin, and Joule.

Straightforward introduction of gravity into thermodynamics leads again to the Jeans instability from a new point of view. It links up with kinetic theory, the cosmic energy equation, and statistical mechanics. All these connections provide quite powerful insights into nonlinear clustering. Their results agree well with N-body simulations and observations of galaxy clustering.

This chapter reviews ideas and relations from ordinary thermodynamics that we will need as we go along. It makes no attempt to be exhaustive, wide ranging, or subtle, for many texts discuss this rich subject in great detail. Some examples are Callen (1985), Fowler (1936), Goodstein (1985), Hill (1986, 1987), and Huang (1987). We follow the discussion in GPSGS (Section 29) closely with some extensions. Our goal here is mainly to establish a clear description and consistent notation. Thermodynamics sometimes seems more complicated than it really is because different sets of dependent and independent variables are useful in different situations. Lack of attention to such details can cause utter confusion.

A basic premise of classical thermodynamics is that systems in equilibrium can be characterized by a finite set of macroscopic parameters. These may be averages of microscopic parameters that are not known in detail and that vary in nature and degree among different types of systems. To discover that a real system is in equilibrium would be to disturb it. Therefore, equilibrium is always an idealization and means that macroscopic disturbances occur on timescales very long compared to microscopic interaction or relaxation timescales.

Macroscopic parameters that describe the system normally include the total internal energy U, the entropy S, the volume V that the system occupies, and the number of objects N it contains. If it contains more than one type of object each species is characterized by its number N_i. Number and volume are obvious parameters. The reason for using the internal energy is that it is a conserved quantity within the system, usually the only one.

Entropy, however, is not so obvious. A second basic premise of thermodynamics is that one can imagine an ensemble of similar systems, each with different values of their macroscopic parameters. If systems in the ensemble are allowed to interact very weakly, just enough to reach equilibrium, then the most probable value of the macroscopic parameters follows from maximizing the entropy. Thus entropy will be a function of the other parameters, $S = S(U, V, N)$. This relation is called the fundamental equation of thermodynamics in the entropy representation. It can be inverted to give the fundamental equation in the energy representation: $U = U(S, V, N)$. It is always important to remember which representation is being used.

Entropy, like the conductor of a symphony orchestra, determines the roles of the other players. The idea that the maximum value of one thermodynamic function can determine the most probable value of other parameters evolved from variational principles in classical mechanics. Gibbs's bold leap from one system to an ensemble of systems created the statistical mechanical foundation of thermodynamics.

A small change in the parameters of a system is represented by the differential of the fundamental equation. In the energy representation this is

$$dU = \left(\frac{\partial U}{\partial S}\right)_{V,N} dS + \left(\frac{\partial U}{\partial V}\right)_{S,N} dV + \left(\frac{\partial U}{\partial N}\right)_{S,V} dN = T\,dS - P\,dV + \mu\,dN. \tag{23.1}$$

Quantities held constant in a partial derivative are often written explicitly for clarity, and as a reminder of the representation used. The three derivatives here are the temperature, pressure (conveniently defined with a minus sign so that energy decreases as volume increases), and the chemical potential. Similarly, in the entropy representation

$$dS = \left(\frac{\partial S}{\partial U}\right)_{V,N} dU + \left(\frac{\partial S}{\partial V}\right)_{U,N} dV + \left(\frac{\partial S}{\partial N}\right)_{U,V} dN = \frac{1}{T}\,dU + \frac{P}{T}\,dV - \frac{\mu}{T}\,dN, \tag{23.2}$$

noting that

$$\left.\frac{\partial S}{\partial V}\right)_{U,N} = -\frac{(\partial U/\partial V)_{S,N}}{(\partial U/\partial S)_{V,N}}. \tag{23.3}$$

Of course, (23.2) could have been derived from (23.1) just by transposing terms, but that would have been neither general nor instructive.

Equation (23.3) is a simple instance of relations among implicit partial derivatives. Since these relations pervade thermodynamics, it is worth pausing to review them. Consider an arbitrary continuous function $A(x, y, z)$ of, say, three variables. We can relate the partial derivatives of A since given values of $A(x, y, z)$ define implicit relations among its independent variables. When A changes, its total differential is

$$dA = \left.\frac{\partial A}{\partial x}\right)_{y,z} dx + \left.\frac{\partial A}{\partial y}\right)_{z,x} dy + \left.\frac{\partial A}{\partial z}\right)_{x,y} dz. \tag{23.4}$$

Since this relation holds for all values of dA, we may choose $dA = 0$ and, in addition, keep one of the variables constant, $dz = 0$, say. The result, dividing through by dx, is

$$\left.\frac{\partial y}{\partial x}\right)_{A,z} = -\frac{(\partial A/\partial x)_{y,z}}{(\partial A/\partial y)_{z,x}}, \tag{23.5}$$

which has the form of (23.3). Similar relations follow by setting $dx = 0$ and $dy = 0$, which is equivalent to cyclic permutation of x, y, and z (i.e., letting $x \to y$, $y \to z$, and $z \to x$). If we divide (23.4) through by dy to give the inverse of (23.5), we see that

$$\left.\frac{\partial x}{\partial y}\right)_{A,z} = \frac{1}{(\partial y/\partial x)_{A,z}}. \tag{23.6}$$

Multiplying the three forms of (23.5) together and using (23.6) shows that

$$\left(\frac{\partial x}{\partial y}\right)_{A,z} \left(\frac{\partial y}{\partial z}\right)_{A,x} \left(\frac{\partial z}{\partial x}\right)_{A,y} = -1. \tag{23.7}$$

Another helpful relation is a chain rule when x, y, and z are themselves functions of a parameter w, so that $dx = (dx/dw)\, dw$, etc. Substituting these three differential relations into (23.4) for $dA = dz = 0$, and comparing the result with (23.5), gives

$$\left.\frac{\partial y}{\partial x}\right)_{A,z} = \frac{(\partial y/\partial w)_{A,z}}{(\partial x/\partial w)_{A,z}}. \tag{23.8}$$

Returning to (23.1) we see that the temperature, pressure, and chemical potential (whose physical justifications are left to the standard textbooks) are all derivatives of the fundamental equation. Thus these variables play a different role than S, V, and N. This shows up in the relation among internal energy, heat, and work.

Heat and work are forms of energy transfer. Unlike internal energy, a system does not contain a certain quantity of work or heat. These are processes. In fact, the amount of heat and work needed to transfer a given quantity of energy depends strongly on how the transfer is made. An early triumph of experimental calorimetry was to show that a small quasi-static transfer (i.e., between states very close to equilibrium) of heat, dQ, can be associated with a quantity whose increase is independent of the transfer method. The trick was to divide the heat change by the temperature,

$$\frac{dQ}{T} = dS, \tag{23.9}$$

and the resulting quantity is the entropy change. The symbol dQ indicates an "imperfect differential," one whose integrated value depends on the path of integration in a thermodynamic $S - V - N$ phase space. Heat is energy transfer into the relative "random" microscopic motions of objects in the system.

The total internal energy transferred quasi-statically is the sum of the heat, mechanical work, and chemical work done on the system:

$$dU = dQ + dW_m + dW_c. \tag{23.10}$$

Substituting (23.9) and comparing (23.1) shows that the mechanical and chemical quasi-static work are

$$dW_m = -P \, dV \tag{23.11}$$

and

$$dW_c = \mu \, dN. \tag{23.12}$$

Since dU and $P \, dV$ can be determined experimentally, (23.10) provides a way to measure the amount of energy transferred as heat. It also makes clear how the choice of fundamental variables determines that the mechanical work is $-P \, dV$. Otherwise, confusion could have arisen from a simple expectation that dU might also contain a $V \, dP$ term.

Another important difference between the fundamental and derivative variables is their extensive–intensive nature. Suppose we imagine two systems that do not interact. Then the values of the fundamental variables, U, S, V, and N, for the combined system are the sum of their values for the individual systems. These quantities are "extensive." On the other hand, combining noninteracting systems in mutual equilibrium does not change the values of their derivative variables, T, P, and μ. They are "intensive." For the special case of perfect gases a stronger definition of extensive is possible. Instead of imagining noninteracting systems, we consider an actual system and subdivide it into two parts with an imaginary wall. Then the extensive and intensive properties apply to interacting regions of actual systems. This applies because the interaction is very weak – just strong enough to create equilibrium. Imperfect gases with long-range correlations, or gravitating systems with long-range forces, do not always satisfy this more restrictive definition of extensive, so one must be a bit careful in drawing consequences from it.

With ordinary thermodynamics, the equality of T, P, or μ between two actual subsystems means they are in thermal, mechanical, or chemical equilibrium respectively. This follows from the entropy maximum principle. For example, consider two subsystems that have a constant total energy $U = U_1 + U_2$ and are separated by a rigid wall that transmits heat but not matter. The total entropy $S = S_1 + S_2$. Maximizing total entropy at constant V_1, V_2, N_1, and N_2 then gives, from (23.2),

$$dS = 0 = dS_1 + dS_2 = \frac{1}{T_1} dU_1 + \frac{1}{T_2} dU_2 = \left(\frac{1}{T_1} - \frac{1}{T_2} \right) dU_1, \quad (23.13)$$

whence $T_1 = T_2$ at equilibrium. Results for P and μ follow similarly. The extensive property of entropy and the conservation of energy are essential to this line of argument.

23.2   Interrelations

Our discussion so far has summarized the framework for describing thermodynamic information. It is useful to emphasize one or another part of this framework in different situations. Therefore we now summarize the interrelations between different descriptions that, however, all contain the same basic information.

From (23.1) we obtain the intensive parameters as functions of the basic extensive variables: $T = T(S, V, N)$, $P = P(S, V, N)$, $\mu = \mu(S, V, N)$. These are the "equations of state." In terms of the basic variables, each equation of state is a partial differential equation. Therefore, it cannot contain as much information about the system as the fundamental equation: A function of integration, at least, is missing.

The set of all equations of state can, however, be equivalent to the fundamental equation in information content. Since the fundamental equation depends on extensive variables, it is first-order homogeneous:

$$U(\lambda S, \lambda V, \lambda N) = \lambda U(S, V, N). \quad (23.14)$$

Differentiating both sides with respect to λ and then selecting the particular value $\lambda = 1$ gives the Euler equation

$$U = TS - PV + \mu N. \quad (23.15)$$

Hence knowledge of $T(S, V, N)$, $P(S, V, N)$, and $\mu(S, V, N)$ enables us to retrieve the fundamental equation $U(S, V, N)$.

An equation that complements (23.1) and involves differentials of the intensive variables can now be derived by taking the differential of (23.15) and subtracting

(23.1) from it. The result is called the Gibbs–Duhem relation:

$$S\,dT - V\,dP + N\,d\mu = 0. \tag{23.16}$$

Thus variations of the intensive parameters are not completely independent. By substituting $S(T, P, N)$, $V(T, P, N)$, and $\mu(T, P, N)$ into (23.16) and integrating we can find a relation between T, P, and μ. Alternatively we can relate T, P, and μ by eliminating S, V, and N from the equations of state. Note that the relation among P, V, T, and N, which for perfect gases is called *the* "equation of state," is a somewhat hybrid representation and does not contain complete information about the thermodynamics.

Everything so far can also be restated in the entropy representation that uses $S(U, V, N)$ as the fundamental equation. Then the Gibbs–Duhem relation, for example, becomes

$$U\,d\left(\frac{1}{T}\right) + V\,d\left(\frac{P}{T}\right) - N\,d\left(\frac{\mu}{T}\right) = 0. \tag{23.17}$$

Some thermodynamic derivatives have an obvious physical meaning and are often used for convenient expressions of relationships. Among these are the isothermal compressibility

$$\kappa_T \equiv -\frac{1}{V}\left.\frac{\partial V}{\partial P}\right)_{N,T}, \tag{23.18}$$

the coefficient of thermal expansion

$$\alpha \equiv \frac{1}{V}\left.\frac{\partial V}{\partial T}\right)_{N,P}, \tag{23.19}$$

the specific heat at constant volume

$$C_V \equiv \frac{T}{N}\left.\frac{\partial S}{\partial T}\right)_{N,V} = \frac{1}{N}\left.\frac{dQ}{dT}\right)_{N,V}, \tag{23.20}$$

and the specific heat at constant pressure

$$C_P \equiv \frac{T}{N}\left.\frac{\partial S}{\partial T}\right)_{N,P} = \frac{1}{N}\left.\frac{dQ}{dT}\right)_{N,P}. \tag{23.21}$$

Each involves the fractional change of an important variable under specified conditions.

Other thermodynamic derivatives can often be conveniently reformulated by making use of the equality of mixed partial derivatives of the fundamental equation. For example, from

$$\frac{\partial^2 U}{\partial V \, \partial S} = \frac{\partial^2 U}{\partial S \, \partial V} \tag{23.22}$$

we readily get

$$\left.\frac{\partial T}{\partial V}\right)_{S,N} = -\left.\frac{\partial P}{\partial S}\right)_{V,N}. \tag{23.23}$$

This is an instance of the "Maxwell relations" of which there are myriads from each basic representation and choice of fundamental variables.

The fundamental variables S, V, and N may not always be best for a particular situation. For example, we may want to rewrite the fundamental equation for two systems in thermal equilibrium in terms of their common temperature T, along with V and N, particularly since T is more readily measurable than their entropy S. Naive substitution of $S(T, V, N)$ into $U(S, V, N)$ to produce $U(T, V, N)$ does not work. Although it produces a perfectly correct relation, $U(T, V, N)$ is a first-order partial differential equation whose integral contains an undetermined function. Thus the information contents of $U(S, V, N)$ and $U(T, V, N)$ are not equivalent.

Finding an equivalent formulation is easy with the Legendre transform. The basic idea is to replace a point-by-point description of a curve (or surface) by a specification of its tangent lines (or planes). A simple illustration with one variable exemplifies this. Suppose the internal energy were just a function of entropy $U(S)$ and we want to replace this with a function of temperature $F(T)$ containing equivalent information. Instead of the point-by-point specification $U(S)$ we consider its tangent whose slope at any point is

$$T(S) = \frac{dU}{dS}. \tag{23.24}$$

But to specify the tangent fully requires its intercept F, as well as its slope. Thus in the U–S plane, the tangent is (see Figure 23.1)

$$T = \frac{\Delta U}{\Delta S} = \frac{U - F}{S - 0}. \tag{23.25}$$

The key idea is to use the intercept to specify the new function of T, or, rearranging (23.25),

$$F = U - TS = U - \frac{dU}{dS}S. \tag{23.26}$$

Finally, using $U(S)$ to eliminate U in (23.26) and substituting $S(T)$ from inverting (23.24) gives $F(T)$. This is the Legendre transform of $U(S)$. Starting with $F(T)$ we can recover $U(S)$ by inverting the process, except at a critical point where

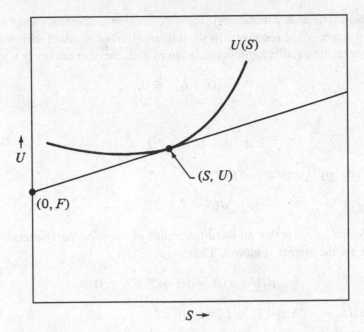

Fig. 23.1. Specification of the curve $U(S)$ by means of the slope and intercept of its tangent.

$d^2 F/dT^2 = 0$. This last condition is related to the requirement that the system be thermodynamically stable; mathematically it ensures that T depends on S.

The symbols of this example were chosen with knowledge aforethought. $F(T, V, N)$ is the free energy and is derived from $U(S, V, N)$ as in (23.26) by keeping the other variables constant. The free energy is one example of a thermodynamic potential. Other Legendre transforms lead to other potentials. If pressure is the relevant variable, then the enthalpy

$$H(S, P, N) = U + PV \qquad (23.27)$$

is useful. If both temperature and pressure are specified, we can use the Gibbs function

$$G(T, P, N) = U - TS + PV \qquad (23.28)$$

and so on. The equivalence of partial second derivatives of the thermodynamic potentials breeds whole new sets of Maxwell relations. Legendre transformations may also be applied in the entropy representation where the results are called generalized Massieu functions rather than thermodynamic potentials. An example is $S(P/T) = S - (P/T)V$, which is a function of U, P/T, and N.

A major role of thermodynamic potentials is to describe systems in equilibrium at constant temperature, pressure, or chemical potential. For example, if a system can exchange energy with a reservoir (another system of such great heat capacity that its temperature remains unchanged), then the equilibrium values of unconstrained internal thermodynamic parameters in the system minimize the free energy at the constant temperature of the reservoir. To see this, recall from standard thermodynamics texts that in stable equilibrium the total system and reservoir energy is a minimum:

$$d(U + U_r) = 0, \tag{23.29}$$

and

$$d^2(U + U_r) = d^2U > 0 \tag{23.30}$$

when the entropy is constant so that

$$d(S + S_r) = 0. \tag{23.31}$$

The reason $d^2U_r = 0$ is that all the differentials of intensive parameters vanish, by definition, for the mighty reservoir. Thus

$$d(U + U_r) = dU + T_r \, dS_r = 0 \tag{23.32}$$

and, since $dS_r = -dS$ and $T_r = T$, we get

$$d(U - TS) = dF = 0. \tag{23.33}$$

Finally (23.30) shows that $d^2F > 0$. Similarly, systems in mechanical contact with a pressure reservoir have constant enthalpy.

23.3   Connections with Kinetic Theory and Statistical Mechanics

Although thermodynamics relates various quantities, it does not tell us what those quantities are for any particular physical system. We need to find the actual equations of state or the fundamental equation. Links with kinetic theory and statistical mechanics make this step possible.

We have already found one link with kinetic theory through the relation (22.28) between gravitational potential energy and the two-point correlation function $\xi(r)$. (This provides an example of confusion when two notations, each standard in a different subject, come together. To help remedy confusion, we shall continue to use U for the total internal energy and now let W become the gravitational correlation energy, rather than the work. We will not have much more use for thermodynamic "work" here, which only occurs in an incomplete differential form anyway.) Through the complete single-particle distribution function we also have a link with the kinetic energy, Equation (22.14), and thus with the total internal energy. For an ordinary homogeneous gas, whose random velocities are represented by a temperature T, the internal energy is

$$\frac{U}{NkT} = \frac{3}{2} + \frac{n}{2kT} \int_0^\infty \phi(r)\xi(r, n, T)\, 4\pi r^2 \, dr, \tag{23.34}$$

where k is Boltzmann's constant. Here $\phi(r)$ is the interparticle potential energy. Generally, the correlation function may depend on density $n = N/V$ and temperature as well as separation. The integral may also be taken over a finite volume.

This result is not a fundamental equation of the system since it is not of the form $U(S, V, N)$. Nor is it an equation of state in the energy representation since it is not of the form $T(S, V, N)$. But it is an equation of state, $T^{-1}(U, V, N)$, in the entropy representation since $T^{-1}(U, V, N) = \partial S(U, V, N)/\partial U$.

To find the fundamental equation and develop the thermodynamics we need two more equations of state. From physical intuition we would expect the pressure to equal that of a perfect gas minus a term proportional to the interaction energy density. Detailed kinetic theory discussions by Fowler (1936, Chapters 8 and 9) show this is indeed the case:

$$\frac{P}{kT} = n - \frac{n^2}{6kT} \int_0^\infty r \frac{d\phi(r)}{dr} \xi(r, n, T) \, 4\pi r^2 \, dr. \qquad (23.35)$$

(Chapter 25 discusses the derivation of this result.) Now we can integrate the Gibbs–Duhem equation (23.17) to find the third equation of state,

$$\frac{\mu}{T} = \frac{\mu}{T}(U, V, N) = \frac{\mu}{T}(u, v), \qquad (23.36)$$

where the equations are all expressed in terms of $u = U/N$ and $v = V/N$. The three equations of state can then be substituted into the entropy representation of the Euler equation,

$$S(U, V, N) = \frac{1}{T}U + \frac{P}{T}V - \frac{\mu}{T}N, \qquad (23.37)$$

to give the fundamental equation. This procedure forges the link between thermodynamics and kinetic theory. As an exercise the reader can apply it when $\xi = 0$ to obtain the entropy of a perfect monotonic gas:

$$S = \frac{N}{N_0}S_0 + NR \ln\left[\left(\frac{U}{U_0}\right)^{3/2}\left(\frac{V}{V_0}\right)\left(\frac{N}{N_0}\right)^{-5/2}\right], \qquad (23.38)$$

where the fiducial entropy is

$$S_0 = \frac{5}{2}N_0 R - N_0\left(\frac{\mu}{T}\right)_0, \qquad (23.39)$$

and $R = 1.986$ cal/mole K.

There are equivalent, alternative paths to the fundamental equation. Two other paths are especially useful. One is to find the specific entropy $s = S/N$ directly by integrating

$$ds = \frac{1}{T} du + \frac{P}{T} dv. \tag{23.40}$$

The second is to calculate the Legendre transform of the entropy. Since all our equations of state are expressed in terms of temperature, the appropriate Legendre transform is the Massieu function

$$S(1/T, V, N) \equiv S(U, V, N) - \frac{\partial S}{\partial U} U = S - \frac{1}{T} U = -\frac{F}{T}. \tag{23.41}$$

This Massieu function is trivially related to the free energy. We call it the free entropy. Since S is a function of U, N, and V, Equation (23.41) gives the relation

$$\left. \frac{\partial (F/T)}{\partial (1/T)} \right)_{N,V} = U. \tag{23.42}$$

Therefore, an integration from $T = T$ to $T = \infty$ using (23.34) for $U(T)$ provides another version of the fundamental equation for the thermodynamics of kinetic systems.

Next we turn to links between thermodynamics and statistical mechanics. These provide other routes toward equations of state, routes that are sometimes more convenient or easier to follow. The basic principles are available in a multitude of texts, so again we just recall the most important ones here.

To each reservoir, and its associated thermodynamic potential, there corresponds an ensemble of statistical mechanics. These ensembles are imaginary collections whose members each replicate the macroscopic properties of a system. Microscopic properties of the system (e.g., which particle or galaxy is where) generally differ from member to member. An ensemble whose members are completely isolated and have well-defined energies, volumes, and numbers of objects is called the "microcanonical ensemble." Its fundamental equation depends just on these variables: $S(U, V, N)$. If the members of an ensemble are in thermal equilibrium with a heat reservoir they define a "canonical ensemble" whose fundamental equation is the free energy $F(T, V, N)$. If the members of the ensemble are in equilibrium with both thermal and particle reservoirs they define the "grand canonical ensemble." Its fundamental equation is the grand canonical potential $U(T, V, \mu) = U(S, V, N) - TS - \mu N$.

Boltzmann and Planck provided the fundamental statistical plank underlying thermodynamics. The entropy of a macrostate of the microcanonical ensemble is related to the number Ω of microscopic configurations the ensemble has for that macrostate by

$$S = k \ln \Omega. \tag{23.43}$$

In this basic equation k is Boltzmann's constant. That something like this relation holds is plausible because entropy is additive while the probabilities of independent microscopic states are multiplicative. The amount of mystery (another variable of statistical mechanics?) in this remark is reduced by considering two independent systems. Their combined entropy is $S_{12} = S_1 + S_2$. Their combined microscopic probability is $\Omega_{12} = \Omega_1 \Omega_2$. If S is to be an extensive function $f(\Omega)$ just of Ω, then $f(\Omega_1 \Omega_2) = f(\Omega_1) + f(\Omega_2)$. The general solution of this functional equation has the form of (23.43).

To apply Boltzmann's equation (23.43) we must learn how to count microstates. Each microstate is assumed to have an equal a priori probability of occurring. Some macrostates of the system, however, are consistent with a greater number of arrangements of microstates. These macrostates have greater entropy and a higher probability of occurring in the ensemble. When the counting is done for a classical gas, standard texts on statistical mechanics show that

$$S = k \ln Z + \frac{U}{T}, \tag{23.44}$$

where Z is the partition function

$$Z = \sum_i e^{\epsilon_i / kT}. \tag{23.45}$$

The sum is taken over all the different energy states ϵ_i of the system. If the number of objects in each energy state is n_i, then $\sum_i n_i \epsilon_i = U$ and $\sum_i n_i = N$. It is now possible, but usually complicated, to express T in terms of U, V, and N to get $S(U, V, N)$. Instead it is more natural to use this approach to deal with the fundamental equation for the free energy

$$F = U - TS = -kT \ln Z \tag{23.46}$$

from the two previous equations. Generally, the energy states depend on the volume of the system. Knowledge of the partition functions, therefore, tells us all the thermodynamic parameters. For example,

$$U = F + TS = -T^2 \frac{\partial}{\partial T} \left(\frac{F}{T} \right)_V = \frac{kT^2}{Z} \frac{\partial Z}{\partial T}, \tag{23.47}$$

using (23.46) for the last step. Combining this result with (23.44) expresses the entropy directly in terms of Z. The rest of thermodynamics follows by the methods described earlier. Of course, the partition functions must be worked out for each specific system.

When working out these details it is often useful to take the "thermodynamic limit" of an infinite system. The reason is that the energy states, which make up the partition function, generally depend on the size and shape of the system. Hidden in (23.45) is an implicit dependence on volume. Thus, the resulting intensive parameters could depend on the shape of the system. Moreover, they may also depend on which ensemble is used for their calculation. The thermodynamic limit avoids these problems, ordinarily, by calculating Z in the limit $N \to \infty$, $V \to \infty$ with $N/V \to$ constant. It is the same general idea we have met before in going from the kinetic description to a fluid description. This limit has been proved rigorously to exist for many, but not all, conditions. Originally the main restrictions were that the interparticle potential decrease faster than r^{-3} as $r \to \infty$ and increase faster than r^{-3} as $r \to 0$. The repulsive core prevents collapse, while the long-range decay prevents global properties from influencing the thermodynamics. Later the proofs were extended to inverse square forces under certain conditions. In Chapters 24 and 25 we shall see why thermodynamics applies to the cosmological many-body problem without having to take the "thermodynamic limit." This may leave some residual dependence on the size and shape of the system. But this will provide information which we can turn to our advantage.

23.4   The Three Laws of Thermodynamics

The law that entropy always increases – the second
law of thermodynamics – holds, I think, the supreme
position among the laws of Nature. If someone
points out to you that your pet theory of the universe
is in disagreement with Maxwell's equations – then so
much the worse for Maxwell's equations. If it is
found to be contradicted by observation – well, these
experimentalists do bungle things sometimes. But if
your theory is found to be against the second law of
thermodynamics I can give you no hope; there is
nothing for it but to collapse in deepest humiliation.

Eddington

As thermodynamic systems evolve quasi-statically through a sequence of equilibrium states, they are constrained by three general principles. (There is a fourth law, sometimes called the "Zeroth Law," that just says temperature is a sensible concept that determines the thermal equilibrium of systems in contact. So if two systems are each in thermal equilibrium with a third, then they are also in equilibrium with each other. This has a natural and more general dynamical interpretation in terms of the kinetic energy of the members of the system.) The three laws can be stated in various ways, some of which are delightfully abstract (e.g., Landsberg, 1961; Buchdahl, 1966). Here I use the standard formulation that emphasizes their essential physical properties.

The First Law we have already met in (23.10). It is the fact that energy is conserved:

$$dU = dQ + dW, \tag{23.48}$$

where dW is the total work done on the system. (The sign before dW depends on a preposition. Some formulations define dW as the work done by the system and therefore precede it by a negative sign.) All types of work such as mechanical, chemical, electromagnetic, and gravitational contribute to energy conservation. Indeed, in the historical development of thermodynamics, when a new type of work was recognized, the concept of energy (due to Leibnitz) was expanded to include it to force the total energy to remain a conserved quantity. More recently, we recognize that energy conservation follows more profoundly from the invariance of the fundamental dynamical laws under time translation $t \rightarrow t + t_0$. If the interaction potential of the members of a system does not depend explicitly on time, energy is conserved. This is a special case of Noether's theorem, which says that there is a conservation law for any continuous dynamical symmetry of the system. Since the fundamental interactions are also translationally and rotationally symmetric in space, linear and angular momentum are conserved. The resulting statistical mechanics and thermodynamics can be generalized to include all these conservation properties, as has long been recognized in stellar dynamics (cf. GPSGS, Sections 39 and 60).

Heat is a different matter, and here thermodynamics stands apart from other formulations of physical theory. It separates a particular macroscopic manifestation of energy, random motion, from all the other dynamics. Initially the reason was historical, stimulated by Benjamin Thompson's caloric research, but ultimately it survived because of its extraordinary success in describing an enormous range of physical systems. (Thompson's personal success as an early scientific entrepreneur and general scamp was equally extraordinary. In 1792, the Elector of Bavaria made him the Imperial Count Rumford of the rapidly decaying Holy Roman Empire. See Brown (1979) for more.) As a system evolves quasi-statically, the small change in its heat, dQ, like the small change in its work, depends on the path the evolution follows in the space of thermodynamic variables such as P, V, and T. However, a small change in energy, and thus its total change $\int dU$ is independent of path, according to the First Law. It only depends on the limits of integration, (i.e., on its initial and final states). This makes dU an exact differential having the form $dz = x(A, B)\, dA + y(A, B)\, dB$ satisfying the condition $\partial x / \partial B = \partial y / \partial A$, for the case of two independent variables A and B. The sum of the work done and the heat produced is the total energy change, independent of how it is apportioned.

From the property that U is exact, we can derive a standard relation among dQ, dT, and dV that is useful for the cosmological many-body problem. Starting with $U(S, V, N)$, we eliminate the entropy using the equation of state $T = T(S, V, N)$ to obtain $U(T, V, N)$. Although this loses some thermodynamic information, as described in Section 23.2, the differential of U remains exact in these variables. If we hold any one of these variables constant, U will remain exact in the other two. Holding N constant,

$$dU = \frac{\partial U}{\partial T}\bigg)_{N,V} dT + \frac{\partial U}{\partial V}\bigg)_{N,T} dV. \qquad (23.49)$$

Substituting this into the First Law and using $dW = -P dV$ gives

$$dQ = \frac{\partial U}{\partial T}\bigg)_{N,V} dT + \left[\frac{\partial U}{\partial V}\bigg)_{N,T} + P\right] dV, \qquad (23.50)$$

in which

$$\frac{\partial}{\partial V}\left[\frac{\partial U}{\partial T}\bigg)_{N,V}\right]_{N,T} = \frac{\partial}{\partial T}\left[\frac{\partial U}{\partial V}\bigg)_{N,T}\right]_{N,V} \qquad (23.51)$$

follows from applying the exactness condition to (23.49). Similar relations follow from $U(P, V, N)$ and $U(P, T, N)$ and from holding other variables constant. These provide a method for calculating dQ provided the relevant derivatives of the energy are known. They also provide constraints on general equations of state because the entropy is also a perfect differential according to the Second Law.

The Second Law of thermodynamics essentially says that if an isolated system is not in equilibrium its entropy will tend to increase: $dS \geq 0$. The equality holds only for equilibrium states which have already undergone all spontaneous entropy-increasing changes. Then their entropy is a maximum and any further infinitesimal spontaneous changes are reversible. Boltzmann's postulate (23.43) implies that if a system evolves so that more and more microscopic states become available to it, its entropy increases. Some of these states may not have been available earlier because initial conditions prohibited them. With evolution, dissipation reduces the memory of these initial conditions as organized motion is converted into heat and more random motions dominate. A simple example (at least conceptually) is the nonlinear mixing of dynamical orbits. As the orbits relax, new microscopic configurations become available.

No real systems are in equilibrium. Gases evaporate, molecules combine, excited atoms decay, and galaxies cluster. The changing availability of states in real systems raises an important question. Why should an essentially equilibrium thermodynamic description apply to real systems? The answer depends on a separation of timescales. If the boundary conditions or the constituents of a system change on a timescale that is long compared to a relaxation timescale, then thermodynamics describes the processes that relax faster. On short relaxation timescales, microscopic states that correspond to long-term nonequilibrium changes are inaccessible to the system. Therefore these microstates do not influence the thermodynamics and can be ignored. Examples are chemical reactions that are very slow unless they are catalyzed (catalysts remove the restraint of an activation energy) and the gravitational formation of singularities in clusters of stars. The latter process is very slow compared to the dynamical crossing time of the cluster and is further restricted by the need to develop states of very low angular momentum. This is why thermodynamics provides a good description of star clusters before singularities form (cf. GPSGS, Sections 40 and 43–48). In other cases this ratio of timescales may itself be a function of the spatial lengthscale. Then a thermodynamic description can apply over certain lengthscales. For the cosmological many-body problem, this includes the shorter lengthscales where clustering is strongest.

Whereas the First Law is exact, the Second is statistical. The property that $dS > 0$ in a nonequilibrium system occurs on average and over time. There might be local regions and short intervals where $dS < 0$, but these are relatively rare. How rare may be seen by inverting Boltzmann's postulate (22.43) and finding the probability ratio $\Omega_2/\Omega_1 = \exp[(S_2 - S_1)/k]$ for a consequent state $S_2 < S_1$. In the classic case when all the molecules of an ideal gas migrate to one half of a room, $S_2 - S_1 = -Nk \ln 2$ and $\Omega_2/\Omega_1 \approx 2^{-N}$, which is tiny for any realistic value of N (even for a small cluster of randomly moving galaxies with $N \approx 10^2$). The increase of entropy also results from the coarse graining of cells that contain microstates in the system's phase space. This is discussed elsewhere (e.g., Chapter 17 earlier; GPSGS, Sections 10 and 38; and many other texts) in detail and is not needed further here.

Entropy, like energy, is an exact quantity and so dS is an exact differential. This has the consequence that dividing (23.50) by T and noting (23.9) gives an exact differential for dS when T and V undergo a small quasi-static change. Applying the exactness condition to the right-hand side, and taking (23.51) into account, gives

$$\left(\frac{\partial U}{\partial V}\right)_T = T\left(\frac{\partial P}{\partial T}\right)_V - P. \tag{23.52}$$

This result, which follows directly from the fact that the First and Second Laws imply the exactness of dU and dS, puts very strong constraints on the form of possible equations of state. We will later apply it to the cosmological many-body problem.

The Third Law of thermodynamics differs in character from the other two. In its original form, due to Nernst, it states that the entropy change is $\Delta S = 0$ for any reversible isothermal process at $T = 0$. Planck's subsequent version is that $S = 0$ at $T = 0$. While this version can be viewed as a thermodynamic consequence of Nernst's statement, it has a more physical interpretation in quantum statistical mechanics. A quantum system in a nondegenerate $T = 0$ ground state has only one microstate, and so $\Omega = 1$ in Boltzmann's relation (23.43), implying $S = 0$.

Unlike the first two laws, the third applies only at a particular temperature, $T = 0$. It has the consequence that a system initially with $T > 0$ cannot be brought to $T = 0$ by any reversible adiabatic process. Considering a change in pressure, the Third Law also implies

$$\left(\frac{\partial S}{\partial P}\right)_T = -\left(\frac{\partial V}{\partial T}\right)_P \rightarrow 0 \text{ as } T \rightarrow 0. \tag{23.53}$$

The second equality follows from the Maxwell relation (23.23). This shows that the coefficient of thermal expansion (23.19) also vanishes as $T \rightarrow 0$. Similarly, for a change in volume

$$\left.\frac{\partial S}{\partial V}\right)_T = \left.\frac{\partial P}{\partial T}\right)_V \to 0 \text{ as } T \to 0, \tag{23.54}$$

using the Maxwell relation from the free energy $F(T, V, N)$.

The main virtue of the Third Law is that the entropy of any state of any system has a unique value. This makes the idea of entropy more readily convertible into forms of information and disorder.

23.5   Fluctuations and Ensembles

Equilibrium thermodynamics deals with average macroscopic quantities. In reality, objects are always buzzing around inside the system. Within any subregion, and from region to region, the macroscopic quantities will fluctuate around their average values. A remarkable aspect of equilibrium thermodynamics is its ability to describe these fluctuations. Although fluctuations are normally small, they can become enormous in gravitational systems. We shall see that to comprehend the architecture of the world and distribution of the galaxies, we must understand these fluctuations.

Fluctuations occur because regions or subsystems – let us call them cells – are not generally isolated. They can exchange extensive quantities such as energy, volume, or numbers of objects with a much larger reservoir. When the cells interact just strongly enough with the reservoir, or with each other, that they are in equilibrium, then they have common values of their corresponding intensive variables such as temperature, pressure, or chemical potential. Idealized cells are imaginary macroscopic replicas with different microscopic states, and the reservoir is a purely mathematical construct. This can often be approximated in a real physical system by dividing it into spatial cells and taking the reservoir for any cell to be the rest of the system excluding that cell. This will clearly work best if the system is statistically homogeneous so that all cells have the same a priori properties.

Fluctuations are a way for real cells to find new states of existence. As cells exchange objects or energy with their surroundings (i.e., the reservoir), they probe different configurations of the entire system. Some of these configurations may become so energetically favorable that the system can escape from them only very slowly, if at all. Escape requires an enormously improbable chance fluctuation or a change in external boundary conditions.

The specific nature of these fluctuations will clearly depend on the constraints imposed by the reservoir. Reservoirs that allow a greater number of extensive variables to be exchanged have more thermodynamic degrees of freedom; any particular variable may behave differently than if the reservoir were more restrictive. The freedom of the reservoir follows directly from the type of ensemble that describes the system. In a microcanonical ensemble where all cells have exactly the same U, V, and N there are obviously no fluctuations. In statistical mechanics, this definition is relaxed to include cells within some small energy range ΔE. Then many texts show that for $\Delta E > 0$, the fluctuations around E tend to zero as $N \rightarrow \infty$ in most physical systems not near phase transitions. In a canonical ensemble, all cells have the same N and V, but their energies differ. These energies can be exchanged with the reservoir. Again, away from phase transitions the fluctuations are usually small and thus the canonical and microcanonical ensembles are thermodynamically equivalent as $N \rightarrow \infty$. The same is generally true for the grand canonical ensemble, which permits cells to exchange both energy and objects.

When fluctuations are large, however, these ensembles are not equivalent. This is the case for nonlinear galaxy clustering. Since both energy and galaxies can move across cell boundaries, the grand canonical ensemble is the one to use. Therefore in this introductory section we illustrate how fluctuations can be calculated simply in the grand canonical ensemble.

First we need to find the general fractional probability for finding a cell with a given energy E_i and number of particles (galaxies) N_i. If the total system (cell plus reservoir) is closed and contains N_{total} particles with total energy E_{total}, of which N_R and E_R are in the reservoir, then let the total number of states available to the total system be $\Omega_{\text{total}} (E_{\text{total}}, N_{\text{total}})$. The total number of states that the reservoir alone can have is $\Omega_R(E_{\text{total}} - E_i, N_{\text{total}} - N_i)$ if just one arbitrary cell, the ith, is not in the reservoir. A basic principle of statistical thermodynamics is that the ratio

$$f_i = \frac{\Omega_R (E_{\text{total}} - E_i, N_{\text{total}} - N_i)}{\Omega_{\text{total}} (E_{\text{total}}, N_{\text{total}})} \qquad (23.55)$$

is the probability of finding a cell in the state (E_i, N_i) in a grand canonical ensemble.

We may write this in terms of the entropy using Boltzmann's relation (23.43) as

$$f_i = \exp k^{-1} \left[(S_R(E_{\text{total}} - E_i, N_{\text{total}} - N_i) - S_{\text{total}} (E_{\text{total}}, N_{\text{total}}) \right]. \qquad (23.56)$$

Let U and $\bar{N}$ be the average values of the energy and number of objects in a cell. All cells have volume V. Since entropy is additive and is contained only in the cell and the reservoir,

$$S_{\text{total}} (E_{\text{total}}, N_{\text{total}}) = S(U, \bar{N}) + S_R(E_{\text{total}} - U, N_{\text{total}} - \bar{N}) \qquad (23.57)$$

and

$$S_R (E_{\text{total}} - E_i, N_{\text{total}} - N_i) = S_R(E_{\text{total}} - U + U - E_i, N_{\text{total}} - \bar{N} + \bar{N} - N_i)$$

$$= S_R(E_{\text{total}} - U, N_{\text{total}} - \bar{N})$$

$$+ \frac{U - E_i}{T} - \frac{\mu(\bar{N} - N_i)}{T}, \qquad (23.58)$$

where the last equation makes use of the fundamental relation (23.1). Now substituting (23.57) and (23.58) into (23.56), the contributions of the reservoir cancel out (as they must physically because altering E_{total} and N_{total} for a sufficiently large reservoir cannot change the thermodynamic properties of the cells), and we are left with

$$f_i = \exp\{[U - TS(U, \bar{N}) - \mu N - E_i + \mu N_i]/kT\}.$$ (23.59)

Since the sum of f_i over all possible states E_i and N_i of the cell must be unity,

$$Z \equiv \sum_i e^{-(E_i - \mu N_i)/kT} = e^{-\Psi/kT},$$ (23.60)

where

$$\Psi \equiv U - TS - \mu \bar{N} = U(T, V, \mu).$$ (23.61)

Here Ψ is gloriously known as "the grand canonical potential," and Z as the "grand canonical partition function." But we see from Section 23.2 that Ψ is just the Legendre transformation of the average energy with respect to both S and N. Thus we have derived the usual statistical mechanics relation

$$\Psi = -kT \ln Z,$$ (23.62)

which gives all the thermodynamic properties of the grand canonical ensemble. For example,

$$-\frac{\partial \ln Z}{\partial(1/kT)} = \frac{\partial(\Psi/kT)}{\partial(1/kT)} = U(T, V, \mu).$$ (23.63)

The power of the grand canonical potential arises from the fact that it provides the normalization of the statistical distribution, whatever the detailed arrangement of microstates. For the canonical ensemble in which N is fixed for each cell forever, we recover analogous results simply by setting $\mu = 0$ in the previous formulae. Thus the canonical potential is just $U - TS$, which we recognize as the free energy from (23.26).

Now to fluctuations. We can find their moments explicitly from the partition function. To find the fluctuation in N, for example, we differentiate the formula for $\bar{N}$ in the form

$$\bar{N} Z = \sum_i N_i e^{-(E_i - \mu N_i)/kT}$$ (23.64)

with respect to μ, the intensive variable corresponding to $\bar{N}$. Using (23.60) and (23.61) we find

$$\frac{\langle (N - \bar{N})^2 \rangle}{\bar{N}^2} = \frac{\langle (\Delta N)^2 \rangle}{\bar{N}^2} = \frac{kT}{\bar{N}^2} \frac{\partial \bar{N}}{\partial \mu}\bigg)_{T,V}.$$ (23.65)

This also gives the density fluctuation $\langle (\Delta n)^2 \rangle / \bar{n}^2$ since $n = N/V$ with V kept constant in (23.65). To put it in a more generally useful form, we note that, at constant T, the Gibbs–Duhem relation (23.16) gives $N d\mu = V dP$ so that

$$\frac{\partial \mu}{\partial (V/N)}\bigg)_T = \frac{V}{N} \frac{\partial P}{\partial (V/N)}\bigg)_T. \qquad (23.66)$$

Consequently,

$$-N^2 \frac{\partial \mu}{\partial N}\bigg)_{T,V} = V^2 \frac{\partial P}{\partial V}\bigg)_{T,N}, \qquad (23.67)$$

and the relative mean square fluctuation in N is

$$\frac{\langle (\Delta N)^2 \rangle}{\bar{N}^2} = -\frac{kT}{V^2} \frac{\partial V}{\partial P}\bigg)_{T,N} = \frac{kT}{V} \kappa_T = \frac{kT}{\bar{N}V} \frac{\partial N}{\partial P}\bigg)_{T,V}, \qquad (23.68)$$

where κ_T is the isothermal compressibility (23.18). In a similar manner, differentiating (23.64) with respect to T and manipulating the thermodynamic derivatives gives the grand canonical ensemble energy fluctuations

$$\frac{\langle (\Delta E)^2 \rangle}{kT^2} = C_V - \frac{\left[(P + \frac{U}{V}) \frac{\partial V}{\partial P})_{T,N} + T \frac{\partial V}{\partial T})_{N,P} \right]^2}{T \frac{\partial V}{\partial P})_{T,N}}, \qquad (23.69)$$

where C_V is the specific heat at constant volume (23.20). Dividing both sides by $\bar{E}^2$ gives the relative fluctuations. If this had been calculated for the canonical ensemble, only C_V would have appeared on the right-hand side; therefore the ensemble does make a difference.

For a perfect gas equation of state, $PV = NkT$, the relative root mean square (r.m.s.) number fluctuations in (23.68) are $1/\sqrt{N}$, usually very small, and characteristic of a Poisson distribution. The perfect gas relative r.m.s. energy fluctuations are also of order $1/\sqrt{N}$ since $C_V = 3Nk/2$ and the second term on the right-hand side of (23.69) is just $+Nk$. Thus for large N, there is little difference between the canonical and grand canonical energy fluctuations of a perfect gas; both are very small. Nonetheless, the extra degree of freedom in the grand canonical ensemble increases the energy fluctuations.

For an interacting gas, however, fluctuations can be much larger. The reason is that when the kinetic and potential energies of interacting objects are nearly in balance, a change in volume may produce a much smaller change in pressure or temperature. Instead of the resulting isobaric or adiabatic change of energy being realized as kinetic motion, much of it is soaked up by the change of potential energy. The system, in a sense, loses its elasticity. This makes it possible for large fluctuations to grow at little cost in total (kinetic plus potential) energy. Large numbers of objects can move relatively freely from cell to cell in the grand canonical ensemble, and the second term on the right-hand side of (23.69) dominates. Formally we may even have $\partial V/\partial P)_{T,N} \to -\infty$ so that the number and energy fluctuations both become very large. They can explore completely new sets of states with little inhibition. Often the system changes its character completely and permanently by falling into a state of much lower total energy and altered symmetry – a phase transition.

23.6   Phase Transitions

A multitude of fluctuations are constantly probing the possibility of a phase transition. Here and there among a thousand fragile variations, some cells find microstates of significantly higher entropy – or lower free energy. Once these cells fall into such a microstate, their earlier microstates become much less probable. In its new state, the cell is trapped. A trapped cell, in turn, modifies the thermodynamic functions of its surrounding cells. Many such cells may form nuclei and their new microstate may propagate throughout the entire system. A phase transition occurs. Its consequences are dramatic and legion. Ice turns to water; water boils to steam. Crystals change structure; magnetic domains collapse.

Thermodynamics cannot portray systems evolving through phase transitions. The situation is too far from equilibrium; fluctuations are as large as their average quantities. Distribution functions of statistical quantities become skew, and their most probable values cease to represent their average values. Consequently, many thermodynamic quantities do not have simple Taylor series expansions; close to a phase transition they become non-analytic. Even so, thermodynamic descriptions contain enough information to predict their own demise. With modern theories of scaling and renormalization, these predictions can be quite detailed. After the transition is over, a new thermodynamic description may apply, perhaps with a modified equation of state. Often, it is possible, moreover, to find equations of state that bridge the transition in an approximate phenomenological way.

Historically, phase transitions were classified (mainly through Ehrenfest's influence) by the continuity of the nth derivative of their Gibbs potential $G[T, P, N] = U - TS + PV$. This would be particularly relevant for a transition occurring at constant pressure so that both phases, represented by G_1 and G_2, would be in equilibrium with a pressure reservoir. The transition would be nth-order if

$$\frac{\partial^n G_1}{\partial T^n} \neq \frac{\partial^n G_2}{\partial T^n} \quad \text{and} \quad \frac{\partial^n G_1}{\partial P^n} \neq \frac{\partial^n G_2}{\partial P^n} \tag{23.70}$$

and all the lower derivatives at this point were equal. Thus in a first-order transition the entropies and volumes of the two phases would differ and there would be a latent heat, $T \Delta S$, of transition. In a second-order transition, the entropies and volumes would be continuous, but the heat capacity $T \partial S / \partial T$ would have a discontinuity, and so on.

Problems arose with this classification, as the higher order derivatives were found not to exist for many transitions such as the Curie point in ferromagnets or the λ transition in liquid helium. Often these higher derivatives have logarithmic or algebraic singularities that become infinite at the transition – a manifestation of their nonanalytic behavior. Moreover, one may want to examine phase transitions in the grand canonical ensemble where $\Psi = U - TS - \mu\bar{N}$ is the relevant potential. Finally, and most profoundly, the higher order transitions often appear to share a number of fundamental properties related to how their basic constituents are arranged over different spatial scale lengths.

As a result, phase transitions are now usually classified into two simple categories: 1. those that are first order and 2. the rest (all called second order). The relevant property of first-order transitions is now that their two states occupy clearly separated regions in a thermodynamic configuration space (having coordinates S, U, V, and N, for example). Solid–liquid–gas transitions are the archetypal examples, as long as they are not close to their critical point. One feature of a first-order transition is that the arrangement of its basic constituents (e.g., molecules, spins, clusters) usually changes its fundamental symmetry. Second-order transitions, in contrast, occur among adjacent states in thermodynamic configuration space. Here the archetype is the liquid–gas transition at the critical point where, at a particular temperature and pressure, the two phases blend into each other continuously. Fluctuations are usually large in both types of phase transitions.

Two additional physical effects associated with large fluctuations are important near phase transitions, especially at a critical point. The first is that correlations of fluctuating basic microscopic properties of the system (e.g., spin, particle density) become very long range. When these correlations become "frozen in" after gas turns to liquid, or a crystal structure changes, the phase transition is over. The second effect is that near a critical point, macroscopic changes occur in slow motion relative to microscopic dynamical timescales. Macroscopic relaxation modes have very long timescales. Both these effects occur because the system is in a state of marginal dynamical stability. Kinetic and interaction energies are nearly balanced. In more direct dynamical terms, the fluctuating bulk forces caused by local thermal pressure gradients are nearly compensated by the interaction forces from local density gradients.

Large, slow, highly correlated fluctuations have the common consequence, only weakly dependent on the specific nature of the interaction forces, that as the correlation lengthscale becomes infinite near a critical point, there are no other lengthscales left to describe the system. Its microscopic structure becomes similar when averaged over any lengthscale in an infinite system. This requires the correlation functions, for example, to become scale-free power laws (18.1). It also provides relations among the critical exponents that represent the algebraic singularities of the heat capacity, the order parameter, the susceptibility, and the equation of state as functions of $(T - T_c)/T_c$, where T_c is the critical temperature. These relations arise essentially because as the system approaches its critical temperature, the only scale length is the correlation length and as it becomes infinite its dimensionality must completely determine the dimensionality of the algebraic singularities of thermodynamic quantities. Actual values of these critical exponents can be calculated using renormalization group theory (e.g., Huang, 1987), which exploits the self-similarity of the system over a wide range of scales.

It is an intriguing question whether a phase transition can occur in the cosmological many-body problem and whether observations of galaxy clustering intimate a role for phase transitions in the future of our Universe. We will return to this question in Chapter 39.

24

Thermodynamics and Gravity

I will not refrain from setting among these
precepts a new device for consideration which,
although it may appear trivial and almost
ludicrous, is nevertheless of great utility
in arousing the mind to various inventions.

Leonardo da Vinci

In the past, there has been a widely held mythology that thermodynamics and gravity are incompatible. The main arguments for that view were threefold. First, thermodynamics applies to equilibrium systems. But self-gravitating systems continually evolve toward more singular states, so they are never in equilibrium. Second, to obtain a thermodynamic or statistical mechanical description it must be possible to calculate the partition function for the relevant ensemble, as in (23.60). But self-gravitating systems contain states where two objects can move arbitrarily close and contribute infinite negative gravitational energy, making the partition function diverge. Third, the fundamental parameters of thermodynamics must be extensive quantities. But self-gravitating systems cannot localize their potential energy in an isolated cell; it belongs to the whole system.

All three of these arguments have a common basis in the long-range nature of the gravitational force and the fact that it does not saturate. By contrast, in the electrostatic case of a plasma, although the Coulomb forces are also long range, the positive and negative charges effectively cancel on scales larger than a Debye sphere where the plasma is essentially neutral, and its net interaction energy is zero. So one can describe plasmas thermodynamically (e.g., Landau & Lifshitz, 1969). Since there are no negative gravitational masses, however, the volume integral of the r^{-2} force or the r^{-1} potential in a infinite homogeneous gravitating system is infinite. Occasionally there have been attempts to get around this by pretending that the gravitational particle potential has an exponential or other form of cutoff. Such models, while entertaining, were unrealistic.

Yet there were clear indications that all was not well with these three arguments. The first evidence came with attempts to apply thermodynamics to finite spherical self-gravitating clouds (Bonnor 1956, 1958) and star clusters (e.g., Antonov, 1962; Lynden-Bell and Wood 1968; Thirring, 1970; Horwitz and Katz, 1978). Thermodynamics worked surprisingly well for describing the instabilities and slow evolution of such systems. Their essential thermodynamic properties could be calculated by imagining each system to be placed in an impenetrable spherical shell and considering a canonical ensemble of them. The results agreed reasonably well with computer simulations. This is discussed in more detail in Sections 30, 31, and 43–50 of *Gravitational Physics of Stellar and Galactic Systems* (GPSGS). Since galaxy clustering is mainly concerned with infinite systems, here I just briefly mention the basic reasons why the three earlier arguments were not important for finite systems when examined more closely. Aspects of these reasons also apply to the cosmological case.

First, the macroscopic evolution of a finite stellar system near equilibrium occurs on a timescale long compared to the dynamical crossing timescale on which the microstates can change. The crossing timescale is $\tau_{\text{cross}} \approx R/v \approx (G\bar{\rho})^{-1/2}$ while the timescale for global changes is approximately of order $N\tau_{\text{cross}}$ as in (13.31). Here N is the total number of stars in a cluster of radius R, average density $\bar{\rho}$, and velocity dispersion v. Since $N \gg 1$, global departures from a previous nearly equilibrium state are slow and the system can pass easily from one nearly equilibrium state to another. It evolves in quasi-equilibrium through a sequence of essentially equilibrium states. At the center, naturally occurring changes may be faster, but it still takes of order $N\tau_{\text{cross}}$ for their effects on temperature and pressure to diffuse throughout the cluster. Microstates that take times $\sim N\tau_{\text{cross}}$ to develop are not available to the system on the shorter timescale τ_{cross}, and thus they do not appear in the partition function when the thermodynamics of these shorter timescales is of interest.

Second, if two stars considered as point masses collide, they will formally introduce an infinite term into the partition function. However, this region of infinite gravitational energy is so localized that it has no significant physical effect on the system. All that happens if the stars coalesce (which we assume to be the case in this idealized example) is that $N \to N - 1$, and the new star's mass is the sum of its component masses. The infinite gravitational potential energy is internalized in the new star and is not an available state for the entire system. Therefore it does not contribute to the partition function. If N is large, an occasional collision will not matter very much. Another possibility is that two or more stars' orbits take them very close to one another without colliding. However, since this occurs with positive kinetic energy comparable to the negative potential energy, it again does not introduce an infinite term into the partition function. Furthermore it occurs for such a short time that its influence on global structure is small, as long as these microstates are rare. More serious is the formation of stable bound binary systems through interactions with other stars, which carry away the excess energy and momentum. Over time, these binaries can transfer energy to other passing stars and contribute very large, but not infinite, terms to the partition function. When these new states become available, they can alter the thermodynamics by acting as local sources or sinks of energy for the orbits of the other stars. Generally their effects take longer than τ_{cross} to diffuse through the cluster. So it remains a good approximation to allow the rest of the cluster to evolve through thermodynamic states in a quasi-equilibrium manner, adding the effects of energy flow through binaries. However, after so many binaries form that they dominate the cluster's dynamics, or when the core becomes so dense that the global thermodynamic state becomes unstable (cf. GPSGS), the cluster can undergo restructuring on a subsequent timescale of order τ_{cross}. During this period the thermodynamic description breaks down. After the dynamical restructuring is over and the stars are redistributed, the system may find another quasi-equilibrium state. This resulting state tends to have a stronger core–halo structure than the original simple density distribution such as $\rho \propto r^{-2}$ for an isothermal sphere. At this stage, a detailed dynamical description is more useful, especially if stars evaporate fairly rapidly from the cluster.

Third, the nonextensivity is not a serious problem for closed finite systems because one is considering a canonical ensemble whose individual systems do not interact strongly. The thermodynamic quantities are averages over all the systems (clusters) in this ensemble, each system contributing a microstate. A thermodynamic instability in the equation of state for this ensemble just indicates that the configuration of the most probable microstate will change significantly.

Additionally, a comment about the effect of gravity on the basic postulate of statistical thermodynamics. This postulate assumes that for given macroscopic constraints on the system (e.g., total U, V, or N), there is an equal a priori probability for the system to be found in any region with the same $6N$-dimensional volume in its position–momentum phase space. In a grand canonical ensemble, for example, this is expressed as (23.55) for the relative probability f_i.

For classical Hamiltonian systems, including gravitational forces, this postulate is the only reasonable assumption (Tolman 1938). Such systems satisfy Liouville's equation (17.14). Therefore there is no inherent tendency for the phase points representing these systems to occupy one part of phase space rather than another; these phase points act like an incompressible fluid. Statistical properties of the system are determined just by the number of microstates available for a given macrostate, not by some nonuniform a priori probability distribution for these microstates. Gravitating systems are more likely to be found in strongly condensed or clustered macrostates simply because these have greater numbers of microstates.

At various times, Maxwell, Boltzmann, and many of their successors hoped that by proving an ergodic (or quasi-ergodic) theorem, they could derive the postulate of equal a priori probabilities. Ergodic hypotheses state that the time average of any statistical quantity in one system equals its ensemble average over all systems in the relevant ensemble. This essentially says that the representative phase point of an isolated system successively passes through all positions in phase space compatible with its conserved quantities before returning to its original position. The weaker quasi-ergodic hypothesis states that over very long times this phase point approaches any compatible phase space position arbitrarily closely.

Despite great mathematical efforts, neither the ergodic nor the quasi-ergodic hypothesis has been proved generally as a theorem from dynamical first principles for physically important classical systems. (It does hold for special classical microcanonical ensembles and for general quantum systems.) Today most physicists are prepared to accept the principle of equal a priori probabilities explicitly as a basic postulate whose a posteriori justification is the agreement of its consequences with observations or experiments. This is the view taken here.

How about infinite gravitating systems? They differ from finite clusters in several fundamental ways. Therefore we should not be surprised that their statistical thermodynamic description is also quite different. If the infinite system is statistically homogeneous, as we shall usually assume for simplicity, then its average properties will have translational and rotational symmetry around any point. Of course these symmetries need not apply to any particular small region. Rather they are expected to hold when averaged over a representative ensemble of regions, regardless of whether the individual regions themselves are large or small. In contrast, the ensemble average over finite spherical clusters only has rotational symmetry around its one central point; none of its other points have any spatial symmetries. This means that finite and infinite systems have different (partially overlapping) sets of available states in their partition functions.

Moreover, the partition function and its consequent thermodynamics will require the grand canonical ensemble for infinite systems, rather than the canonical ensemble used for finite systems. This is because, as already remarked, the cells representing galaxy clustering are permeable to galaxies as well as to energy. Each cell is in contact with a reservoir of galaxies, a reservoir that can be depleted.

These differences already imply that the reasons a thermodynamic description can be useful will differ for finite and infinite gravitating systems. But there is another overriding difference: Infinite systems evolve in an expanding cosmological background. Finite bound systems are detached from the expansion of the Universe, as the cosmic energy equation (22.31) shows for the case of local virial equilibrium. We shall see that this difference is the essential additional feature that makes a thermodynamic description of the cosmological many-body problem possible.

In the next chapter, we examine the detailed quantitative conditions for a thermodynamic description of gravitational clustering in an infinite system. Here I give a more qualitative brief physical description of these conditions in relation to the three major concerns discussed previously. Some of the basic discussion for finite systems also applies to infinite systems.

First, quasi-equilibrium can apply to the infinite cosmological case for the same basic reason as in finite slowly evolving systems. Average correlations and clustering change on a global Hubble expansion timescale $\sim (G\bar{\rho})^{-1/2}$, or longer, whereas microstates change on the local crossing timescale $\sim (G\rho)^{-1/2}$. On lengthscales where the local density $\rho > \bar{\rho}$, these timescales will be distinct. Relaxation of the system may therefore be a function of lengthscale. On smaller lengthscales, thermodynamic quantities such as temperature and pressure have well-defined values at any given time, but these values will change slowly on the longer timescale. This is the standard basic approximation of nonequilibrium thermodynamics (e.g., de Groot & Mazur, 1962). Its procedure is to calculate equations of state and other thermodynamic properties as though the system were in equilibrium, with given values of P, $\bar{n}$, T, and U, and then let these values change slowly compared to the time on which detailed local configurations change. Thus the system on these lengthscales evolves through a sequence of equilibrium states – the definition of quasi-equilibrium. The separation between timescales needed for this to be a good approximation can be estimated a priori and justified a posteriori.

Second, the unavailability and unimportance of singular states in the partition function of infinite systems also occurs for similar reasons as in finite systems.

Any colliding points, which may now represent galaxies, merge and their mutual gravitational energy is internalized. The expansion of the Universe makes such collisions less likely than they would be for similar size objects in finite nonexpanding systems. Moreover, the eventual formation of many bound clusters, which become singular or acquire a large fraction of the system's total energy, takes so long that these states are essentially unavailable on the thermodynamic timescales of interest (cf. Chapter 30).

Third, extensivity is a good approximation in infinite systems, but for a different reason than in finite clusters. Clearly for a grand canonical ensemble of cells whose size is greater than the extent of correlations, the thermodynamic functions will be essentially extensive. One might argue that correlations across cell boundaries could be important. However, this is mitigated by two factors. First, the ratio of cross-surface to volume correlation energies decreases as the cell size increases and its surface area to volume ratio decreases. Second, relatively few of the cells in the grand canonical ensemble are contiguous with each other. Therefore the properties (microstates) of one cell are mostly uncorrelated with those of other cells. Thermodynamic properties, averaged over all cells, will have little residual correlation in the ensemble. This last result also applies to smaller cells. For any size cell, the most stringent requirement for extensivity is that the correlation energy between cells be much less than the correlation energy within an average cell. We will see that this holds if the two-point correlation function is sufficiently steep, as is generally the case on relaxed scales. Thus extensivity will usually be a good approximation over a wider range of scales than might be expected initially.

Properties of infinite systems that diminish the usefulness of a thermodynamic description also become apparent from this discussion. Strong initial large-scale coherence is one. If this coherence is not near equilibrium, the system may not relax to a thermodynamic state, nor will its properties be extensive on smaller scales. Systems in a very early stage of relaxation are another. Their timescales for changes of local microstates and global macrostates may not separate sufficiently for quasi-equilibrium to apply over long distances. Determining which sets of initial conditions are attracted toward thermodynamic states and which ones evolve in other directions is an interesting unresolved problem.

Finally, the role of expansion for infinite systems. It is intuitively obvious (after thinking about it for some time!) that the expansion removes the mean field – because the expansion is the mean field. At least this is true for the Einstein–Friedmann models. We demonstrate it explicitly in the next chapter. The result of removing the mean field is that only the fluctuations remain. The thermodynamics is a description of these fluctuations, their averages, and distributions around the average. Consequently, the long-range nature of the gravitational force, which gives rise to the mean field, does not lead to difficulties in the cosmological case. This differs from most other many-body problems, which start with a mean field and perturb around it, to higher and higher and higher orders. By comparison, the cosmological case is delightfully simple.

25

Thermodynamic Formulation of the Cosmological Many-Body Problem

Many shall run to and fro,
and knowledge shall be increased.
Daniel 12:4

We start with the simplest form of the cosmological many-body problem: the one Bentley posed to Newton (see Chapter 2). In its modern version it asks "If the universe were filled with an initial Poisson distribution of identical gravitating point masses, how would their distribution evolve?" This is clearly an interesting physical problem in its own right, and it provides the simplest model for galaxy clustering in the expanding universe. If we can solve it, other complications such as a range of masses, different initial conditions, galaxy merging, and inhomogeneous intergalactic dark matter may be added to account for observed galaxy clustering. Actually, we will find that these complications are secondary. The simple gravitational cosmological many-body problem contains the essential physics of the observed galaxy distributions.

To formulate the problem thermodynamically in a solvable way, we first examine why the mean gravitational field is not dynamically important and why extensivity is a good approximation. This leads to a derivation of the form of the energy and pressure equations of state. In turn, this derivation provides insight into the requirements for quasi-equilibrium. To complete the basic description, equivalent to finding the third equation of state, Chapter 26 develops very general physical properties of these gravitating systems. This closely follows the analyses by Saslaw and Fang (1996).

25.1 Expansion Removes the Mean Gravitational Field
from Local Dynamics

The result which titles this section has long been known for Einstein–Friedmann cosmological models. Physically it is obvious because the mean field in these models is homogeneous and exerts no local force. It determines only the global expansion $R(t)$. To see this more precisely, consider the field equations for the Einstein–Friedmann models given by $\Lambda = 0$ in (16.4) and (16.5). At late times, after galaxies have formed and begun to cluster, the pressure P and kinetic energy of their motion is negligible compared to the average rest mass energy density $\epsilon = \bar{\rho}c^2$. Under these conditions, (16.4) and (16.5) give

$$\ddot{R} = -\frac{4}{3}\pi G \bar{\rho} R \tag{25.1}$$

for all values $(-1, 0, +1)$ of the curvature k. As usual, two particles with constant separation x in comoving coordinates expanding with the Universe will have a separation

$$\mathbf{r} = R(t)\mathbf{x} \tag{25.2}$$

in local proper coordinates. Substituting (25.2) into (25.1) shows that the background acceleration $\ddot{\mathbf{r}}$ in proper coordinates is the gradient of a potential,

$$\phi = -\frac{2}{3}\pi G\bar{\rho}r^2. \tag{25.3}$$

(This gradient also gives the local background force $-GM(r)/r^2$ used in Newtonian cosmology. See the discussion before Equation 14.48.) Therefore the gravitational potential in proper coordinates, ϕ_{proper}, is related to the potential in comoving coordinates, ϕ_{comov}, by the transformation

$$\phi_{\text{proper}} = \phi_{\text{comov}} + \frac{2}{3}\pi G\bar{\rho}r^2 = \phi_{\text{comov}} - \frac{1}{2}\frac{\ddot{R}}{R}r^2 = \phi_{\text{comov}} - \frac{1}{2}R\ddot{R}x^2. \tag{25.4}$$

If there were no comoving potential, then two massless test particles, with no initial relative velocity, would just accelerate away from each other in proper coordinates.

In a statistically homogeneous system, the density around any galaxy is [see (14.6)]

$$\rho(\mathbf{r}) = \bar{\rho}[1 + \xi(\mathbf{r})], \tag{25.5}$$

where $\bar{\rho}$ is the average unconditional density. Local Newtonian motions of the galaxies in proper coordinates are governed by Poisson's equation (conventionally defined in this context to have a positive sign)

$$\nabla_{\mathbf{r}}^2\phi_{\text{proper}} = 4\pi G\bar{\rho}[1 + \xi(\mathbf{r})]. \tag{25.6}$$

Note that $\nabla_{\mathbf{r}}^2$ symbolizes derivatives with respect to proper coordinates, since the local matter distribution is generally irregular. Substituting (25.4) into (25.6) and using (25.1) and (25.2) shows explicitly that the expansion of the Universe cancels the long-range part of the gravitational field, $4\pi G\bar{\rho}$, leaving

$$\frac{1}{R^2(t)}\nabla_{\mathbf{x}}^2\phi_{\text{comov}} = \nabla_{\mathbf{r}}^2\phi_{\text{comov}} = 4\pi G\bar{\rho}\xi(r). \tag{25.7}$$

The left side represents the forces that produce the local peculiar velocities of the galaxies. These are seen to be related to the short-range departures $\xi(r)$ of the conditional correlations from the average unconditional density. These departures can be arbitrarily nonlinear, unlike the linear analyses (see Chapter 16) used to follow the growth of small density perturbations in the expanding universe.

Therefore the local dynamics depend only on the local fluctuations of the gravitational force. Since thermodynamics is essentially an average over these local
dynamical states, it too will depend just on local fluctuations, as we will see shortly.

25.2   Extensivity and Gravity

Although the division of thermodynamic quantities into extensive and intensive variables is not a fundamental requirement (and modified versions of thermodynamics
exist that allow for nonextensivity directly), it is a great simplifying convenience.
Strictly speaking, it does not hold exactly for any real system whose particles interact
over a finite range, but it is usually a good approximation.

To see why extensivity is also a good approximation in the cosmological many-
body problem (Sheth & Saslaw, 1996), consider dividing a region into two cells. They
could have equal size, or one could be part of the other. Naturally their volumes and
total numbers of particles remain extensive since $V = V_1 + V_2$ and $N = N_1 + N_2$
despite gravity. However, because there is a gravitational correlation energy, U_{corr},
between the cells the internal energy is not extensive:

$$U = U_1 + U_2 + U_{corr}. \tag{25.8}$$

As the entropy is a function of U, V, and N, gravitating systems are not completely
extensive, since U and S are not additive. Note that U_1 and U_2 contain the extensive
kinetic energy as well as the internal correlation energy of the volumes (i.e., the
part of the correlation energy contributed by interactions among all particles within
a volume). For extensivity to be a very good approximation, we require $U_{corr} \ll$
$U_1 + U_2$.

This approximation holds for several reasons in an ensemble of comoving cells
containing gravitating particles in an expanding universe. First, consider an ensemble containing only cells larger than the correlation lengthscale. This ensemble is
nearly extensive because the universal expansion effectively limits the dynamical
and consequent thermodynamic effects of gravity to about the correlation lengthscale, as (25.7) and the results of the next Section 25.3 indicate. Then the correlation
energy between two cells is small compared to their internal correlation energy and
U_{corr} is negligible in (25.8). Second, most cells in this or any other ensemble are
separated from each other by more than a correlation length and therefore do not
interact strongly.

Third, consider an ensemble of cells that are smaller than the correlation length.
If the correlation function is sufficiently steep, then the internal correlation energy
within a cell is much greater than the external correlation energy contributed by
other cells in the ensemble. External interactions may be neglected and extensivity is
again a good approximation. As a specific example, consider a power-law correlation
$\xi(r) = \xi_0 r^{-\gamma}$. Then the average correlation energy within a spherical cell of volume

$V = 4\pi R^3/3$ is

$$W_V = -\bar{n}V \int_0^R \frac{Gm^2}{2r}\xi(r)\, 4\pi\bar{n}r^2\, dr = -2\pi Gm^2\bar{n}^2 V\xi_0 \frac{R^{2-\gamma}}{(2-\gamma)}. \qquad (25.9)$$

Similarly, on average, the correlation energy within a spherical cell of volume $2V$ (i.e., of radius $2^{1/3}R$) is $-2\pi Gm^2\bar{n}^2 2V\xi_0(2^{1/3}R)^{2-\gamma}(2-\gamma)^{-1}$. To justify the approximation of extensivity requires that $W_{2v} \approx 2W_V$. This means that $|W_{2v}/2W_V| = 2^{(2-\gamma)/3} \approx 1$, which is satisfied when $\gamma \gtrsim 1$. In particular, it holds to within a few percent for the observed value $\gamma \gtrsim 1.7$. Analyses of the ratio of cells' interaction correlation energy to their internal correlation energy for other geometries leads to similar results. Note that, for this example, the criterion is independent of R because both the correlation function and the potential energy are scale-free power laws. Furthermore, in many systems $\xi(r)$ decreases faster than a simple power law near the edge of a sufficiently large cell, further reducing the relative importance of external interaction energy.

For these reasons, although extensivity is not exact, it is usually a reasonable approximation even on small scales.

25.3   The Energy Equation of State

To describe the thermodynamics of a particular physical system, we must specify its three equations of state. These may be represented by $U(T, V, N)$, $P(T, V, N)$, and $\mu(T, V, N)$, or by an equivalent set.

We start with the cosmological many-body equation of state for the internal energy U. It has three contributions: the average kinetic energy of peculiar velocities, the potential energy of the gravitational interactions, and the potential energy of the global expansion. Since the gravitational potential does not depend on velocity, the kinetic energy separates from the potential energy in the Hamiltonian. Therefore the partition function is also separable. If we have a unimodal distribution of peculiar velocities throughout the entire system, including all degrees of clustering from isolated field galaxies to the richest bound clusters, then there is a single global kinetic temperature

$$T = \frac{m\langle v^2 \rangle}{3}. \qquad (25.10)$$

This temperature is related to the total peculiar velocity dispersion $\langle v^2 \rangle$ in the usual way (except that we set Boltzmann's constant $k = 1$ throughout). For simplicity, all particles have the same mass. Later we will examine how a range of masses modifies the results. The kinetic energy is therefore

$$K = \frac{3}{2}\bar{N}T. \qquad (25.11)$$

We will confirm the monotonic nature of the global velocity distribution a posteriori, both from N-body simulations and observations. From place to place there will naturally be fluctuations around the averages. It is just these fluctuations that describe the detailed clustering.

Next, consider the average gravitational potential energy in a volume, which we take to be spherical for simplicity. The number of galaxies in a shell around any specified central galaxy at $r = 0$ is $\bar{n}[1 + \xi(r)]4\pi r^2 dr$ on average. Therefore the gravitational potential energy is

$$\frac{\bar{N}m\bar{n}}{2} \int_0^R \phi_{\text{proper}}(r)[1 + \xi(r)]\, 4\pi r^2\, dr, \tag{25.12}$$

since there are $\bar{N}$ possible central galaxies in the cell and each pair interaction contributes once. With $\phi_{\text{proper}}(r) = -Gm/r$, the first term of (25.12) is

$$-\pi G\bar{\rho}M(R)R^2 = -\frac{3}{4}\frac{GM^2(R)}{R}, \tag{25.13}$$

which diverges as R^5.

The third contribution to the energy equation of state is from the global expansion whose potential is given by (25.4) as

$$\phi_{\text{proper}} - \phi_{\text{comov}} = \frac{2}{3}\pi G\bar{\rho}r^2 = \frac{1}{2}\bar{N}\frac{Gm}{r}. \tag{25.14}$$

The corresponding potential energy,

$$\frac{\bar{\rho}\bar{N}}{2} \int_0^R \frac{Gm}{r} 4\pi r^2\, dr = \frac{3}{4}\frac{GM^2(R)}{R}, \tag{25.15}$$

exactly cancels the divergent contribution (25.13) of the mean field to the gravitational potential energy in (25.12).

Therefore, only the correlation term of the potential energy survives in the equation of state, consistent with our dynamical expectation from (25.7). The resulting energy equation of state may be written in the form

$$U = K + W = \frac{3}{2}\bar{N}T(1 - 2b), \tag{25.16}$$

where

$$b = -\frac{W}{2K} = \frac{2\pi Gm^2\bar{n}}{3T} \int_0^R \xi(\bar{n}, T, r)r\, dr \tag{25.17}$$

represents the average departure from a noninteracting ($\xi = 0$ or perfect gas) system for cells of radius R.

In the grand canonical ensemble, the two-point correlation function depends on $\bar{n}$ and T as well as on r. For ensembles at a higher temperature, the correlations will generally be less, all else being equal. Thus $b = b_R(\bar{n}, T)$. Very small cells, as $R \to 0$, contain either zero or one particle, and so they have no internal gravitational potential energy and $\lim_{R \to 0} b = 0$. A grand canonical ensemble of such cells would behave like a perfect gas. In the opposite limit of cells so large that $r\xi(\bar{n}, T, r) \to 0$, the value of b_R tends to a constant b independent of volume. We will find that this volume-independent value of b is often a good approximation for a fairly wide range of observations. Outside this range, it provides a useful method to measure $\xi(r)$ from counts in cells of different volumes.

Although simple spherical volumes were used to develop (25.17), the value of b can readily be calculated (at least numerically) for volumes of any shape, and in general

$$b_V(\bar{n}, T) = -\frac{W}{3\bar{N}T} = \frac{Gm^2\bar{n}^2}{6\bar{N}T} \int_V \int_V \frac{\xi(\bar{n}, T, |\mathbf{r}_1 - \mathbf{r}_2|)}{|\mathbf{r}_1 - \mathbf{r}_2|} d\mathbf{r}_1 d\mathbf{r}_2. \quad (25.18)$$

It is averaged over a grand canonical ensemble of cells, all having the same comoving volume and shape. For different ensembles whose cells have the same volume but different shapes, the values of b will depend on shape. This is because statistically homogeneous correlations contribute more to the internal interactions in directions where the cell boundaries are wide than where they are narrow. For most reasonably shaped cells (e.g., spheres, cubes, rectangular solids with axis ratios $\lesssim 5:1$), this is a small effect. In some circumstances the shape dependence of b can also provide useful information about the correlation function.

There are general limits on the value of b. Since the average net effect of gravitational fluctuations produces a negative potential energy, (25.18) shows that $b \geq 0$. If the system were virialized at all levels, so that it could be considered to be an isolated, bound distribution in equilibrium, then it would have $b = 1$. This would be an extreme case, since (25.18) shows that b is not what is usually meant by the virial ratio $-W/2K$. This usual virial ratio applies *only* to finite bound equilibrium systems in which W is the total potential energy obtained from $\bar{n}(r)$, not the correlation energy from $\xi(r)$. Only in the extreme limit where the entire universe is virialized would these two descriptions become equivalent. From another point of view, we shall see in the next section that the physical requirement of a nonnegative pressure also implies $b \leq 1$. Therefore we expect that in general

$$0 \leq b_V(\bar{n}, T) \leq 1. \quad (25.19)$$

25.4 The Pressure Equation of State

We derive the form of the pressure in the cosmological many-body system from two points of view, each providing its own insight. The first is from the force per unit area across a mathematical surface, and the second is directly from the dynamics. These follow standard analyses for interacting systems, except that now we include the

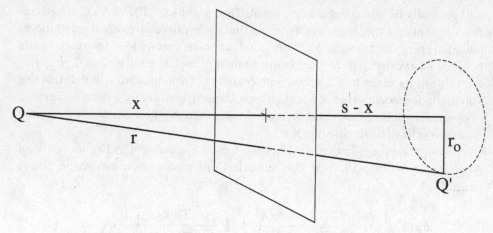

Fig. 25.1. Calculation of pressure as force per unit area in an interacting system.

contribution of cosmological expansion. This produces a simplification by removing the effect of the mean field, as in the energy equation of state. Another simplification occurs because gravity is a binary interaction; the basic potential involves only pairs of particles (not triplets or higher multiples). This implies that the interaction pressure depends only on the two-particle correlation function directly (not the three-particle or higher correlations). A final simplification will arise from the inverse square nature of the gravitational force.

To calculate pressure from the force across a surface, we follow Hill (1960) who follows Fowler (1936 and 1966) who in turn follows Rayleigh and Laplace, all for the nonexpanding case. Then we add expansion and specify the gravitational interaction. Figure 25.1 shows a representative surface. Galaxies moving through it will contribute the usual kinetic momentum transport NT/V to the pressure since this is independent of the interaction potential in the Hamiltonian. Here the temperature, as in the kinetic energy, refers to the local velocity dispersion. This dispersion may differ from place to place, depending on the local degree of gravitational clustering. Including all the local velocities in the entire ensemble gives a global temperature from the dispersion of the total peculiar velocity distribution. Local velocity dispersions then represent fluctuations from this average.

A galaxy at Q exerts a force $-m\, \partial\phi(r)/\partial r$ on another galaxy at Q' across the surface, with a component $-m\phi'(r)s/r$ normal to the surface. On average there are $\bar{n}[1 + \xi(r)]2\pi r_0 dr_0 ds$ galaxies in a cylindrical shell between r_0 and $r_0 + dr_0$ and between s and $s + ds$. Thus the total normal force that galaxy Q exerts on this differential shell is

$$F_{\text{shell}} = -\phi'(r)\frac{s}{r}m\bar{n}[1 + \xi(r)]2\pi r_0\, dr_0 ds = -2\pi m\bar{n}\phi'(r)[1 + \xi(r)]s\, dr\, ds$$

$$(25.20)$$

using $r^2 = s^2 + r_0^2$ from Figure 25.1. Integrating this force over a volume $x \leq s \leq R$, $s \leq r \leq R$ on the other side of the surface from Q gives the normal force exerted by Q on all the galaxies in this volume:

$$F_{\text{volume}} = -2\pi m\bar{n} \int_x^R s\, ds \int_s^R \phi'(r)[1 + \xi(r)]\, dr$$

$$= \pi m\bar{n}x^2 \int_x^R \phi'(r)[1 + \xi(r)]\, dr - \pi m\bar{n} \int_x^R r^2 \phi'(r)[1 + \xi(r)]\, dr.$$

$$(25.21)$$

The second equality results from an integration by parts with "$dv = s\, ds$," "$u = \int_s^R$," and recalling Leibnitz's theorem.

The interaction pressure from all galaxies in a cylinder of unit cross-sectional area normal to the surface and extending from $x = 0$ to $x = R$ is their combined force. Since this cylinder contains $\bar{n}\, dx$ galaxies between x and $x + dx$, its interaction pressure is

$$\bar{n} \int_0^R F_{\text{volume}}\, dx = -\frac{2\pi m\bar{n}^2}{3} \int_0^R r^3 \phi'(r)[1 + \xi(r)]\, dr \qquad (25.22)$$

using (25.21) and integrating each resulting term by parts. Adding the kinetic part of the pressure gives the total static pressure of galaxies in the volume:

$$P = \bar{n}T - \frac{\bar{n}^2 m}{6} \int_0^R r \frac{d\phi}{dr}[1 + \xi(\bar{n}, T, r)] 4\pi r^2\, dr. \qquad (25.23)$$

In summary, the first term in (25.23) is the usual kinetic contribution from the momentum of particles crossing a mathematical surface in the system. The second term contains the force per unit area across this surface exerted by interacting particles having the average uniform density, and also the corresponding force caused by correlated departures from the average density. Note that this second term involves the volume integral of $r\phi'[1+\xi]$ over a sphere. In the usual case of an interacting molecular gas or fluid, the forces and correlations are short range and so this integrand becomes negligible after very short distances and R may be replaced by infinity. In the cosmological case, global expansion cancels the long-range contribution of gravity.

The expanding universe contributes an additional effective pressure to (25.23) from the expansion force per unit area acting on the surface. For a uniform density sphere of given radius R (not to be confused with the expansion scale of the universe) around any arbitrary point in the expanding universe, the Newtonian analog of (25.1) for all $r \leq R$ is

$$\ddot{r} = -\frac{4}{3}\pi G\bar{\rho}r. \qquad (25.24)$$

Multiplying by the average density and integrating over the sphere gives the effective total expansion force acting on the surface:

$$F_{exp} = -\frac{4}{3}\pi G\bar{\rho}^2 \int_0^R r\, 4\pi r^2\, dr = -\frac{4}{3}\pi^2 G\bar{\rho}^2 R^4. \qquad (25.25)$$

The absolute value of this force divided by the surface area $4\pi R^2$ gives an effective expansion pressure

$$P_{exp} = \frac{1}{3}\pi G\bar{\rho}^2 R^2, \qquad (25.26)$$

which exactly cancels the homogeneous interaction term in (25.23).

Two other essentially equivalent alternative derivations provide further intuition into this result from slightly different points of view. First, since all points on the surface $4\pi R^2$ are within a distance $r < R$ of some point within the sphere, and all points are equivalent in the uniform universe, an average of the acceleration $\ddot{r}$ over the sphere, multiplied by the total mass within the sphere, and divided by the surface area, also yields (25.26). Second, one can define an expansion force per unit mass, $F_{exp}/m = \partial(\phi_{comov} - \phi_{proper})/\partial r = 4\pi G\bar{\rho}r/3$, from (25.4). Averaging this over a sphere of radius R and then multiplying by the total mass per unit area of the sphere, $\bar{\rho}R/3$, also gives (25.26).

Thus the pressure equation of state in the expanding case is obtained by adding (25.26) to (25.23) yielding

$$P = \frac{\bar{N}T}{V}(1 - b), \qquad (25.27)$$

with b given by (25.17) or (25.18). Like the energy equation of state (25.16), this form is exact. It does not depend on correlation functions of higher order than ξ_2. (Nor is it the lowest order term of a "virial" or "cluster" expansion as in most interacting molecular systems.) This is because gravity interacts with a pairwise potential, as mentioned earlier. Moreover, the simple similar forms of these two equations of state derive from the particular simple r^{-1} form of the gravitational potential. Equation (25.27) also confirms the limit $b_V \le 1$ for $P \ge 0$.

These kinetic derivations of the energy and pressure equations of state make no mention of the ensemble they represent. Therefore their *form* is valid for both the canonical and the grand canonical ensembles. However, the detailed nature of $\xi(\bar{n}, T, r)$ and thus of b_V will differ for each of these ensembles. This is most readily seen by recalling from Section 23.5 that fluctuations will differ in the different ensembles and that these fluctuations are related to ξ, for example by (14.16). Furthermore, grand canonical ensembles that differ just in the size (or shape) of their cells will have different values of b_V.

As an alert reader, you will have noticed an implicit assumption in the derivations of U and P. The kinetic and interaction pressure (or internal energies) in any given local region must refer to the same macrostate. This means, for example, that during the time a galaxy imparts momentum to the surface in Figure 25.1 by moving from Q to the surface, the correlation function $\xi(\bar{n}, T, r)$ does not change significantly. Since ξ is averaged over the ensemble (or system), this ensures that on average there will be another galaxy that has moved to Q so that the interaction energy across the surface remains essentially unchanged. Thus the local macrostate must not change significantly during the period that the local microstate changes. This is the basis of the quasi-equilibrium approximation, to be discussed in more detail shortly. First we rederive the equations of state directly from the dynamical equations of motion, since this makes the quasi-equilibrium assumption explicit.

25.5 Dynamical Derivation of the Equations of State

We sketch how the pressure equation of state (25.27) follows directly from the equation of motion of the galaxies by calculating their second moment or virial. (The "virial" should not be confused with the approximate "virial expansion" of the equation of state for nonpairwise potentials, nor with the "virial theorem" for finite relaxed bounded systems. There are many uses of virials in the dynamical broth.) This approach just modifies the standard derivation for a stationary imperfect gas (Fowler, 1936 and 1966, Section 9.7). A similar analysis applies to the energy equation of state (25.16).

Consider Newton's law in proper coordinates. A galaxy at $\mathbf{r}$ interacts with the gravitational potential $\phi_{\text{proper}}(r)$ and has a velocity $\mathbf{v}$. In comoving coordinates Newton's second law becomes

$$\frac{d\mathbf{v}}{dt} = \frac{d\mathbf{v}_{\text{pec}}}{dt} + \ddot{R}\mathbf{x} + \dot{R}\dot{\mathbf{x}} = -\nabla_{\mathbf{r}}\phi_{\text{proper}} \tag{25.28}$$

using (25.2), with $\mathbf{v}_{\text{pec}}$ the galaxy's peculiar velocity relative to the universal expansion. The uniform smoothed mass distribution contributes a smooth potential to ϕ_{proper}, which determines the expansion deceleration. From (25.1)

$$\ddot{R}\mathbf{x} = -\frac{4}{3}\pi G\bar{\rho}R\mathbf{x} = -\nabla_{\mathbf{r}}\left(\frac{2}{3}\pi G\bar{\rho}r^2\right) = -\nabla_{\mathbf{r}}\left(-\frac{1}{2}R\ddot{R}x^2\right). \tag{25.29}$$

Therefore, the smooth potential is $-\frac{1}{2}R\ddot{R}x^2$, as found in (25.4). The transformation (25.4) removes the uniform mass distribution leaving the potential ϕ_{comov}, which arises from the correlated mass distribution centered around a galaxy. This potential produces the peculiar motion of the galaxy:

$$\frac{d\mathbf{v}_{\text{pec}}}{dt} + \frac{\dot{R}}{R}\mathbf{v}_{\text{pec}} = -\frac{1}{R}\nabla_{\mathbf{x}}\phi_{\text{comov}}. \tag{25.30}$$

In these comoving coordinates, the change of the moment of inertia for all N galaxies in a comoving volume is

$$\frac{R^2}{4}\frac{d^2}{dt^2}\left(\sum_{i=1}^{N}m_i x_i^2\right) + \frac{R\dot{R}}{2}\frac{d}{dt}\left(\sum_{i=1}^{N}m_i x_i^2\right)$$

$$= \frac{1}{2}\sum_{i=1}^{N}m_i\,(R\dot{x}_i)^2 - \frac{1}{2}\left(\sum_{i=1}^{N}m_i\mathbf{x}_i\cdot\nabla_{\mathbf{x}}\phi_i\right). \tag{25.31}$$

Taking the time average (indicated by a bar) of this equation over the interval 0 to τ gives

$$\frac{1}{\tau}\left[\frac{R^2}{4}\frac{d}{dt}\left(\sum_{i=1}^{N}m_i x_i^2\right)\right]_0^{\tau} = \frac{1}{2}\overline{\sum_{i=1}^{N}m_i\,(R\dot{x}_i)^2} - \frac{1}{2}\overline{\sum_{i=1}^{N}m_i\mathbf{x}_i\cdot\nabla_{\mathbf{x}}\phi_i}. \tag{25.32}$$

The first term on the right-hand side is the kinetic energy of peculiar velocities and is $(3/2)NT$ as indicated in (25.16). The second term is the virial for the peculiar motions of galaxies. There are two contributions to the virial. One is from the peculiar momentum transport across a mathematical surface in the system, and it gives rise to the usual macroscopic pressure

$$-\frac{P}{2}\oint \mathbf{x}\cdot d\mathbf{s} = -\frac{3}{2}PV. \tag{25.33}$$

The other contribution comes from the gravitational interactions between the correlated particles. Using the formal solution of (25.7), this contribution is

$$-\frac{1}{2}\int m\bar{n}\mathbf{x}\cdot\nabla_{\mathbf{x}}\phi\,d^3x = \frac{GR^2m^2\bar{n}^2}{2}\int\int\mathbf{x}\cdot\frac{(\mathbf{x}'-\mathbf{x})}{|\mathbf{x}'-\mathbf{x}|^3}\xi\,d^3x\,d^3x'$$

$$= -\frac{GR^2m^2\bar{n}^2}{4}\int\int\frac{1}{|\mathbf{x}'-\mathbf{x}|}\xi\,d^3x\,d^3x'. \tag{25.34}$$

Here the last equality comes from the antisymmetry of $\mathbf{x}$ and $\mathbf{x}'$ in the previous equality. Changing the integral to the relative proper separation, since ξ is isotropic, and combining with the momentum transport, we have

$$P = \frac{\bar{N}T}{V}\left(1 - \frac{2\pi Gm^2\bar{n}}{3T}\int_V \xi(r)r\,dr\right) - \frac{2}{3V\tau}\left[\frac{R^2}{4}\frac{d}{dt}\left(\sum_{i=1}^{N}m_i x_i^2\right)\right]_0^{\tau}. \tag{25.35}$$

Now comes the quasi-equilibrium approximation. The scale factor $R(t)$ changes with time as the universe evolves. However, on lengthscales where the overall clustering has relaxed sufficiently that the total moment of inertia changes slowly compared to the expansion timescale $\tau \approx R/\dot{R}$, the last term on the right-hand side of (25.35) is small. This is either because the time changes are small or because the average configurations are similar at $t = 0$ and $t = \tau$. These are the dynamical conditions for quasi-equilibrium evolution, and they again yield (25.27) as the equation of state for the cosmological many-body system. Thus, in an expanding universe, the relaxed local gravitational distribution can be close to an equilibrium state for periods less than about a Hubble time. Then the equilibrium forms of the equations of state are a good approximation. Over longer timescales, if the system evolves through a sequence of these equilibrium states (quasi-equilibrium evolution) a thermodynamic description remains useful. When will it apply?

25.6 Physical Conditions for Quasi-Equilibrium

In the derivations of U and P, there are two implicit timescales. One is the macroscopic timescale for U and P to alter through changes of $\bar{n}$, T, and $b_V(t)$ or $\xi(\bar{n}, T, \mathbf{r}, t)$. The other is the microscopic timescale over which the specific detailed configuration of individual galaxy positions and velocities (microstate) changes from one nearly relaxed state to another. If the macroscopic timescale exceeds the microscopic timescale, quasi-equilibrium holds. This means that the system can adjust locally relative to slowly changing average conditions. In the two previous derivations of the pressure equation of state, for example, momentum transport over the lengthscale of ξ must occur faster than the timescale for ξ to change. Thus the detailed local configuration of particles changes faster than ξ (or the average moment of inertia) changes.

The macroscopic timescale, τ_{macro}, for ξ or U to change is given by the cosmic energy equation (22.31) and may be written as

$$\tau_{\text{macro}}^{-1} \equiv \left| \frac{\dot{U}}{K} \right| = 2\frac{\dot{R}}{R}(1 - b). \tag{25.36}$$

Since $0 \le b \le 1$ and $\dot{R}/R$ is essentially the inverse of the Hubble timescale, we have $\tau_{\text{macro}} \approx \tau_{\text{Hubble}}$, not surprisingly. As b approaches unity, macroscopic evolution becomes even slower, as the distribution of more galaxies over larger scales becomes closer to virial equilibrium.

The microscopic timescale will involve a lengthscale L, and the peculiar velocity v (for a unit mass $m = 1$):

$$\tau_{\text{micro}} \approx \frac{L}{v} \approx \frac{L}{K^{1/2}}. \tag{25.37}$$

The microscopic lengthscale over which a change in the detailed configuration is most relevant to U and P is the lengthscale of the correlation function that enters through the functional $W[\xi]$. Thus we may consider L to be the lengthscale over which most of the contribution to W comes. This will generally be less than the radius R of the volume V, particularly if the correlations are strongly nonlinear, since W is the r^{-1} moment of $\xi(r)$ over the volume. If we define L by the requirement

$W[L]/W[R] = 1/2$, then the power law $\xi(r) \propto r^{-\gamma}$ implies

$$\frac{L}{R} = \frac{1}{2^{1/(2-\gamma)}}, \tag{25.38}$$

which is about 0.03 for $\gamma = 1.8$ in the nonlinear regime. In this nonlinear regime, the dynamical crossing time $L/v \ll \tau_{\text{Hubble}}$, and consequently $\tau_{\text{micro}} \ll \tau_{\text{macro}}$. Even in the linear regime where $L/v \approx \tau_{\text{Hubble}}$, growing correlations will produce $\tau_{\text{micro}} \lesssim \tau_{\text{macro}}$ over increasing timescales. Therefore quasi-equilibrium is a reasonable initial approximation and becomes better as the clustering increases. It enables local fluctuations to be averaged over a timescale between the fluctuation timescale and τ_{macro}, making a thermodynamic description possible. Unstable large-scale coherent structure over the entire system would shorten τ_{macro}, and if the instability were fast enough, thermodynamics would not apply.

It is also interesting to examine the quasi-equilibrium assumption from the point of view of Liouville's equation (17.16) for the N-particle distribution function in $6N$-dimensional phase space,

$$\frac{df^{(N)}}{dt} = \frac{\partial f^{(N)}}{\partial t} + \sum_{i=1}^{3N} \left(\frac{\partial f^{(N)}}{\partial q_i} \dot{q}_i + \frac{\partial f^{(N)}}{\partial p_i} \dot{p}_i \right) = 0. \tag{25.39}$$

This represents all members of the grand canonical ensemble that contain exactly N galaxies. The distribution $f^{(N)}$ is an average over a small volume element $d^{3N}q\, d^{3N}p$ in phase space, and its total time derivative along the Hamiltonian trajectories is zero. This means that $f^{(N)}$ is itself an integral of the motion of all N galaxies. Therefore, $f^{(N)}$ depends only on other integrals of the motion such as the total energy $U(\mathbf{p}, \mathbf{q})$. It does not depend on linear momentum because a translational motion of the whole system would not affect the distribution. Nor does it depend on angular momentum because there is rotational symmetry around every point in the statistically homogeneous system. Arbitrary mass distributions with no constraints imposed on the gravitational potential lead to $f^{(N)} = f^{(N)}(U)$ (e.g., GPSGS, Section 39). For strict equilibrium, $\partial f^{(N)}/\partial t = 0$ in any given region of phase space. Now if U varies slowly on a timescale $\sim \tau_{\text{Hubble}}$, then the partial time derivative of $f^{(N)}(U)$ in (25.39) will be small compared to the coordinate and momentum terms, which nearly cancel to give an approximate steady state for the whole system. To estimate how accurate the quasi-equilibrium approximation is, we may therefore use (25.36) for $\dot{U}$. As clustering becomes stronger and clusters become more stable and bound, b increases, $\dot{U}$ decreases, τ_{macro} increases, and quasi-equilibrium becomes more accurate.

The existence of quasi-equilibrium evolution and the applicability of thermodynamics to a good approximation does not depend on the specific form of $b(\bar{n}, T, t)$. However, strong quasi-equilibrium will hold if the functional form of $b(\bar{n}, T)$ remains invariant as the system evolves. For this we must determine how general physical requirements constrain the form of $b(\bar{n}, T)$. We shall also need this form to specify the full thermodynamic description.

26

The Functional Form of $b(\bar{n}, T)$

From the general, unfold the particular.

26.1 $b(\bar{n}, T) = b(\bar{n}T^{-3})$

Our analysis so far is incomplete. One way of looking at it is that we have found two out of the three equations of state necessary to specify the cosmological many-body system. Another way is that we need to complete the temperature and density dependence of the two equations of state (25.16) and (25.27) for $U(\bar{n}, T, b)$ and $P(\bar{n}, T, b)$. From the first point of view we would directly derive the equation of state for the chemical potential μ. From the second, adopted here, we would use general physical principles to determine $b(\bar{n}, T)$. This has the advantage of yielding an exact form of $\mu(\bar{n}, T)$ quite simply.

Nothing is more general than the First and Second Laws of thermodynamics. Their combination (23.52) must be satisfied by any equation of state. Substituting the forms of the equations of state (25.16) and (25.27) into (23.52) shows that the First and Second Laws imply that $b(\bar{n}, T)$ must satisfy (Ahmad, 1996)

$$3\bar{n}\left(\frac{\partial b}{\partial \bar{n}}\right)_T + T\left(\frac{\partial b}{\partial T}\right)_{\bar{n}} = 0. \tag{26.1}$$

This delightfully simple linear partial differential equation has the general solution

$$b(\bar{n}, T) = b(\bar{n}T^{-3}) \tag{26.2}$$

for any functional form of b. This solution follows from (26.1) by inspection, or more formally by integrating along its characteristics $d\bar{n}/3\bar{n} = dT/T$. Therefore the dependence of b on $\bar{n}$ and T can only occur in the combination $(\bar{n}T^{-3})$. Physically this can be traced to the requirement that the entropy of a system in an equilibrium state (recall that the states defining quasi-equilibrium evolution are all equilibrium states) does not depend on the path by which the system reaches that state and so the entropy must be a total differential.

We can also derive (26.2) from a scaling property (Saslaw and Hamilton, 1984; Saslaw and Fang, 1996) of the gravitational partition function, which is a special case of more general scaling (Landau and Lifshitz, 1980, p. 93). From (23.60), the grand canonical partition function is (with Boltzmann's constant $k = 1$)

$$Z(T, V, \mu) = \sum_j e^{\mu N j/T} z_j(T, V), \tag{26.3}$$

where

$$z_j(T, V) = \int \exp\left[-\left(\sum_{i=1}^{j} \frac{p_i^2}{2m_i} - \frac{1}{2} \sum_{i \neq k}^{j} \frac{Gm_im_k}{r_{ik}} \right) T^{-1} \right] d^{3N}p \, d^{3N}r \quad (26.4)$$

is the canonical partition function for a system with j particles.

Now consider a scale transformation of length, momentum, temperature, and chemical potential:

$$r \rightarrow \lambda r,$$
$$p \rightarrow \lambda^{-1/2} p,$$
$$T \rightarrow \lambda^{-1} T,$$
$$\mu \rightarrow \lambda^{-1} \mu. \quad (26.5)$$

This leaves the exponentials in z_j and Z invariant. Thus it is a transformation that leaves the ratio of the kinetic and gravitational interaction energy to the average energy independent of scale. But it changes the phase-space volume by a factor $\lambda^{3N/2}$. Therefore $z \rightarrow \lambda^{3N/2}z$. The general solution of this functional transformation is $z_j(T, V) = T^{-3N/2}\eta(\bar{n}T^{-3})$, where η is an arbitrary function of just $\bar{n}T^{-3}$, since $V \rightarrow \lambda^3 V$ and $n = N/V \rightarrow \lambda^{-3}n$ for a fixed N. Consequently, all modifications of the thermodynamic functions due to the potential energy term in (26.4) will depend on the thermodynamic variables only in the combination $\bar{n}T^{-3}$. This results just from the fact that the ratio of the gravitational potential energy to the temperature in the canonical and grand canonical ensemble is invariant under the transformation (26.5). Such invariance clearly holds whether or not the statistically homogeneous ensemble expands with the universe. Expansion removes those terms that contribute to the mean field in the sum over the potential.

Similar scaling arguments can also be used to derive the pressure equation of state (25.23) in interacting systems (e.g., Hill, 1956; Balescu, 1975, GPSGS, Sections 30.1, 30.2). Here, I emphasize scaling arguments less because they are less physically intuitive than the earlier kinetic and thermodynamic derivations. The scaling derivations also confirm that these equations of state are exact for pairwise interacting potentials.

26.2   The Specific Function $b(\bar{n}T^{-3})$

Finding the actual function $b(\bar{n}T^{-3})$ for the cosmological many-body system requires a deeper analysis of its physical properties (Saslaw and Fang, 1996). One of the most fundamental of these properties is the pairwise nature of the gravitational interaction. The last chapter derived its well-known implication that the equations of state and other thermodynamic quantities depend only on the two-particle correlation function ξ_2. Even though some properties such as the distribution of fluctuations may depend on higher order correlations, these higher order correlations can be reexpressed in terms of ξ_2, as we shall see. This basic dependence on just ξ_2 will lead us to the form of b.

In the grand canonical ensemble, the average density $\bar{n}$ and temperature T are fixed, but individual densities and energies fluctuate among members of the ensemble. From Section 26.1, the gravitational contribution to thermodynamic quantities such as U, P, and S depends on the combination of $\bar{n}$ and T only in the form

$$x \equiv b_0 \bar{n} T^{-3}, \tag{26.6}$$

where b_0 may depend on time, but not on the intensive variables $\bar{n}$ or T. Therefore the volume average $\bar{\xi}_2$ is also a function of x. Consequently, the relation (14.17),

$$\langle [N(r_1) - \bar{N}][N(r_2) - \bar{N}] \rangle = \langle (\Delta N)^2 \rangle$$

$$= \bar{N} + \frac{\bar{N}^2}{V} \int_V \xi_2(\bar{N} T^{-3}, r) \, dV$$

$$= \bar{N} + \bar{N}^2 \bar{\xi}_2(x), \qquad \bullet \tag{26.7}$$

between the variance of the numbers of galaxies among subsystems, all having volume V, and the volume-averaged correlation $\bar{\xi}_2$ in the grand canonical ensemble constrains the dependence of $\bar{\xi}_2$ on x. The left-hand side of (26.7) depends on powers of the density up to and including second order. Therefore $\bar{N}^2 \bar{\xi}_2(x)$ may depend on powers of x up to and including x^2. Since $\bar{N}$ is a given constant for any particular grand canonical ensemble, we may generally express $\bar{N}\bar{\xi}_2$ as

$$\bar{N}\bar{\xi}_2 = a_0 + a_1 x + a_2 x^2, \tag{26.8}$$

where one factor of $\bar{N}$ has been absorbed into the a_i coefficients for a given ensemble.

In obtaining (26.8), it is helpful to recognize the difference between the volume average $\bar{N}$ and the grand canonical ensemble average $\langle N \rangle$, which averages over energy states. The general relation between $\bar{\xi}_2$ and the fluctuations in the grand canonical ensemble is (e.g., Goodstein, 1985, Equation 4.2.54)

$$V(1 + \bar{\xi}_2) = \frac{1}{\bar{n}^2 Z} \sum_N \frac{N(N-1)e^{\mu N/T}}{N! \Lambda^{3N}} \int \cdots \int e^{-U\{N\}/T} d\{N-1\}$$

$$= \frac{1}{\bar{n}^2 V} \langle N(N-1) \rangle, \tag{26.9}$$

where Z is the grand canonical partition function, Λ is the phase-space volume of a cell, Boltzmann's constant is again unity, and the interaction energy $U\{N\}$ depends on N coordinates. Therefore

$$\bar{N}^2 \bar{\xi}_2 = \langle N^2 \rangle - \langle N \rangle - \bar{N}^2, \tag{26.10}$$

where the grand canonical ensemble average (angle brackets) is essentially an average over all energy configurations.

For equilibrium systems in the limit $N \to \infty$ (which is not a necessary condition for a thermodynamic description to apply), the relative fluctuations usually become so small that the grand canonical ensemble becomes equivalent to the canonical ensemble and $\langle N \rangle \to \bar{N}$. Then the average over all spatial volumes becomes the same as the average over all energy configurations. However, for cells with a finite N, even though $\langle N \rangle$ and $\bar{N}$ may be related by a simple numerical factor for any particular grand canonical ensemble, they will generally have different functional dependences as the system evolves from one ensemble to another. For example, clustering may change the complexions within some volume V and alter the energy states over which the $\langle N \rangle$ average is taken, but it need not alter $\bar{N}$ for the entire system. A Poisson distribution in the limit $N \to \infty$ satisfies

$$\langle N^2 \rangle = \overline{N^2} = \bar{N} + \bar{N}^2. \tag{26.11}$$

However, usually it is necessary to make use of the more general relation

$$\langle N^2 \rangle = \alpha_0 + \alpha_1 \langle N \rangle + \alpha_2 \langle N \rangle^2, \tag{26.12}$$

where the αs are constants. There are no higher order terms since both sides must be consistent to order N^2. From (26.3), the thermodynamic properties of $\bar{\xi}_2$ under a change of grand canonical ensemble depend on N only in the combination $\langle N \rangle V^{-1} T^{-3}$, which is related to x in (26.6) by a coefficient that absorbs the numerical difference between $\langle n \rangle$ and $\bar{n}$ for any particular grand canonical ensemble with a given constant $\bar{N}$. Substituting (26.12) into (26.10) and combining all the constant coefficients gives (26.8).

Next, we see how basic physical properties of the cosmological many-body system determine the coefficients in (26.8). In the limit of vanishing gravitational interaction as $x \to 0$ (effected either by $b_0 \sim G^3 \to 0$ or by $T \to \infty$), the system reduces to a perfect gas with $\bar{\xi}_2 \to 0$, and so

$$a_0 = 0. \tag{26.13}$$

Physical constraints on $b(x)$ will determine the coefficients a_1 and a_2. The variance of number fluctuations is related to the equation of state by (23.68):

$$\langle (\Delta N)^2 \rangle = \frac{T\bar{N}}{V} \frac{\partial N}{\partial P}\Big)_{T,V} = \frac{\bar{N}}{\left[1 - b - x\frac{db}{dx}\right]}. \tag{26.14}$$

The second equality follows for any equation of state with the form (25.27), whatever the function $b(x)$ is. Substituting (26.8), (26.13), and (26.14) into (26.7) gives

$$\frac{db}{dx} = -\frac{b}{x} + \frac{a_1 + a_2 x}{1 + a_1 x + a_2 x^2}. \tag{26.15}$$

Consider a sequence of ensembles in which x increases from zero so that $\bar{\xi}_2$ and therefore b increases as the ensemble becomes more clustered. Since b (like all thermodynamic variables away from a phase transition) must be an analytic function of x in this limit and $b \to 0$ as $x \to 0$, we must have $b \to \gamma x + O(x^2)$. But the coefficient γ may be absorbed into the definition of x through b_0, and so we may set $\gamma = 1$ to obtain

$$\lim_{x \to 0} \frac{db}{dx} = 1. \tag{26.16}$$

This would also follow from the physical requirement that db/dx remains finite, positive, and nonzero as $x \to 0$. Therefore from (26.15) we find

$$a_1 = 2. \tag{26.17}$$

In the limit of very strong clustering, as $x \to \infty$, the solution of (26.15) is

$$\lim_{x \to \infty} b = 1 - \frac{1}{x}, \tag{26.18}$$

which will constrain the general solution of (26.15).

This general solution is

$$b = 1 + \frac{a_3}{x} - \frac{1}{x} \int \frac{dx}{1 + 2x + a_2 x^2}, \tag{26.19}$$

where a_3 is a constant of integration. As $x \to \infty$, the integral term becomes of order x^{-2}, and therefore (26.18) requires

$$a_3 = -1. \tag{26.20}$$

Only a_2 remains to be determined. The character of the integral in (26.19) depends on whether $a_2 < 1$ or $a_2 > 1$ or $a_2 = 1$, and we examine each case.

If $a_2 < 1$, we may write

$$\alpha \equiv (1 - a_2)^{1/2} \tag{26.21}$$

to obtain

$$b = 1 - \frac{1}{x} - \frac{1}{2\alpha x} \ln \left(\frac{1 - \alpha + a_2 x}{1 + \alpha + a_2 x} \right)$$

$$= -\left[1 + \frac{1}{2\alpha} \ln \left(\frac{1 - \alpha}{1 + \alpha} \right) \right] \frac{1}{x} + O(x), \tag{26.22}$$

where the last equality applies near the noninteracting (perfect gas) limit as $x \to 0$. In order that b not diverge in this limit, the coefficient of x^{-1} must be zero, implying that

$$1 = \frac{1}{2\alpha} \ln\left(\frac{1+\alpha}{1-\alpha}\right) = 1 + \frac{\alpha^2}{3} + \frac{\alpha^4}{5} + \cdots, \qquad (26.23)$$

which has no solution for $a > 0$ or $a_2 < 1$. It is consistent only in the limit $a_2 \to 1$. If $a_2 > 1$, we may write

$$\beta \equiv (a_2 - 1)^{1/2} \qquad (26.24)$$

to obtain

$$b = 1 - \frac{1}{x} - \frac{1}{\beta x} \tan^{-1}\left(\frac{1 + a_2 x}{\beta}\right)$$

$$= -\left(1 + \frac{1}{\beta} \tan^{-1} \frac{1}{\beta}\right) \frac{1}{x} + O(x), \qquad (26.25)$$

with the last equality again applying as $x \to 0$. In this limit there are no divergence-free solutions for positive values of $\tan^{-1}(1/\beta)$. Moreover, the exact solution in (26.25) is inconsistent with the boundary condition (26.18) for large x. So $a_2 > 1$ is also inconsistent.

Therefore the physical boundary conditions force us to the special value of

$$a_2 = 1. \qquad (26.26)$$

For this value the solution of (26.19) is

$$b = 1 - \frac{1}{x} + \frac{1}{x(1+x)} = \frac{x}{1+x} = \frac{b_0 \bar{n} T^{-3}}{1 + b_0 \bar{n} T^{-3}}. \qquad (26.27)$$

This also satisfies the condition (25.19) that $0 \le b \le 1$, since b is bounded by the perfect gas ($x = 0$) and by the fully clustered ($x \to \infty$) regimes.

Thus we have derived $b(\bar{n} T^{-3})$ from the general functional form of $\bar{\xi}_2(\bar{n} T^{-3})$, its relation to thermodynamic fluctuations through the general form of the equation of state for a pairwise interacting gravitational system, and its physical boundary conditions. The result is the same as adopted earlier (Saslaw and Hamilton, 1984) as the simplest mathematical expression of the condition (25.19).

More complicated relations than (26.27) have been postulated from time to time, but one can show (Ahmad, 1996; Saslaw and Fang 1996) that they lead to undesirable physical consequences such as distribution functions having negative probabilities, violations of the Second or Third Laws of thermodynamics, or gravitational clustering, which decreases as the universe expands. For example, a form of $b(\bar{n} T^{-3})$ that differs from (26.27) and leads to the negative binomial distribution for $f(N)$ fails to satisfy the Second Law of thermodynamics in the highly clustered expanding universe.

26.3 Minimal Clustering

We can also derive the form (26.27) of $b(x)$ from a different physical point of view: minimal clustering. Alternatively we could use the previous derivation of (26.27) to establish the principle of minimal clustering. This principle states that as $x(t) = b_0(t)\bar{n}(t)T^{-3}(t)$ increases in quasi-equilibrium, the system tends to minimize its macroscopic nonequilibrium response. This implies that in a representative part of the universe (or in a quasi-equilibrium sequence of grand canonical ensembles) gravitational clustering develops as slowly as possible subject to general physical constraints.

Physically, minimal clustering occurs because clustering and expansion are nearly balanced in the linear perturbation regime of the Einstein–Friedmann models, and in the nonlinear regime the increasing virialization reduces the rate of evolution (see Sections 18.4 and 22.2). Faster evolution, which may occur in some non-Friedmann models (Saslaw, 1972), would decrease the macroscopic timescale and weaken the quasi-equilibrium condition $t_{micro} \ll t_{macro}$ (Sections 25.5 and 25.6). For similar reasons, minimal clustering will not hold in systems having significant power in large-scale nonequilibrium modes (or structures) which are rapidly unstable. It is most applicable to an initial Poisson state and will apply approximately to initial states "near" the Poisson state. The precise meaning of "near" is a questions of dynamics rather than thermodynamics and is not yet understood. It may be related to the existence of dynamical attractors in the equations of motion, which cause the system to evolve in a quasi-equilibrium manner. Clues from numerical N-body simulations (Saslaw, 1985b; Itoh, 1990; Suto, Itoh, and Inagaki, 1990) suggest that minimal clustering applies reasonably well for initial power-law perturbation spectra with $-1 \lesssim n \lesssim 1$ (where n is defined in Equation (14.35)). This includes the Poisson case, $n = 0$, but its full range of applicability needs further exploration.

More precisely, minimal clustering implies that at $x = 0$, the value of $\bar{\xi}_2(x) = 0$ must be an absolute minimum and

$$\frac{d^n \bar{\xi}_2}{dx^n} = 0 \quad \text{for } n \geq 3 \tag{26.28}$$

so that this minimum is as flat as possible. Moreover, (26.28) must continue to hold further along the sequence of increasing x to minimize the difference between each member of the ensemble and its predecessor along this sequence. As a consequence of (26.28), $\bar{\xi}_2$ has the form of (26.8) and the analysis of the previous section again leads to $b(\bar{n}T^{-3})$ given by (26.27).

Another way to see how (26.27) follows from minimal clustering is to examine constraints on the more general relation

$$b(x) = \frac{x}{1+x} g(x), \tag{26.29}$$

where $g(x)$ is arbitrary. The requirement

$$0 \le \frac{db}{dx} \le 1, \tag{26.30}$$

which follows from the fundamental results (26.14)–(26.18) noting that $\langle (\Delta N)^2 \rangle$ is greater than 0, then implies

$$g + x(1+x)g' = (1+x)^2 - y(x), \tag{26.31}$$

where $y(x)$ satisfies

$$0 \le y(x) \le (1+x)^2. \tag{26.32}$$

Minimal clustering requires that, for all x, the function $y(x)$ has a maximum value consistent with $y(0) = 0$ and that $y(x)$ joins smoothly onto $(1+x)^2$ for large x so that $b'(x)$ is minimal. Therefore

$$y(x) = 2x + x^2. \tag{26.33}$$

Substituting (26.33) into (26.31) and solving for $g(x)$ shows that the only physical solution that remains finite as $x \to 0$ is $g(x) = 1$ for all x, again giving (26.27).

An important sequel follows immediately by inserting (26.27) into (26.14) to obtain the variance of fluctuations for the number of particles in any given size volume:

$$\langle (\Delta N)^2 \rangle = \frac{\bar{N}}{(1-b)^2}, \tag{26.34}$$

with $b = b_V$ from (25.17) or (25.18). When $b = 0$, these reduce to the Poisson fluctuations of a noninteracting system. As $b \to 1$ the fluctuations become infinite; they intimate a phase transition. If $b[x(t)]$ increases minimally when $x(t)$ increases in a given system, or in a sequence of ensembles, then the dynamical fluctuations will grow slowly, preserving the quasi-equilibrium nature of the evolution.

27

Derivation of the Spatial Distribution
Function $f(N)$

... it is with these attributes that this my small book
will be interwoven, recalling to the painter by what
rules and in what way he ought by his art to imitate
all things that are the work of nature and the
adornment of the world.

Leonardo da Vinci

Having completed the equation of state, we can begin to explore its consequences. They are tremendously rich and far reaching; only a few have been examined in detail. The three main ones are the spatial distribution function $f(N)$, the velocity distribution function $f(v)$, and the evolution $b(t)$. All of these results can be compared directly with N-body simulations and, most importantly, with observations. We take them up in turn here and in the following chapters.

27.1   Entropy and Chemical Potential

The first step in deriving the spatial distribution function is to calculate the entropy and chemical potential for the cosmological many-body system. These quantities also provide interesting insights in their own right. To find them, it is simplest to work with the specific energy

$$u(s, v) = N^{-1} U(S, V, N) \tag{27.1}$$

in terms of the specific entropy $s = S/N$ and the specific volume $v = V/N = \bar{n}^{-1}$. Its change is generally $du = (\partial u/\partial s)_V \, ds + (\partial u/\partial v)_s \, dv$ with $(\partial u/\partial s)_V = (\partial U/\partial S)_{V,N} = T$, and similarly $(\partial u/\partial v)_s = -P$. This gives the standard relation $du = T \, ds - P \, dv$. Changes in v (or $\bar{n}$) in members of an ensemble can be produced by variations of either N or V, keeping the other one constant. Substituting the equations of state (25.16) and (25.27) for U and P gives for the entropy

$$d\left(\frac{S}{N}\right) = \frac{1}{T} d\left(\frac{U}{N}\right) + \frac{P}{T} d\left(\frac{V}{N}\right)$$

$$= d \ln (\bar{n}^{-1} T^{3/2}) - 3 \, db + b \, d \ln (\bar{n} T^{-3}), \tag{27.2}$$

where $\bar{n}$ is the average density in V. Substituting (26.27) into (27.2) and integrating [note that b_0 is independent of $\bar{n} T^{-3}$ although it may depend separately on time and on the given volume of the integral in (25.18)] yields

$$\frac{S}{N} = \ln \bar{n}^{-1} T^{3/2} - 3b - \ln(1-b) + s_0 \tag{27.3}$$

with s_0 an arbitrary constant.

For $b = 0$, the entropy reduces to that of a perfect classical gas. For fixed $\bar{n}^{-1} T^{3/2}$, the gravitational contribution $S_{\text{grav}}/N = -3b - \ln(1-b)$ decreases the total entropy for small b but causes it to become infinitely positive as $b \to 1$. This is because mild clustering introduces more structure (more correlation information if you like) into the $b = 0$ Poisson distribution when $\bar{n}^{-1} T^{3/2}$ is constrained to be constant. However, as $b \to 1$ the system becomes virialized on all scales, clusters become dominant, and the entropy becomes infinite. The transition from $S_{\text{grav}} < 0$ to $S_{\text{grav}} > 0$ occurs at $b \approx 0.95$. Infinite entropy is another manifestation of the infinite density fluctuations (26.34) as $b \to 1$. If $\bar{n}^{-1} T^{3/2}$ is not constrained, but allowed to change consistently, then S/N always increases monotonically with increasing b, as in (30.16), and again becomes infinite as $b \to 1$.

Chemical potential, μ, measures the change of internal energy if one particle is added to the system while keeping all other thermodynamic quantities such as volume and temperature constant. To obtain μ and to simplify the algebra in the following derivation of $f(N)$, it is more convenient to use an entropy representation (cf. discussion after (23.28)) of the grand canonical potential in (23.61). This is readily done by letting

$$-\frac{\Psi(23.61)}{kT} \to \Psi, \tag{27.4}$$

and as usual setting $k = 1$. With this new Ψ, we write (23.61) as

$$\Psi = S - \frac{U}{T} + \frac{N}{T}\mu, \tag{27.5}$$

and (23.62) becomes

$$\Psi = \Psi\left(\frac{1}{T}, V, \frac{\mu}{T}\right) \ln Z, \tag{27.6}$$

along with $U = T^2 \, \partial\Psi/\partial T$ for (23.63), which is just (23.47). Combining (27.5) with the Euler equation (23.15) and the pressure equation of state (25.27), we see that

$$\Psi = \frac{PV}{T} = \bar{N}(1-b). \tag{27.7}$$

Now (27.5), (27.7), (27.2 with constant T), and the energy equation of state (25.16) give the general chemical potential

$$\frac{\mu(n,b)}{T} = \ln(nT^{-3/2}) - b - \int n^{-1} b \, dn \tag{27.8}$$

for any form of $b(n, T)$. Substituting the specific form (26.27) for b gives

$$\frac{\mu}{T} = -b + \ln\left(\frac{b}{b_0 T^{-3/2}}\right). \tag{27.9}$$

The chemical potential is negative since adding a particle with the average energy of the system decreases the total gravitational binding energy.

27.2   The Gravitational Quasi-Equilibrium Distribution $f(N)$

Now we use our earlier results to derive the gravitational quasi-equilibrium distribution (GQED) $f(N)$ for the cosmological many-body system (Saslaw and Hamilton, 1984). Collect an ensemble of subregions, all of the same shape and volume, which are much smaller than the total system. Both the number of galaxies and their mutual gravitational energy will vary among the members of the ensemble. This describes a reasonable way to sample the universe. Thermodynamically, each subregion can be regarded as being in contact with a large reservoir. Across the boundaries flow particles (galaxies) and energy. This is our grand canonical ensemble characterized by a given chemical potential μ, temperature T, and value of $b = b_V(\bar{n}T^{-3})$ from (25.18) and (26.27). It describes systems with variable N and U.

Statistical thermodynamics gives us the probability (23.59) of finding N_j particles in the energy state $U_j(N_j, V)$:

$$P_{Nj}(V, T, \mu) = \frac{e^{-U_j/T + N_j\mu/T}}{Z(V, T, \mu)}, \tag{27.10}$$

where

$$Z(V, T, \mu) = \sum_j e^{-U_j/T + N_j\mu/T} \tag{27.11}$$

is the grand partition function, now explicitly summing over U_j and N_j (in units with $k = 1$). Substituting P_{Nj} into the general definition of entropy,

$$S = -\sum_j P_{Nj} \ln P_{Nj}, \tag{27.12}$$

directly returns (27.5) and (27.6) with $N = \bar{N}_j$.

The probability of finding a particular number N of particles in a volume V of any given shape (which we shall see also applies to a volume such as a cone projected onto an area of sky, as long as the appropriate b_V is used) follows from summing (27.10) over all energy states:

$$f(N) = \sum_j P_{Nj}. \tag{27.13}$$

Helpfully, we do not have to do the sum explicitly to find $f(N)$. Instead, we calculate its generating function from the observation that e^Ψ is the partition function, which normalizes this sum over all energy states. The generating function is the quantity

$$\langle z^N \rangle \equiv \sum_N z^N f(N), \tag{27.14}$$

where z is an arbitrary variable. Thus the term multiplying z^N in the power series expansion of the average $\langle z^N \rangle$ is just $f(N)$. To determine this average, let $z \equiv e^q$ and use (27.11) and (27.6) in (27.14) to give

$$\langle e^{qN} \rangle = Z^{-1} \sum_{N,j} e^{-U_j/T + (q+\mu/T)N} = \exp\left\{ \Psi\left(\frac{\mu}{T} + q \right) - \Psi\left(\frac{\mu}{T} \right) \right\}. \tag{27.15}$$

Here we regard the arguments T^{-1} and V of Ψ as being understood since they do not enter explicitly in the following analysis.

At this stage it simplifies the algebra to regard $\Psi(x)$ as a function of e^x rather than of x. In terms of z, Equation (27.15) then becomes

$$\sum_N z^N f(N) = \exp\{ \Psi(ze^{\mu/T}) - \Psi(e^{\mu/T}) \} = e^{-\Psi} e^{\Psi(ze^{\mu/T})}. \tag{27.16}$$

Now expand the last exponential in (27.16) in a Taylor series around $z = 0$ and equate the coefficients of powers of z on both sides of the equation to find

$$f(N) = e^{-\Psi} \frac{e^{\mu N/T}}{N!} (e^\Psi)_0^{(N)}, \tag{27.17}$$

with

$$(e^\Psi)_0^{(N)} = \left[\left(\frac{d}{d(ze^{\mu/t})} \right)^N \exp \Psi(ze^{\mu/T}) \right]_{ze^{\mu/T}=0}. \tag{27.18}$$

This is the basic result.

We can apply it straightaway to the distribution of voids. For $N = 0$ and, consequently, $U = 0$, we see from (27.11) and (27.6) that $\Psi(ze^{\mu/T} = 0) = 0$ and therefore

$$f(0) = e^{-\Psi(\mu/T)} = e^{-\bar{N}(1-b)} = e^{-4\pi r^3 \bar{n}(1-b)/3}, \tag{27.19}$$

using (27.7) The last expression, in particular, is the probability of finding an empty spherical volume of radius r.

In the limit when gravity is unimportant so that $b \to 0$, this reduces to the standard result for a Poisson distribution. As $b \to 1$, the galaxies become more and more clustered, and so the probability of finding an empty region of any size approaches unity in the idealized limit. Notice that the void distribution does not depend on the specific thermodynamic functional form of $b(nT^{-3})$. Nor does the functional form of $f(0)$ depend on the volume of the cells in the grand canonical ensemble. However, the value of $b = b_V$ does depend on this volume. So, in general, there will be an additional volume dependence of (27.19) through b_V in (25.17) or (25.18). This additional dependence will be small for large cells over the range where $\xi(r)$ contributes little to the volume integral in b. For very small cells, $b_V \to 0$ if $\xi(r)$ is less singular than r^{-2} in (25.17), and the distribution will appear Poisson. This is because such small cells almost always contain either one or zero particles, and their internal interaction energy vanishes. Later we will see that N-body computer simulations accurately confirm this result.

To find the full set of distributions $f(N)$ we rewrite (27.18) in a more specific form. Replacing the dummy variable $ze^{\mu/T}$ by $e^{\mu/T}$ (alternatively taking the limit $z \to 1$), we note that

$$\frac{d}{de^{\mu/T}} = e^{-\mu/T} \frac{db}{d(\mu/T)} \frac{d}{db}. \tag{27.20}$$

To obtain $db/d(\mu/T)$, we use (27.7):

$$\frac{d\Psi}{d(\mu/T)} = \frac{d[Vn(1-b)]/db}{d(\mu/T)db} = nV, \tag{27.21}$$

where the last equality follows from the standard result

$$N = T \left(\frac{\partial \Psi}{\partial \mu} \right)_{V,T}, \tag{27.22}$$

which follows in turn from (27.11) and (27.6), or from (27.5) at constant V. Substituting (27.20) and (27.21) into (27.18) gives

$$(e^{\Psi})_0^{(N)} = \left[\left(ne^{-\mu T} \frac{db}{d[n(1-b)]} \frac{d}{db} \right)^N e^{nV(1-b)} \right]_{n=0}. \tag{27.23}$$

Here $n = n(b, T)$ and the complete operator in parenthesis is applied N times.

Two features of (27.23) are especially noteworthy. First, the properties of a distribution for this gravitationally interacting system are determined by its infinitely diffuse limit, $n = 0$. This also occurs in the description of an imperfect gas by its virial expansion in powers of the density. Second, there is a form of $b(n)$ that simplifies the operators in (27.23) enormously, namely $n(1-b) \propto b$. Remarkably, this is just the form (26.27) to which basic physical principles led us earlier.

With (26.27) for $n(b)$, and (27.9) for μ/T, it is straightforward to calculate $(e^{\Psi})_0^{(N)}$ for the first few values of N to determine its general form, which may be proved by induction (see Saslaw and Hamilton, 1984). Then substituting (27.7) for Ψ and combining with (27.17) gives the general formula for the GQED distribution functions:

$$f(N) = \frac{\bar{N}(1-b)}{N!}[\bar{N}(1-b) + Nb]^{N-1}e^{-\bar{N}(1-b)-Nb}. \tag{27.24}$$

We will often use this basic result in subsequent discussions. Notice that this form of $f(N)$ is the same for grand canonical ensembles with any given value of V or T, which determines b_V. All we have to do to determine how the distribution changes for different V or T is to select another ensemble with larger or smaller, or hotter or colder, cells. The spatial distribution depends only on $\bar{N} = \bar{n}V$ and $b = b_V$, both of which can be determined easily if the particles' positions and velocities are known. So there are no free parameters in the basic theory.

28

Properties of the Spatial GQED

Here lyeth muche rychnesse
in lytell space.
Heywood

The spatial gravitational quasi-equilibrium distribution (27.24) is replete with fascinating physical and mathematical properties. Many of these have been explored; many more remain to be discovered. In this chapter, I summarize some important results, deriving those whose derivations are simple and quoting others whose origins are more esoteric. These properties unfold from the mainly physical to the mainly mathematical, allowing for logical precedence and similarities. Theoretical insights, computer simulations, and agreement with observations have all stimulated these developments.

28.1   Physical Limiting Cases and Self-Organized Criticality

Since the GQED depends on both N and V, we can fix a value for N and regard the distribution as a continuous function of volume, $f_N(V)$. This is the probability that a cell with exactly N particles (galaxies) has a volume between V and $V + dV$. (We can also make this a description for shapes by considering the shape dependence of b_V.) Alternatively, we can fix the value of V to obtain the discrete probability $f_V(N)$ that a given size volume in the system contains exactly N particles. Both these distributions are orthogonal cuts through the generic distribution function denoted by $f(N, V)$ or sometimes more simply by $f(N)$, for given values of b and $\bar{n}$. Although $f_N(V)$ and $f_V(N)$, for all N and V, contain the same information as $f(N, V)$, each emphasizes different aspects of this information. Different emphases are often useful in different contexts.

For example the $f_V(N)$ distribution is often called "counts in cells." It is easily related to observations either in space or projected onto the sky and is especially suitable for representing $f(N, V)$ when N is large. The distribution $f_N(V)$, in contrast, makes it easier to recover information about the N particles nearest to a randomly chosen spatial point in the system. This is an especially suitable representation when N is small. In particular,

$$f_0(V) = e^{-\bar{N}(1-b_V)} \tag{28.1}$$

is the probability that no particles are in a randomly positioned volume of size $V = \bar{N}/\bar{n}$. This is often called the "void distribution" (not to be confused with the merely underpopulated regions that observers sometimes call "voids." Theoretical voids are guaranteed empty – or you get your money back!)

333

Let us look at four limits. The simplest is $b = 0$, for which (27.24) reduces to a Poisson distribution – the expected noninteracting case. The opposite fully clustered case with $b \to 1$ is more subtle. The void probability with fixed $\bar{n}$ tends to unity for any size volume with $b_V \to 1$. Thus the particles tend to cluster within smaller and smaller regions. This is the origin of the formally infinite fluctuations in $\langle (\Delta N)^2 \rangle$ from (26.34) and the infinite entropy in (27.3) – all signs of a phase transition.

As the system approaches this phase transition, it also shows signs of self-organized criticality on larger and larger scales. Substituting Stirling's approximation for $N!$ with $N > 0$ into the GQED (27.24) gives

$$f(N) = \frac{\bar{N}(1 - b)}{\sqrt{2\pi}} b^{N-1} e^{-\alpha/12N} e^{(N-\bar{N})(1-b)} \left[1 + \frac{\bar{N}(1 - b)}{Nb} \right]^{N-1} N^{-3/2}, \quad (28.2)$$

where α is a number from the higher order asymptotic expansion satisfying $0 < \alpha < 1$. As $b \to 1$, $f(N) \sim N^{-3/2}$. This power-law behavior of the fluctuation spectrum characterizes self-organized critical systems (see Bak, 1996). But for the cosmological many-body system there is a difference. Normally, self-organized criticality applies to an entire system. Here it may apply only on those spatial scales where the $N^{-3/2}$ power law is a good approximation. This is generally the case on small spatial scales where $N < 1/(1 - b)$. So as $b \to 1$, the scales distinguished by self-organized criticality can grow. The implications of self-organized criticality for the cosmological many-body system have not yet been examined in detail.

The third limit (e.g., Saslaw & Sheth, 1993) considers $f(N)$ for very large volumes. For fixed $\bar{n}$, this is equivalent to very large $\bar{N}$. In cells so large that correlations do not contribute significantly as the volume in the integral of (25.17) or (25.18) is increased, the value of $b_V = b$ becomes independent of scale. For such large relatively uncorrelated cells, we expect their departures $\epsilon \equiv (\bar{N} - N)/\bar{N}$ from the average value of $\bar{N}$ to be small. Writing the GQED (27.24) in terms of ϵ for any value of ϵ,

$$f(N) = \frac{\bar{N}^N}{N!} e^{-\bar{N}} (1 - b)(1 - \epsilon b)^{N-1} e^{\epsilon \bar{N} b}, \quad (28.3)$$

we see that it has a Poisson component modified by ϵ and b. If $b \to 0$ there are no correlations and we regain the purely Poisson distribution as mentioned earlier. However, in the limit $\epsilon \to 0$ for large $\bar{N}$, this Poisson distribution is renormalized by a factor $(1 - b)$. What does this factor mean?

We get some insight from the related limit $\epsilon \to 0$, $\bar{N} \to \infty$, but $\bar{N}|\epsilon| \gg 1$. (For a fixed volume, $\bar{N} \to \infty$ would be a "fluid" limit. Here, however, the last inequality implies that this is just a large volume limit with small fractional fluctuations and large numerical fluctuations of N.) Using Stirling's approximation the GQED becomes

$$f(N) = \frac{1}{\sigma (2\pi)^{1/2}} \exp\left[-\frac{(\bar{N} - N)^2}{2\sigma^2} + \frac{(1 + 2b)}{2} \frac{(\bar{N} - N)}{N} \right.$$

$$\left. - \frac{(1 + 2b)(1 - b)^2}{6} \frac{(\bar{N} - N)^3}{\bar{N}^2} \right] \quad (28.4)$$

to order $\bar{N}^{-1/2}$ with $\sigma^2 = \bar{N}/(1-b)^2$. Since $(\bar{N}-N)^2 \sim \bar{N}$, the first term in the exponential is of order unity, while the other two terms are each of order $\bar{N}^{-1/2}$. Thus, in the limit of large cell size as $\bar{N} \to \infty$ and b_V becomes independent of V, the GQED becomes Gaussian with a variance that depends on b. Departures from Gaussian introduce a skewness of order $\bar{N}^{-1/2}$, which also depends on b.

In the dominant Gaussian term of (28.4), the variance depends on the factor $(1-b)^{-2}$. If we consider an effective number of particles $N_{\text{eff}} = N(1-b)^2$, with a similar relation for $\bar{N}_{\text{eff}}$, then to lowest order (28.4) becomes a renormalized Gaussian for N_{eff}. This indicates that ensembles with larger values of b, and therefore greater clustering, effectively have a factor of $(1-b)^2$ fewer independent particles than in a completely unclustered ensemble. This represents the fact that more particles are in clusters, reducing the effective number of independent objects in the distribution on all scales. Instead of a simple Gaussian distribution of the fundamental particles (galaxies) in the system, there is now a distribution of bound and partly bound clusters that is approximately Gaussian on very large scales.

The fourth limit (Fry, 1985) is a genuine continuum limit in which $b \to 1$, $\bar{N} \to \infty$, and $\bar{N}(1-b)^2 = \bar{\xi}^{-1}$ from (14.17) remains finite. Setting $x \equiv \rho/\bar{\rho} = \lim_{\bar{N}\to\infty} N/\bar{N}$ the GQED becomes

$$f(\rho)d\rho = \frac{1}{(2\pi\bar{\xi}x^3)^{1/2}} \exp - \left[\frac{(x^{1/2} - x^{-1/2})^2}{2\bar{\xi}} \right] dx. \qquad (28.5)$$

In this fully clustered continuum limit, all the $\bar{\xi}_n$ are independent of volume, so in this sense the distribution becomes hierarchical (Fry, 1985).

28.2 Normalizations, Underdense Regions, and the Shape of the GQED

For a given volume, the $f_V(N)$ distribution of (27.24) is already normalized when summed over all values of N:

$$\sum_{N=0}^{\infty} f_V(N) = 1. \qquad (28.6)$$

This follows directly from setting $z = 1$ in (27.14), or from (27.13). Therefore, the probability that such a volume has a number of objects in some range, N_1–N_2, is just the summation of $f(N)$ over that range. In particular, the probability that a volume is underdense by a factor α, say, is just the sum from zero to the largest integer less than $\alpha\bar{N}$. Observationally, underdense regions with $\alpha \lesssim 0.2$ are sometimes melodramatically called "voids." However, we shall reserve the use of "void" for $N = 0$.

For a given number of objects, the $f_N(V)$ distribution is not automatically normalized. Moreover, since b_V will usually also depend on volume (for a given shape) the normalization

$$\int_0^\infty f_N(V)\, dV = 1 \tag{28.7}$$

will generally depend differently on b for each value of N. Numerical normalization is usually simplest. In the case where a constant value of b independent of volume is a good approximation, the normalization of $f_N(N, \bar{n}V)$ over $\bar{n}V$ gives the normalized distribution (Crane & Saslaw, 1986)

$$f_N^* = \frac{N!(1-b)e^{Nb}}{\Psi(N)} f(N, V), \tag{28.8}$$

where the first few values of Ψ are

$$\Psi(0) = 1,$$
$$\Psi(1) = 1,$$
$$\Psi(2) = 2(1+b),$$
$$\Psi(3) = 9b^2 + 12b + 6,$$
$$\Psi(4) = 8(8b^3 + 12b^2 + 9b + 3),$$
$$\Psi(5) = 5(125b^4 + 200b^3 + 180b^2 + 96b + 24),$$

and in general (E. Weisstein, 1997, private communication)

$$\Psi(N) = b(bN)^N - (1-b)e^{bN}\gamma(1+N, bN), \tag{28.9}$$

where γ is the standard incomplete gamma function. These often give useful approximate fits to the probabilities for finding objects at various distances from a random point in space if the average density $\bar{n}$ is known. More precise fits follow by determining b_V from the variance of counts in cells of volume V using (26.34) and normalizing numerically. We will see examples later when comparing the theory with N-body computer simulations.

Figures 28.1–28.5 (kindly provided by Dr. Eric Weisstein) illustrate the shapes of $f_V(N)$, $f_N(V)$, and $f_b(N, V)$ for a range of parameters. In each case, $\bar{N} = \bar{n}V$ is plotted to scale the volume, giving $f_{\bar{N}}(N, b)$ and $f_N(\bar{N}, b)$. As b increases, $f_{\bar{N}}(N, b)$ becomes increasingly flat and skew compared to a Poisson. Its peak moves to smaller N until, as $b \to 1$, it becomes unity for $N = 0$ and zero otherwise. The normalized graphs of $f_N(\bar{N}, b)$ show the peak becoming shallower and shifting to larger volumes (i.e., larger $\bar{N}$) as b increases until it again becomes zero for all $N > 0$ when $b = 1$. The peak also moves to larger volumes as N increases. Finally, the graphs of $f_b(N, \bar{N})$ for fixed b show the complementary nature of f_N and $f_{\bar{N}}$ with each emphasizing different aspects of the full distribution.

Figures 15.1 and 15.2 (kindly provided by Dr. Fan Fang) have already shown higher resolution illustrations for several distributions that are particularly relevant

$\bar{N} = 1$

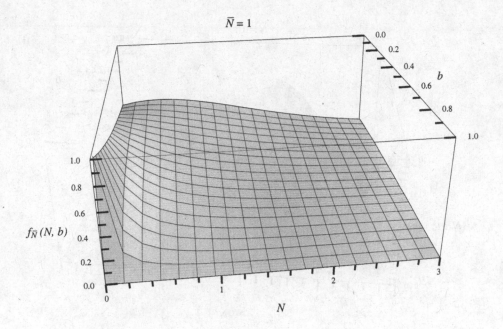

$\bar{N} = 5$

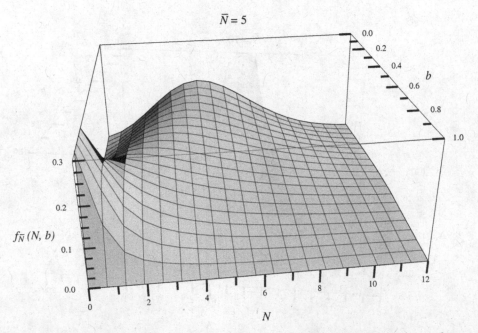

Fig. 28.1. Three-dimensional distributions $f_V(N, b) = f_{\bar{N}}(N, b)$ for counts in given volumes $V = \bar{N}/\bar{n}$ having a fixed average number density $\bar{n}$.

$\bar{N} = 10$

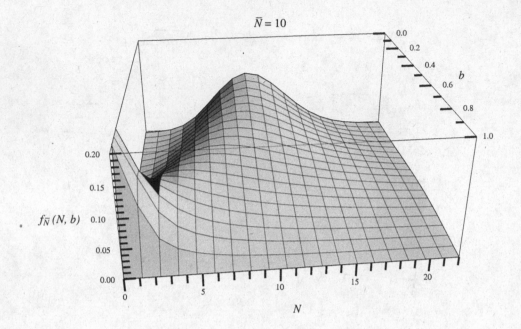

$\bar{N} = 100$

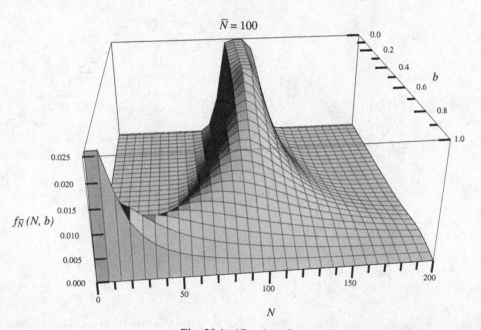

Fig. 28.1. (*Continued*)

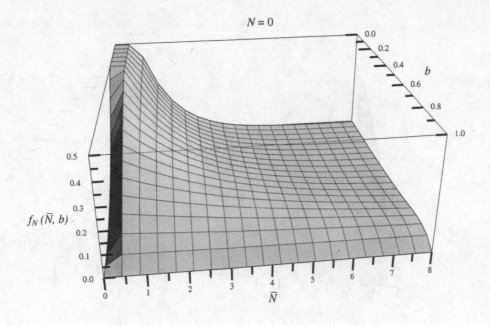

$$N = 0$$

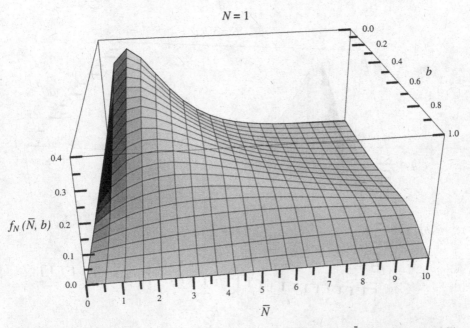

$$N = 1$$

Fig. 28.2. Normalized three-dimensional distributions $f_N(V, b) = f_N(\bar{N}, b)$ for the probability of finding N objects within a volume $V = \bar{N}/\bar{n}$ around a random point in space for $N = 0$ and $N = 1$.

N = 2

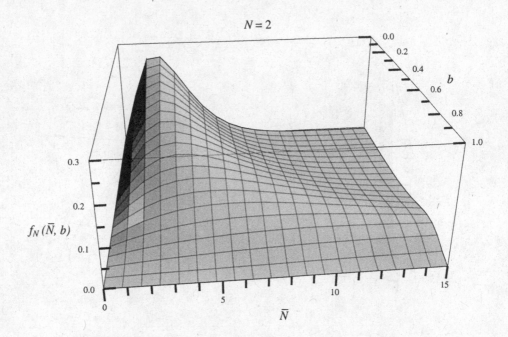

N = 5

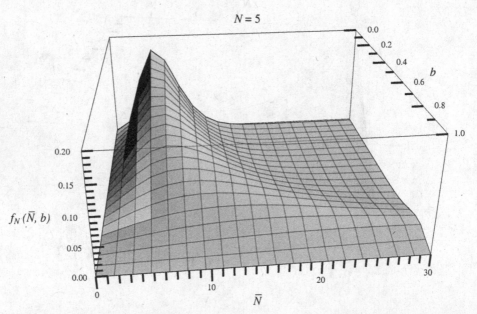

Fig. 28.3. Normalized three-dimensional distributions $f_N(V, b) = f_N(\bar{N}, b)$ for the probability of finding N objects within a volume $V = \bar{N}/\bar{n}$ around a random point in space for $N = 2$ and $N = 5$.

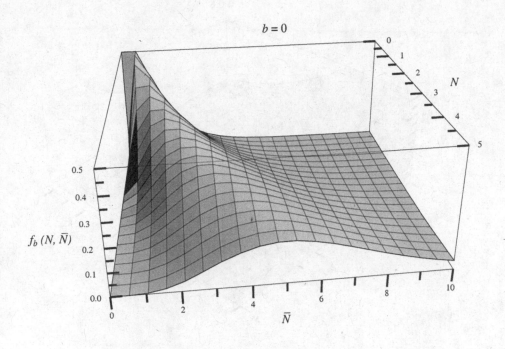

$b = 0$

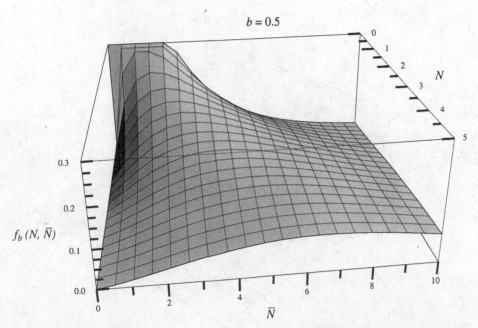

$b = 0.5$

Fig. 28.4. Distributions $f_b(N, \bar{N} = \bar{n}V)$ of both counts in cells and of N objects within an arbitrary random volume V for given values of $b = 0$ and $b = 0.5$.

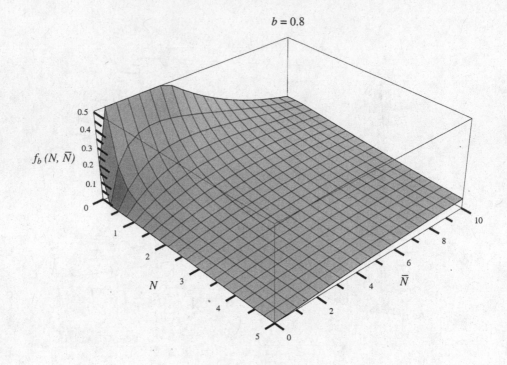

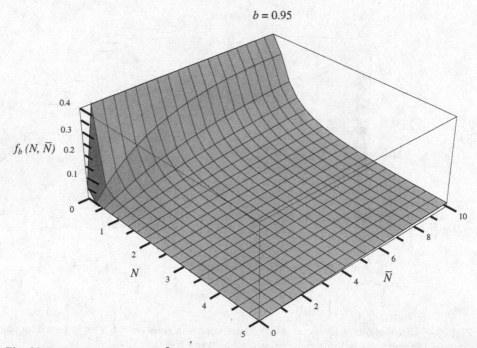

Fig. 28.5. Distributions $f_b(N, \bar{N} = \bar{n}V)$ of both counts in cells and of N objects within an arbitrary random volume V for given values of $b = 0.8$ and $b = 0.95$.

to the range of observations. Figure 15.2 gave $f_V(N)$ for a volume containing $\bar{N} = 10$ galaxies and for three different values of b. Figure 15.1 gave $f_N(V)$ for $b = 0.8$ and $b = 0.4$ with each case having the three values $N = 0, 3$, and 6 in order of their decreasing peak amplitude. Here the spatial density is normalized to unity so that $V = \bar{N}/\bar{n} = \bar{N}$. Generally curves with values of b that differ by about 0.05 can be readily distinguished from one another. Comparison of Figures 15.1 and 15.2 with Figures 28.1–28.5 illustrates several aspects of the complete spatial distribution function.

28.3   Specific Heats, Compressibility, and Instability

Changes of the specific heat, from that of a perfect gas for $b = 0$ to that of a fully virialized system for $b = 1$, provide illuminating physical insights into clustering. Inserting the equations of state (25.16) and (25.27) for U and P along with (26.26) for $b(\bar{n}T^{-3})$ into the specific heat at constant volume (23.20) gives

$$C_V = \frac{1}{N}\frac{\partial U}{\partial T}\bigg)_{V,N} = \frac{3}{2}(1 + 4b - 6b^2). \qquad (28.10)$$

For $b = 0$, $C_V = 3/2$, which is the value for a monotonic perfect gas. As b increases, so does C_V, reaching its maximum value of $5/2$ at $b = 1/3$. It then decreases smoothly, passing through a critical value $C_V = 0$ at

$$b_{\text{crit}} = \frac{2 + \sqrt{10}}{6} = 0.8604 \qquad (28.11)$$

until reaching $C_V = -3/2$ at $b = 1$. For the fully virialized limit, $U = -K$ and $K = 3NT/2$, so this gives the expected result. (Figure 30.3 in Section 30.2 illustrates C_V.)

When C_V becomes negative, instability occurs. So many galaxies have joined clusters that adding energy causes the galaxies, on average, to rise out of cluster potential wells. Consequently they lose kinetic energy and cool, thus producing the negative overall value of C_V. We shall soon see explicitly how the numbers of galaxies in different clusters depend on b.

Quite generally for any thermodynamic system, the necessary criteria for the thermodynamic stability of fluctuations are (e.g., Callen, 1985)

$$C_P \geq C_V \geq 0 \qquad (28.12)$$

and

$$\kappa_T \geq \kappa_S \geq 0, \qquad (28.13)$$

where C_P is the specific heat at constant pressure, κ_T is the isothermal compressibility, and κ_S is the isentropic compressibility (which for gravitating systems is not equivalent to the adiabatic compressibility). To calculate these quantities, we first combine (26.27) for $b(nT^{-3})$ with the equation of state (25.27) for P to give

$$b = b_0 P T^{-4}. \tag{28.14}$$

Then from (25.16), (25.27), (26.27), and (28.14) we find the coefficient of thermal expansion

$$\gamma = -\frac{1}{V}\frac{\partial V}{\partial T}\bigg)_{P,N} = \frac{1}{T}\left(\frac{1+3b}{1-b}\right). \tag{28.15}$$

Next we entreat a little help from the standard thermodynamic identities

$$C_P = C_V + \frac{TN\gamma^2}{V\kappa_T}, \tag{28.16}$$

$$\frac{\kappa_S}{\kappa_T} = \frac{C_V}{C_P} \tag{28.17}$$

along with the equations of state to obtain

$$C_P = \frac{dQ}{dT}\bigg)_{P,N} = \frac{5}{2} + 12b, \tag{28.18}$$

$$\kappa_T = -\frac{1}{V}\frac{dV}{dP}\bigg)_T = \frac{V}{NT(1-b)^2}, \tag{28.19}$$

and

$$\kappa_S = -\frac{1}{V}\frac{dV}{dP}\bigg)_S = \frac{3}{5}\frac{V}{NT(1-b)^2}\frac{(1+4b-6b^2)}{[1+(24b/5)]}. \tag{28.20}$$

As b increases, a large change in volume with constant pressure produces a smaller change in temperature because more particles are in clusters less affected by the expansion. So γ increases. Adding heat to a clustered system at constant pressure causes less of a temperature rise than for an unclustered system because some of the heat energy goes into expanding and thus "cooling" the clusters, as well as into heating the unclustered particles. For $b < 1$, increasing the volume at constant temperature would decrease the pressure, and since $\kappa_T > 0$ the system is locally mechanically stable according to this criterion. However, from (28.11), for $b > b_{\text{crit}}$, the value of $\kappa_S < 0$; thus the system is not mechanically stable to volume fluctuations at constant entropy.

The nature of these instabilities in gravitationally interacting systems is not generally the same as in imperfect gases because semistable clustering adds extra degrees of freedom. Moreover, the isotherms of the $P(V)$ relation in the equations of state are single valued, unlike a van der Waals gas for example, so these instabilities do not lead to a standard simple phase transition. The statistically homogeneous cosmological many-body system has a unique thermodynamics of its own. Its instability leads to ever more complex clustering on increasing scales. When $b = b_{crit}$, particularly interesting extended structures may develop.

28.4 Fluctuations

Earlier we saw (26.34) how the variance of counts in cells diverges $\sim \bar{N}(1 - b)^{-2}$. And we just saw in the last section that the coefficient of thermal expansion and the compressibilities also diverge as $b \to 1$. Higher order fluctuations are also informative and can be calculated (Saslaw and Hamilton, 1984) straightforwardly in the manner described in Callen (1960), or from the generating function in Section 28.8. We give some examples here but caution that a few fluctuation moments do not adequately describe a distribution that is highly non-Poisson, as for the GQED at large values of b:

$$\langle (\Delta N)^2 \rangle = \frac{\bar{N}}{(1 - b)^2}, \tag{28.21a}$$

$$\langle (\Delta N)^3 \rangle = \frac{\bar{N}}{(1 - b)^4}(1 + 2b), \tag{28.21b}$$

$$\langle (\Delta N)^4 \rangle = \frac{\bar{N}}{(1 - b)^6}(1 + 8b + 6b^2) + 3\langle (\Delta N)^2 \rangle^2, \tag{28.21c}$$

$$\langle (\Delta U)^2 \rangle = \frac{3}{4}\frac{\bar{N}T^2}{(1 - b)^2}(5 - 20b + 34b^2 - 16b^3), \tag{28.21d}$$

$$\langle (\Delta N)(\Delta U) \rangle = \frac{3}{2}\frac{\bar{N}T}{(1 - b)^2}(1 - 4b + 2b^2), \tag{28.21e}$$

$$\langle (\Delta N)^2(\Delta U) \rangle = \frac{3}{2}\frac{\bar{N}T}{(1 - b)^4}(1 - 6b + 2b^2). \tag{28.21f}$$

All these fluctuations diverge and become formally infinite when $b = 1$, and large dominant bound clusters form. The higher the moment, the stronger its divergence. Correlated fluctuations $\langle (\Delta N)(\Delta U) \rangle$ are especially interesting. For small b, their ensemble average is positive, indicating that a region of density enhancement typically coincides with a region of positive total energy. Its perturbed kinetic energy exceeds its perturbed potential energy. Similarly, an underdense region has negative total energy since it has preferentially lost the kinetic energy of the particles that have fled. At a critical value of $b = 1 - \sqrt{2}/2 = 0.293$, a positive (ΔN) is just as likely to be associated with a positive or a negative (ΔU). And for larger values of b, overdense regions typically have negative total energy; underdense regions usually have positive total energy. The critical value of b where this switch occurs is very close to the value of $b = 1/3$, at which C_V peaks, consistent with clusters being important but not dominant.

28.5   Projection

One of the most useful observational properties of the GQED is that its projection onto the sky has essentially the same functional form as the three-dimensional spatial distribution. The value of b differs in the two cases but usually this difference is small (Itoh et al., 1988; Sheth and Saslaw, 1996). The reason is that the derivation of the GQED applies to cells of any nonpathological shape, including long cones centred on an observer and projected as circles (or rectangles) onto the celestial sphere. By numerically integrating ξ over the appropriate volume in (25.18) to find b and then expressing the results in terms of the projected area of the volume, we obtain $f(N, V, b_V) \rightarrow f(N, \theta, b_\theta)$ for an average projected surface density of sources (Lahav and Saslaw, 1992).

Generally the value of b_V for a cone will be slightly smaller than for a sphere of the same volume since the correlations contribute less within the narrow apex of the cone. Usually this affects the value of b by a few percent or less. Its importance diminishes for shapes whose dimensions all exceed the correlation lengthscale.

Applying the GQED to projected counts in cells of area A on the sky is actually much easier than integrating ξ to get b_V. All that is necessary is to measure the variance $\langle (\Delta N)^2 \rangle$ in cells of a chosen size and shape. Then b_V follows from the first expression in (28.21). Now $\bar{N} = \bar{n}A$, where $\bar{n}$ is the average surface density in the sample. If all is self-consistent, the full distribution function should satisfy the GQED with no free parameters. More statistical information can then be extracted by using cells of other sizes and shapes.

28.6   Random Selection

Often we select a subsample from a clustered population and ask whether that subsample is more clustered than if it had been selected randomly. Examples could be the elliptical galaxies, or the brightest galaxies, chosen from a more general population. The GQED has the useful property that the distribution of a subsample chosen randomly from it has a form very close to the GQED form, but with a smaller value of $b = b_{\text{sample}}$ (Saslaw, Antia, and Chitre, 1987; Saslaw, 1989; Lahav and Saslaw, 1992; Sheth, 1996a). Thus it is easy to compare any subsample of a GQED with a randomly chosen one of the same size.

To chose a random subsample, suppose that the original parent population contains an average of $\bar{N}$ objects in a cell of given size. Each object has a constant binomial probability, p, to belong, or $q = 1 - p$ not to belong to the subsample. A cell of the subsample then has an average of $\bar{N}_{\text{sample}} = p\bar{N}_{\text{parent}}$ objects. Since the distribution of the subsample is very close to a GQED, its variance will satisfy (28.21a) combined with (14.17):

$$\frac{1}{(1 - b_{\text{sample}})^2} = 1 + \bar{N}_{\text{sample}} \bar{\xi}_{\text{sample}} \qquad (28.22)$$

to an excellent approximation (usually within a couple percent). For the parent GQED

$$\frac{1}{(1 - b_{\text{parent}})^2} = 1 + \bar{N}_{\text{parent}}\bar{\xi}_{\text{parent}}. \tag{28.23}$$

Now random selection leaves the correlation functions, and ξ_2 in particular, un-changed. The simplest way to see this is to note that random selection reduces the number of objects uniformly throughout the system. Therefore the numerator and denominator of $\xi(r)$ in (14.10) are reduced similarly, leaving $\xi(r)$ unchanged. Thus $\bar{\xi}_{\text{parent}} = \bar{\xi}_{\text{sample}}$ and the last two equations give

$$(1 - b_{\text{sample}})^2 = \frac{(1 - b_{\text{parent}})^2}{1 - (1 - p)(2 - b_{\text{parent}})b_{\text{parent}}}. \tag{28.24}$$

Figure 28.6 shows this relation between b_{parent} and b_{sample} for a range of selection probabilities, p. A subsample with the same p but a larger b than a random subsample

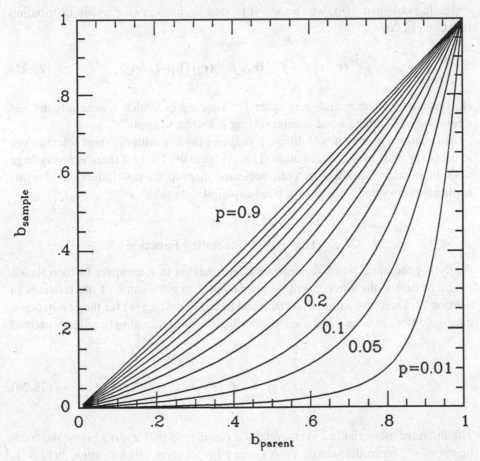

Fig. 28.6. The relation between b_{parent} and b_{sample} for random selection for a range of values of probability $p = 0.01, 0.05, 0.1, 0.2, \ldots, 0.8, 0.9$. (From Lahav and Saslaw, 1992.)

would be more intrinsically clustered than the parent population. Equation (28.86) gives an alternative derivation of this result.

28.7   Recovery of $\xi(r)$ from b_V

The relation (28.23) holds for cells of any size volume, with $b_{\text{parent}} = b_V$, $\bar{N} = \bar{n}V$, and $\bar{\xi}$ depending on volume. Of course, all the cells in a given grand canonical ensemble have the same volume, and so properties of cells with different volumes represent different grand canonical ensembles with different equations of state, and therefore with different fluctuations. Since each ensemble is self-consistent, we may obtain the spatial dependence of $\xi(r)$ from the values of $b(r)$ in the GQEDs that best fit the complete counts-in-cells distribution for cells of different sizes (Itoh, 1990). This is sometimes easier than trying to measure $\xi(r)$ directly, especially if the data are noisy for $\xi(r) \lesssim 1$. Alternatively, we can use (14.17) to determine b_V, or $\xi(r)$, from the variance of counts in cells of different volumes.

Having obtained $\xi(r)$, we can use it to step up to the two-particle distribution function (14.26)

$$f^{(2)}(\mathbf{r}_1, \mathbf{r}_2) = f^{(1)}(\mathbf{r}_1) f^{(1)}(\mathbf{r}_2) [1 + \xi_2(r_{12})]. \qquad (28.25)$$

However, the bootstrap ends here since $f^{(3)}$ requires ξ_3, which is generally not just a function of ξ_2 unless some special scaling is known to apply.

Fluctuations in volumes of different shapes, possibly with different orientations, may also provide useful information. This is especially likely if there is a very large scale filamentary or otherwise coherent component to the distribution. So far this application does not seem to have been explored.

28.8   The GQED Generating Function

Analytic generating functions represent distributions in a compact fashion that is handy to deal with. Since it will be very helpful to have some of the relations in Section 21.3 here, we summarize them and then determine $g(s)$ for the gravitational quasi-equilibrium case. For a discrete distribution, the generating function is defined as

$$g(s) = \sum_{N=0}^{\infty} f_V(N)s^N. \qquad (28.26)$$

The intermediate variable s is essentially a counter, so that $f_V(N)$ is just the coefficient of s^N. Since the sum of $f_V(N)$ over all N is normalized to unity, $g(1) = 1$. When $g(s)$ is available in closed analytic form, as for the GQED, it is easy to find any particular $f_V(N)$ from

$$f_V(N) = \frac{1}{N!} \frac{d^N g(s)}{ds^N} \bigg|_{s=0}. \tag{28.27}$$

Among the general statistical properties derivable from generating functions are the kth factorial moment

$$m_{(k)} \equiv \sum_{N \geq k} \frac{N!}{(N-k)!} f_V(N) = \frac{d^k g(s)}{ds^k} \bigg|_{s=1}, \tag{28.28}$$

which is also given by

$$g(1+s) = \sum_{N=0}^{\infty} f_V(N)(1+s)^N = \sum_{k=0}^{\infty} \frac{m_{(k)}}{k!} s^k \tag{28.29}$$

using the binomial expansion of $(1+s)^N$. For those distributions uniquely specified by all their factorial moments, (28.28) may be inverted to give

$$f_V(N) = \sum_{k=N}^{\infty} (-1)^{k-N} \frac{m_{(k)}}{N!(k-N)!}. \tag{28.30}$$

(Lognormal and Cauchy distributions are common exceptions.) Thus the mean or expectation $E(N)$ of N is

$$\bar{N} = E(N) = \sum_{N=0}^{\infty} N f_V(N) = g'(1), \tag{28.31}$$

the second factorial moment is

$$E[N(N-1)] = \sum N(N-1) f_V(N) = g''(1), \tag{28.32}$$

the second moment is

$$E(N^2) = \sum N^2 f_V(N) = g''(1) + g'(1), \tag{28.33}$$

and therefore the variance is

$$E[(N - \bar{N})^2] = g''(1) + g'(1) - [g'(1)]^2. \tag{28.34}$$

Higher order moments are generated by

$$\sum_k \frac{E[N^k]}{k!} t^k = g(e^t). \tag{28.35}$$

Central moments,

$$\mu_k = \sum_N (N - \bar{N})^k f_V(N), \tag{28.36}$$

are generated by

$$\sum_k \frac{\mu_k}{k!} t^k = e^{-\bar{N}t} g(e^t), \tag{28.37}$$

and the cumulants of κ_k are generated by the expansion

$$\sum_k \frac{\kappa_k}{k!} t^k = \log g(e^t). \tag{28.38}$$

Sometimes, as when dealing with clusters of galaxies, we may want the tail of the distribution for the probability, q_N, that N is greater than a given value:

$$q_N = f_V(N+1) + f_V(N+2) + \cdots, \tag{28.39}$$

which has the generating function

$$Q(s) = q_o + q_1 s + q_2 s^2 + \cdots = \frac{1 - g(s)}{1 - s}. \tag{28.40}$$

(To obtain the last equality, multiply through by $1 - s$ for $|s| < 1$.) The complementary probability that N is less than or equal to a given value has the generating function

$$R(s) = f_V(0) + [f_V(0) + f_V(1)] s + [f_V(0) + f_V(1) + f_V(2)] s^2 + \cdots$$

$$= \frac{g(s)}{1 - s} = g(s) e^{\sum s^k / k} \tag{28.41}$$

with the sum over $k = 1, 2, 3 \ldots$. Another useful form of generating function for a statistically homogeneous and isotropic point distribution is related to the volume averages of the N-point correlation functions (Daley and Vere-Jones, 1988; Balian and Schaeffer 1989):

$$g(s) = \exp \left[\sum_{N=1}^{\infty} \frac{\bar{N}^N (s-1)^N}{N!} \bar{\xi}_N \right], \tag{28.42}$$

where the volume integrals

$$\bar{\xi}_N = \frac{1}{V^N} \int_V \xi_N(\mathbf{r}_1, \ldots, \mathbf{r}_N) \, d^3\mathbf{r}_1 \ldots d^3\mathbf{r}_N, \tag{28.43}$$

and $\bar{\xi}_1 \equiv 1$. This explicitly shows that the distribution functions depend on the correlation functions of all orders. Any distribution function thus contains much more information than just the low order correlations. In particular, the derivatives of the void distribution $f_V(0) = g(0)$ with respect to $\bar{n}$ taken at constant values of $\bar{\xi}_N$ and V formally give the other $f_V(N)$ (White, 1979). To use this rigorously, one must already know all the $\bar{\xi}_N$, so there is no information gain. For the GQED, as well as for other distributions, the requirement that $\bar{\xi}_N$ remain constant is equivalent to a random dilution, which provides an especially interesting formulation (Sheth, 1996a) of the relation between $f_V(0)$ and the other $f_V(N)$.

If we were to obtain a distribution function directly from a generating function such as (28.42), we would need to know all the $\bar{\xi}_N$. One approach would be to try to solve a BBGKY hierarchy for the $\bar{\xi}_N$, but this is mathematically prohibitive. Another approach would be to guess some scaling relation for the $\bar{\xi}_N$ in terms of $\bar{\xi}_2$ and see where it leads, but this has a less obvious physical connection (Balian and Schaeffer, 1989). However, if we already have a distribution function, we can determine its generating function directly from (28.26).

To determine $g(s)$ for the GQED (Saslaw, 1989), we substitute its form (27.24) into (28.26) to get

$$g(s) = ve^{-vb} \sum_{N=0}^{\infty} \frac{[v+N]^{N-1}}{N!} (be^{-b}s)^N, \qquad (28.44)$$

where

$$v \equiv \frac{\bar{N}(1-b)}{b}. \qquad (28.45)$$

We can rewrite this in a simpler and more transparent form, which effectively sums the series, by starting with Abel's (1826) generalization of the binomial identity

$$(x+y)^n = \sum_{k=0}^{n} \binom{n}{k} x(x-kz)^{k-1}(y+kz)^{n-k}. \qquad (28.46)$$

Letting $x \to t$, $y \to 0$, $z \to u$ and reordering the series gives a generalization of the Taylor series expansion for any formal series:

$$h(t) = \sum_{n=0}^{\infty} a_n t^n = \sum_{k=0}^{n} \frac{t(t-ku)^{k-1}}{k!} h^{(k)}(ku), \qquad (28.47)$$

where $h^{(k)}(ku)$ is the kth derivative of $h(t)$ evaluated at ku. This result has long been known (e.g., Pincherle, 1904). For $u = 0$ it reduces to the ordinary Taylor expansion of a function (as Equation [28.46] reduces to the binomial identity for $z = 0$). Equation (28.47) shows how a function can be determined from its kth derivative at the kth lattice point multiple of u, for all values of k and arbitrary u.

Now to specialize this to the case of the GQED, let $t = vb$ and substitute $e^{vb} = h(t)$ into (28.47) with $u = -b$, thereby obtaining the identity

$$e^{vb} = \sum_{N=0} \frac{v(v+N)^{N-1}}{N!} b^N e^{-Nb}.$$
(28.48)

This looks like the summation in (28.44) except for the factor s^N. To improve the resemblance, we write (28.48) as

$$e^{vb*} = \sum_{N=0} \frac{v(v+N)^{N-1}}{N!} b_*^N e^{-Nb*},$$
(28.49)

where the function $b_* = b_*(b, s)$ is defined by the relation

$$b_*^N e^{-Nb*} = (bs)^N e^{-Nb}.$$
(28.50)

Since this holds for any value of N, we must have

$$b_* = b_*(be^{-b}, s) = be^{-b} se^{b*}.$$
(28.51)

Thus from (28.44), (28.45), (28.49), and (28.50), we find

$$g(s) = e^{-v(b-b*)} = \exp\left\{ -\bar{N} + \bar{N}\left[b + \frac{(1-b)}{b} b_*(b, s) \right] \right\},$$
(28.52)

where b_* satisfies (28.51).

We can go further and solve (28.51) for $b_*(b, s)$ by applying Lagrange's expansion (see Whittaker and Watson, 1927, p. 133):

$$f(\zeta) = f(a) + \sum_{n=1} \frac{t^n}{n!} \frac{d^{n-1}}{da^{n-1}} \{ f'(a) [\phi(a)]^n \},$$
(28.53)

where

$$\zeta = a + t\phi(\zeta).$$
(28.54)

The result follows directly as

$$b_* = \sum_{N=1} \frac{N^{N-1}}{N!} (be^{-b}s)^N.$$
(28.55)

Therefore we may finally write the generating function of the GQED as

$$g(s) = e^{-\bar{N}+\bar{N}h(s)}.$$
(28.56)

with

$$h(s) \equiv b + \frac{(1-b)}{b}b_* = b + \frac{(1-b)}{b} \sum_{N=1}^{\infty} \frac{N^{N-1}}{N!} b^N e^{-Nbs^N} \equiv \sum_{N=0}^{\infty} h_N s^N,$$

(28.57)

where we define $h_0 \equiv h(s = 0) = b$. The result $g(s) = 1$ implies $h(s) = 1$ from (28.56), and since $h_N \geq 0$, it also generates a distribution.

Note that raising $g(s)$ to any power $1/m$ where m is a positive integer gives another generating function whose only difference is that the average number $\bar{N}$ is renormalized. The resulting distribution has the same form as the GQED. This shows that the GQED belongs to the restricted class of distribution functions that are infinitely divisible (see Section 21.5). The Poisson, normal, gamma, Cauchy, and negative binomial distributions are other examples. The general form of (28.56) is also known as the compound Poisson distribution and has been discussed extensively (e.g., Feller, 1957; Consul, 1989).

A property of infinitely divisible distributions is that the distribution of the sums from their independent distributions in two different volumes has the same form as the individual distributions. This arises (e.g., Feller, 1957) from the standard result (cf. (28.72) below) that if two mutually independent integral-valued random variables X and Y have generating functions $g_1(s)$ and $g_2(s)$, then the distribution of the sum $S = X + Y$ is the convolution of X and Y and it has the generating function $g_1(s) g_2(s)$. So if we divide a volume V into two mutually exclusive volumes with $V = V_1 + V_2$, then since $\bar{N} = \bar{n}V = \bar{n}(V_1 + V_2)$, Equation (28.56) shows that $g_{V_1+V_2}(s) = g_{V_1}(s)g_{V_2}(s)$. Naturally this generalizes to an arbitrary number of mutually independent volumes. Distribution functions with this property form a semigroup, since $g(s)$ generally has no inverse.

28.9   Scaling and Moments

The relation between the volume integrals $\bar{\xi}_N$ of the N-particle correlation functions and b_V determines the nature of the scaling relations $\bar{\xi}_N(\bar{\xi}_2)$ in (14.44) and (18.1). These can also be used to find the higher order moments of the density contrast in a simple closed form.

To obtain these results (Saslaw and Sheth, 1993) we equate the factorial moments (28.28) derived from the two forms of the generating function (28.42) and (28.56). From (28.42), it is clear that the kth factorial moment depends upon all the $\bar{\xi}_i$ for $1 \leq i \leq k$, and similarly from (28.56)–(28.57) the kth factorial moment of the GQED depends upon all the ith factorial moments of the h_N distribution for $1 \leq i \leq k$. Thus (Zhan and Dyer, 1989)

$$\bar{N}^{k-1}\bar{\xi}_k = \frac{d^k}{ds^k}h(s)\Big|_{s=1} = \frac{1-b}{b} \sum_{M=k}^{\infty} \frac{M^{M-1}}{(M-k)!} b^M e^{-Mb}.$$

(28.58)

Solving this for $\bar{\xi}_k$ and using (28.55) shows that

$$\bar{\xi}_N = \frac{1-b}{b} \bar{N}^{1-N} \frac{d^N b_*}{ds^N}\bigg|_{s=1}. \tag{28.59}$$

Since $b_* = b$ for $s = 1$, it is easy to evaluate these derivatives. As examples, the first few $\bar{\xi}_N$ exact to all orders of b are:

$$\bar{\xi}_1 = 1, \tag{28.60}$$

$$\bar{\xi}_2 = \frac{b}{(1-b)^2} \frac{2-b}{\bar{N}}, \tag{28.61}$$

$$\bar{\xi}_3 = \frac{b^2}{(1-b)^4} \frac{9 - 8b + 2b^2}{\bar{N}^2}, \tag{28.62}$$

$$\bar{\xi}_4 = \frac{b^3}{(1-b)^6} \frac{64 - 79b + 36b^2 - 6b^3}{\bar{N}^3}, \tag{28.63}$$

$$\bar{\xi}_5 = \frac{b^4}{(1-b)^8} \frac{625 - 974b + 622b^2 - 192b^3 + 24b^4}{\bar{N}^4}. \tag{28.64}$$

For $b \to 1$, all these correlation volume integrals, except $\bar{\xi}_1$, become infinite. The higher the order of $\bar{\xi}_N$, the faster its divergence. This illustrates again why the basis set of correlation functions does not provide a particularly useful representation of nonlinear clustering. In the weak clustering limit ($b \ll 1$), these $\bar{\xi}_N$ satisfy the relation

$$\bar{\xi}_N = \left(\frac{N}{2}\right)^{N-1} (\bar{\xi}_2)^{N-1}. \tag{28.65}$$

This is the scaling property (14.44) since the coefficients of $(\bar{\xi}_2)^{N-1}$ do not depend on scale.

Generally, we have seen that $b = b_V$ does depend on scale, except over a range where ξ_2 does not contribute significantly to the integrals defining b in (25.17) or (25.18), or to $\bar{\xi}_2$. Thus for large values of b outside this invariant range, the $\bar{\xi}_N$ will not satisfy the particular scaling property (14.44). This is because gravitational relaxation in an expanding universe is never complete on all scales if discrete galaxies dominate the gravitational forces. Other approaches (e.g., Bouchet and Hernquist, 1992), where a dark matter fluid dominates the gravity, proceed by postulating some scaling relation for the coefficients S_m in (14.44) as the end result of clustering, attempting to relate it to the underlying dynamics, and exploring its consequences. Generally, such scaling relations will not describe clustering where inhomogeneities over different lengthscales evolve on different timescales. However, they may apply, as in the GQED, over restricted ranges of lengthscales.

Density moments of a distribution may be related to the $\bar{\xi}_N$ for the density contrast

$$\delta(V) = \frac{N - \bar{N}}{\bar{N}} \tag{28.66}$$

in cells of volume V by (Lahav et al., 1993)

$$\langle \delta^2 \rangle = \frac{1}{\bar{N}} + \bar{\xi}_2, \tag{28.67}$$

$$\langle \delta^3 \rangle = \frac{1}{\bar{N}^2} + \frac{3}{\bar{N}}\bar{\xi}_2 + \bar{\xi}_3, \tag{28.68}$$

$$\langle \delta^4 \rangle = \frac{1}{\bar{N}^3} + \frac{1}{\bar{N}^2}(3 + 7\bar{\xi}_2) + \frac{6}{\bar{N}}(\bar{\xi}_2 + \bar{\xi}_3) + 3\bar{\xi}_2^2 + \bar{\xi}_4. \tag{28.69}$$

The first of these is already familiar from (14.16). Substituting (28.61)–(28.64) into (28.67)–(28.69) we recover (28.21a–c), which were previously derived thermodynamically and can also be obtained directly from the GQED generating function using (28.37).

From the density moments, we can form the usual normalized skewness and kurtosis for the GQED:

$$S \equiv \frac{\langle \delta^3 \rangle}{\langle \delta^3 \rangle^2} = 1 + 2b \tag{28.70}$$

and

$$K \equiv \frac{\langle \delta^4 \rangle - 3\langle \delta^2 \rangle^2}{\langle \delta^2 \rangle^3} = 1 + 8b + 6b^2. \tag{28.71}$$

Evidently these ratios will depend on scale when the scale dependence of b is significant. Many attempts to measure these quantities in observed catalogs, and even in simulations, have led to uncertain results. Unfortunately, small uncertainties and irregularities in an empirical distribution are magnified when estimating its moments and combining these to obtain a skewness or kurtosis. This often makes it difficult to know whether results for a small number of realizations really represent the moments of a true sample. Numerical experiments can provide many realizations, but observations provide only one. Therefore it is usually best to deploy the entire distribution rather than to characterize it by skewness, kurtosis, or other combinations of low order moments. This especially holds for strongly nonlinear distributions that are very non-Gaussian, as in galaxy clustering.

28.10   Bias and Selection

Section 28.6 described how simple random selection decreases the value of b relative to a complete sample while retaining the GQED form of $f(N)$ to a very good approximation. More generally, bias and selection will change the $f(N)$ distribution, altering all the correlation functions. Bias between the two-point correlations of galaxies and dark matter is an example (14.40). Other examples are distributions of subsamples of galaxies selected from the same parent distribution by luminosity, size, morphology, mass, or local density. Different degrees of clustering in these subsamples may be related to astrophysical processes involved in the formation, merging, stripping, and gravitational segregation of galaxies.

Two general methods are often useful to relate parent and selected distributions. The first (cf. Saslaw, 1989) modifies the generating function by the selection process. The second (cf. Lahav and Saslaw, 1992) uses a conditional probability to calculate the distribution functions directly. While this second approach always works, the first is sometimes easier and provides more insight. We will describe some examples of each approach.

Homologous selection is one of the simplest cases. A homologous subsample with $\bar{N}_{\text{subsample}} = p\bar{N}_{\text{parent}}$ for $0 < p < 1$ is also a GQED with the same value of b as the parent population. Substituting $p\bar{N}$ for $\bar{N}$ in the GQED generating function (28.56) does not affect $h(s)$, and therefore it gives the same form for the subsample's generating function. If $p^{-1} \equiv m$ happens to be an integer, the parent distribution $f(N)_{\text{parent}}$ represents the distribution of the sums of m of these independent subsamples. This is because $g(s)_{\text{parent}}$ is the mth convolution power of $g(s)_{\text{subsample}}$. For example, if $m = 2$ the parent population contains two independent subsamples. Each subsample contributes $N^{(1)}$ and $N^{(2)}$ galaxies with probability $p^{(1)}$ and $p^{(2)}$, respectively, to the total number $N = N^{(1)} + N^{(2)}$ of galaxies in that cell. Therefore the probability of finding exactly N galaxies in that cell is

$$p_N^{(1)} p_0^{(2)} + p_{N-1}^{(1)} p_1^{(2)} + \cdots + p_0^{(1)} p_N^{(2)}, \tag{28.72}$$

which is the coefficient of s^N in the product $g_1(s)g_2(s)$ of the two generating functions. Thus the probability distribution of $N^{(1)} + N^{(2)}$ has the generating function $g_1(s)g_2(s)$, which is called the "convolution" of the distributions of $N^{(1)}$ and $N^{(2)}$. Its cumulants (28.38) are clearly the sums of the cumulants of $f^{(1)}(N)$ and $f^{(2)}(N)$, illustrating the convenience of cumulants. Generalization to $m > 2$ is straightforward.

Although homologously selected distributions retain the form of their parent distribution, the amplitudes of their correlation functions exceed the parent correlations. From the preceding discussion and (28.56), it is evident that the generating functions for homologously selected and parent distributions are related by

$$g(s)_{\text{homologous}} = [g(s)_{\text{parent}}]^p. \tag{28.73}$$

Applying (28.73) to (28.42) yields

$$\sum_{N=1}^{\infty} \frac{(p\bar{N})^N (s-1)^N}{N!} \bar{\xi}_N(\text{homologous}) = \sum_{N=1}^{\infty} \frac{p\bar{N}^N (s-1)^N}{N!} \bar{\xi}_N(\text{parent}) \tag{28.74}$$

and hence

$$\bar{\xi}_N(\text{homologous}) = p^{1-N}\bar{\xi}_N(\text{parent}). \tag{28.75}$$

In fact, this holds for any compound Poisson distribution, independent of the form of $h(s)$. Since b (or more generally $h(s)$) is unchanged, the pattern of correlations in the numerator of $\bar{\xi}_N$ remains the same, but the normalization of $\bar{\xi}_N$(homologous) $\propto$ $(p\bar{N})^{-N}$. This is essentially why the higher order correlations have greater amplitude increases for homologous selection.

A simple nonhomologous case occurs when the total population is the sum of homogeneous subsamples that are all GQEDs but with different values of b. So these subsamples have the same basic distribution with different degrees of clustering. The subsamples need not be independent, and their values of $\bar{N}$ and b may have probability distributions $p(\bar{N})$ and $p(b)$. To keep the illustration simple, suppose they all have the same $\bar{N}$ but different values of b. Then

$$f_{\text{total}}^{(N)} = \int f(N, \bar{N}, b)\, p(b)\, db \tag{28.76}$$

integrated over the range of b. This is a "sum of distributions" rather than the "distribution of sums" discussed previously. Its generating function is

$$g_{\text{total}}(s) = \sum_N f_{\text{total}}(N)\, s^N = \int g(s, b)\, p(b)\, db. \tag{28.77}$$

The integral of a distribution over another probability distribution for one of its parameters is usually called a compound distribution. If b can take on just two values b_1 and b_2, for example, with probabilities p and $1 - p$, respectively, then

$$g_{\text{total}}(s) = p\, g(s, b_1) + (1 - p)\, g(s, b_2). \tag{28.78}$$

More generally, $p(\bar{N})$ and $p(b)$ may be correlated as $p(\bar{N}, b)$ and (28.76) becomes a double integral over $\bar{N}$ and b. This may describe a variety of models, including those where galaxies in different mass ranges cluster with different values of b (Itoh et al., 1993).

More complicated forms of selection may be represented by a "galaxy biasing function" $\beta(N \mid M)$. This is the conditional probability of selecting N galaxies with specified properties (e.g., magnitudes, masses, morphological types, or diameters) from a parent population of M "objects." Let $\Psi(M)$ be the distribution function of the parent population. Then the distribution of the selected sample is

$$f(N) = \sum_{M \geq N}^{M_{\text{max}}} \beta(N \mid M)\Psi(M). \tag{28.79}$$

The biased generating function for $f(N)$ is

$$g_B(s) \overset{\cdot}{=} \sum_{N=0}^{\infty} \sum_{M \geq N} \beta(N \mid M) \Psi(M) s^N$$

$$= \sum_{M \geq 0} g_\beta(s, M) \Psi(M), \tag{28.80}$$

where

$$g_\beta(s, M) = \sum_{N=0}^{M} \beta(N \mid M) s^N. \tag{28.81}$$

At least four general biases may contribute to β: random selection, gravitational segregation, environmental biasing, and observational selection. All biases act together and depend on the history of galaxy formation and clustering, none of which is very certain.

   With this apparatus, we can now have a more detailed look at the special case of random selection (Saslaw, Antia, and Chitre 1987), which Section 28.6 started to examine. If the probability that any one galaxy is selected at random is $p \equiv 1 - q$, then the probability that N galaxies are selected from a sample of M galaxies is given by the binomial distribution

$$\beta(N \mid M) = \binom{M}{N} p^N q^{M-N}. \tag{28.82}$$

Substituting this into (28.81) and (28.80), using (28.56) and the binomial expansion (perhaps more accurately called the binomial contraction in this case), gives the biased generating function

$$g_B(s) = \sum_{M=0} (q + ps)^M \Psi(M) = g(q + ps)$$

$$= \exp\left\{-\bar{N} + \bar{N}\left[b + \frac{(1-b)}{b} \sum_{N=1} \frac{N^{N-1}}{N!} b^N e^{-Nb}(q + ps)^N\right]\right\}, \tag{28.83}$$

which again has the compound Poisson form. Equations (28.31) and (28.34) readily give the mean and variance of the randomly selected GQED:

$$E(N) = p\bar{N}, \tag{28.84}$$

$$E[(\Delta N)^2] = \frac{p\bar{N}}{(1-b)^2}[1 - (1-p)(2-b)b]. \tag{28.85}$$

   The mean is just what we expect since p is the probability of selecting a galaxy from the sample. The variance shows how, for a given b, the dispersion is reduced from the $p = 1$ case of no selection. For a value of p less than 1, there will be a value of b, call it b_{parent}, that will give the same relative variance $E[(\Delta N)^2]/p\bar{N}$ as in the unselected population characterized by a given value of $b = b(p = 1)$. Equating these normalized variances gives

$$\frac{1}{(1 - b_{sample})^2} = \frac{1 - (1 - p)(2 - b_{parent})b_{parent}}{(1 - b_{parent})^2}. \tag{28.86}$$

In other words, for the values of b_{parent} and $p < 1$, we get the same normalized variance as observed in the random sample. This provides an alternative derivation of (28.24). The full randomly biased distribution $f(N)$ will differ from the GQED $\Psi(M)$ but the scaled difference will generally be small. This may be seen by comparing the ratio of their generating functions $g(s, \bar{N}, b)$ and $g(q + ps, p\bar{N}, b')$ using (28.55)–(28.57). The important difference involves s and is that of $b_*(b, s)$ and $b_*(b', q + ps)$. From (28.51) this difference essentially depends on the logarithm of their ratio, which usually varies slowly. Thus random selection provides a good reference comparison to estimate the relative correlations of subsamples of data.

Seldom, however, is observational selection purely random. Properties for which a sample is selected are usually correlated with other, often unknown, properties. For example, suppose biasing is modified by the environment so that the total chance of selecting a galaxy is linearly proportional to the total number of galaxies in the region:

$$\beta(N \mid M) = \frac{M}{\bar{M}} \binom{M}{N} p^N q^{M-N}. \tag{28.87}$$

Here M is the "weighting function." The constant of proportionality, $1/\bar{M}$, is determined by the requirement that $g_B(s = 1) = 1$, where the biased generating function is

$$g_B(s) = \frac{1}{\bar{M}} \sum_{M \geq 0} M \Psi(M) (q + ps)^M = \frac{(q + ps)}{\bar{M} p} \frac{d}{ds} \left[\sum_{M \geq 0} \Psi(M) (q + ps)^M \right]$$

$$= \frac{(q + ps)}{\bar{M} p} \frac{dg(q + ps)}{ds}. \tag{28.88}$$

If $\Psi(M)$ is the GQED, then $g(q + ps)$ is again given by (28.56) with s replaced by $q + ps$. By evaluating the first derivative at $s = 1$ we get the mean of the selected sample:

$$\bar{N} = g'_B(s = 1) = p\bar{M} + \frac{p}{(1 - b)^2}. \tag{28.89}$$

Random selection contributes the first term. Weighting by the surrounding number modifies this by an amount that increases rapidly as $b \to 1$ and clustering becomes stronger.

We can use the variance to define a biasing function B_*^2 between the sample and the parent populations (including the Poisson contribution) by

$$\left\langle \left(\frac{\Delta N}{\bar{N}}\right)^2 \right\rangle_V = B_*^2 \left\langle \left(\frac{\Delta M}{\bar{M}}\right)^2 \right\rangle_V. \qquad (28.90)$$

Here $\bar{N}$ and $\bar{M}$ are the mean number of objects in the sample and parent populations, respectively, in a given volume V. This is analogous to the definition in (14.40), to which (28.90) reduces when the fluctuations are strongly nonlinear (recall 14.17). Often it is simplest to deal with the variances directly. Sometimes it is also convenient to subtract off the Poisson contribution in (14.17), and this would just leave (28.90) as the definition for B_*^2 with the Poisson contribution removed. An example considers ΔM to be the fluctuations in the matter (especially for models dominated by dark matter) and ΔN to be the fluctuations in the galaxies selected in some manner from the matter.

Originally B_*^2 was considered to be a positive constant (with $B_*^2 < 1$ representing relative antibias) but then it was allowed to depend on scale. Despite many attempts to measure its value observationally, it remains obscure. The measurements are subtle, different galaxy samples give different average values $0.5 \lesssim B_*^2 \lesssim 2$, and B_*^2 may indeed depend on scale. Plausible physical reasons related to galaxy formation or clustering can be adduced for almost any observed value of B_*^2. Occasionally there is a tendency to use B_*^2 as an arbitrary degree of freedom to reconcile incomplete theories with uncertain observations. Moreover, because B_*^2 only characterizes the difference between the second moments of the distributions, its information, although useful, is restricted. One could use higher moments to describe more refined biases, but usually it is best to use the distribution functions themselves when they are available.

For the record, and to show that B_*^2 can easily be quite complicated, we give the result for the relatively simple example of (28.87). Substituting (28.88) into (28.34) yields after some algebra (Lahav & Saslaw, 1992)

$$B_*^2 = \frac{1}{p\left(1 + \delta_M^2\right)}[1 - (1-p)(2-b)b] + \frac{\delta_M^2}{\left(1 + \delta_M^2\right)^2}(2b - 1), \qquad (28.91)$$

where $b = b_V$ for the parent population and $\delta_M^2 \equiv \langle(\Delta M/\bar{M})^2\rangle = [\bar{M}(1-b)^2]^{-1}$ (including the Poisson contribution). Thus the nonrandom density biasing of (28.87) introduces a dependence of B_*^2 on the variance of the parent population and on scale through both $\bar{M} = \bar{m}V$ and b_V. On very large scales $\bar{M} \to \infty$, $\delta_M \to 0$, and B_*^2 in (28.91) reduces to the result for random selection (28.82), which depends only on p and b_V. Also, for a Poisson parent population with $b = 0$, one gets $\langle(\Delta N/\bar{N})^2\rangle = B_*^2 \delta_M^2 = (\bar{M}+1-p)/[p(\bar{M}+1)^2]$. When $p = 1$ and $\bar{M} \gg 1$, this gives $\langle(\Delta N/\bar{N})^2\rangle \approx 1/\bar{M}$, as expected.

These results can be extended straightforwardly to bias functions $\beta(N \mid M)$ proportional to any function of M. Expanding the weighting function in powers of M

introduces higher order derivatives into (28.88), which can be used to obtain more general versions of (28.90). It is also easy to use other distributions $\Psi(M)$ for the parent population.

Another form of environmental biasing generalizes random selection by supposing that the probability p in (28.82) for finding a galaxy in any region depends on the number and properties of other galaxies in that region. This is not quite the same as (28.87) since $q(M) = 1 - p(M)$ would also depend on M. There are many possibilities, depending on specific models. For example, galaxies of a particular type (say ellipticals) may be produced by intergalactic gas stripping of galaxies that originally had another type. If the amount of intergalactic gas responsible for stripping is proportional to the total number of galaxies of all types, it might be modeled by $p(M) \propto M$. However, if collisions ("merging"), rather than stripping, produce a particular type of galaxy, we might expect $p(M) \propto M^2$. These and related possibilities are represented by

$$p(M) = \bar{p}(1 + \mu) \left(\frac{M}{M_{\max}} \right)^{\mu}, \tag{28.92}$$

normalized so that the observed fraction of galaxies is

$$\bar{p} = \frac{1}{M_{\max}} \int_0^{M_{\max}} p(M)\, dM. \tag{28.93}$$

Similarly, deviations from random selection may be represented rather generally by

$$p(M) = \bar{p} \frac{(1 + \alpha)}{(2 + \alpha)} \left[1 + \left(\frac{M}{M_{\max}} \right)^{\alpha} \right] \tag{28.94}$$

or by a first-order Taylor expansion in the number of galaxies in the parent population:

$$p(M) = \left(\bar{p} - \frac{C}{2} \right) + C \left(\frac{M}{M_{\max}} \right). \tag{28.95}$$

Clearly when $\alpha = 0$, $\mu = 0$, or $C = 0$ we are back to random selection ($\bar{p} = p$). Therefore models such as these can be used to test whether physical models of galaxy selection and bias are consistent with observed samples. Chapter 33 discusses some comparisons.

28.11 The Multiplicity Function and Related Interpretations

Any probability distribution has many interpretations. Therefore it is no surprise that the GQED is related to a variety of statistical models and stochastic processes. Each of these provides interesting additional insights into the cosmological many-body problem. Some of these insights are spawned by the multiplicity function.

To find the multiplicity function of the GQED (Hamilton, Saslaw, and Thuan, 1985; Saslaw, 1989) most simply, we use the generating functions (28.56) and (28.57). If we have a sum of independent random variables,

$$\sum_M = x_1 + \cdots + x_M, \qquad (28.96)$$

then the number M of random variables may itself be a random variable whose distribution has some generating function $g(s)$. Suppose the random variables each have the same distribution function whose generating function is $h(s)$. Then the generating function of the sum is shown in standard probability texts to be the compound distribution $g[h(s)]$. When M has a Poisson distribution, for which $g(s) = \exp(-\bar{N} + \bar{N}s)$, then the sum is a compound Poisson distribution with the form of (28.56). In a Poisson distribution, the values of M extend to infinity, although the probability for finding large values of M decreases rapidly.

Therefore we may interpret the generating function (28.56) and (28.57) as representing an infinite number of clusters whose centers have a Poisson distribution in large volumes of space and whose probability, $h_N = h(N)$, for containing N galaxies is given by the generating function $h(s)$ in (28.57) with $h(0) = b$. Since a cluster center with no galaxies is not distinguishable, we may drop $h(0)$ from the sequence of $h(N)$ and renormalize the remaining terms so that their sum is unity. This gives the multiplicity function

$$\eta(N) = \frac{h(N > 0)}{(1 - b)} = \frac{(Nb)^{N-1}}{N!} e^{-Nb} \quad \text{for} \quad N = 1, 2, 3, \ldots. \qquad (28.97)$$

This is familiar in probability theory as the Borel distribution.

In the limit of very large N, we see that $\eta(N + 1)/\eta(N) \to be^{1-b}$. The value of this limit is 0.945 for $b = 0.7$, 0.977 for $b = 0.8$, and tends to unity as $b \to 1$, and so the curve becomes very flat near $\eta(N) = 0$. For $b = 1$, the only nonzero value of $h(N)$ is $h(0) = 1$. In this limit, all the Poisson centers, except for a set of measure zero, contain no galaxies. If $\bar{n}$ is to remain constant as $b \to 1$, this set of measure zero must contain an infinite number of galaxies. For an actual system, this effectively means that all the galaxies have condensed into one huge cluster, and all the rest of space is empty of galaxies, as we also expect from $\lim f(0) = e^{-\bar{N}(1-b)} = 1$ as $b \to 1$. In the opposite limit $b \to 0$, we see $h(N) \to \delta_{N,1}$, which is the standard Poisson distribution where each Poisson center has just one galaxy.

We may gain further insight into the multiplicity function by decomposing the generating function into multiplets. Writing

$$\bar{n} = \sum_N \bar{n}h_N \equiv \sum_N \bar{n}_N \qquad (28.98)$$

and recalling that $\bar{N} = \bar{n}V$, we may rewrite (28.56) using the last equality of (28.57) as

$$g(s) = \exp\left(-V\sum\bar{n}_N + V\sum\bar{n}_N s^N\right)$$

$$= [\exp(-\bar{n}_1 V + \bar{n}_1 Vs)][\exp(-\bar{n}_2 V + \bar{n}_2 Vs^2)][\exp(-\bar{n}_3 V + \bar{n}_3 s^3)]\cdots$$

$$= \prod_N \exp(-\bar{n}_N V + \bar{n}_N Vs^N). \tag{28.99}$$

Each factor in (28.99) is the generating function for a Poisson distribution with an average $\bar{N}_N = \bar{n}_N V$ but with a variable that is N times the standard (singlet) Poisson variable (cf. 21.95). Thus the $N = 1$ factor represents a Poisson distribution of single particles with density $\bar{n}h_1$, the $N = 2$ factor is a Poisson distribution of doublets with density $\bar{n}h_2$, and so on. Since the generating function of a sum of independent random variables is the product of the generating function of each random variable, the total distribution $f(N)$ is the sum of all N-tuples, each with its Poisson distribution and mean density $\bar{n}_N$.

Multiplicity functions contain less information than the complete $f(N)$ distribution. The complete distribution additionally requires the form of the generating function with which the multiplicity function is compounded. For a statistically homogeneous system or an ensemble of independent systems, this generating function may be Poisson. If statistical homogeneity breaks down on some high level of clustering, the compound Poisson nature of the distribution will also break down.

Shapes of clusters comprise another type of information beyond the scope of multiplicity functions. However, the independence of cluster centers (or multiplets), which characterize a compound Poisson distribution, does provide a formal constraint in the limiting case when b is independent of scale (Sheth and Saslaw, 1994). From (14.17) and (15.10), one b for all scales implies a single value of $\bar{\xi}$ for all scales. But this holds rigorously only if $\xi(\mathbf{r}) = \delta(\mathbf{r})$, with $\delta(\mathbf{r})$ the Dirac delta function. This implies that all the particles in each cluster are located at its center. Such a model of point clusters, while often providing an excellent simple approximation over a range of scales where b changes slowly (Sheth and Saslaw, 1996), is incomplete. Incorporating the scale dependence for b_V, as in its derivation in Chapter 25, gives more realistic models of extended clusters. It is just this scale dependence which contains information about the correlation function (as in Section 28.7) and therefore about the shapes of clusters.

With the scale dependence of b_V and the GQED form of the multiplicity function, it becomes possible to develop cluster models that synthesize the observed galaxy distribution (Sheth and Saslaw, 1994). This essentially completes the program proposed by Neyman and Scott in the 1950s (see Chapter 6) and solves the problem of connecting the forms of clusters to the underlying dynamics for the cosmological many-body system. The basic answer is that a Poisson distribution of nonoverlapping spherical clusters whose radial density profiles closely resemble isothermal spheres, and whose probability of containing N galaxies is given by the multiplicity function (28.97), agrees very well with the observed $f(N)$ for all N including $f(0)$. It also agrees with the observed galaxy–galaxy correlation functions and percolation properties. Various aspects of this agreement can be improved somewhat, at the expense of others, by tinkering with the clusters' ellipticities, their ratio of size to separation, and their individual density profiles.

It was pointed out (Hamilton, Saslaw, and Thuan, 1985) that the GQED multiplicity function is closely related to the multiplicity function for bound virialized clusters proposed by Press and Schechter (1974) and subsequently extended and modified (e.g., Epstein, 1983; Bond et al., 1991; Lacey and Cole 1994; Jedamzik, 1995: Sheth, 1995a,b, 1996c,d). In this approach, the distribution function for the mass, or number of objects, that eventually composes a bound system, is derived from the distribution of isolated overdense regions in the initial Gaussian, or Poisson, density field. This essentially assumes that the condensations typically evolve spherically and that nonlinear dynamics or tides do not transfer significant material from one cluster to another. Results may also depend somewhat on the window function used to define the initial overdense regions. For an initial Poisson distribution, any random point in space has an equal a priori probability of being a cluster center, and this may lead to a GQED distribution function (Sheth, 1995a). Such clustering can also be expressed in the formalism of a Galton–Watson branching process (Sheth, 1996d). From a physical point of view, the basic reason that these approaches are essentially equivalent is the quasi-equilibrium nature of the system's evolution. Moreover, although clusters in the GQED may have any shape and degree of virialization, bound clusters are usually approximately spherical and unbound clusters with sufficient density contrast to be identified usually do not stray far from the virial criterion. The latter is also a consequence of quasi-equilibrium evolution.

28.12   Relation to Multifractals

Back in Chapter 12 we saw that a spatial distribution could be characterized by its fractal structure. Borgani (1993) has worked this out for several distribution functions and finds that the GQED has an interesting multifractal behavior. Its fractal scaling is a function of density and therefore of position. This is not unexpected, since the local gravitational forces are density dependent. For values of $q > 1/2$ in (12.8) the Renyi dimension (12.11) is

$$D_q = 3 - \gamma \tag{28.100}$$

(where γ is the usual exponent of the two-particle correlation function). This is independent of q and is just the correlation dimension. In this regime, which would include most of the overdense regions, the distribution is a simple fractal.

However, for $q < 1/2$, which is weighted toward large underdense regions (especially for $q < 0$), the Renyi dimension is

$$D_q = 3 - \frac{\gamma q}{1 - q},$$
(28.101)

which has a fairly strong dependence on q. Corresponding values of α in (12.14) are $\alpha = 3 - \gamma$ for all values of q and, in (12.15), $f(\alpha) = 3 - \gamma$ for $q \geq 1/2$ and $f(\alpha) = 3$ for $q < 1/2$. Although these fractal properties can be found once $f(N)$ is known, they have not been derived ab initio from fundamental physical properties of the system.

29

The Velocity Distribution Function

To and fro they were hurried about!
And to and fro, and in and out,
The wan stars danced between.

Coleridge

Spatial distribution functions, for all their usefulness, describe only half the phase space. Velocities are the other half. In the cosmological many-body problem, with galaxies interacting just through their mutual gravity, these velocities must be consistent with the galaxies' positions. On nonlinear scales where dissipation prevails, only weak memories of initial conditions or spatial coherence remain.

One measure of galaxy velocities, much studied in the 1980s (see Peebles, 1993 for a detailed summary), is the relative velocity dispersion of galaxy pairs as a function of their projected separation on the sky. For small separations this has an approximately exponential distribution. Simple dynamical models, including projection effects, can relate this pairwise distribution to the galaxy two-point correlation function in redshift space. Today, with increasing ability to measure redshifts and better secondary distance indicators, more accurate radial peculiar velocities are becoming available. So, with an eye to the future, this chapter considers the three-dimensional and radial peculiar velocity distributions predicted for the cosmological many-body problem. In Chapters 32 and 34, we will see that these predictions agree well with relevant N-body experiments and with the first observational determinations.

The peculiar velocity distribution function $f(v)\, dv$ is just the probability for finding a galaxy in the system with a peculiar velocity between v and $v + dv$ (relative to the Hubble expansion). It is the analog for galaxy clustering of the Maxwell–Boltzmann distribution for a perfect gas. Unlike a homogeneous perfect gas, however, $f(v)$ represents the entire range of clustering, from the minor motions of field galaxies to the largest velocities in the richest bound clusters. All these velocities must be consistent with the spatial clustering from nonlinear to linear scales.

In the early linear stages of clustering, velocities may be calculated from the BBGKY hierarchy in the form (17.58) and (17.59). (See also GPSGS (Section 23) and integrate over configuration space rather than over velocities.) As with spatial correlations, mathematical complexity makes the velocity distributions impossible to extract rigorously from the BBGKY hierarchy for nonlinear clustering. Instead, a simple physical assumption that relates $f(v)$ to $f(N)$ provides a good shortcut (Saslaw, Chitre, Itoh, and Inagaki, 1990).

The basic assumption is that in quasi-equilibrium, the fluctuations in potential energy over a given volume (caused by correlations among particles) are proportional to the fluctuations of local kinetic energy, so that

$$GmN\left(\frac{1}{r}\right) = \alpha v^2. \tag{29.1}$$

In this equation, $\langle 1/r \rangle = \Gamma(2/3)\bar{N}^{1/3} = 1.35\,\bar{N}^{1/3}$ (see GPSGS, Equation 33.23 with $a = 0$) is the ensemble average inverse particle separation expected for a uniform Poisson distribution. Therefore α incorporates all the detailed information about the local configuration. In particular, α represents the local correlations that depart from a uniform distribution, including the correlations between N and r^{-1}. It may be thought of as a local "form factor" relating the kinetic and potential energy fluctuations. Smaller values of α imply a more condensed, nonuniform region of particles with higher local velocity dispersion v^2. Generally α will vary from volume to volume. We shall henceforth make the simplifying approximation that α has a constant value, which is its average over the entire system. The computer experiments in Chapter 32 show that this is quite a good approximation. If we then average (29.1) over the entire system to obtain the mean square velocity dispersion $\langle v^2 \rangle$, treating each particle's peculiar velocity individually so that $\langle N \rangle = 1$, we have the relation

$$\alpha = \left(\frac{1}{r}\right)\langle v^2 \rangle^{-1} \tag{29.2}$$

in natural units with $G = m = R = 1$ (see (19.7)–(19.9) for conversion into physical units), taking α to denote its average value. This is an approximate relation that may be used as the definition of α, neglecting higher order fluctuations.

With these assumptions, the first step to finding the velocity distribution is to rescale $f(N)$ in (27.24),

$$f(N) = \frac{\bar{N}(1-b)}{N!}[\bar{N}(1-b) + Nb]^{N-1}e^{-\bar{N}(1-b)-Nb}, \tag{29.3}$$

from density fluctuations to kinetic energy fluctuations by replacing N with $N\langle 1/r \rangle$, and replacing $\bar{N}$, the average number in a given size volume, with $\bar{N}\langle 1/r \rangle$. (Equivalently, one can rescale $\alpha \to \alpha/\langle 1/r \rangle$ and then use N and $\bar{N}$ in (29.3) directly.) The second step is to substitute αv^2 for $N\langle 1/r \rangle$ from (29.1) and $\alpha \langle v^2 \rangle$ for $\bar{N}\langle 1/r \rangle$, also recognizing that $N! = \Gamma(N+1)$. The third step is to convert from kinetic energy fluctuations to velocity fluctuations using the Jacobian $2\alpha v$. The result is

$$f(v)\,dv = \frac{2\alpha^2\beta(1-b)}{\Gamma(\alpha v^2 + 1)}[\alpha\beta(1-b) + \alpha b v^2]^{\alpha v^2 - 1}e^{-\alpha\beta(1-b)-\alpha b v^2}v\,dv, \tag{29.4}$$

where

$$\beta \equiv \langle v^2 \rangle \tag{29.5}$$

and Γ is the usual gamma function. This is the basic velocity space complement to the configuration space distribution (29.3).

On the high-velocity tail of the distribution, for $\alpha v^2 \gg 1$ and $v^2 \gg \beta$, we obtain

$$f(v)\,dv \sim \left(\frac{2}{\pi}\right)^{1/2} \alpha^{3/2} \frac{\beta(1-b)}{b} (\alpha v^2)^{-1} (be^{1-b})^{\alpha v^2}\,dv. \qquad (29.6)$$

In the limit of very high velocities, this satisfies the differential equation

$$\frac{df}{dv} = 2\alpha \ln (be^{1-b})\,vf. \qquad (29.7)$$

Note that for values of b in the range of physical interest, $0 < b < 1$, the logarithm in (29.7) is negative, so that the high-velocity tail of $f(v)$ always behaves like a Maxwellian. However, the limit of $f(v)$ as $b \to 0$ will not be a Maxwell–Boltzmann distribution because a Maxwellian does not satisfy the relation $N \propto v^2$ of (29.1) in this limit. The Maxwell–Boltzmann limit of $f(v)$ requires both $b \to 0$ and no correlation between N and v.

When gravitational clustering is very nonlinear, the galaxies will have high peculiar velocities, and they deflect the orbits of each other only slightly. Fokker–Planck equations describe this situation. We would therefore expect $f(v)$ to be a solution of the stationary Fokker–Planck equation whose general form is

$$\frac{\partial}{\partial v}[a(v)f(v)] = \frac{1}{2}\frac{\partial^2}{\partial v^2}[\sigma^2(v)f(v)], \qquad (29.8)$$

when the time derivative of $f(v)$ is small enough to be ignored, as in the quasi-equilibrium evolution we consider here. In gravitating systems, the high-velocity limits of the functions $a(v)$ and $\sigma^2(v)$, which characterize dynamical friction and stochastic acceleration, have the form (e.g., GPSGS, Equations 3.24, 5.15, 5.16)

$$a(v) = -\frac{\delta}{v^2}\,;\; \sigma^2(v) = \frac{\gamma}{v^3}, \qquad (29.9)$$

where δ and γ are positive constants. Substituting these gravitational functions into (29.8) and retaining only the dominant terms for large v gives the simple result

$$\frac{\partial f}{\partial v} = -2\frac{\delta}{\gamma}vf, \qquad (29.10)$$

which is identical to the earlier thermodynamic result (29.7) if we identify

$$\frac{\delta}{\gamma} = -\alpha \ln (be^{1-b}). \qquad (29.11)$$

Thus the asymptotic form of the GQED $f(v)$ satisfies the gravitational Fokker–Planck equation with an essentially Maxwellian tail.

Just as it is possible to obtain a good approximation to the gravitational quasi-equilibrium spatial distribution (29.3) from a detailed model of the Borel distribution of clusters that compose it (Sheth and Saslaw, 1994), one can also obtain an approximation to the velocity distribution (29.4) from a detailed model of the galaxy velocities in these clusters (Sheth, 1996b). Such cluster models provide alternative derivations of these distributions.

For direct comparison with observations, we also require the radial velocity distribution function $f(v_r)$. This is easily related to the general three-dimensional velocity distribution $F(\mathbf{v})$, and to the distribution $f(v)$ in (29.4) for speeds (Inagaki et al., 1992). Let the spherical coordinates (r, θ, ϕ) in configuration space correspond to the coordinates (v, Θ, Φ) in velocity space where

$$v^2 = v_r^2 + v_\theta^2 + v_\phi^2 \equiv v_r^2 + v_\perp^2. \tag{29.12}$$

In general

$$f(v) = \int_0^\pi \int_0^{2\pi} F(\mathbf{v}) v^2 \sin\Theta \, d\Phi \, d\Theta. \tag{29.13}$$

If the velocity distribution $F(\mathbf{v})$ is isotropic so that

$$f(v) = 4\pi v^2 F(v^2), \tag{29.14}$$

then the radial velocity distribution is

$$f(v_r) = \int_{-\infty}^\infty \int_{-\infty}^\infty F(v^2) \, dv_\theta \, dv_\phi = 2\pi \int_0^\infty v_\perp \frac{f(v)}{4\pi v^2} \, dv_\perp. \tag{29.15}$$

Thus when $f(v)$ has the form of (29.4) we obtain

$$f(v_r) = \alpha^2 \beta (1 - b) e^{-\alpha\beta(1-b)} \int_0^\infty \frac{v_\perp}{\sqrt{v_r^2 + v_\perp^2}}$$

$$\times \frac{\left[\alpha\beta(1 - b) + \alpha \left(v_r^2 + v_\perp^2\right) b\right]^{\alpha\left(v_r^2+v_\perp^2\right)-1}}{\Gamma\left[\alpha \left(v_r^2 + v_\perp^2\right) + 1\right]}$$

$$\times e^{-\alpha b\left(v_r^2+v_\perp^2\right)} \, dv_\perp \tag{29.16}$$

for the radial velocity distribution function. So far, this integral must be done numerically. The usual Maxwellian factor $\exp(-\alpha v_r^2)$ is substantially modified by an enhanced probability for finding galaxies near $v_r = 0$; these are mostly field galaxies whose peculiar velocities are diminished by the adiabatic expansion of the universe. There is also an enhanced probability, relative to the Maxwellian, for finding galaxies with very high peculiar radial velocities; these are the galaxies in rich condensed clusters. Equation (29.16) accounts for the entire range of galaxy peculiar radial velocities in the cosmological many-body problem. Chapter 15 showed

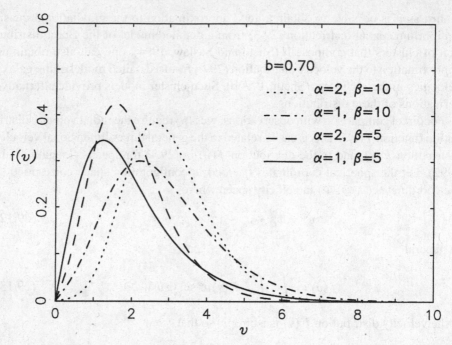

Fig. 29.1. The velocity distribution function (29.4) for $b = 0.70$ and several values of α and β. Velocities are in natural units.

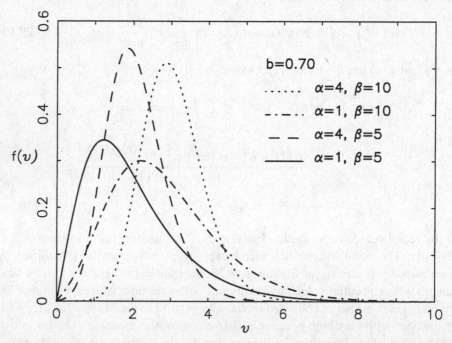

Fig. 29.2. The velocity distribution function (29.4) for $b = 0.7$ and several values of α and β. Two sets of values from the previous graph are repeated for direct comparison. Velocities are in natural units.

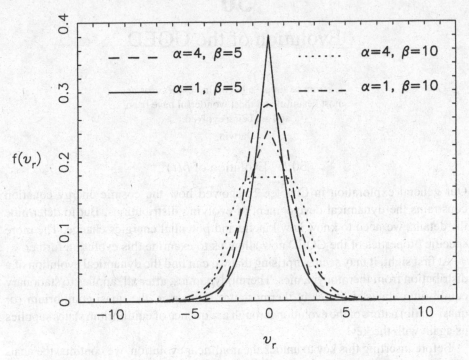

Fig. 29.3. The radial velocity distribution function (29.16) for $b = 0.7$ and the same values of α and β as in the previous figure.

some comparisons with computer experiments, and we will examine more detailed comparisons with experiments in Chapter 32 and with observations in Chapter 34.

Figures 29.1–29.3 along with Figure 15.3 (kindly provided by Dr. Fan Fang) illustrate $f(v)$ and $f(v_r)$ for a range of parameters. All these velocities are in natural units, which can be converted to physical units by (19.8). As b increases, $f(v)$ becomes increasingly skewed relative to a Maxwellian. The peak migrates to smaller v and the tail grows. Two physical effects cause this migration: the decreasing velocities of field galaxies (including those galaxies ejected from clusters) as the universe expands and the increasing number of low-velocity (but high total energy) galaxies in the outer regions of clusters. The increasing tail represents the increased number of high-velocity galaxies in massive clusters for larger b. For given values of b and α, a larger velocity dispersion, β, broadens the distribution and shifts its peak to greater velocities. For given β and b, a smaller value of α broadens the distribution, shifts the peak to smaller velocities, and extends the tail. This represents a state where, on average, the galaxies are separated by larger distances because the clusters are separated by larger distances. Similar properties may also be seen in the radial velocity distribution function, which is more sharply peaked for larger b. Since the universe becomes more clustered with time, we should now explore how b evolves.

30

Evolution of the GQED

... from so simple a beginning endless forms
most beautiful and most wonderful have been
and are being evolved.

Darwin

30.1   Evolution of $b(t)$

Our general exploration in Chapter 22 showed how the cosmic energy equation constrains the dynamical development of evolving distributions. But to determine the details, we need to know how kinetic and potential energies change. The more specific properties of the GQED now allow us to examine this evolution further.

At first sight, it may seem surprising that we can find the dynamical evolution of a distribution from thermodynamics. Thermodynamics, after all, applies to stationary equilibrium states. It is really thermostatics. However, the quasi-equilibrium (or quasi-static) nature of the evolution through a sequence of equilibrium states supplies us again with the key.

Before inserting this key to unlock the nonlinear evolution, we continue the analysis of Chapter 22 to follow linear evolution away from a Poisson state toward the GQED. Expanding the general form of the generating function (22.17) in powers of b, and comparing with a similar expansion of the GQED generating function (28.56), shows that they are identical to order b. Therefore the earliest dynamical evolution away from the Poisson state is toward the GQED. Comparing these two expansions to order b^2 shows they are also identical to this order if

$$Q = \frac{3}{4} \quad \text{and} \quad \alpha = \frac{3}{4}. \tag{30.1}$$

Thus the GQED $f(N)$ distribution is consistent with a hierarchical three-particle correlation function having this value of Q. Interestingly, observations in redshift space for the three-dimensional second Southern Sky Redshift Survey give $Q = 0.61 \pm 0.04$ (Benoist et al., 1997), while earlier optical surveys, mainly in two-dimensions projected onto the sky, gave $Q = 1.0 \pm 0.2$ (Peebles, 1993). There are significant differences among samples depending on luminosity, color (the *IRAS* infrared catalog gives a smaller Q), measurements in two or three dimensions and in real or redshift space, as well as on details of the analysis. Although a precise observational value of Q is difficult to extract, it seems reasonably consistent with (30.1).

Further evolution is nonlinear. Rather than pursue it with higher and higher order developments of linear style analyses, it is usually more useful to change our point of view and follow a simpler macroscopic description (Saslaw, 1992). As long as the evolution is quasi-equilibrium, we can use thermodynamics. And the aspect of thermodynamics that enables us to move forward is that the expansion of the system of galaxies is, to a good approximation, adiabatic.

If the Universe were strictly homogeneous (unclustered and filled with an ideal fluid), then its expansion observed locally in comoving coordinates would be adiabatic and satisfy the first law of thermodynamics

$$dQ = dU + P \, dV, \tag{30.2}$$

with the net heat flow into or out of the volume $dQ = 0$ (see Equation 16.6). For nonrelativistic velocities in volumes small compared to the scale of the Universe, the internal energy U can exclude the rest mass energy. Gravitational clustering destroys this strict homogeneity, although it may preserve statistical homogeneity, and it generates local entropy (acting somewhat like an internal degree of freedom). Clustering produces motions of galaxies across the boundary of a volume. Because the motions are correlated, there will generally be a net heat flow dQ. This term represents the departure from an adiabatic expansion. Its importance will depend on the ratio of the linear gravitational clustering timescale, τ_{grav}, to the expansion timescale τ_{expand}.

We can find this ratio directly from the Einstein–Friedmann equation (16.14), using (16.18):

$$\left(\frac{dR}{dt}\right) = \frac{8 \pi G \rho_0 R_0^3}{3} \frac{1}{R} + c^2, \tag{30.3}$$

where a 0 subscript denotes quantities at the present radius. With

$$\tau_{expand}^{-2} \equiv \frac{\dot{R}^2}{R^2} = H^2 \tag{30.4}$$

and

$$\tau_{grav}^{-2} \equiv \frac{8}{3}\pi G \rho = \Omega H^2 \tag{30.5}$$

we may use (30.3) to write the "adiabatic parameter" of the expansion similarly to (16.62) for open models as

$$\alpha^2 \equiv \frac{\tau_{grav}^2}{\tau_{expand}^2} = 1 + \frac{(1 - \Omega_0)}{\Omega_0} \frac{R}{R_0} \tag{30.6}$$

(note that this is not the same α as in 30.1; we no longer need the earlier α), where $\Omega_0 \equiv \rho_0/\rho_{critical}$ as usual. When $\alpha > 1$, the characteristic timescale for gravitational clustering to grow and for correlated galaxies to move across the boundaries of comoving volume elements exceeds the global expansion timescale and the evolution is approximately adiabatic. The adiabatic approximation becomes better for models where α is greater (i.e., Ω_0 is smaller). This is because lower density universes expand faster relative to their gravitational timescale. For models with the critical density $\Omega_0 = 1$, Equation (30.6) shows that their expansion timescale always equals their gravitational timescale. This property is also nearly true at very early times $R/R_0 < \Omega_0/(1 - \Omega_0)$ when low-density universes expand as though they had the critical density (cf. 16.20).

Adiabatic processes have no net heat flow across the boundary of a system (which may be a subregion of a larger system). This does not necessarily imply that the process is reversible, or isentropic, or quasi-static, or quasi-equilibrium. Various combinations of these properties may apply to different systems in different circumstances (e.g., Tolman, 1934; Callen, 1985). For example, adiabatic quasi-static expansion of a perfect gas in a piston is isentropic and reversible, but adiabatic free expansion of a perfect gas is not reversible or isentropic because it is not quasi-static. In contrast, the expansion of a homogeneous unenclosed perfect gas in the expanding universe can be adiabatic, isentropic, and reversible, even if it is not quasi-static. A gas with two chemically (or nuclearly) reacting components in these same circumstances, however, can expand adiabatically but may not be isentropic or reversible since energy may be redistributed among the degrees of freedom of the interaction. Quasi-equilibrium evolution occurs through a sequence of states, each described by an equilibrium distribution, but whose parameters evolve. It is more general than quasi-static evolution, which requires the microscopic relaxation time of the system to be shorter than the macroscopic rate of change. Some systems can evolve through equilibrium states even though their relaxation timescale is long. This can occur if the timescale for microscopic configurations to change is shorter than the timescale for macroscopic changes, as discussed in Section 25.6.

Gravitational clustering may be adiabatic but, as Chapters 26, 27, and the next section show, it will be neither isentropic nor reversible. As the Universe expands, clustering generates entropy, which is reflected in the increase of b.

The adiabatic increase of b is equivalent to the cosmic energy equation (22.31) and provides another derivation of it (our fourth). Recall the equations of state for the internal energy (25.16) and pressure (25.27):

$$U = \frac{3}{2}NT(1 - 2b), \tag{30.7}$$

$$P = \frac{NT}{V}(1 - b), \tag{30.8}$$

which along with the temperature and density dependence (26.27) for b,

$$b = \frac{b_0 \bar{n} T^{-3}}{1 + b_0 \bar{n} T^{-3}}, \tag{30.9}$$

describe the system. Using the general definitions $U = W + K$ and $b = -W/2K$, we can eliminate b and U in (30.7) and (30.8) to obtain $P = (2K + W)/3V$. This resembles the standard virial theorem with a surface term $3PV$ (see Equation 43.6 of GPSGS) but differs in that W is the correlation energy of fluctuations rather than the total energy of a finite bound system. Substituting these expressions for P and $dU = d(W + K)$ into the First Law of thermodynamics (30.2) with $dQ = 0$, and noticing that $dV/V = 3\,dR/R$, directly gives $d(W+K)/dt + \dot{R}(2K+W)/R = 0$, which is the cosmic energy equation (22.31). More generally, when the adiabatic approximation does not apply, the cosmic energy equation will also need to be generalized to include correlations across boundaries.

Scales to which this analysis applies best will clearly be those over which the transfer of correlation energy is least. These are the scales on which $\xi(r)$ is small and its contribution to W and b_V is negligible, that is, in the asymptotic regime where $b_V \approx$ constant. Even on smaller scales, however, this analysis is often a good approximation since the correlation energy or heat transferred over the boundary of a volume will generally be small compared to the correlation energy or heat within the volume, for reasons similar to those discussed in Section 25.2. A second approximation that simplifies the nonlinear calculation of $b(t)$ is the use of $dW = P\,dV$ to go from the more general form of the First Law of thermodynamics to (30.2). This is strictly true only for reversible processes, and it is a better approximation for larger values of α. We shall see later that the results of these approximations agree well with N-body simulations.

The trick that simplifies the calculation of $b[R(t)]$ is to regard U and b as functions of P and T since (30.8) and (30.9) give

$$b = b_0 P T^{-4}. \tag{30.10}$$

Then we obtain

$$dQ = 0 = dU + P\,dV = \left(\frac{\partial U}{\partial T}\right)_P dT_P + \left(\frac{\partial U}{\partial P}\right)_T dP_T + P\,dV$$

$$= \frac{3}{2}N(1 - 2b)\,dT_P + \frac{\partial U}{\partial b}\left[\left(\frac{\partial b}{\partial T}\right)_P dT_P + \left(\frac{\partial b}{\partial P}\right)_T dP_T\right] + P\,dV$$

$$= \frac{3}{2}N(1 - 2b)\,dT_P + \frac{\partial U}{\partial b}\,db + P\,dV$$

$$= -\frac{9}{4}NT\left(1 + \frac{1}{6b}\right)db + NT(1 - b)\frac{dV}{V}, \tag{30.11}$$

which integrates to

$$R = R_* \frac{b^{1/8}}{(1 - b)^{7/8}}, \tag{30.12}$$

giving $R(b)$ since $V \propto R^3$. Initially after the system has relaxed, if $R = R_i$ when $b = b_i \neq 0$ then $R_* = R_i(1 - b_i)^{7/8}b_i^{-1/8}$. This specifies the adiabat. Inverting

(30.12) yields $b(t)$ for a model with given $R(t)$ (e.g., for $\Omega = 1$, $R(t) \sim t^{2/3}$. However, since $R(t)$ is an implicit function of time for $\Omega_0 < 1$, it is simplest to use (30.12) directly for the nonlinear development of b. Equation (30.12) is the basic result for the evolution of b.

Nonlinear gravitational clustering in the cosmological many-body problem strongly tends to evolve along an adiabat in the equation of state that includes gravitational interactions. The faster the universe expands relative to its gravitational clustering timescale, the better this adiabatic approximation becomes. This generalizes the standard result that a homogeneous unclustered perfect gas in an expanding universe evolves along an adiabat of the perfect gas equation of state.

Another way to understand the usefulness of the adiabatic approximation and the reversible approximation for dW is to note that departures from this state will tend to cancel out in a more exact version of (30.11). For a nonadiabatic but reversible change, $dQ = T\,dS$. Irreversible clustering, however, increases the entropy further so that $T\,dS = dQ + \mathcal{O}(\alpha^{-1})$. Thus dQ is less than it would be if the situation were nonadiabatic but reversible. This same irreversibility increases the work done and so $dW = -P\,dV + \mathcal{O}(\alpha^{-1})$. Therefore in the most general form of the First Law of thermodynamics, $dQ = dU - dW$, the two irreversible contributions tend to cancel. For large values of Ω_0, and consequent small values of a, we would expect this cancellation to break down slowly as the clustering evolves.

It is interesting that the adiabatic approximation breaks down for a special class of models with $\Omega_0 = 1$ and dominated by perturbed collisionless dark matter. These models have a scale-free similarity solution for the growth of correlations in the fluid limit (Davis and Peebles, 1977; Peebles, 1980, Equation [75.6]). This would lead to a constant value of $b = (7 + n)/8$, where n is the exponent of an initial power-law spectrum; $n = 0$ and $b = 7/8$ for the Poisson case. However, self-similarity does not apply to the cosmological many-body problem where the graininess of the gravitational field introduces new lengthscales, not present in the fluid limit, that evolve with time. These are the scales on which local nonlinear evolution has caused the galaxy distribution to relax. The growing presence of such scales destroys the self-similar nature of the solution and causes b to increase from its Poisson value. Moreover, the self-similar models are rather special in that they assume a particular decomposition of the BBGKY hierarchy and a velocity distribution that is symmetric (i.e., not skewed). Both these assumptions apply, even to dark matter dominated models, only in special situations (e.g., Hansel et al., 1986; Yano and Gouda, 1997).

For the cosmological many-body problem, however, Figure 15.7 previously illustrated that b grows in good agreement with (30.12). More detailed comparisons with N-body experiments await Chapters 31 and 32. First we shall draw some useful implications from adiabatic evolution.

When evolution is essentially adiabatic, nonlinear clustering generally grows more slowly than the universe expands and eventually, when $\alpha^2 > 1$, the clustering pattern ceases to grow significantly. It "freezes out." This happens at higher redshift for lower Ω_0 models, as (30.6) indicates. To estimate this we set $\alpha^2 = 2$ in (30.6) giving

$$\frac{R_{\text{freeze}}}{R_0} = \frac{1}{1 + z_{\text{freeze}}} \approx \frac{\Omega_0}{1 - \Omega_0}. \tag{30.13}$$

This is essentially the previous expression (16.20) for the redshift at which the open universe starts expanding as though it were effectively empty. In the case $\Omega_0 = 1$, the pattern never freezes out. It always keeps evolving. For $\Omega_0 = 0.1$, this gives $R_{\text{freeze}} \approx 0.1 R_0 \approx 3$ for $R_0 \approx 30$ suggested by N-body simulations. If $\Omega_0 = 0.01$, $R_{\text{freeze}} \approx 0.01 R_0$, but this case leads to such a small current value of b that it is inconsistent with the presently observed galaxy distribution. If $\Omega = 0.2$ in our universe and galaxies clustered mostly in this manner, we should expect to see little evolution in $f(N)$ for $z \lesssim 2$.

After the pattern freezes out, it is a useful approximation to regard it as retaining its general form but expanding with the universe – stretching where unbound. In this case, the value of b_{pattern} obtained by fitting the GQED distribution function (29.3) for $f(N)$ to the observed spatial distribution will differ from the physical value $b_{\text{physical}} = -W/2K$, which incorporates velocity information. This is because many velocities decay adiabatically, but the clustering pattern does not change significantly in the late stages of an open universe when $z < z_{\text{freeze}}$. Therefore $b_{\text{pattern}} \leq b_{\text{physical}}$, a symptom of weaker clustering. In these terms, the result (30.12) for $b(R)$ refers, of course, to b_{physical} on scales where adiabatic expansion is a good approximation. These are generally the scales over which b_V remains essentially independent of scale. If the difference between b_{pattern} and b_{physical} could be determined observationally, especially on large scales where both these quantities lose their scale dependence, it would provide a method for estimating Ω_0. The difficulty lies in determining accurate peculiar velocities for b_{physical}, even if they are isotropic. An indirect, but potentially effective, approach would be to determine b_{physical} from the radial velocity distribution (29.16).

30.2   Evolution of Energy, Entropy, and Specific Heat

Once $b(t)$ is known, the evolution of the spatial distribution (29.3) is determined since $\bar{N}$ remains constant in comoving coordinates. It then becomes easy to find how other thermodynamic quantities evolve (Saslaw and Sheth, 1993). For example, the temperature and kinetic energy follow by using $b = -W/2K$ to eliminate W from the cosmic energy equation (22.31) and with (30.12):

$$K = \frac{3NT}{2} = K_* b^{-1/4} \tag{30.14}$$

for $T(b)$ and hence $T(R)$, with K_* as a fiducial value. Then substituting for $\bar{n} \propto R^{-3}$ and T in (30.9), along with $b(R)$, gives the evolution of $b_0(R)$.

This result may be understood as follows. Adiabatic expansion would cause the temperature of a freely expanding homogeneous system to decrease according to

$K \propto R^{-2}$ as in (22.57) or (22.25). Equations (30.12) and (30.14) show that, in fact, $K \propto R^{-2}(1 - b)^{-7/4}$. The $(1 - b)^{-7/4}$ factor represents the effect of clustering on the kinetic energy. As the clustering characterized by b increases, most of the kinetic energy is in partially or totally bound clusters, which are not much affected by the expansion. Thus (30.14) shows that for large b at late times, the kinetic energy becomes nearly constant. This agrees with the behavior seen in N-body simulations (Fig. 8.5 of Itoh et al., 1988) for low-density cosmologies ($\Omega_0 = 0.1$ and 0.01). When $\Omega_0 = 1$, the adiabatic hypothesis is less exact and does not describe the evolution of K quite as well. Using (30.14) immediately gives the approximate dependence of the specific internal energy on b:

$$\frac{U}{N} = \frac{3T}{2}(1 - 2b) = \frac{K_*}{N}b^{-1/4}(1 - 2b), \tag{30.15}$$

which decreases steadily as b increases.

Similarly, one calculates the specific entropy by substituting $\bar{n}(R) = 3N/4\pi R^3$ and (30.14) for T into (27.3):

$$\frac{S}{N} = s_0 + s_* - 3b - \frac{29}{8}\ln(1 - b), \tag{30.16}$$

where $s_*(N, K_*, R_*)$ is a fiducial value of the entropy. The entropy increases steadily as b increases. Thus the gravitational clustering process tends to decrease the internal energy and maximize the entropy.

Evolution of the specific heat C_V can now be understood in terms of changes in clustering associated with increasing b (Saslaw and Sheth, 1993). Figure 30.1 shows

$$C_V = \frac{1}{N}\frac{\partial U}{\partial T}\Big)_{V,N} = \frac{3}{2}(1 + 4b - 6b^2), \tag{30.17}$$

derived in (28.10), as a function of b. Clustering is related to the multiplicity function (28.97),

$$\eta(N) = \frac{(Nb)^{N-1}}{N!}e^{-Nb}, \tag{30.18}$$

which is the probability of having a cluster or multiplet with $N = 1, 2, 3 \ldots$ galaxies. This is associated with the generating function (28.56) of the GQED. The centers of each cluster have a Poisson distribution throughout space. Figure 30.2 plots $\eta(N)$ as a function of b for $N = 1 - 5$. The curves are strictly nested with $\eta(N) < \eta(N-1)$ for all values of b. The maximum of $\eta(N)$ for any value of N occurs at $b = (N - 1)/N$, and $\eta(1)$ decreases exponentially with b. The difference, $\eta(2) - \eta(3)$ of binaries relative to triplets is greatest for $b = 0.312$, which is close to the value of $b = 1/3$ where C_V peaks at $5/2$. Figure 30.3 plots $\eta(2) - \eta(N)$ for $N = 3, 4, 5$ as a function of b. These curves also peak near $b = 1/3$.

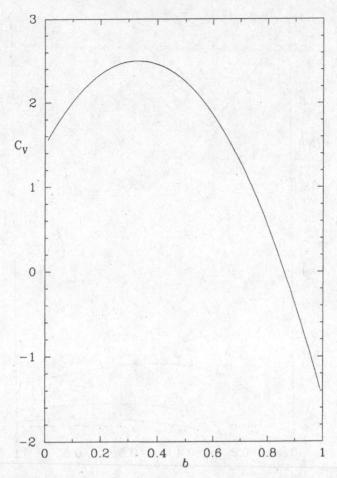

Fig. 30.1. Specific heat at constant volume, C_V. (From Saslaw & Sheth, 1993.)

This suggests an approximate physical interpretation for the behavior of the specific heat C_V. If the galaxies are initially uncorrelated, the system behaves as though it were a perfect monatomic gas ($C_V = 3/2$). When the system begins to relax, the gravitational interactions cause galaxies to cluster. Binary pairs form, which tends to push the specific heat upward toward the diatomic regime ($C_V = 5/2$). However, once a galaxy is firmly bound into a larger multiplet, the effect of adding energy to the system is to make the galaxy rise out of its cluster potential well, resulting in a loss of kinetic energy equivalent to lowering the temperature. Thus there is an interplay between small clusters, which tend to push the specific heat up, and larger bound systems, which tend to decrease it. The net specific heat is an average over all these clusters (or multiplets). As long as the number of binary clusters increases rapidly compared to the number of larger bound clusters, the effect of adding energy to the entire system will be to increase the specific heat. At about $b = 1/3$ the number of binaries relative to triplets is maximal. As b continues to increase, the number of

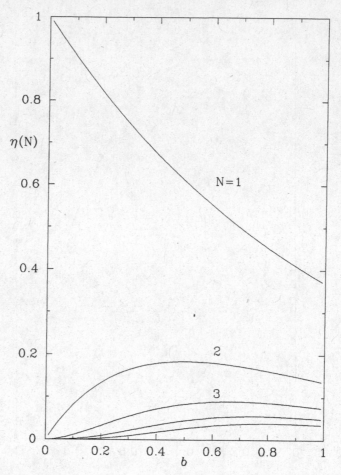

Fig. 30.2. Multiplicity function for $N = 1$–5. Only the first three curves are labeled. (From Saslaw and Sheth, 1993.)

binaries relative to triplets decreases. For $b > 1/3$, the effects of larger clusters on the specific heat begin to dominate, and C_V decreases steadily as b increases further. For $C_V < 0$ the large galaxy clusters determine the energetics of the system. The limit of $C_V = -3/2$ as $b \to 1$ results from the system becoming virialized on all scales, as shown after (28.11).

The regime $b \gtrsim 1/3$ where large clusters dominate the energetics is also the regime where linear theory becomes less valid. In the notation of the linear theory (Section 16.3) this corresponds to $\delta(t)/\delta(t_i) = 1.5$ in an $\Omega_0 = 1$ universe, where t_i is the time of the initial perturbation. With the quasi-equilibrium growth of $b[R(t)]$, we can now follow the nonlinear evolution of the correlations and compare it with the linear evolution.

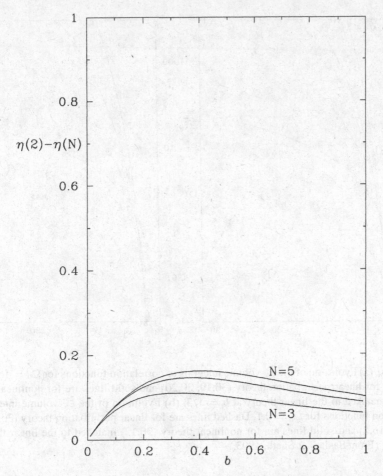

Fig. 30.3. The difference of $\eta(2) - \eta(N)$ for $N = 3, 4, 5$. (From Saslaw & Sheth, 1993.)

30.3 Evolution of Correlations

Combining the relations (28.60)–(28.64) for the volume integrals of the correlation functions, $\bar{\xi}_N(b)$, with the evolution of b, and recalling that $\bar{N}$ remains constant in a comoving volume, gives the evolution of the $\bar{\xi}_N$. Initially we may use a linear growth rate to approximate $\bar{\xi}_N$ until the early distribution settles more completely into the GQED form (Zhan, 1989). Similarly, the linear growth of fluctuations provides the linear growth of b. For an initial smooth fluid state at $R(t) = R_i$ with $\delta(t_i) \equiv 1$ the fluctuations (26.34) for $\delta = \langle(\Delta N^2)\rangle^{1/2}/\sqrt{\bar{N}}$ may be written simply as

$$b = \frac{\delta(t) - 1}{\delta(t)} \tag{30.19}$$

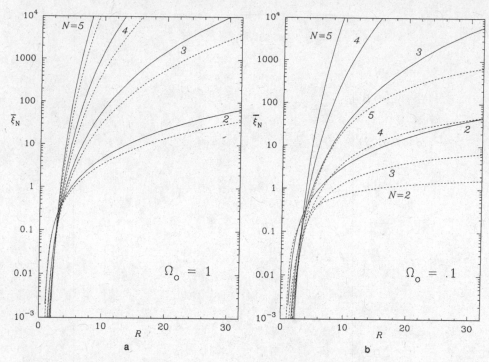

Fig. 30.4. (a) Evolution of the $\bar{\xi}_N$ volume integrals of correlation functions for $\Omega_0 = 1$. Dashed lines are for linear perturbation theory (30.19, 30.20) and solid lines are for nonlinear theory (30.12) matched to the linear theory at $b = 1/3$. (b) Evolution of the $\bar{\xi}_N$ volume integrals of correlation functions for $\Omega_0 = 0.1$. Dashed lines are for linear perturbation theory (30.19, and Section 16.2) and solid lines are for nonlinear theory (30.12) matched to the linear theory at $b = 1/3$. (From Saslaw & Sheth, 1993.)

since $b_i = 0$. In the $\Omega = 1$ Einstein–de Sitter case, (16.59) gives

$$\delta(t) = \frac{3}{5}\frac{R}{R_i} + \frac{2}{5}\left(\frac{R_i}{R}\right)^{3/2} \quad \text{and} \quad R \propto t^{2/3}, \qquad (30.20)$$

for initially cold particles with $d\delta(t_i)/dt = 0$. This also applies to models with $\Omega_0 < 1$ at $1 + z \gtrsim (1 - \Omega_0)/\Omega_0$ according to (16.20) when their expansion resembles the $\Omega_0 = 1$ case. Otherwise for $\Omega_0 \neq 1$, the linear growth (which is not affected by changing the time-independent normalization of δ) must generally be solved numerically using the equations in Section 16.2.

The reason the linear approximation (30.19)–(30.20) works better than might be expected is twofold. First, much of the contribution to the volume integral of $\bar{\xi}_2$, which determines b through the fluctuations, comes from large scales that evolve linearly (Inagaki, 1991). Second, from (30.19) and (30.20) we have approximately that $R \propto \delta \propto (1-b)^{-1}$, while from (30.12) $R \propto (1-b)^{-1}[b^{1/8}(1-b)^{1/8}]$. Therefore linear and nonlinear evolution are not very different. This means that the linear growth matches smoothly onto the nonlinear growth as b increases. For $\Omega_0 < 1$, this accord continues until $R = R_{\text{freeze}}$ in (30.13). Then the linear evolution continues to give a good approximation to b_{pattern}, which grows slowly. The nonlinear evolution continues to give a good approximation to $b_{\text{physical}} > b_{\text{pattern}}$, which increases adiabatically (Saslaw and Sheth, 1993, Sheth & Saslaw, 1996).

Figure 30.4 illustrates the growth of the low order $\bar{\xi}_N$ with both the linear and nonlinear evolution rates for two representative values of Ω_0. For large values of Ω_0 these growth rates are reasonably close, although nonlinear evolution is always faster than linear evolution after several initial expansion timescales. For small Ω_0, the effects of freezing on the linear growth become apparent and its disparity with nonlinear growth increases rapidly. In both cases it is clear that the higher order $\bar{\xi}_N$ increase very rapidly and diverge from one another after $\bar{\xi}_N \approx 1$. This occurs after only several expansions and shows again how the use of $\bar{\xi}_N$ to represent the nonlinear development becomes rather problematical.

Figure 30.5 (kindly provided by James Edgar) explicitly shows the evolution of b with redshift $z = (R_0/R) - 1$. For $\Omega_0 = 1$, this uses (30.20), and for $\Omega_0 < 1$ it numerically integrates (16.13), (16.14), (16.55), and (30.19). This is directly applicable to the pattern value of b, which can be observed by fitting the GQED (27.24) to counts in cells at high redshifts. It represents a straightforward prediction of cosmological many-body clustering, since the appropriate redshift surveys are just beginning.

It is clear that $b(z)$ depends strongly on Ω_0 and, to a lesser extent, on its current value $b(z = 0)$. Current observations suggest $b(z = 0) \approx 0.75$ (see Chapter 33). Thus if b can be determined with reasonable accuracy at the present epoch and also at an earlier epoch with $z \gtrsim 1$, and if the high-redshift distribution has the GQED form, we will have a reasonably good measure of Ω_0. If galaxy merging is moderately important at high redshifts (see Chapter 36), the distribution will tend to retain the GQED form but with a somewhat lower value of b. This would suggest that the actual simple many-body clustering value of b would be higher than the observed value, which includes merging, and so the value of Ω_0 derived from Figure 30.5 would be an upper limit.

The graphs of Fig. 30.5 also illustrate how lower values of Ω_0 produce greater clustering, at a given redshift, as Section 30.1 discussed. Although observing the complete spatial galaxy distribution function at high redshifts is clearly the best procedure, it is also possible to estimate an upper limit to Ω_0 from observations of overdense regions in the tail of the distribution. Larger values of b at moderate redshifts enhance the probabilities of finding galaxy clusters (which need not be bound) at these redshifts, because the tail of $f(N)$ is particularly sensitive to $b(z)$ and therefore to Ω_0. Unfortunately, it is also sensitive to $\bar{N}$ and to $b(z = 0)$, as well as to the a priori description of a "cluster" for these purposes. High-redshift clusters are beginning to be found (e.g., Steidel et al., 1998) and their interpretation will open an exciting new area.

A further consequence of $b(z)$ indicated in Fig. 30.5, is that the values of $b(z = 0)$ and Ω_0 imply an initial value of z at which $b \approx 0$ and clustering effectively began. This provides an estimate of the epoch when galaxies formed as coherent dynamical

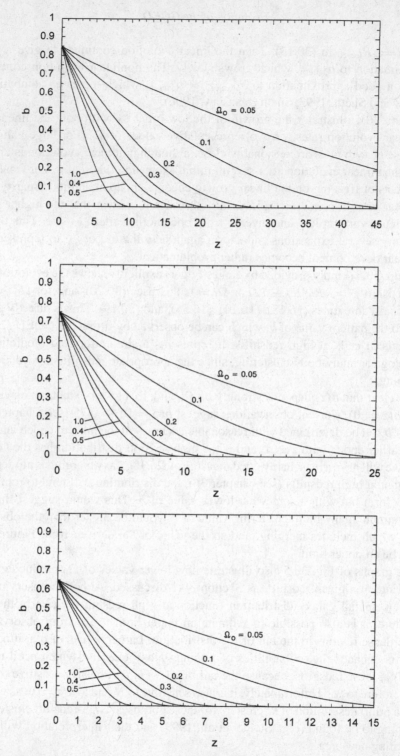

Fig. 30.5. The linear perturbation theory evolution of b with redshift for values of the cosmo-logical density parameter Ω_0 between 0.05 and 1.0 and for present values of $b(z = 0)$ of 0.65, 0.75, and 0.85.

entities, though it is subject to caveats concerning the initial conditions, which could have been "preclustered" to some extent. Even so, this evolution suggests that galaxies had to form at redshifts $\gtrsim 5$ for $\Omega_0 = 1$ and at considerably higher redshifts for $\Omega_0 \lesssim 0.5$ in these models.

Having worked out the theory of spatial and velocity distribution functions and their evolution for the cosmological many-body problem, we can next compare these results with computer experiments. This will start us along the way to judging their applicability in our observed Universe.

PART V

Computer Experiments for Distribution Functions

Many motives spur astronomers toward numerical simulations. These computer experiments test well-defined theories, display outcomes of complex interactions, elicit quantitative comparisons with observations, and provoke new insights. Moreover, they are almost always guaranteed to lead to a publishable result. What could be finer and more delightful!?

Streams of simulations have therefore poured forth in abundance. They differ mainly in their assumptions about the amount, nature, and role of dark matter, and in their initial conditions. Most agree with some aspects of observations, but none with all. None, so far, are generally accepted as complete descriptions of galaxy clustering.

As computing power expands, each new generation essentially repeats these simulations with more complicated physical interactions, greater detail, higher resolution, and added parameters. While this development continues, it seems to me wiser not to discuss the latest examples here, for they will soon be as obsolete as their predecessors. Instead, we concentrate on the simplest case: the cosmological many-body problem. Even this reveals a richness of behavior that surpasses current understanding. Understanding is more than simulation, for it embeds the simulations in a much richer conceptual context.

31

Spatial Distribution Functions

Order and simplification are the first steps toward
the mastery of a subject – the actual enemy is the
unknown.

Thomas Mann

Since our main aim here is to examine the basic physics of gravitational clustering, we simplify the system in order to ease its physical interpretation. The simplest system containing enough physics to be interesting for galaxy clustering is the cosmological many-body problem. Point particles represent clustering galaxies in the expanding universe. Integration of their equations of motion, as described in Section 15.3 and Chapter 19, shows how they evolve. Section 15.3 gave examples and a brief summary of some important results. Here we use the conceptual structure built up in the last few chapters to explore those results in more detail. We start with Poisson initial conditions, minimalist in the sense of having no initial information other than the interparticle separation. Then we consider possible effects of non-Poisson initial conditions and of dark matter between galaxies, even though its amount and distribution are unknown.

31.1   Poisson Initial Conditions

31.1.1   The Simplest Case: $m = \bar{m}$, $\Omega_0 = 1$

All galaxies have the same mass and the $\Omega_0 = 1$ Einstein–de Sitter cosmology has the simplest evolution. Its gravitational timescale always equals its expansion timescale (30.6). Returning to the 4,000-body computer simulation of Figure 20.1 now helps us appreciate the nature of the quasi-equilibrium approximation. Most of the strong clusters at an expansion factor of $a/a_0 = 15.62$ are seen to have their progenitors at earlier times. Some can be traced back to chance enhancements in the initial Poisson distribution. This correspondence is not exact since merging, ejection, and projection all play a role. But those roles are secondary.

If quasi-equilibrium does indeed dominate, we should find the predicted spatial $f(N)$ distribution (27.24). Figure 15.4 (Section 15.3) for a similar experiment but with 10,000 particles showed this is indeed the case. The agreement of the GQED with these experiments is equally good at earlier times. However, at times much later than the putative age of our Universe in these simulations the experimental distributions become more ragged (at expansion factors $a/a_0 \gtrsim 15$ for $N = 4,000$ or $\gtrsim 30$ for $N = 10,000$). Typically their peak broadens relative to the GQED; sometimes they have a double peak. The reason for this is that inhomogeneous structures eventually form on the scale of the entire system. It loses statistical homogeneity and the simulation is no longer a representative volume of space.

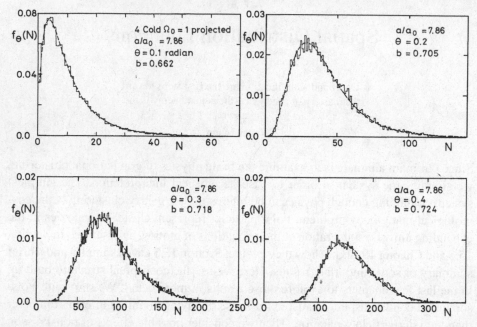

Fig. 31.1. The projected distribution functions (solid histograms) $f_\theta(N)$ for combined data of four initially cold $\Omega_0 = 1$ simulations with $N = 4{,}000$ as in Fig. 20.1. The expansion factor is 7.86 and θ is the radius of the circles on the sky in radians. Dashed lines show the best fit to the GQED with the resulting value of b given in the diagrams. (From Itoh et al., 1988.)

Projected experimental distributions $f(N)$ also agree well with the GQED. Figure 31.1 shows $f(\theta)$ for $N = 4{,}000$ as seen by an observer at the center of the sphere looking at counts on the sky in circles of radius 0.1, 0.2, 0.3, and 0.4 radians. These correspond to the top right projection in Figure 20.1, averaged over four experiments with expansion factors $a/a_0 = 7.86$. As a reminder, this is the expansion factor for $\Omega_0 = 1$ when ξ_2 has about the same slope and amplitude as the presentobservations.

In accord with the discussion of Section 28.5, values of b for the projected distributions are about 5–10% lower than for the three-dimensional spatial distributions on similar scales. This is because the volume integral for b in (25.18) over a cone is less than over a sphere whose radius is the same as the radius of the base of the cone. This difference becomes greater as the correlations become stronger. Therefore we expect this difference to be greater for larger Ω_0 and for longer evolution. Simulations bear this out, although for $\Omega_0 \lesssim 0.1$ there is so little evolution in the clustering pattern after an expansion factor of $\sim$10 that any evolutionary effect is small. In principle, we could use differences between projected and three-dimensional values of b to determine Ω_0. But, as we will see in Section 31.3, the effects are small and unlikely to be disentangled using current observations.

Changing the value of Ω_0 affects the rate of evolution as Section 31.3 will show, but the form of the experimental $f(N)$ continues to agree with the GQED. This is because, except perhaps for very small values of $\Omega_0 < 0.01$, the system relaxes to a GQED before $z \approx \Omega_0^{-1}$ when the Universe stops behaving as though it were closed (see 30.13). Subsequent evolution retains the "frozen" form of the distribution function, although its values of $\bar{N}$ and b_{physical} continue to evolve.

31.1.2 Effects of Initial Peculiar Velocities

Most of the cosmological N-body experiments start with a cold system. All galaxies participate fully in the Hubble flow with no initial peculiar velocities. This is more for convenience than from any conviction that we know their initial peculiar velocities. Indeed, if galaxies formed from the random agglomeration of smaller protogalaxies, we would expect them to have residual peculiar velocities of order $\langle v_{\text{protogalaxies}} \rangle / \sqrt{N}$, where N is the average number of protogalaxies per galaxy.

Unless these initial peculiar velocities are very high (which is physically unlikely), they have little lasting effect on the distribution. The reason is that the few-body gravitational interactions of near neighbors cause the velocity distribution to relax to its self-consistent form if $v_{\text{initial}} \lesssim \bar{r} H_i$, where $\bar{r}$ is an average initial galaxy separation and v is the r.m.s velocity. Simulations confirm (Itoh et al., 1988) that this relaxation occurs rapidly, after only one or two initial Hubble expansion timescales H_i^{-1}. If the velocities do start off very hot, $v_{\text{initial}} \gtrsim 3\bar{r} H_i$, then the system cools roughly adiabatically with $v \approx v_{\text{initial}}(R_i/R)$ as in (22.25) until $v \lesssim \bar{r} H$, when it starts to cluster significantly.

Thus, for a wide range of initial peculiar velocities, the cosmological many-body system quickly settles down into a state where the velocities and spatial clustering are mutually self-consistent and have the GQED form. (Velocity distribution functions will be described in Chapter 32.) The memory of initial velocities on scales that become nonlinear is rapidly forgotten.

31.1.3 Two Mass Components

The next simplest case has two components with different masses. This introduces two new parameters: the mass ratio m_2/m_1 of two types of particle and the ratio N_2/N_1 of their total numbers (or total masses). Although galaxies of two different masses are the obvious interpretation of this system, we can stretch our imagination to consider two other interpretations in the case of very high mass ratios. One is to think of the massive particles as crudely representing cores of dark matter structures, with the less massive particles as galaxies attracted to these cores. The other is for the massive particles to represent galaxies, and the less massive ones to represent particles in dark matter halos around the galaxies. Both these are very crude models of their respective situations. Simplicity, rather than realism, is their main virtue. The literature abounds with much more sophisticated models, but whether their increased sophistication is matched by equally increased realism is still being debated.

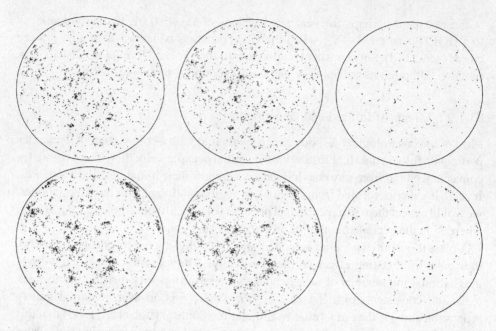

Fig. 31.2. Projected positions at expansion factor $a/a_0 = 7.86$ for the $m_2/m_1 = 5$ (top) and $m_2/m_1 = 100$ (bottom) simulations described in the text. From left to right they show all ($\sim 2{,}000$) projected particles in one hemisphere, the less massive ($\sim 1{,}750$) particles, and the most massive (~ 250) particles. (From Itoh et al., 1990.)

It is instructive to examine the two cases $m_2/m_1 = 5$, where effects of mass differences begin to show clearly, and $m_2/m_1 = 100$, where they dominate. (For more details see Itoh et al. (1990), which we follow here.) Both are again 4,000-body systems starting from cold Poisson initial conditions with direct integrations of the orbits for $\Omega_0 = 1$. Each contains 500 heavy particles of mass m_2 and 3,500 of mass m_1. Thus in the first case $M_2/M_1 = 0.71$ and the massive particles almost dominate the total gravitational field, while in the second case $M_2/M_1 = 14$ and the massive particles are completely dominant (with the less massive particles essentially acting as test particles).

Figure 31.2 shows the projected configurations that a central observer would see at an expansion factor of 7.86, when the amplitude and slope of ξ_2 are similar to present observations. In the top row $m_2/m_1 = 5$; in the bottom row $m_2/m_1 = 100$. Projections on the left contain all the particles, those in the middle contain just the less massive particles, and those on the right contain just the massive particles.

The first thing to notice, especially in contrast with Figure 20.1 for $m_2/m_1 = 1$, is the enhanced visual clustering as m_2/m_1 increases. Filamentary structures appear to become longer and more conspicuous. More massive particles tend to collect less massive satellites. However, within a particular simulation there are often regions that are exceptions to this general trend. Moreover, as Chapter 8 pointed out, appearances can be deceiving.

Therefore to begin to quantify this clustering, we look at its two-point correlation functions. Figures 20.2 and 20.3 showed comparable results when all particles have the same mass. Figures 31.3a and 31.3b now show how the two-point correlation functions evolve for the mass ratio $m_2/m_1 = 5$ at four values of the expansion factor a/a_0 (sometimes denoted by R/R_0). Correlations for all the galaxies are shown in Figure 31.3a; those for the massive and less massive components are shown separately in Figure 31.3b. Figures 31.4a and 31.4b show comparable results for the case $m_2/m_1 = 100$. So here we are looking at the effects of varying just one parameter.

We can look next at the spatial distribution functions for these two cases and then discuss some of their combined implications. (Figure 15.4 showed the case $m_2/m_1 = 1$, albeit for $N = 10,000$, which provided a smoother histogram.) Figure 31.5 shows $f_V(N)$ at $a/a_0 = 7.86$ for $m_2/m_1 = 5$ using randomly placed spheres of radius $r = 0.1, 0.2, 0.3$, and 0.4 (whose centers are at least a distance r from the edge of the simulation). The normalized comoving radius $R = 1$ for the whole simulation. Results are given for all particles together, and separately for the less massive and the massive particles. Figure 31.6 gives comparable results for the $m_2/m_1 = 100$ case. The dotted lines are the best-fit GQED, and the figures give their value of b.

What do these results mean? First we gain some insight into the speed of the system's evolution. Thirty years ago the rapid clustering caused by the local graininess of the gravitational field, even for identical mass particles, came as a great surprise. Now we take it for granted. Two mass components enhance clustering further. This is apparent from visual comparison of the projected galaxy positions in Figures 20.1 and 31.2. However, it is not so clear from comparing the two-point correlation functions for all particles in Figures 20.2, 31.3a, and 31.4a at the same expansion times. The reason is partly that less massive particles dominate ξ_2 for all three cases, and partly that ξ_2 is intrinsically not very discriminatory. Its discriminatory power needs to be increased by measuring ξ_2 for the two components separately. Then Figures 31.3b and 31.4b show that the more massive component generally has a steeper two-particle correlation that grows faster.

Increasing m_2/m_1 accentuates the difference between ξ_2 for the two components until m_2/m_1 becomes very large. Then, while ξ_2 for the massive component continues to grow, the growth of less massive correlations is curtailed. This is an indication that many of the less massive particles have been caught up as bound satellites of their more massive companions. The resulting groups continue to cluster.

It might seem curious that ξ_2 becomes roughly equal for both components on scales $r \gtrsim 0.05$. This similarity occurs sooner for smaller m_2/m_1. Spheres of this radius initially contain about four particles, of which one is typically more massive. Thus the distribution on larger scales becomes dominated by groups; positions of particles on these scales depend less on their individual masses.

Scales on which correlations become independent of mass signify that the nature of the relaxation is essentially collective. It depends on particles interacting with groups, and on groups interacting with groups. Not on particles interacting just with

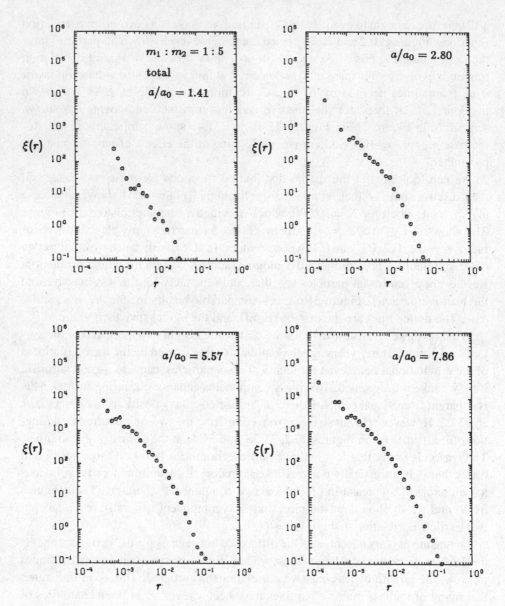

Fig. 31.3(a). Two-point correlation functions for all particles with $m_2/m_1 = 5$, $a/a_0 = 1.41$–7.86 in the simulation of Fig. 31.2.

their two or three nearest neighbors. Once a particle interacts mainly with a much larger collective mass, the particle's own mass becomes essentially irrelevant to its orbit – as Galileo pointed out.

Collective relaxation is a harbinger of quasi-equilibrium evolution. Averaging evolution over more particles means their behavior is more likely to represent the

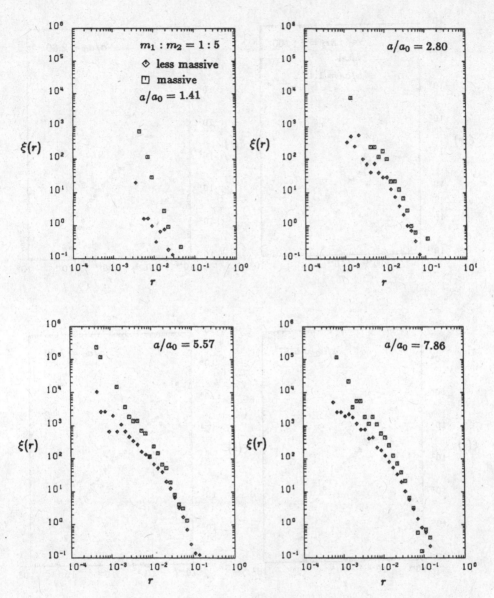

Fig. 31.3(b). Two-point correlation functions for the less massive and massive particles with $m_2/m_1 = 5$. (From Itoh et al., 1990.)

whole system. Consequently, as an example, local values of $b = -W/2K$ on small scales should be close to their ensemble average for this scale. In other words, the refinement produced by taking a distribution of values of b in the spatial GQED $f(N)$, rather than its average for a given scale, is small. The evolution of a Poisson distribution into the GQED discussed in Section 30.1 suggests that this collective

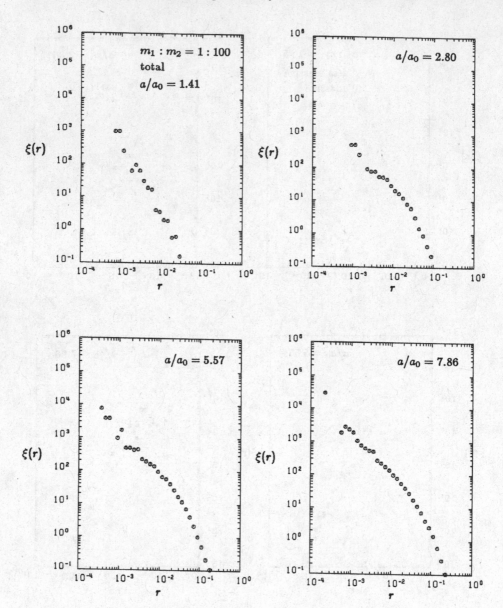

Fig. 31.4(a). Two-point correlation functions for all particles with $m_2/m_1 = 100$, $a/a_0 = 1.41$–7.86 in the simulation of Fig. 31.2.

relaxation dominates over near neighbor two-body relaxation quite quickly. This is also indicated by the result for $m_2/m_1 = 1$ that the GQED fits well after only about one expansion time (Itoh et al.,1988). For instance, already at $a/a_0 = 2.8$ the value of $b = 0.45$ is quite large in spheres of radius 0.3 for the $N = 4,000$ initially cold Poisson simulations with $\Omega_0 = 1$.

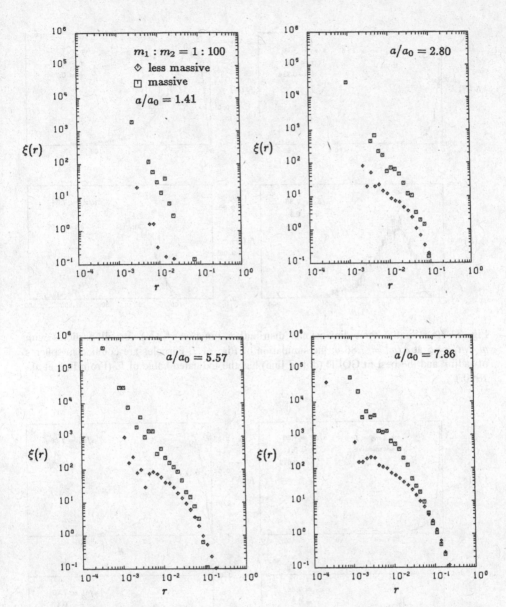

Fig. 31.4(b). Two-point correlation functions for the less massive and more massive particles with $m_2/m_1 = 100$. (From Itoh et al., 1990.)

As the ratio m_2/m_1 increases, the value of b for the massive particle distribution becomes increasingly greater than the value it would have from (28.24) for random selection (Itoh et al., 1990). This is especially noticeable on smaller scales ($r \lesssim 0.2$) where the clustering is prominent. For $m_2/m_1 = 5$, for example, the difference may be 20%. Larger scales spatially smooth out this effect, and by $r \approx 0.4$ it is only a

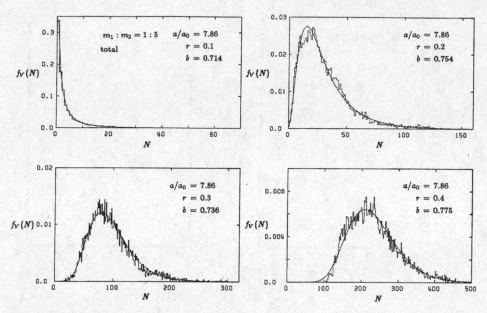

Fig. 31.5(a). Spatial three-dimensional distribution function $f_V(N)$ for all particles with $m_2/m_1 = 5$ at $a/a_0 = 7.86$ in the simulation of Fig. 31.2. Particles are counted in spheres of radii r and the best fit GQED (dotted line) has the indicated value of b. (From Itoh et al., 1990.)

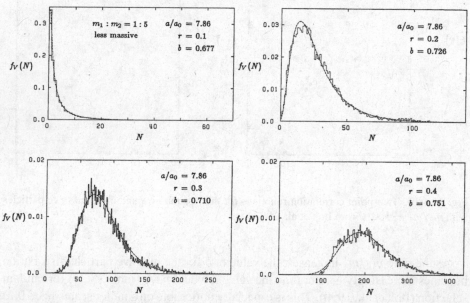

Fig. 31.5(b). $f_V(N)$ for the less massive particles with $m_2/m_1 = 5$. (From Itoh et al., 1990.)

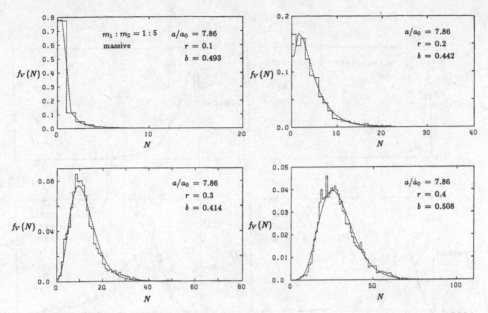

Fig. 31.5(c). $f_V(N)$ for the more massive particles with $m_2/m_1 = 5$. (From Itoh et al., 1990.)

few percent. Different mass components fit the GQED well enough to measure these differences easily.

When $m_2/m_1 = 100$, the 500 massive particles dominate the total gravitational field. Less massive particles mostly shadow the clustering, often as satellites of their more forceful companions. This introduces a new effect, exemplified by Figure 31.6. On larger scales ($r \gtrsim 0.3$), the $f_V(N)$ distributions become more ragged. This is particularly true for the less massive particles, and consequently for the total population, but not for the massive particles. Massive particles continue to cluster quite smoothly. The irregularities become pronounced on scales initially containing a few massive particles that subsequently cluster together. Eventually the contiguity of these satellite systems can give the visual impression of filaments.

31.1.4 Continuous Mass Distributions

Although the simplicity of two-component systems provides useful insight, continuous mass distributions are more realistic. These models (Itoh et al., 1993) have the same overall parameters as before, except that $N = 10,000$ and their mass spectrum is a truncated gamma distribution:

$$dN \propto \left(\frac{m}{m_*}\right)^{-p} e^{-m/m_*} d\left(\frac{m}{m_*}\right). \tag{31.1}$$

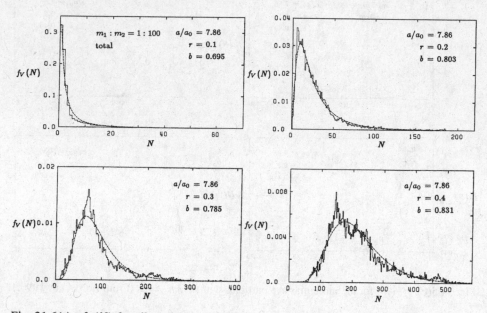

Fig. 31.6(a). $f_V(N)$ for all particles with $m_2/m_1 = 100$ at $a/a_0 = 7.86$ in the simulation of Fig. 31.2. (From Itoh et al., 1990.)

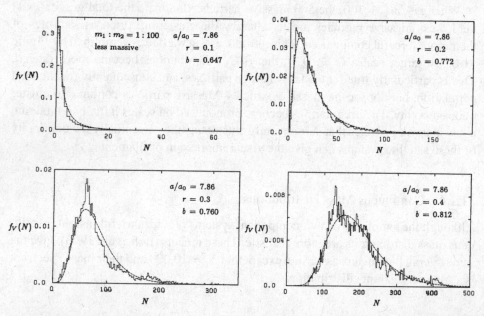

Fig. 31.6(b). $f_V(N)$ for the less massive particles with $m_2/m_1 = 100$. (From Itoh et al., 1990.)

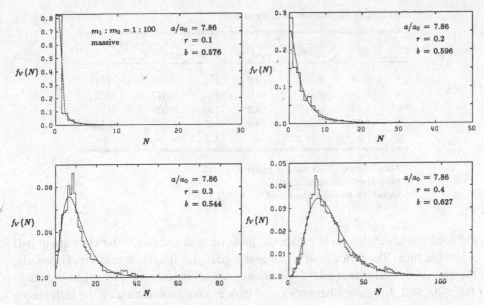

Fig. 31.6(c). $f_V(N)$ for the more massive particles with $m_2/m_1 = 100$. (From Itoh et al., 1990.)

The reason for using this form is that if $m \propto L$ it becomes the observed galaxy luminosity function determined by Schechter (1976) in which $1.1 \lesssim p \lesssim 1.25$. Of course the simple linear proportionality between mass and luminosity might not generally hold, but (31.1) is a reasonable example. With $N = 10,000$ this distribution is naturally discrete but reasonably smooth. It is specified by its values of p, $m_{\max}/m_*$, and $m_{\min}/m_*$.

Table 31.1 gives the properties of the five models first described. The last two columns give the total mass fractions of the 2,000 least massive and 2,000 most massive particles. The statistics illustrated here were averaged over 6, 5, 4, 5, and 4 realizations each for Models A, B, C, D, and S. In Model S all particles have the same mass for direct comparison.

Figure 20.4 showed projections of Models S and B at $a/a_0 = 8$, which were described earlier in Section 20.1. Many compact clusters arise in the single-mass model, while larger size clusters and some apparent sheetlike structures also grow in the multimass model. Multimass models form more clusters more quickly in the early expansion. Multimass clusters generally have larger radii and more members than in corresponding single-mass models. Comparing the positions of the 20% most massive and 20% least massive particles in Model B shows that most of the massive particles are in clusters, but the least massive ones are found in both clusters and the general field.

Figure 20.5 showed ξ_2 separately for the three mass groups in Model B as well as for all particles. The 20% most massive particles, containing almost three-quarters of

Table 31.1. *Parameters of Simultations*

Model	p	m_{min}/m_*	m_{max}/m_*	Ω_0	M_{LM}/M_T[a]	M_M/M_T[b]
A	1.00	0.01	10.0	1.0	0.013	0.716
B	1.25	0.01	10.0	1.0	0.017	0.741
C	1.50	0.01	10.0	1.0	0.025	0.733
D	1.25	0.10	10.0	1.0	0.058	0.536
S[c]	—	—	—	1.0	—	—

[a] Mass fraction of less massive galaxies.
[b] Mass fraction of massive galaxies.
[c] Model S is a single-mass model.

the total mass, develop an approximate power-law correlation after only about half a Hubble time. Then they drag less massive particles into their clusters. Gradually, the amplitude and extent of $\xi_2(r)$ increases for all masses. Slopes are in the range $1.5 \lesssim \gamma \lesssim 2$. Irregular distributions of dark matter should modify the differences of ξ_2 for various mass groups, but this has not been studied systematically. In the extreme case where clustering of dark matter dominates the total gravitational field, there should be little distinction between different masses – all become test particles (except perhaps on very small scales).

Figure 31.7 shows the void distribution $f_0(r)$ and $f_5(r)$, as well as $f_V(N)$ for $r = 0.1 - 0.4$ in Model B at $a/a_0 = 8.0$. These are for all the particles. In $f_0(r)$ and $f_5(r)$, best fits are found using one value of b on all scales (more on the scaling of b in Section 31.2). Comparing with Figure 15.4 for Model S at $a/a_0 = 8$, we see that the distributions in Model B are more irregular and agree less well with the GQED, especially for $r \gtrsim 0.20$. This is a direct effect of the satellite systems that are not accounted for in the GQED. They produce inhomogeneities on larger scales, rather similar to what happens in Model S at $a/a_0 = 32$. Interestingly, on smaller scales where these inhomogeneities are essentially filtered out (by a spherical window function if you like), the agreement with the GQED remains excellent. This agreement also holds for the void distribution, even though it is approximated using only one value of b for all scales.

The sense of the discrepancy with GQED, both for Model S at late times and Model B near the "present" time, is that large volumes contain too many particles. Other models (Itoh et al., 1993) with values $0 \lesssim p \lesssim 5$ show better agreement than Model B with the GQED. This is because the value $p \approx 1.3$ maximizes the contribution of massive galaxies to the total gravitational field and therefore exacerbates formation of satellite systems.

Having described some basic properties of these simulations, we can turn to their scale dependence and evolution of b, before exploring non-Poisson initial conditions and adding dark matter.

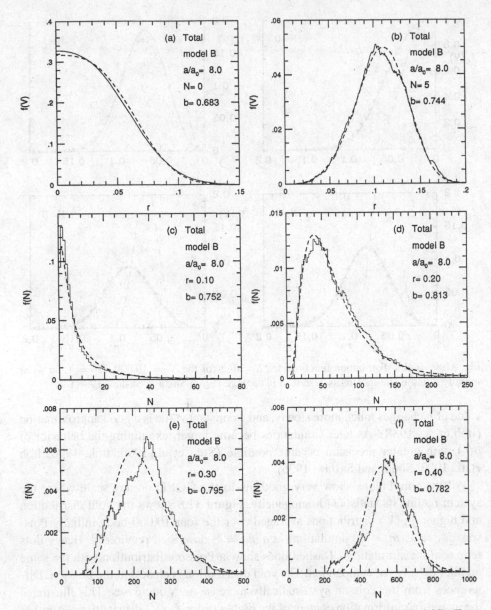

Fig. 31.7. The void distribution $f_0(r)$, and $f_5(r)$, and $f_V(N)$ for $r = 0.1$–0.4 in Model B. (From Itoh et al., 1993.)

31.2 Scale Dependence of b

Derivations of the GQED in Chapters 25 and 26 concluded that its form is independent of scale but its value of b (25.18) is not. Simulations can check this. Early experiments (Saslaw, 1985b, Figure 8.4) found this variation, especially at small scales containing only a few particles, near the expected Poisson limit. On larger scales the

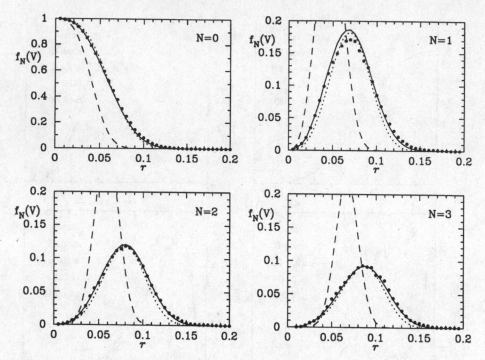

Fig. 31.8. $f_N(V)$ distribution functions for $N = 0$–3 for the $N = 10,000$, $\Omega_0 = 1$, $m = \bar{m}$ initially Poisson simulations as described in the text. (From Sheth & Saslaw, 1996.)

value of b changes much more slowly, and a constant value is a good approximation (Itoh et al., 1988). As later simulations became larger, examining the behavior of $b(r)$ with greater precision became possible (Suto et al., 1990; Itoh, 1990; Itoh et al., 1993; Sheth and Saslaw, 1996).

All the simulations show very good fits to the GQED on any scale where the system retains its statistical homogeneity. Figure 31.8 shows the void distribution and higher $f_N(V)$ distributions averaged over the four 10,000-body initially Poisson, $\Omega_0 = 1$, $m = \bar{m}$ simulations at $a/a_0 = 8$ described previously. Heavy dots represent the simulations. Dashed lines show a Poisson distribution with the same number density for comparison. The void distribution is closest to the Poisson. Differences from the Poisson systematically increase as N increases. This illustrates the increasing information content of the higher order $f_N(V)$ distributions. Another way to see this is to notice that the Poisson contribution to the generating function (28.56) would involve just the form $h(s) = s$ in (28.57) and that the GQED void distribution $f_0(V) = e^{-\bar{N}(1-b)}$ closely resembles the Poisson $f_0(V) = e^{-\bar{N}}$. The Poisson resemblance fades as N increases and (28.57) adds more information.

Attempting to fit the simulations using one value of $b = 0.7$ as a free fitting parameter independent of scale in the GQED gives the dotted lines in Figure 31.8. Not bad, but not perfect either. In the void distribution, the fit is systematically high at small scales, because b is really smaller on these scales, and systematically low at large scales where b is really larger. Since $f_N(V)$ is systematically biased toward larger scales for larger values of N, the fits fall slightly below the simulation values.

We can incorporate the scale dependence of b into the GQED directly using (26.34) for the variance of counts in cells of volume v. The self-consistent average $\bar{N}$ is

$$\bar{N} = \frac{1}{N_{\text{cells}}} \sum_{i=1}^{N_{\text{cells}}} N_i, \tag{31.2}$$

where N_i is the number of particles in the ith cell. (This may differ slightly from $\bar{N} = \bar{n}V$ if there is statistical inhomogeneity in the cells that are sampled.) Then the self-consistant variance is related to $\bar{\xi}_2$ and b by (14.17) and (26.34) is

$$\frac{1}{N_{\text{cells}} - 1} \sum_{i=1}^{N_{\text{cells}}} (N_i - \bar{N})^2 = \bar{N}(1 + \bar{N}\bar{\xi}_2) = \frac{\bar{N}}{(1-b)^2}. \tag{31.3}$$

This determines $b = b_V = b(r)$ on any scale simply from the variance of counts in cells on that scale. Then this self-consistent value of b_V is substituted into the GQED $f_N(V)$ on that scale. This is the solid line in Figure 31.8. It fits the simulations almost perfectly to 98% accuracy. (There is a decrement in the simulations at the peak of $f_1(V)$ caused by exceptional chance structure in one of the four averaged simulations.) This is one of the rare examples where a decrease in the number of free parameters (now zero) results in better agreement between theory and simulations. This improved agreement also extends to the observations, as we will see later.

For these same simulations, Figure 31.9 shows $b(r)$ determined from (31.3) at different expansion factors $2.86 \leq a/a_0 \leq 16$. For the most part these are quite smoothly increasing functions; occasionally there is a glitch when unusual structure forms by chance as at $r \approx 0.015$ when $a/a_0 = 11.3$. On scales containing an average of more than one or two particles, the scale dependence of b is slight. As the simulation expands, the maximum scale-invariant value of b increases. Section 31.3 presents more on evolution.

So far we have been concerned with b_{pattern}. Now we turn to $b_{\text{physical}} = -W/2K$. These are nearly the same in $\Omega_0 = 1$ initially Poisson models, but they may differ for others, as Section 30.1 discusses and the experiments referenced earlier confirm. Figure 31.10 shows b_{physical}, again for the same simulations as in the previous two figures. Extracting b_{physical} is a bit more subtle since at these high levels of precision we need to include the effect of the softening parameter, ϵ, on the correlation energy W. Details of how this is done are described in Sheth and Saslaw (1996). The results are that b_{physical} generally increases with scale in the manner of b_{pattern}. However, b_{physical} is more sensitive to statistical inhomogeneity because it involves both the correlation inhomogeneity and the inhomogeneity in kinetic energy.

Finally, we look at $b_{\text{pattern}}(r)$ for the models with more continuous mass spectra described in Section 31.2 (Itoh et al., 1993). Figure 31.11 portrays this for Models A, B,

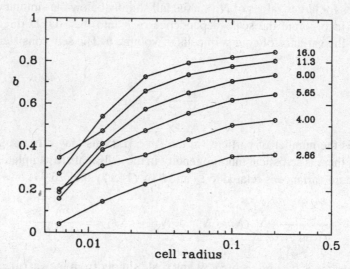

Fig. 31.9. Scale dependence of b_{pattern} (for the simulations in the previous figure) determined from the variance of counts in cells as a function of comoving cell radius at six values of the expansion factor a/a_0. (From Sheth & Saslaw, 1996.)

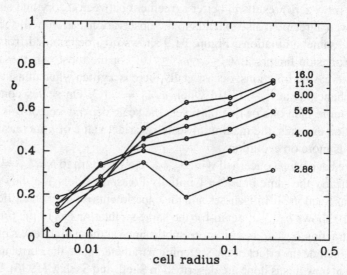

Fig. 31.10. Scale dependence of b_{physical} (for the simulations of Fig. 31.8) as a function of comoving cell radius at six expansion factors a/a_0. Arrows show the softening length, ϵ, for the force when the expansion factors are 2.83, 4.00, and 5.65. In comoving coordinates, ϵ shrinks as the simulation expands. (From Sheth & Saslaw, 1996.)

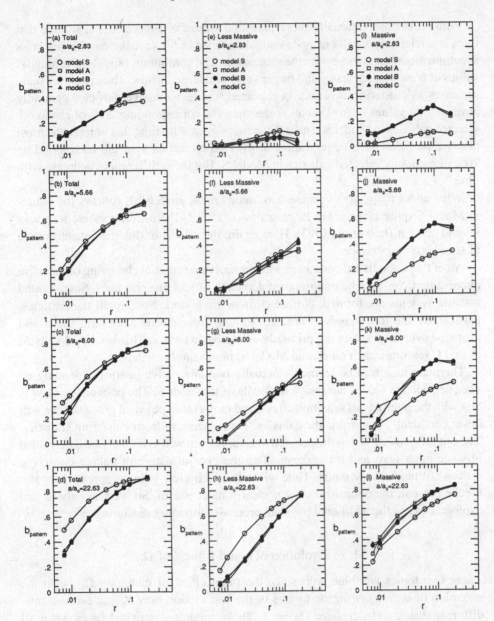

Fig. 31.11. Scale dependence of b_{pattern} for models in Table 31.1. The total 20% least massive and 20% most massive particles are each shown at four expansion epochs. (From Itoh et al., 1993.)

C, and S in Table 31.1. Since all particles in Model S have the same mass, their 20% "less massive" and 20% "most massive" members are just selected randomly. This provides a useful comparison. All these models, apart from S, evolve in pretty much the same way. In fact, for these models it takes rather drastic values of $p \gtrsim 3$, with $M_{\text{LM}}/M_{\text{T}} \gtrsim 0.11$ and $M_{\text{M}}/M_{\text{T}} \lesssim 0.43$, to give substantially different $b(r)$ (see Itoh et al., 1993).

Within the observationally more plausible values of Figure 31.11, we see that there is a relatively small but systematic difference in $b(r)$ for the continuous mass spectrum models and those with the same mass. The continuous models have smaller values of b on smaller scales and larger values on larger scales. The crossover scale increases as clustering evolves. As expected, the less massive particles generally have smaller values of $b(r)$ on all scales than the corresponding 20% of randomly selected galaxies in Model S. These differences grow with time. In contrast, the most massive particles have larger values of $b(r)$ on all scales than the corresponding 20% of randomly selected galaxies in Model S. But these differences decrease with time.

What about $b_{physical}(r)$ in these continuous mass models? It follows the values in Model S quite closely. On large scales ($r \gtrsim 0.3$) the differences are less than several percent (Itoh et al., 1993). Here again, the effects of the mass spectrum are of secondary importance.

All of these results are consistent with the early preferential clustering of the more massive particles and their subsequent accretion of less massive ones. Some bound satellite systems are formed. Not all clusters are bound. Nor are all the particles, particularly the less massive ones, in clusters. Interactions after long times and averaged over large scales are primarily collective in nature. This is why Models A, B, and C resemble each other, and Model S, increasingly closely.

The main lesson to be learned – actually two alternative lessons depending on your perspective – does not reside in details of the models. The pessimistic view is that with the interplay of selection effects and many other physical processes we will never confidently determine the galaxies' mass spectrum from clustering statistics. The optimistic view is that the detailed mass spectrum will eventually be determined in some other way, and for understanding the general nature of galaxy clustering it does not matter very much. Time will tell which view is more appropriate. But for the present, the optimistic view is clearly more useful. So we press ahead and explore some of the additional physical processes affecting clustering.

31.3   Evolution of b and Effects of Ω_0

These two topics combine well since the main effect of changing Ω_0 is on the evolution of b. We saw in the figures of the last section how $b_{pattern}$ evolved over different scales and times for $\Omega_0 = 1$. Those models contrasted cases when all masses were equal and when they ranged over three orders of magnitude.

Looking at the evolution of $b_{pattern}$ on large scales, where b_V approximates its asymptotic value, removes most of the effects of "spatial filtering." It facilitates clearer comparisons with different values of Ω_0 and with the adiabatic theory of Chapter 30. Adiabatic theory should describe both $b_{pattern}$ and $b_{physical}$ approximately when $\Omega_0 = 1$, since they are nearly equal. When Ω_0 is significantly less than unity, adiabatic theory should continue to describe $b_{physical}$, but not $b_{pattern}$ for expansion factors greater than $R_{freeze}/R_0 \approx \Omega_0/(1 - \Omega_0)$ from (30.13).

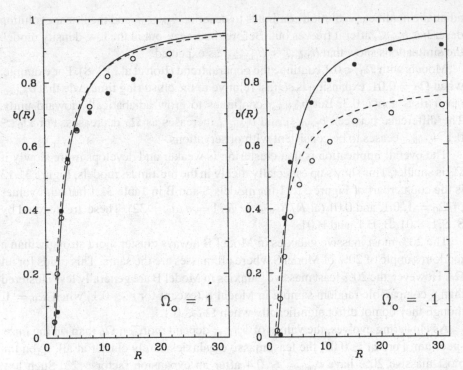

Fig. 31.12. Linear (dashed lines) and nonlinear (solid lines) evolution of $b_{pattern}$ (open circles) and $b_{physical}$ (filled circles) for cosmologies with $\Omega_0 = 1$ and Ω_0 ($R_0 = 26$) = 0.1. (From Sheth & Saslaw, 1996.)

Figure 31.12 illustrates these effects for two 4,000-body experiments with $\Omega_0 = 1$ and $\Omega_0(R_0 = 26) = 0.1$. Their particle masses are all the same and their softening parameters are $\epsilon = 0.03$. The open circles show how $b_{pattern}$ evolves for large cells as the radius of the universe $R(= a)$ expands. Filled circles show $b_{physical}$. Initially the distributions are Poisson and for $\Omega_0 = 1$ the values of $b_{pattern}$ and $b_{physical}$ agree closely until $R_0 \approx 8$ (roughly the present time). Later values of $b_{physical}$ fall short by 10%. Although this difference appears to be affected by the numerical softening parameter, it is not fully understood (Sheth and Saslaw, 1996). For $\Omega_0 = 0.1$ at $R_0 = 26$, roughly the present time as judged by the presently observed slope and amplitude of ξ_2 for this lower density universe, $b_{pattern}$ is about 20% less than $b_{physical}$. Since correlations are less in the lower Ω_0 universe, the effects of the softening parameter are diminished. Indeed, the two forms of b begin to diverge in Figure 31.12 at $R \approx 3$ for $\Omega_0 = 0.1$, as expected from (30.13).

The dashed line in Figure 31.12 is the linear evolution of b from (30.19) and (30.20). The solid line is the adiabatic evolution for $b_{physical}$ on large scales (30.12) with the adiabat specified by $R_* = 2.12$ and 2.38 for the $\Omega_0 = 1$ and 0.1 cases, respectively. For $\Omega_0 = 1$, the linear and adiabatic nonlinear evolutions agree closely, and both describe $b_{pattern}$ well for the reasons discussed in Section 30.3. For $\Omega_0 = 0.1$, adiabatic nonlinear evolution continues to describe $b_{physical}$, and the linear evolution describes $b_{pattern}$ after it freezes out. Before the freeze-out in the low-density model, the simulations show that $b_{pattern} \approx b_{physical}$ as expected.

Models with $\Omega_0 < 0.1$ continue the general trend (Itoh et al., 1988). For example, when $\Omega_0 = 0.01$, expansion is so fast relative to the clustering timescale that $b_{pattern}$ never rises above 0.3. But $b_{physical}$ continues to grow adiabatically toward unity. The difference between $b_{physical}$ and $b_{pattern}$ increases as Ω_0 decreases. For $\Omega_0 \lesssim 0.1$, $b_{pattern}$ ceases to be consistent with observations.

The overall implication is that clustering is weaker and develops more slowly if Ω_0 is smaller. This shows up especially nicely in the multimass models. Figure 31.13 is the counterpart of Figure 31.11 for models S and B in Table 31.1 but with values of $\Omega_0 = 1, 0.1$, and 0.01 (at $R/R_{initial} = R/1 = a/a_0 = 32$). These are denoted by S, S.1, S.01, B, B.1, and B.01.

The 20% most massive galaxies in Model B always cluster more strongly than a random sample of 20% of Model S where all masses are the same. This holds for all Ω_0. However, the 20% least massive galaxies in Model B are generally less clustered than a comparable random sample in Model S (except for $r \gtrsim 0.1$) when $\Omega_0 = 1$, though they do not differ significantly when $\Omega_0 \lesssim 0.1$.

As clustering evolves, the values of $b_{pattern}$ depend more on Ω_0 than on the mass spectrum. For $\Omega_0 \lesssim 0.05$, the least massive galaxies barely cluster at all. Even the most massive 20% have $b_{pattern} \lesssim 0.4$ after an expansion factor ~ 23. Such low values of Ω_0 are clearly inconsistent with the observed values of $b_{pattern} \gtrsim 0.7$. Models with $\Omega_0 \approx 0.1$ agree with observed clustering only after expansion factors $\gtrsim 10$, depending on their mass spectrum. If galaxies clustered in this manner from Poisson initial conditions, then they generally had to start clustering as dynamical entities before redshifts $z \approx 10$. However, if $\Omega_0 \approx 1$, galaxies could have started clustering in this way at $z \approx 5$.

Alternatively, particular non-Poisson initial conditions might be carefully contrived so that clustering starts at small redshifts and yet agrees at present with the gravitational quasi-equilibrium distribution and therefore with observations. We will examine some non-Poisson initial conditions shortly.

For the moment, there are two main points that emerge from these simple models. The first is that four properties may each contribute significantly to the presently observed value of $b_{pattern}$: the value of Ω_0, the redshift when galaxies essentially start to cluster as dynamical entities, their spatial distribution when clustering began, and their mass spectrum. All these are connected to the mostly unknown earlier conditions for galaxy formation. Even without additional components of dynamically important dark matter, there is considerable complexity. Details of this complexity may become difficult to disentangle.

The second main point is that if $\Omega_0 \gtrsim 0.2$ then the details of these other properties appear to be secondary. A reasonable range for these properties becomes consistent with the observed value of $b_{pattern}$ and the form of $f(N)$. So if economy of explanation is at least a starting point, then these models commend themselves for further exploration.

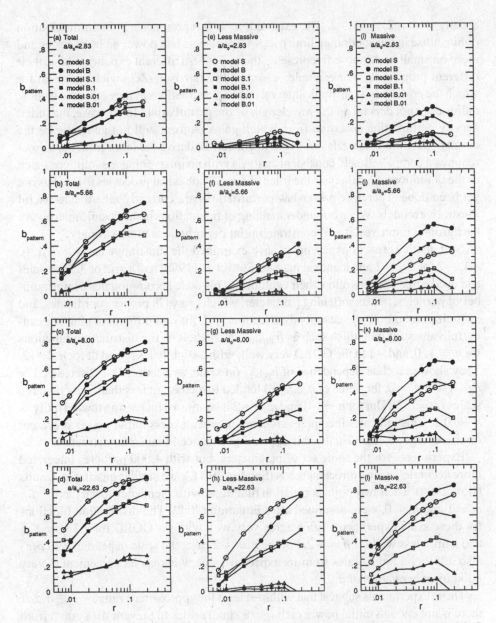

Fig. 31.13. Scale dependence of b_{pattern} at four expansion epochs as in Fig. 31.11 for Models S and B of Table 31.1 except that here results are compared for three values of $\Omega_0 = 1, 0.1$, and 0.01, denoted S, S.1, S.01 etc. (From Itoh et al., 1993.)

31.4   Non-Poisson Initial Conditions

The next level of exploration is to drop the initial Poisson distribution. Simplicity was its original motivation. What shall replace it? For lack of anything better, people have often replaced it by a power-law spectrum of perturbations (14.35) with $\langle |\delta_k|^2 \rangle \propto k^n$, usually with $-2 \lesssim n \lesssim 2$. The exponent $n = 0$ represents a Poisson distribution (white noise in the continuum limit); $n > 0$ represents less power on large scales and more on small scales; $n < 0$ represents the opposite. Different exponents have their different proponents. For example, $n = 1$ is the Harrison–Zeldovich spectrum for which the power per logarithmic interval of k in perturbations as they enter the Hubble radius does not depend on the wavelength of the perturbation. In any case, the initial density perturbation spectrum from which galaxies form will not generally be the density spectrum when galaxies begin to cluster. To derive the latter from the former requires a completely self-consistent analysis with no intervening assumptions (such as the definition of a galaxy or the roles of specific physical processes). This has not yet been done. Therefore power-law perturbations are currently regarded as useful illustrative models. When our understanding of fluctuations in the cosmic microwave background improves, it will constrain initial perturbations more strongly.

Systematic series of experiments have examined the illustrative cases $n = 1, 0, -1, -2$ for $\Omega_0 = 1$ and identical masses. Suto et al. (1990) had a grid of 32^3 particles and used a tree code to follow their evolution. This code accommodates a large number of particles, but its softening parameter, which grows in proper coordinates, and its multipole smoothing, described in Section 19.2, make it difficult to obtain meaningful values for quantities such as b_{physical}. Nonetheless, the distribution functions for $n = 1, 0$, and -1 fit the GQED very well, with a slightly less good fit for $n = -2$. They showed a clear dependence of b_{pattern} on scale, as Section 31.2 discusses. For $n = 1$ and $n = 0$, the value of $b_{\text{pattern}}(r)$ tended to a constant less than unity on large scales ($r = 0.2$). But for $n = -1$ and $n = -2$ it continued increasing toward unity as $r \to 0.2$. This indicates incomplete relaxation when excess initial power is present on these large scales. Visually this appears as enhanced filamentary structure.

Experiments for the same set of parameters, but with 4,000 particles integrated more accurately using direct methods (Section 19.1), gave similar qualitative results (Itoh, 1990). Their two-point correlation functions, with a tendency to vary as r^{-2} for $n = 1$ and $n = 0$, were described after Equation (20.2). The distribution functions for these same experiments again agree very well with the GQED for $-1 \lesssim n \lesssim 1$ and a little less well for $n = -2$. For the $n = -2$ case, the scale dependence of $b(r)$ also differs on large scales from its expected value, whereas its agreement is very good for other values of n.

These experiments suggest that within at least the approximate range $-1 \lesssim n \lesssim 1$ there is not enough initial power on large or small scales to prevent the system from being attracted to the GQED as it relaxes. It is easy to imagine that for sufficiently negative values of n, for spikey power spectra, or for other forms of initial large-scale coherence such as cosmic strings, the distribution would not be attracted to a GQED on all scales. To determine which initial distributions would not evolve toward a GQED would require either a large and varied library of simulations or a general theory of the basins of attraction to the GQED. Neither of these has yet been developed.

There are, however, specific experiments that clearly depart from a GQED and might indicate more general classes of such models. One is a case where 10^4 particles of identical mass in an $\Omega_0 = 1$ universe were first placed on the vertices of a cubic lattice. Then they were artificially perturbed by a superposition of small-amplitude plane waves with random phases and orientations, and with the cube of the wavenumber distributed uniformly over the range 0.6–1.0. This produced pronounced initial filamentary structure throughout the distribution, which was then evolved by direct integration of the orbits (Dekel and Aarseth, 1984). The filaments became stronger and eventually fragmented, especially at their intersections, into clusters of various shapes. But filaments persisted until at least the "present," when the experiment stopped.

Analysis (Saslaw, 1985b) of this and related experiments showed that its two-point correlation relaxed to the usual power law. Thus ξ_2 lost nearly all memory of its initial conditions, except perhaps on scales $\gtrsim 0.25$ where its amplitude was smaller than for a similarly evolved initial Poisson case. By contrast, the $f(N)$ distribution for this evolved perturbed lattice case departed noticeably from the GQED. On scales $\gtrsim 0.1$ it had many more volumes with a small number of galaxies (points) than the best fitting GQED. It also had somewhat fewer volumes with an intermediate number of galaxies. Moreover, its scale dependence of b was different, having lower values on small scales than in the evolved Poisson case.

The moral of this experiment is that although filamentary structure may be a good thing insofar as the observations imply it, models can easily produce too much. Analysis of distribution functions over a range of scales provides a quantitative check of filamentary structure.

31.5   Models with Dark Matter

All these have never yet been seen –
But Scientists who ought to know,
Assure us that they must be so . . .
Oh! let us never, never doubt
What nobody is sure about!

Belloc

Models with dark matter have many more degrees of freedom than those we have considered so far. If most of the dark matter is in or closely around each galaxy while it clusters, then point masses can still approximate the clustering. Such dark matter may change the timescale of clustering through its effects on the global expansion and on the graininess of the local gravitational field, but it will have little effect on the statistical form of clustering. At the other extreme, a nearly uniform distribution of radiation or hot particles that mainly contribute to the mean field will also leave the statistical distribution of galaxies little changed from that of point masses. It is the intermediate case with substantial dark matter irregularly placed between the clustering galaxies that can strongly affect their distribution.

Unfortunately nobody knows where the dark matter was when the galaxies clustered. We do not even know where most of it is now, what it consists of, or how much there is. The only evidence for it outside galaxies is in clusters, partly from X-ray emission, partly from gravitational lensing, and mainly from peculiar velocities. The sum total of dark matter observed so far suggests that $0.1 \lesssim \Omega_0 \lesssim 0.3$.

Although we may not know which models of dark matter best describe our Universe, galaxy distribution functions provide useful insights into a model's applicability and dynamics. This is best done for simulations of the same style, to minimize differences caused by smoothing, resolution, boundaries, and other techniques of computation. To maximize physical differences, we consider an analysis (Bouchet and Hernquist, 1992) that contrasts two models that were popular in the late 1980s and early 1990s: cold dark matter and hot dark matter. Their popularity waned as subsequent models had to introduce more and more ad hoc complications such as bias, various initial perturbation spectra, arbitrary rules to decide where galaxies form, and hybrid combinations of both types of dark matter with different distributions. How much closer all this brings us to a true understanding of observations is still widely debated.

The original cold dark matter models were based on the view that inflation in the early universe required an initial power-law perturbation spectrum with $n = 1$. (Subsequently, the number of different inflationary models has itself inflated, and other initial perturbation spectra are possible.) As the universe evolved, the interplay of particle processes, radiation, transition to a matter-dominated universe, decoupling between various forms of matter and radiation, and the expansion of the universe would modify this initial perturbation spectrum. On large scales where the initial perturbation amplitudes were very small and evolved slowly, they would retain their $n = 1$ form. On small scales where amplitudes were larger and evolution faster, the spectrum would become approximately $n = -3$, provided it still remained in the linear regime. On intermediate scales, its form could be calculated for the physical processes in the model and then fit to a simple algebraic function, which varied smoothly between these limits (see, e.g., Peebles, 1993 for details). This would then set the initial perturbation spectrum for galaxy formation and subsequent clustering in such cold dark matter models.

The hot dark matter models were a different story. They are gravitationally dominated by weakly interacting particles such as massive neutrinos whose peculiar velocities are relativistic when their mass density exceeds that of the nonrelativistic component (such as baryons). When the mass densities become comparable, the relative motions of the hot matter smooth out earlier perturbations on scales less than the Hubble length. This leaves long wavelengths $\lambda \gtrsim 10 \, \Omega_0^{-1} h^{-2}$ Mpc dominant. Consequently, cluster size inhomogeneities form first and then fragment into galaxies. Later variants included more massive dark matter particles, which decayed into radiation whose energy faded away in the adiabatic expansion of the universe, and cosmic strings. These modifications helped assist galaxy formation on smaller scales.

The original hot and cold dark matter models interest us here because they illustrate galaxy clustering for initial power spectra that do not follow a simple power law. Their evolution was followed and analyzed by Bouchet and Hernquist (1992) using a hierarchical tree code (see Section 19.2) that included the monopole and dipole gravitational contributions of the more distant cells. They also evolved an $n = 0$ white noise power spectrum similarly for comparison. All cases had $\Omega_0 = 1$.

After ten initial expansion factors, the spatial distribution function of the white noise model agreed very well with the gravitational quasi-equilibrium distribution, using (31.3) for b_V, over a range of about four orders of magnitude in $f(N)$ and 2–3 orders in N. For large values of N, there was some discrepancy. However, since there was only one example for each of these simulations, this discrepancy could have been caused by the finite size of the simulation and by inhomogeneities that form over the scale of the system, as mentioned in Section 31.1.4 also for the direct integration experiments. For good statistics on the low probability tail, one needs a large ensemble of simulations.

The cold dark matter simulation, again after ten expansions, agreed very well on large scales but poorly on small scales. For small cells, the probability of finding an intermediate number of galaxies was less than the GQED prediction. This is consistent with the result in Section 31.4 that initial power spectra with $-1 \lesssim n \lesssim 1$ are attracted toward the GQED. The $n \approx -3$ spectrum for cold dark matter on small scales would therefore not be expected to evolve toward a GQED, and it doesn't. It has too much anticorrelation on these small scales, reminiscent of the lattice model described earlier.

True to form, the corresponding hot dark matter simulation, which has even less initial power on small scales, does not agree at all with the GQED on these scales. The sense of the disagreement is the same as in the cold dark matter case: Too few simulated cells are populated with intermediate numbers of galaxies. However, on large scales where the initial spectrum satisfies $-1 \lesssim n \lesssim 1$, the agreement is much better.

These results, along with those of Section 31.4, indicate that in relatively simple models, on scales dominated by initial power spectra with $-1 \lesssim n \lesssim 1$, the relaxed spatial distribution of particles can be approximated very well by the gravitational quasi-equilibrium distribution. Naturally we would expect more complicated classes of models with multiple components, completely different types of initial perturbations, unstable matter, biases between galaxies and dark matter clumps, etc. to have different combinations of conditions (or perhaps none at all) that lead to the GQED. This has not been systematically explored because it takes too much computer time to simulate complicated models over wide ranges of possible parameters.

In addition to spatial distributions, simulations can test theoretical velocity distributions for models, and we turn to them next, as we look further along the path toward that ultimate arbiter of models – the observations.

32

Velocity Distribution Functions

> If you ask the special function
> Of our never-ceasing motion
> We reply, without compunction
> That we haven't any notion.
> Gilbert and Sullivan

To the denizens of Iolanthe, we can only say with wonder that their motions gravitational are very much more rational, and easier to understand. Not individually, but statistically. To test the special function (29.4) for many-body motions in the context of cosmology we return to simulations in Sections 31.1–31.4.

Figures 15.5 and 15.6 in Section 15.3 illustrated some results of these simulations, which we now examine in more systematic detail. As in Section 31, we start with the simplest $\Omega_0 = 1$ case and identical masses. Then we explore the effects of smaller Ω_0 and of components with different masses. Unlike the spatial distributions, velocity distribution functions are just beginning to be computed for experiments with dark matter and non-Poisson initial conditions. This is partly because the definition of which particles constitute a galaxy is still unsettled for such cases and partly because observations of $f(v)$ for representative samples are just starting to be analyzed. Both these situations should improve. Then velocity distribution functions will become very valuable because they are more sensitive than the spatial distributions to some of the basic cosmological parameters.

Figure 32.1 shows the velocity distribution functions at four different expansion factors for the 4,000-body, $\Omega_0 = 1$, initially cold Poisson simulations with particles of identical masses. These are the same cases whose spatial distributions are shown in Figures 31.1 and 20.1. The solid histograms are the experimental distributions. The dashed lines through the histograms, barely distinguishable from them, are the best least-squares fits of the gravitational quasi-equilibrium velocity distribution (29.4), and their parameters are given in the graphs. Maxwell–Boltzmann distributions with the same velocity dispersion are shown to the right of each histogram; they don't fit at all.

The fits to the GQED are remarkably good. Their fitted velocity dispersions, β, are very close to their directly measured values, $\langle v^2 \rangle$; their fitted values of b are close to their values from the spatial distribution function in large cells, and the average values of $\alpha\beta$ agree with the theoretical relation (29.2). We expect the value of b for the velocity distribution to equal the peak value on large scales from the spatial distribution since the velocity distribution is determined by interactions over the entire scale of correlations.

For similar models with $\Omega_0 = 0.1$, the fits remain very good. Since clustering does not develop as rapidly, their velocity dispersions are smaller than in the $\Omega_0 = 1$

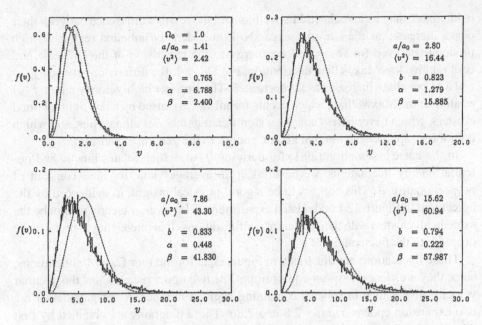

Fig. 32.1. Peculiar velocities from simulations averaged over four $\Omega_0 = 1$ experiments with cold Poisson initial conditions for $N = 4,000$ at different epochs a/a_0. Solid histograms are the experimental distribution and the dashed lines through them are the best fits of Equation (29.4) with the parameters listed. The dashed line to the right of each histogram is a comparison Maxwell–Boltzmann distribution with the same velocity dispersion $\langle v^2 \rangle$ as the simulation. (From Saslaw et al., 1990.)

case. At expansions $a/a_0 = 2.8$, 7.86, 15.62, and 31, the values of $\langle v^2 \rangle$ are 10.4, 18.2, 20.2, and 21.9. Not only are these values lower than in Figure 32.1, but their growth is slower. Moreover, the best-fit value of b for $f(v)$ is 0.86 at $a/a_0 = 7.86$ and it only grows to 0.89 at $a/a_0 = 31$. These larger values are consistent with the larger values of b_{physical} for $\Omega_0 = 0.1$, as in Figure 31.12.

In the similar experiments with very small $\Omega_0 = 0.01$, the fits of the form of the GQED to $f(v)$ are again very good, but now the values of β and $\langle v^2 \rangle$ become discordant, and $\alpha\beta$ differs greatly from $\langle 1/r \rangle$ in (29.2). This system has not achieved full quasi-equilibrium, as we expect from the freezing of the spatial clustering pattern at high redshift and the subsequent adiabatic cooling of velocities discussed in Chapter 30. Here the velocity dispersions are much smaller: $\langle v^2 \rangle = 2.3$, 2.9, and 2.6, and $\beta = 2.1, 2.2$, and 2.2 when $a/a_0 = 2.8$, 7.86, and 15.62. Therefore measurements of $\langle v^2 \rangle$ or β in the nonlinear regime are sensitive indicators of Ω_0. This is because for low Ω there are many more loosely clustered and field galaxies, and their peculiar velocities, never very high, decay rapidly in the adiabatic expansion.

The full velocity distribution function contains a great deal of information about all the orbits of galaxies in all their stages of clustering. Comparison with the Maxwellian in Figure 32.1 shows that the gravitational distribution always has a colder peak than its associated Maxwellian and the relative difference between their peaks increases as the Universe expands. Lower density unbound regions, which predominate even for $\Omega_0 = 1$, lose energy to the expansion of the Universe and cool relative to a Maxwellian distribution. For $\Omega_0 < 1$, the difference between $f(v)$ and a Maxwellian increases as Ω_0 decreases. The stronger high-velocity tail of $f(v)$ relative to the Maxwellian, which occurs for all Ω_0, is caused by a few tightly bound clusters, which nevertheless contain a significant number of all galaxies, and which do not lose appreciable internal kinetic energy to the general expansion.

In the basic assumption (29.1) for deriving $f(v)$, which relates kinetic and potential energy fluctuations, we have taken an average value for their constant of proportionality, α. This appears to be a good physical insight, as evidenced by the agreement in Figure 32.1 and related experiments. However, α clearly cannot be the same in all clusters with their different configurations. It is interesting to probe more deeply into the fluctuations of α.

These fluctuations are illustrated in Figure 32.2 for the four $\Omega_0 = 1$ simulations, since they are closest to quasi-equilibrium. The two upper panels show the peculiar velocity dispersions in spheres containing a given number, N, of galaxies at the two expansion epochs $a/a_0 = 2.8$ and 7.86. These diagrams are obtained by first choosing 25 random spatial points within a radius $R = 0.6$ of the center of each simulation. Then we find the N nearest galaxies to each random point and calculate

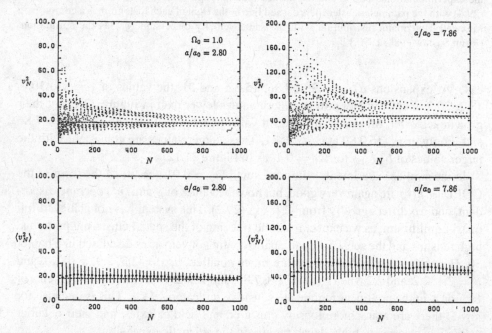

Fig. 32.2. Velocity dispersions, v_N^2, in spheres with a given number of galaxies, as described in the text. The horizontal dashed line is the average velocity dispersion of all galaxies within the normalized comoving radius $R = 0.8$. The lower two panels give the average and standard deviation for all spheres of a given N. (From Saslaw et al., 1990.)

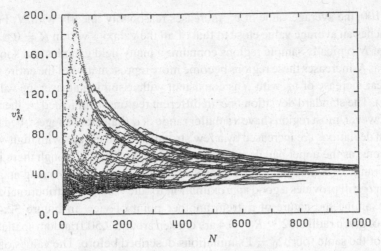

Fig. 32.3. A high-resolution version of one of the simulations in the upper right-hand panel of Fig. 32.2. It shows v_N^2 for spheres of different N around 100 points randomly selected throughout the entire $\Omega_0 = 1$ simulation. Each value of N is plotted. All 4,000 particles are used. (From Saslaw et al., 1990.)

the velocity dispersion of their peculiar motions. This is plotted for every twentieth value of N. Results for each of the four simulations start at values of $N = 20, 25, 30$, and 35, respectively, for clarity. Thus for each value of N in the upper panels, every point represents the velocity dispersion in a sample sphere containing N galaxies. The dashed horizontal line shows the average velocity dispersion within $R = 0.8$ for the entire sample.

Initially the galaxies move just with the Hubble expansion and $v_N^2 = 0$ for all N. The velocity dispersion builds up very rapidly, particularly for small groups, which form quickly. Large values of N tend to average over the velocities of many weakly clustered galaxies, thus giving lower and more representative values of v_N^2. At later times, interesting network patterns develop in the individual diagrams. They portray the correlations between velocity dispersion and spatial clustering. Figure 32.3 shows an example at higher resolution for a single simulation. The thick streaks whose v_N^2 decreases with increasing N arise from the random spatial points that happen to fall near the centers of strongly bound clusters where the velocity dispersion is large. It decreases toward the average with increasing distance from the cluster as more loosely bound and field galaxies are counted. The thin chains whose v_N^2 increases with increasing N arise from spatial points outside the cores of strong clusters, but whose spatial sphere begins to encompass a core as N increases. The points at small N with very low v_N^2 are those that happen to be in underdense regions. So this provides a rather nice pictorial representation of both position and velocity clustering, which could eventually be compared with observations.

Returning to Figure 32.2, its two lower panels give the average and the standard deviation of v_N^2 for each value of N in the upper panels. For values of N between about

20 and 100, the average value of v_N^2 increases reasonably linearly with N. For large N it reaches an average value close to that of all the galaxies within $R = 0.8$. Larger values of N typically sample regions containing many field galaxies and some clusters. So as N increases these regions become more representative of the entire system. The linear increase of v_N^2 with N is consistent with assuming an average value of α in (29.1). The standard deviation of α in different regions is indicated by the vertical bars. However, most regions have a smaller range of α than these suggest, because the standard deviations are increased by a few strongly bound groups with high-velocity dispersions, as the upper panels and Figure 32.3 show. Thus, although there is a significant range of α in regions having the same number of galaxies, using an average value for α still provides a good approximation for the velocity distribution function.

We examine the nature of α from another point of view in Figure 32.4. Here spheres of given radii $0.02 \leq R \leq 0.4$ are chosen around 9,500 random spatial points in each of the same four $\Omega_0 = 1$ simulations described before. The values of N and the velocity dispersions $\langle v_R^2 \rangle$ for each of these spheres give the overall averages and standard deviations for $\langle v_R^2 \rangle$. As the system evolves, there is an increasing absolute dispersion around the average value of $\langle v_R^2 \rangle$, especially for small values of R. Despite this scatter, one sees a fairly linear relation between $\langle v_R^2 \rangle$ and N, again most clearly for small values of R and N. For $R \geq 0.3$, this linear relation has a break at larger values of N. Moreover, the slope, which is the value of α, differs for the different

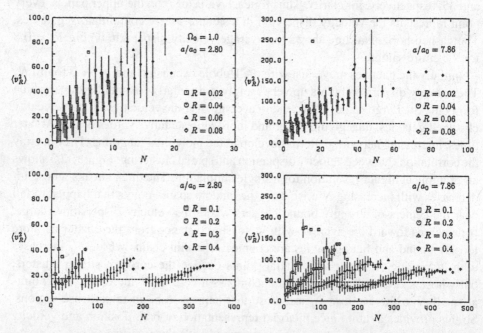

Fig. 32.4. Velocity dispersion in spheres of given radius, averaged over four $\Omega_0 = 1$ simulations shown at expansion epochs of 2.8 and 7.86. Radii R of spheres range from 0.02 to 0.4 as a fraction of the total radius of the simulation, normalized to $R = 1$. (From Saslaw et al., 1990.)

radii. Therefore the average value of α determined by (29.2) represents a weighted average over these different radii. Eventually it should become possible to obtain comparable diagrams for large observational samples to help determine their detailed degree of relaxation and estimate Ω_0.

For more than one mass component, Figure 15.5 already showed that if 500 particles each have five times the mass of the other 3,500, then the GQED describes $f(v)$ for both components. This remains true even for systems with mass ratios as high as 100:1. Recall from the last chapter that these high mass ratio simulations were not described well by $f(N)$ because they tended to form satellite systems. However, the motions of both components separately, as well as their combination, continue to satisfy (29.4) for $f(v)$.

The evolution of these velocity distributions when $N = 4,000$ and $\Omega_0 = 1$ have been followed systematically for a range of mass ratios $m_2:m_1 = 1:1, 2:1, 3:1,$ 4:1, 5:1, 8:1, 15:1, 30:1, and 100:1, leading to several conclusions (Inagaki et al., 1992).

First, there are systematic trends in the values of $b, \alpha,$ and β as the mass ratio $m_2:m_1$ of the two components increases. Simulations with larger mass ratios generally have larger values of b, larger values of β, and smaller values of α. These differences are small, at about the 10% level, for b but for α and β they may range over nearly an order of magnitude. Thus the velocity dispersion, which depends mostly on the relative graininess of the gravitational field, is particularly sensitive to the mass ratio.

Second, as these systems evolve, the values of b and β initially increase faster and reach asymptotic values that are greater for higher mass ratios. Values of α decrease faster for higher mass ratios. These results indicate that higher mass ratios promote formation of more compact clusters.

Third, the spread of values in $b, \alpha,$ and β produced by different mass ratios is about the same whether one examines the distributions of all the particles or those for just the more massive or the less massive components.

Fourth, time fluctuations in the values of $b, \alpha,$ and β are greater for the distributions of massive components than for either the less massive or the total distributions. This is mostly a reflection of the smaller number (500) of galaxies in the more massive sample.

Fifth, although the values of β increase rapidly as the system evolves, and the values of α decrease rapidly, their product $\alpha\beta$ tends toward the same value after about $a/a_0 \approx 10$ in all cases. Initial departures from this value for the total sample are greatest for larger ratios of m_2/m_1, but these cases relax rapidly.

Sixth, even though the value of β increases by more than a factor of ten during the evolution, the value of $\beta/\langle v^2 \rangle$ always remains close to unity. Its variance among different simulations tends to decrease slightly with time. Thus β, which is the fitted value, is nearly equal to $\langle v^2 \rangle$, which is the value measured directly from the velocities.

Therefore the general quasi-equilibrium principles that guide the evolution of two-component systems are similar to those of one-component systems. Details

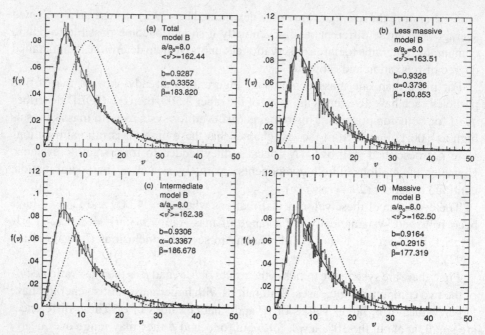

Fig. 32.5. (a) The velocity distribution function for Model B of Section 31.1.4 at expansion epoch $a/a_0 = 8$. (b) The same for the 20% least massive galaxies. (c) The same for the 60% of galaxies with intermediate masses. (d) The same for the 20% most massive galaxies. (From Itoh et al., 1993.)

naturally differ, with heavier components promoting the more rapid formation of more compact clusters.

Turning to velocity distributions for more realistic continuous mass spectra, we return to Model B. Section 31.1.4 described its spatial distribution as an illustration. The four panels of Figure 32.5 show its velocity distribution functions at $a/a_0 = 8$ for the total, 20% least massive, 60% intermediate, and 20% most massive galaxies in this 10,000 body simulation with $\Omega_0 = 1$. As usual, histograms are experimental velocity distribution functions, solid curves are the best least-square fit of the theoretical distribution (29.4), and dotted lines are the closest Maxwell–Boltzmann distribution (with the same experimental velocity dispersion $\langle v^2 \rangle$). Each panel shows the best-fit values of b, α, and β as well as $\langle v^2 \rangle$.

It is clear that the theoretical $f(v)$ from (29.4) agrees closely with the experimental distributions for continuous mass spectra, although $\langle v^2 \rangle$ is about 10% less than β. This agreement is better than for the spatial distribution $f(N)$ for cells with $r \gtrsim 0.3$, illustrated in Figure 31.7. Thus the spatial inhomogeneity of this model, shown in Figure 20.4, does not give rise to irregularity in $f(v)$. This is because the limited and centrally weighted sampling characteristic of large spatial cells exacerbates fluctuations in $f(N)$, but all galaxies are sampled for $f(v)$. This sampling may also affect the value of b, which is about 10% higher for $f(v)$

than for $f(N)$. Since the velocity dispersions of galaxies in the three different mass groups are almost equal, equipartition has not been established. Another way of expressing this is that the temperature of a component is proportional to its mass. It is another sign that collective interactions have dominated the dynamics.

Evolution of $b_{velocity}$ (the best-fit value of b in $f(v)$), α, and β for the systematically changing properties of models in Table 1 are shown in Figures 32.6 and 32.7. Results

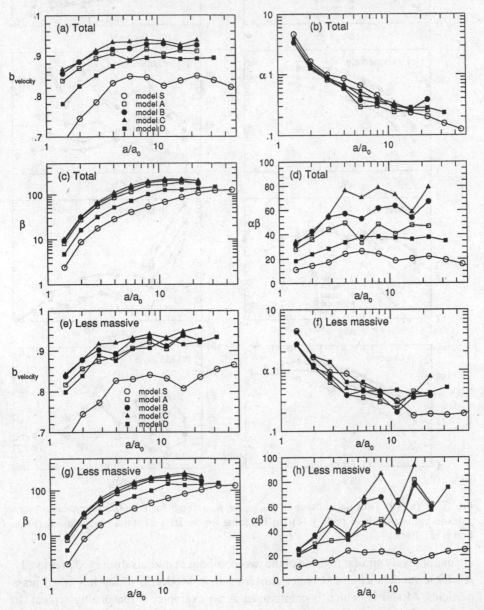

Fig. 32.6. (a)–(d) Time dependence of $b_{velocity}$, α, β, and $\alpha\beta$ for all galaxies in the models of Table 1. (e)–(h) The same for the 20% least massive galaxies. (From Itoh et al., 1993.)

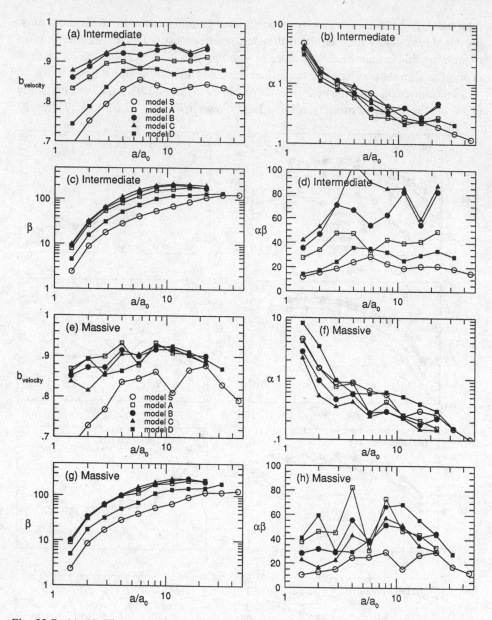

Fig. 32.7. (a)–(d) Time dependence of $b_{velocity}$, α, β, and $\alpha\beta$ for the 60% intermediate mass galaxies in the models of Table 1. (e)–(h) The same for the 20% most massive galaxies. (From Itoh et al., 1993.)

are qualitatively similar to those of the two-component models discussed previously. All these models exhibit stronger clustering than Model S, which has equal mass particles. Model B, which is emphasized as an example, is reasonably typical. In it, the 20% most massive galaxies contain about 75% of the total mass. Massive galaxies cluster by their own self-gravity and form the satellite systems visible in Figure 20.4. For models in which massive galaxies contribute less to the total mass, their self-gravity is not as strong. Therefore the satellite systems cannot grow as rapidly and deviations from the gravitational quasi-equilibrium distribution are smaller. The main property on which the features of clustering depend is the total mass fraction of massive components, rather than the value of p in the mass spectrum (31.1).

Regarding the dependence on Ω_0, the reduced clustering in models with lower Ω_0 also reduces the effects of different mass spectra. Massive galaxies affect clustering properties less in lower-density universes. There is significant interplay between the mass spectrum and the cosmology in producing details of galaxy clustering. Basic features of clustering in cosmological many-body systems, however, are common to a wide range of conditions. How well these results apply to the clustering observed in our own Universe is the next question.

PART VI

Observations of Distribution Functions

The heavens themselves, the planets, and this centre,
Observe degree, priority, and place,
Insisture, course, proportion, season, form,
Office and custom, in all line of order

Shakespeare

Physics tries to discover the pattern of
events which controls the phenomena we observe.

Jeans

Not explanation, but prediction is our most stringent test of understanding. Many are the explanations after the fact; few are the predictions that agree with later observation. This holds especially for complex phenomena, whether of astrophysics, cosmology, economics, history, or social behavior.

When statistical thermodynamics first yielded distribution functions for cosmological many-body systems, neither simulations nor observations had been analyzed for comparison. Only Hubble's galaxy counts in the limit of large cells were known, though not to us at the time. Correlation functions then dominated studies of large-scale structure.

Old simulations, after their spatial distribution functions were analyzed, first showed that the theory was reasonable. As a result, many new simulations, sketched in Part V, began to explore the range of conditions to which the theory applies; this exploration continues.

It took a couple of years to persuade observers to analyze distribution functions for modern catalogs. Would the distributions created by the great analog computer in the sky agree with those of digital computers here on Earth? Is the pattern of the galaxies dominated by simple gravitational clustering? The answer is still a continuing saga. Although observed spatial distribution functions are now moderately well studied, velocity distributions are just beginning to be determined, and predictions of past evolution remain to be observed in our future.

33

Observed Spatial Distribution Functions

He had bought a large map representing the sea
Without the least vestige of land:
And the crew were much pleased when they found it to be
A map they could all understand.

Lewis Carroll

33.1 Basic Questions

Early analyses (e.g., Gregory and Thompson, 1978; Kirshner et al., 1981; Fairall et al., 1990; Fairall, 1998) of modern magnitude-limited galaxy catalogs revealed large empty regions, originally called voids. These regions filled in somewhat as more sensitive surveys found fainter galaxies (e.g., Kirshner et al., 1987), but they remained underpopulated. Cosmological many-body simulations gave a probability $f_0(V)$ for finding such empty regions (Aarseth and Saslaw, 1982), even before their theoretical distribution function was calculated. Soon it became clear that these voids and underdense regions were part of a more general distribution function description.

In retrospect, we may regard many analyses of observed spatial distribution functions as attempts to answer several basic questions. Although these questions are not yet fully answered, and relations among them are not always apparent, we will use them to guide our discussion here.

1. Is the observed form of $f(N, V)$ generally consistent with gravitational quasi-equilibrium clustering? If so, does it rule out other possibilities such as particular dark matter distributions or initial conditions?
2. Do the two-dimensional distribution functions for projections onto the sky give a good estimate of $f(N, V)$ or do we need the full three-dimensional distribution function?
3. How does $b(r)$ depend on spatial scale? Can this dependence restrict possible models significantly?
4. What is the magnitude scaling of the distribution function as the limiting magnitude of the sample changes? Is it consistent with random selection independent of environment?
5. Does the form of $f(N, V)$ within individual clusters or superclusters agree with results for larger more statistically homogeneous samples? What is the size of a fair sample?
6. Do galaxies of different mass, morphology, and luminosity have significantly different distribution functions?
7. What are the distribution functions for radio galaxies and for clusters of galaxies?

33.2   Catalogs: A Brief Sketch

The ideal catalog for answering these seven basic questions does not exist, and perhaps it never will. Such a catalog would be complete, containing all the dynamically important objects of every type in a volume large enough to be statistically homogeneous and to represent even larger volumes. Peculiar velocities of all the objects would be known with high precision, to determine distances as well as velocity distribution functions. Morphological classifications, closely related to underlying physical properties rather than just to appearance in the visible spectral band, would be determined. Luminosities, diameters, masses, and starburst features of individual galaxies would all be observed. Bound companions and subgroups would be noted. Finally, the dark matter would be located. And all these abundant observations would reach back to high redshifts.

Fortunately the real catalogs we deal with at present have already revealed many basic properties of galaxy clustering. Future catalogs will expand and refine these results; there is still time for robust predictions.

Here I summarize the essential features of some current catalogs and mention some catalogs in the making, which are relevant to analyses of distribution functions. References describe other important details of catalog construction.

The Zwicky catalog (Zwicky et al., 1961) contains the positions and magnitudes for 28,891 galaxies found on the Palomar Sky Survey plates. The catalog is limited by apparent magnitude $m_{zw} = 15.5$ and is reasonably complete to $m_{zw} = 14.5$. It covers a large representative part of the sky. Originally it contained a small number of redshifts for scattered bright galaxies. Subsequently it spawned many more complete redshift surveys (e.g., the CfA survey of Geller and Huchra, 1989) over limited areas but increasing depth. From the same Palomar Sky Survey plates, Abell derived his catalog of clusters described in Chapter 7.

The UGC and ESO catalogs are not selected primarily by magnitude, but by angular diameter. Their galaxies have angular major diameters $\geq 1'$. The UGC (Nilson, 1973) covers the declination range $\delta > -2°.5$ while ESO (Lauberts, 1982) covers $\delta < -17°.5$. Although both claim to be complete to $1'$, their diameter systems are different, and actually ESO is ~15% deeper (e.g., Lahav et al., 1988). It has also been shown (e.g., Hudson and Lynden-Bell, 1991) that they both suffer incompleteness at small diameters. However, these selection effects are not serious for determining distribution functions, provided they are isotropic over the sky, since we are comparing projected cells with equal volumes. To avoid systematic effects from galactic obscuration, it is best to restrict the analysis to galactic latitudes $|b^{II}| \geq 30°$. This leaves 8,757 galaxies in the UGC, averaging about 0.8 galaxies per square degree, and 7,906 galaxies in ESO (excluding members of our local group), averaging about 1.2 galaxies per square degree.

An advantage of the UGC and ESO catalogs is that their galaxies are classified by morphological type. This is not straightforward since the classification of galaxies by eye is clearly subjective and becomes less certain for fainter members. Moreover,

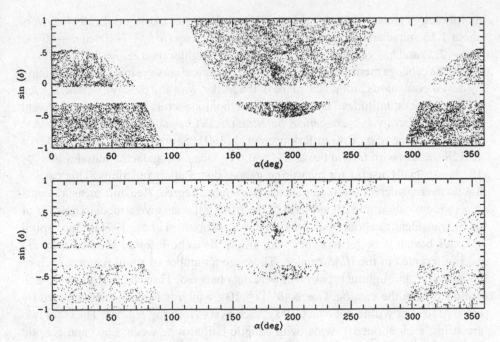

Fig. 33.1. (a) The distribution of all UGC and ESO galaxies with diameters larger than $1'$, projected in equatorial coordinates for Galactic latitudes $|b^{II}| = 30°$. The empty strip ($-17°5 < \delta < -2°5$) is not surveyed by these catalogs. (b) The distribution of E + S0 galaxies in the UGC and ESO catalogs. (From Lahav & Saslaw, 1992.)

in diameter-limited catalogs, the more compact elliptical galaxies may be underrepresented relative to large low surface brightness spirals and irregulars. Therefore it is usually best not to split morphological samples too finely in such catalogs. In fact (Lahav and Saslaw, 1992) a coarse division combining E and S0 type galaxies is consistent for both catalogs, containing 16% of UGC and 18% of ESO. Similarly, combining Sa, Sb, and Sb/c gives a group with 26% of UGC and 33% of ESO, A third less well-defined group combining Sc, "S," irregular, and dwarf spirals has 42% of UGC and 37% of ESO. So these make three reasonably comparable groups, though they do not include all types.

Figure 33.1a shows these UGC and ESO galaxies with diameters larger than $1'$, in equatorial coordinates for galactic latitudes $|b| = |b^{II}| \geq 30°$. The empty strip ($-17°5 < \delta < -2°5$) is not covered by these catalogs. Figure 33.1b shows the distribution of E + S0 galaxies in these catalogs. Visual inspection shows a number of filaments and voids with an overall impression of reasonable statistical homogeneity.

The Southern Sky Redshift Surveys, SSRS and SSRS2, systematically sample galaxy redshifts over smaller areas of the sky. The SSRS (da Costa et al., 1988) draws from the UGC and ESO catalogs for $b^{II} < -30°$ and $\delta < -17°5$ and contains redshifts for 1,657 galaxies with angular diameters (whose definitions depend on morphology) greater than 1.1 arcmin. The SSRS2 (da Costa et al., 1994) covers about 1.13 steradians around the southern galactic cap ($b^{II} < -40°$ and $-40° < \delta < -2°5$) and has velocities for 3,592 galaxies brighter than $m_B = 15.5$.

All the catalogs mentioned so far are based on optical surveys in which obscuration by dust is continuous around the plane of the Milky Way and patchy elsewhere, even at high galactic latitudes. To obtain a more homogeneous catalog for a different sample, the *Infrared Astronomical Satellite* (*IRAS*) has surveyed most of the sky. An *IRAS* catalog (e.g., Rowan-Robinson et al., 1991; Saunders et al., 1990, 1992) of 13,128 sources with 60 μm fluxes $S_{60} \geq 0.6$ Jy located at galactic latitudes $|b^{II}| \geq 10°$ is especially useful for measuring galaxy distribution functions. This catalog has an average density of 0.43 galaxies per square degree. Redshift measurements of a random subsample of one sixth of the *IRAS* galaxies give a median redshift of 0.03, equivalent to about 90 h^{-1} Mpc (Rowan-Robinson et al., 1991). Since spiral galaxies have a stronger tendency than ellipticals to be infrared sources, they are overrepresented in the *IRAS* catalog. There are a number of small regions, each of about 1 deg^2, throughout the sky that were not observed. Three large regions are also excluded from the catalog. One is $|b^{II}| \leq 10°$, which is heavily contaminated by infrared sources within our own Galaxy, such as star-forming regions. The other two are strips, each about 10° wide, with ecliptic latitudes between ±60° and ecliptic longitudes from 150–160° and 330–340°. These could not be observed because the satellite ran out of coolant.

Figure 33.2 shows this *IRAS* catalog as well as the unobserved regions. It uses an equal area projection with polar coordinates $r\cdot = 90 - |b|$ and $\theta = l$. When l_i and b_i are the galactic coordinates of the ith galaxy, the Cartesian coordinates using this projection are $x_i = f(b_i) \sin l_i$ and $y_i = f(b_i) \cos l_i$ with $f(b_i) = b_i - 90$ for galaxies in the northern hemisphere and $f(b_i) = b_i + 90$ for galaxies in the southern hemisphere. The outer circles drawn on the maps of excluded regions show $b = ±20°$ in the northern and southern hemispheres; the inner circles show $b = ±40°$. For $b \geq 40°$ in the north, nearly all the excluded regions were caused by the coolant problem.

Apart from the excluded regions, the *IRAS* distribution looks fairly homogeneous to the eye. The main exception is a long dense filament in the north. This is the supergalactic plane. Can you find it? The distribution function can.

The largest current modern catalog used the Automatic Plate Measuring (APM) machine in Cambridge, U.K. to scan 185 UK Schmidt survey plates. This APM catalog contains over two million galaxies brighter than magnitude $B_j = 20.5$, about twice as many as the Lick catalog described in Chapter 6. Moreover its information – positions, magnitudes, image profiles, surface brightness, and position angles – is more precise than for the Lick catalog. Among other results, the APM catalog provided improved measures of the angular two-particle correlation function, especially on large angular scales (e.g., Maddox et al., 1996, shown in Figure 14.1).

All the examples mentioned previously aim to catalog large representative volumes or their projections on the sky. There are also many catalogs of smaller regions,

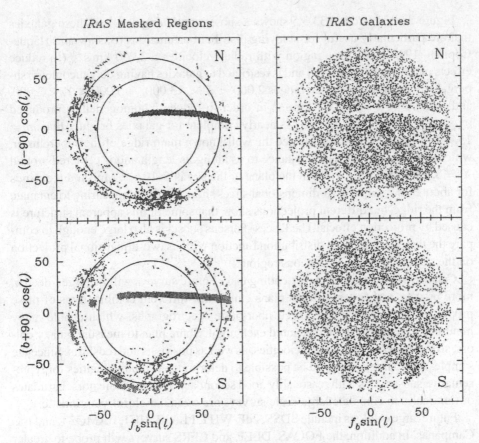

Fig. 33.2. Equal area projection showing the distribution of excluded regions (left-hand side) and *IRAS* galaxies (right-hand side) in the northern (top) and southern (bottom) Galactic hemispheres. On the left, the inner circles show $b^{II} = \pm 40°$ and the outer circles show $b^{II} = \pm 20°$. (From Sheth, Mo & Saslaw, 1994.)

interesting for their own sake as well as for their contrast with statistical homogeneity. The Pisces–Perseus region has been especially well studied for more than a decade (e.g., Giovanelli and Haynes, 1993 and earlier references therein). It is reasonably nearby ($z \approx 0.02$) and covers an area extending 6^h on the sky ($0^h < \alpha < 4^h$ and $22^h < \alpha < 24^h$) in the declination band $0 < \delta < +50$. The catalog provides equatorial coordinates, galaxy major and minor diameters, and magnitudes (many from the Zwicky catalog); but its major feature is $\sim 10^4$ redshifts, mostly measured from the 21 cm line of neutral hydrogen in the spiral galaxies. The catalog contains galaxies to apparent blue magnitude $+19$ and is essentially complete to $m = 15.5$ and angular size greater than 0.8 arcmin. The goal is to complete it to $m = 15.7$ and sizes >1.0 arcmin using both radio and optical redshifts. The dominant structure in this region is the Pisces–Perseus supercluster. It also contains several rich Abell clusters such as A262, A347, and A426.

Figure 33.8 in Section 33.3.5 shows a subset containing 4,501 of these galaxies that are particularly useful for studying distribution functions (Saslaw and Haque-Copilah, 1998). It is for a region with radial velocities $>2,000$ km s^{-1} (to reduce effects of bulk flow velocities) and is restricted to galaxies having de Vaucouleurs supergalactic coordinates in the ranges $2,000 < X < 13,000$, $-5,000 < Y < 2,000$, and $-7,000 < Z < 7,000$ km s^{-1} in order to reduce inhomogeneities produced by sparse sampling. This subset is nearly complete for galaxies brighter than $m = 15.7$. In Figure 33.8 we readily see the well-known main ridge of the supercluster, which is roughly at the same distance to us along its length and is centered around $X = 5,000$ km s^{-1}. Its width in the plane of the sky is 5–$10 h^{-1}$ Mpc and it extends for about $50 h^{-1}$ Mpc. Two-dimensional slices show filaments appearing to emanate from the ridge, but different projections show that some of this apparent structure is caused by projection effects. The Pisces–Perseus subset is also large enough to compare the three-dimensional distribution function with its two-dimensional projection on the sky for this inhomogeneous region.

Cataloging galaxies is a never-ending task. Future surveys will be larger, deeper, more precise, and faster. Various plans emphasize different combinations of these properties. Most rely on advanced fiber-optic spectrographs, which enable many hundreds of redshifts to be measured each night. Some plan to measure galaxy colors, sizes, shapes, and internal velocities, as well as positions and redshifts. Since the samples are so large, it should be possible to determine peculiar velocities for representative subsamples with reasonably good secondary distance indicators. Estimates of timescales to complete large-scale surveys are often delightfully optimistic.

Future large catalogs include SDSS, 2dF, WHT, FLAIR, HET, LAMOST, and Las Campanas. In addition, the FOCAS, DEEP, and CFRS surveys will probe to greater redshifts. DENIS 2 and SIRTF will catalog infrared galaxies, and AXAF will explore X-ray sources, although their redshifts will require optical identifications. Here I just mention the optical programs behind these acronyms. Summaries in Maddox and Aragón-Salamanca (1995) and other more recent progress reports contain further details.

The Sloan Digital Sky Survey (SDSS) uses a dedicated 2.5-meter telescope with a $3°$ diameter field to measure CCD color images in five bands (around 3,540Å, 4,760Å, 6,280Å, 7,690Å, and 9,250Å) and redshifts. The goal is to obtain about 10^6 galaxy and 10^5 quasar redshifts complete to about 18th magnitude at 6,280Å in a north galactic cap of π steradians. The median redshift of this galaxy sample is $z \approx 0.1$. Image catalogs several magnitudes fainter than the redshift catalog are planned for parts of both hemispheres. The larger Anglo-Australian telescope, using a 400-fiber spectrograph with a 2 degree field, known as the 2dF, has been undertaking a very large survey similar to that of the SDSS. Another such survey is being done on the 4.2-meter William Herschel Telescope (WHT) on La Palma with the 160 fibre Autofib-2 spectrograph. FLAIR-II is a smaller ~ 90 fiber spectrograph that operates with the larger 40 square degree field of the United Kingdom Schmidt Telescope in Australia. It has already obtained tens of thousands of redshifts. Plans for the segmented constant zenith angle, 8.5-meter Hobby-Eberly Telescope (HET) at McDonald Observatory in Texas include a multifiber wide-field spectrograph. Its large diameter and wide field of view and dark sky there will make it especially useful for measuring redshifts of faint galaxies and clusters fairly uniformly distributed over large fields. LAMOST is a plan by Chinese astronomers for a Large Sky Area Multi-Object Fiber Spectroscopic Telescope that has an effective aperture of four meters. It is a Schmidt telescope with a meridian mounting and a $5.2°$ field of view. The goal is to concentrate on spectra and cover about 2×10^5 square degrees of sky. The Las Campanas 2.5-meter telescope has measured over 25,000 redshifts with a median $z \approx 0.1$ and isophotal magnitudes brighter than about 17.5 for several long ($80°$) thin ($1.5°$) strips in both galactic hemispheres.

Among fainter surveys, the Deep Extragalactic Evolutionary Probe (DEEP) is using the 10-meter Keck telescope to obtain ~15,000 spectra brighter than about 24.5 B magnitude, giving a median $z \approx 0.6$ corresponding to a lookback time of several billion years (depending on Ω_0). This may be enough to begin to measure significant evolutionary effects for both clustering and internal galaxy evolution. The Canada–France Redshift Survey (CFRS) has already measured a sample of ~10^3 galaxy redshifts selected in I band with a median $z \approx 0.55$ to search especially for evolution of the galaxy luminosity function. Results show that evolution may depend on galaxy color in complicated ways, perhaps resulting from mergers and starbursts. The Faint Object Camera and Spectrograph (FOCAS) for the Japanese 8.2-meter SUBARU telescope will be able to measure spectra brighter than about $m_v = 24$ (depending on the spectral dispersion). In addition, about a half dozen smaller catalogs plan to probe to $z \approx 1$ or beyond.

All these projects, including the infrared and X-ray catalogs, are massive long-term undertakings involving huge teams. Each approach has its strengths and weaknesses, but taken together they will provide a far more precise understanding of the distribution of matter in our Universe and how it evolved.

33.3   Partial Answers to Basic Questions

Samples and catalogs described in the last section have already provided rough answers to our earlier questions and pointed the way to further refinements.

33.3.1   Gravitational Quasi-Equilibrium Distributions Agree with Observations

A couple of years after predicting the GQED (27.24), the first observational analyses (Crane and Saslaw, 1986) for both $f_N(V)$ and $f_V(N)$ projected on the sky were made using the Zwicky catalog. The observations agreed very well with the theoretical GQED giving values of b in the range $0.6 \lesssim b \lesssim 0.75$, depending somewhat on the size and less on the shape of the cells that were used. Uncertainties in the effects of galactic obscuration at different latitudes, selection properties of Zwicky galaxies, random positioning or tesselation for cell centers, and treatment of the sample boundary all contributed to the uncertainty in b. (It is usually best and simplest to just ignore all cells that cross any boundary.) Also, the dependence of b on scale was not fully taken into account. Subsequent analyses showed that all these uncertainties, while interesting, were of secondary importance. The representative values of $b = 0.70 \pm 0.05$ that were initially found turned out to fit other samples as well, when corrected for random or biased selection (Chapter 28). The UGC and ESO distribution functions in Figure 15.8 are examples of this, and will be discussed in Section 33.3.6.

An extensive two-dimensional analysis (Sheth et al., 1994) of the *IRAS* catalog, much less strongly affected by the uncertainties mentioned above, agrees excellently with the GQED. Figure 33.3 shows the void distribution $f_0(\theta)$ for galaxies in Figure 33.2. The circles are the observations and they are threaded very nicely by the solid line for the self-consistent GQED with $b(r)$ determined directly from the variance of counts in cells for the data using Equation (31.3). This is the same procedure used to examine the simulations in Figure 31.8. With no free parameters, the theory agrees with the observations as well as it does with the simulations. Similar analyses using negative binomial, compounded lognormal, or any other distributions nearly identical to the GQED in this regime will of course fit almost as well (see also Bouchet et al., 1993). However they will not generally satisfy the Second and Third

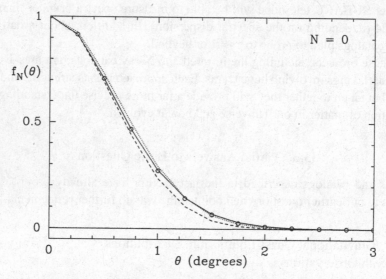

Fig. 33.3. Void distribution function as a function of angular cell radius θ for *IRAS* galaxies. The circles are observed for the region with Galactic latitude $>40°$ in the northern hemisphere, which is relatively uncontaminated by excluded regions, apart from the coolant strip, shown in Figure 33.2. The dashed line is a Poisson distribution with the same mean as the data on each scale, the dotted line shows a GQED with a single best value of $b = 0.15$ independent of scale, and the solid line shows the fully self-consistent fit of the GQED to the data, with no free parameters. (From Sheth, Mo & Saslaw, 1994.)

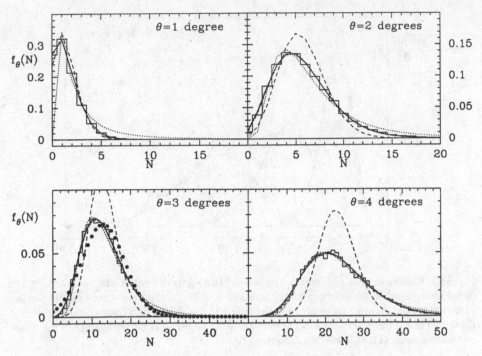

Fig. 33.4. Counts-in-cells distribution functions (histograms) of *IRAS* galaxies in the northern hemisphere with Galactic latitude >40° for cells of various angular radii. Histograms show the probability that a randomly placed disk contains N *IRAS* galaxies. Solid lines show the best GQED, dashes are the best Poisson, light dots are the best discrete lognormal, and heavy dots for $\theta = 3°$ are the best Gaussian. (From Sheth, Mo & Saslaw, 1994.)

Laws of thermodynamics and the other physical requirements described in Chapter 26 for gravitational galaxy distribution functions.

In Figure 33.3, the dotted line shows the best GQED fit using a single value of b independent of scale. It is just slightly worse than the self-consistent GQED, and both are better than the best Poisson fit (dashed line). Thus the void distribution of *IRAS* galaxies is consistent with simple gravitational clustering.

On larger scales, Figure 33.4 shows the *IRAS* counts-in-cells distribution function, $f_\theta(N)$, for circular cells of radius θ degrees on the sky ($1° \approx 1.5\,h^{-1}$ Mpc at the median distance of the sample and is somewhat larger at the average or maximum distance of the sample). Again the agreement of the observed galaxy distribution (solid histogram) with the GQED is excellent. The dashed Poisson distributions fit progressively worse as the scale increases, because correlations contribute more to $f(N)$ on larger scales. Similarly, Gaussian distributions (heavy dots) fit poorly on these scales, though they become much better representations on scales of $\theta \approx 13°$. For $\theta \gtrsim 15°$, there are too few representative regions to obtain useful sampling. The dotted line in Figure 33.4 shows the best fit of a discrete lognormal distribution such as Hubble (1934, 1936a) originally used. It works for $\theta \gtrsim 4°$ but fails progressively

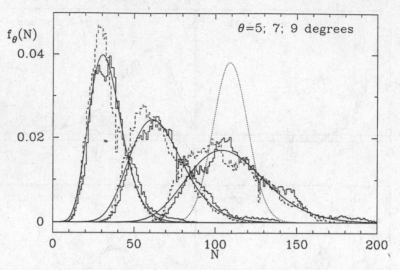

Fig. 33.5. Counts-in-cells distribution functions of *IRAS* galaxies with Galactic latitude $\geq 20°$ (solid histograms) in the northern hemisphere and $\leq -20°$ (dashed histograms) in the south. From left to right these are for cells of angular radius 5°, 7°, and 9°. As a comparison for $\theta = 9°$, the dotted line is the best Poisson. Solid lines show the GQED that fits the average of the north and south counts. (From Sheth, Mo & Saslaw, 1994.)

for smaller θ (a result that Hubble appears not to have noticed). This illustrates the importance of examining $f(N, V)$ over a wide range of scales.

We pick up a new property of the clustering on scales of 5–9 degrees. Figure 33.5 shows those scales where the counts are shown separately for the northern and southern galactic hemispheres with $|b^{II}| \geq 20°$ in each case. Differences between the two hemispheres give an idea of the statistical fluctuations on these scales. These fluctuations tend to fall on opposite sides of the GQED. More interestingly, at $\theta = 5°$ but not on other scales there is a double peak in the northern hemisphere only. This is caused by the concentration of galaxies near the supergalactic plane, alluded to in the discussion of Figure 33.2. In general, broad, and especially double, peaks in $f_\theta(N)$ are signatures of density inhomogeneities and show that the distribution function statistic is sensitive to large-scale filamentary structure. The cell size that picks out this structure corresponds well with visual impressions of its angular scale. Figures 33.4 and 33.5 show that the double peak is filtered out on scales $\lesssim 4°$ and smoothed over on scales $\gtrsim 7°$. Apart from the galaxies near the supergalactic plane (our own local clustering), the GQED fits the *IRAS* distribution from scales $0.2° \lesssim \theta \lesssim 13°$. Beyond this, the small number of independent cells give ragged histograms.

It is clear that the simplest explanation of this agreement between the GQED and observations is that the dark matter has not significantly affected galaxy clustering. This may be either because it was nearly uniformly distributed or, more likely in my view, because it clustered similarly to the galaxies. In the latter case most of the dark matter would have been closely associated with galaxies as they clustered, either internally or in halos around them. Another possibility is that the dark matter might be in the form of clumps or dark (low surface brightness) galaxies that clustered similarly to the visible galaxies. After strong clustering, tidal debris would accumulate in the centers of clusters. Generally, specific numerical simulations are necessary to find spatial and velocity distribution functions for these possibilities and compare them with the observations. However, in Section 33.3.3 we shall see how some classes of models may be ruled out by comparing their predicted scale dependence of $b(r)$ with the observations. This will also illustrate the observed $b(r)$ for the *IRAS* distribution functions in the previous three figures. But first we check the relation between observations of the two-dimensional and the three-dimensional distribution functions.

33.3.2 Projected and Volume Distribution Functions Are Observed to Be Similar

Earlier we saw how both the gravitational quasi-equilibrium theory (Section 28.5) and the relevant N-body experiments (Section 31.3.3) gave the same form for the projected and volume distribution functions. Only their values of b differed modestly, depending on the different shapes of the cells (projected cones versus spheres or cubes) relative to the scale length of the two-particle correlations. This similarity arises essentially because the correlations, averaged over large enough systems, are statistically homogeneous.

How about the observations? Early accurate determinations of $f(N)$ were for projected distributions because few redshifts were known. The Southern Sky Redshift Survey was the first optical catalog to have its three-dimensional distribution functions determined (Fang and Zou, 1994). They also fit the GQED very well with a value of $b = 0.70 \pm 0.05$ on scales of about 5–$12\, h^{-1}$ Mpc. This was very similar to the earlier two-dimensional result for the Zwicky catalog. Subsequent analysis (Benoist et al., 1997) of the complete magnitude-limited SSRS2 survey showed similar good fits to the GQED, essentially confirming the earlier result.

For the *IRAS* galaxies, volume-limited subsamples of galaxies with $60\,\mu$m flux densities greater than 1.2 Jy and velocities in ranges of $2,400$ to $12,000$ km s^{-1} are also in good agreement with the GQED (Bouchet et al., 1993).

As future catalogs acquire more completely sampled velocities, they may find interesting discrepancies between the projected and volume distribution functions. But at present these seem consistent with each other and with cosmological many-body clustering.

33.3.3 The Observed Scale Dependence of $b(r)$ Restricts Possible Models

Examples of catalogs for which $b(r)$ has been determined explicitly are Zwicky (Saslaw and Crane, 1991), ESO and UGC (Lahav and Saslaw, 1992), and IRAS

(Lahav and Saslaw, 1992; Sheth et al., 1994). Determining $b(r)$ is simple to do either by fitting the GQED to counts in cells of different sizes, or by just measuring the variance of the counts for cells of a given size and repeating for different sizes using (28.21a). Both techniques give the same result to within a few percent. In all cases, the general behavior of $b(r)$ is to start near the $b = 0$ Poisson value on small scales and to rise rapidly with scale until it levels off on large scales, as expected from (25.18). It may decrease on very large scales if $\xi_2(r)$ makes a substantial negative contribution to $\bar{\xi}_2$ (see Figures 8.5 and 8.6 of Sheth et al., 1994). This decrease has not yet been seen, although there may be hints of it in the SSRS catalog (Fang and Zou, 1994). Searches in future catalogs for decreasing $b(r)$ will be exciting since they will determine $\xi_2(r)$ for very small negative values, which are hard to measure directly.

Heavy dots in Figure 33.6 show $b(r)$ for the *IRAS* distribution functions in the preceding figures. Since *IRAS* galaxies are mainly spirals, they do not form a complete sample; a roughly random selection would reduce their value of b relative to a more complete sample (see Section 28.6) such as the Zwicky catalog. The lower *IRAS* value of b may also be related to a lower average mass for those galaxies, reducing their dynamical clustering slightly. Here again, future more informative catalogs will be helpful.

Thin lines in Figure 33.6 show $b(\theta)$ for some illustrative cold dark matter models. Their initial perturbations have power spectra with the form

$$P(k) = \frac{Ak}{\{1 + [ak + (bk)^{3/2} + (ck)^2]^\nu\}^{2/\nu}} \tag{33.1}$$

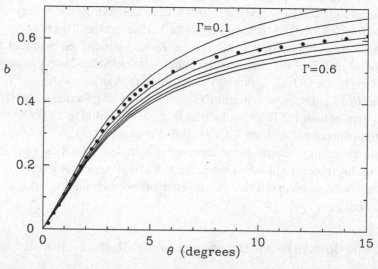

Fig. 33.6. Values of b as a function of angular scale for the *IRAS* galaxies (heavy dots) and for illustrative cold dark matter models discussed in the text. (From Sheth, Mo & Saslaw, 1994.)

with A, a, b, and c positive constants. These five-parameter models are the Fourier transform of a class of correlation functions that are positive on small scales, cross zero only once, and then remain negative even on the very largest scales. The volume integral of their correlation function becomes zero only for infinite volumes. When $a = (6.4/\Gamma)h^{-1}$ Mpc, $b = (3.0/\Gamma)h^{-1}$ Mpc, $c = (1.7/\Gamma)h^{-1}$ Mpc, and $\nu = 1.13$, Equation (33.1) can describe a wide class of theoretical models (Efstathiou, Bond, and White, 1992). For $\Gamma = h$, this power spectrum is similar to that of the once-popular "standard" cold dark matter model (with $\Omega_0 = 1$ and very small baryon density). Then $\xi_2(r)$ is characterized by two parameters: the amplitude A and the value of Γ.

To compare these or other models with the observed $b(\theta)$ one first normalizes the amplitude of $P(k)$ to the observations. This is often done by matching σ_8, the r.m.s. density variation, in spheres of radius $8\,h^{-1}$ Mpc and assuming a constant bias factor B relating the variances of the dark and luminous matter on all scales (Section 8.3). More generally, as in Figure 33.6, one can match the amplitude of the linearly biased model correlation function with that of the observations at a given scale, say r_0 in $\xi(r/r_0)$, which is $3.8\,h^{-1}$ Mpc for this *IRAS* sample. This gives $B_{\text{IRAS}}\,\sigma_8 \approx 0.65$ for $0.1 < \Gamma < 0.6$ and determines A for a given Γ. As Γ increases, the amplitude of the correlation function on large scales decreases and becomes significantly less than that observed for $\Gamma \gtrsim 0.5$, whereas $\Gamma = 0.2 \pm 0.1$ agrees with the APM correlation function (Efstathiou et al., 1992).

Next, one Fourier transforms the normalized $P(k)$ to get $\xi(r)$ and integrates over a volume along the line of sight to get $W(\theta)$. This integral involves the selection function for the catalog, which in turn requires a knowledge (or estimate) of the galaxy luminosity function and of how density evolves with redshift. Then $W(\theta)$ gives $b(\theta)$ from (14.37) and (28.21a). It is possible of course to compare $W(\theta)$ directly with the observations, but it is often easier to determine $b(\theta)$ more accurately, especially on large scales. Details may be found in Sheth et al. (1994).

Figure 33.6 shows that these models clearly disagree with the *IRAS* catalog. They do not fit the observed strength of clustering consistently on both small and large scales. Different normalizations of the power spectrum do not help. The only way to get agreement with these models would be to add an ad hoc bias function whose dependence on spatial scale is contrived to remove the disagreement. Alternatively, one could try to retrodict (or, better, predict) such a bias function from the underlying physics of galaxy formation or clustering, but this has not yet been done.

Any models comparing themselves to our Universe must not only reproduce the observed $f(N, V)$, but also the observed $b(\theta)$, and soon the observed velocity distribution function as well. These are quite strong constraints and may eliminate whole classes of models, as in this illustration, as well as help determine the best value of Ω_0 for those models that remain. These constraints will become even more powerful when they are applied to $b(r)$ in three dimensions rather than in projection.

33.3.4   Observed Distribution Functions Scale Essentially Randomly with Limiting Magnitude

This is a test of two basic properties. One is statistical homogeneity. The other is that the luminosity of a galaxy is essentially independent of the number of galaxies in its environment. Random scaling holds reasonably well for Zwicky's catalog, the only case studied so far. If a brighter limiting magnitude randomly selects a smaller sample of the original population, then their values of b should be related by (28.24). Scaling the values of b in this way should then produce agreement between $b(\theta)$ for the different magnitude-limited samples.

Figure 33.7a illustrates b for counts in cells of increasing area using galaxies of the Zwicky catalog with apparent magnitude limits 15.7, 15, and 14.5. All are above galactic latitude $30°$ to minimize obscuration by interstellar dust in our Galaxy. The general trend of b with scale is similar to that of the *IRAS* galaxies. When the two smaller magnitude-limited subsamples are rescaled by (28.24) for their relative fractions of the total sample, the result is Figure 33.7b.

The fact that these rescaled subsamples have approximately the same amplitude and form for b (area) as the full Zwicky sample, indicates that magnitude sampling is essentially random. There is little, if any, overall correlation between the luminosity of a galaxy and its position on the sky. This suggests that average environmental effects on galaxy luminosities are small. However, it does not preclude strong effects of local interactions such as mergers or tides.

33.3.5   Superclusters Are Surprisingly Close to Fair Samples

Having found that the GQED describes statistically homogeneous galaxy catalogs covering large volumes, we naturally ask if it also applies to smaller regions such as superclusters. This might tell us something about how superclusters form, whether by fragmentation of larger perturbations or by accumulation of smaller ones. Although the physics of fragmentation is not well understood, it is unlikely to lead directly to the observed GQED. Accumulation, in contrast, could remain close to a GQED until local relaxation is nearly complete (as in clusters such as Coma). Nearly relaxed clusters would resemble an isothermal sphere with $\rho \propto r^{-2}$ and roughly Maxwellian velocities, though they might still have subcondensations.

One of the most interesting regions to analyze is the Pisces–Perseus supercluster. Using the previously mentioned radial velocity catalogs of Giovanelli, Haynes, and their collaborators gives the three-dimensional velocity space view of these galaxies shown in Figure 33.8. This is a reasonably uniformly sampled region, in de Vaucouleurs' supergalactic coordinates, and contains 4,501 galaxies brighter than $m = 15.7$. This and related samples are relatively unaffected by bulk flows of galaxies either nearby or on larger scales and their distribution functions have been explored quite extensively (Saslaw and Haque-Copilah, 1998).

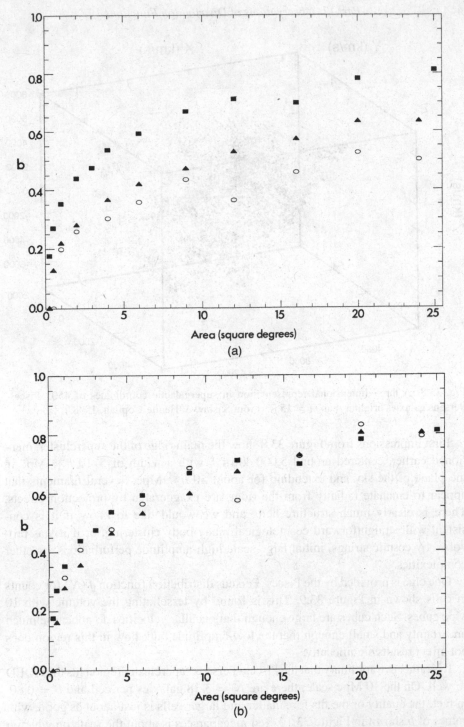

Fig. 33.7. (a) Observed variation of b with cell area for the Zwicky galaxy catalog with different apparent magnitude limits. Squares are for the entire catalog above Galactic latitude $30°$, triangles are for brighter galaxies with $m_{zw} \leq 15$, and ovals for those with $m_{zw} \leq 14.5$. (b) The same results rescaled for random sampling using Equation (28.24). (From Saslaw & Crane, 1991.)

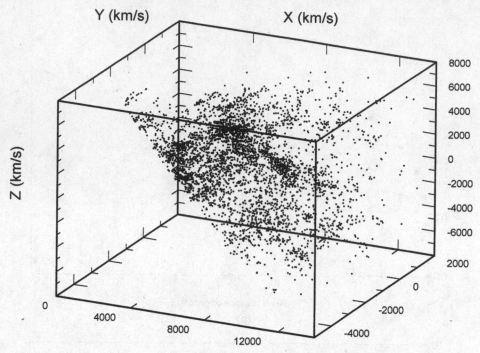

Fig. 33.8. A three-dimensional representation in supergalactic coordinates of 4501 Pisces–Perseus galaxies brighter than $m = 15.7$. (From Saslaw & Haque-Copilah, 1998.)

First impressions from Figure 33.8 show the main ridge of the supercluster, mentioned earlier, centered around 5,000 $km\,s^{-1}$ with a width of 5–10 h^{-1} Mpc in the plane of the sky and extending for about 50 h^{-1} Mpc. Several filaments that appear to emanate radially from the ridge are exaggerated by projection effects. There is clearly much structure here, and we would like to know if it is consistent with straightforward cosmological many-body clustering, or if it was provoked by cosmic strings, initial large-scale high-amplitude perturbations, or other complexities.

One clue is provided by the Pisces–Perseus distribution function $f(N)$ for counts in cells shown in Figure 33.9. This is found by tessellating the volume with 10 Mpc cubes. Such cubes are large enough that peculiar velocities do not cause much uncertainty and small enough that the low-amplitude bulk flow in this region does not affect results significantly.

It is rather extraordinary that the Pisces–Perseus supercluster region fits the GQED so well. On the 10 Mpc scale, there are $\bar{N} = 3.38$ galaxies per cell and $b = 0.80$. In fact, the quality of the fits for smaller and larger cells is just about as good, with values of b shown in Figure 33.10. Ten megaparsecs is about the scale on which b reaches its asymptotic value.

All the inhomogeneities of Pisces–Perseus – the voids and the filaments, the clumps and the ridges – come under the purview of the GQED. There seems to be

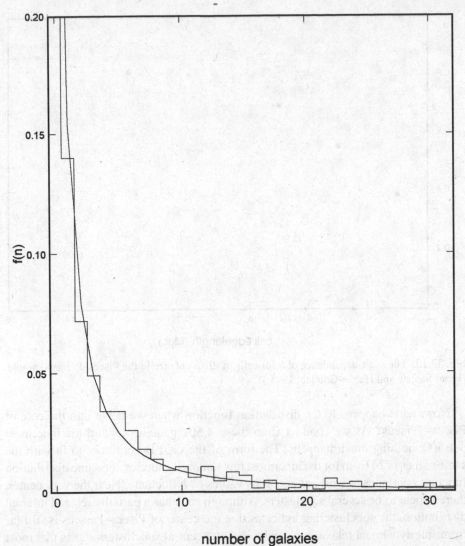

Fig. 33.9. The observed three-dimensional distribution function for counts in cells of length 10 mpc (histogram) for the Pisces–Perseus galaxies in the previous figure. The smooth curve is the GQED, with $b = 0.80$. (From Saslaw and Haque-Copilah, 1998.)

no necessity for non-gravitational clustering nor for strong large-scale structure in initial conditions. (See also Chapter 35 for the probability of finding such a large overdense region.)

Weak large-scale initial perturbations could have facilitated superclustering in the Pisces–Perseus region, but even their existence may not be required. They will be hard to ferret out. In any case, the numerical simulations described in Chapter 31 showed that such large-scale initial perturbations evolve toward the GQED for power-law power spectra with indices $-1 \lesssim n \lesssim 1$.

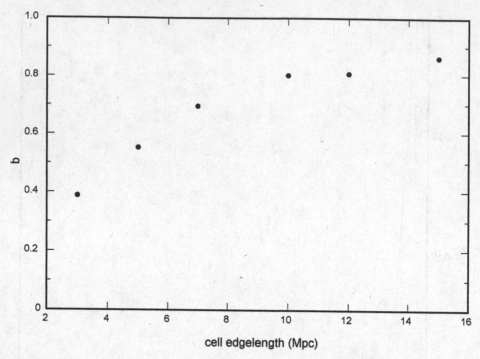

Fig. 33.10. The scale dependence of b for cells of different size in the Pisces–Perseus sample. (From Saslaw and Haque-Copilah, 1998.)

Now, what happens to the distribution function when we delve into the core of Pisces–Perseus? As we contract from these 4,501 galaxies toward the innermost ~1,500, nothing much happens. The form of the GQED continues to fit with the same values of b. Even for the innermost few hundred galaxies, the same distribution function continues to work, although its values of b fluctuate. Near the very center, there appear to be several subclusters. Although each has a partially relaxed internal distribution, the subclustering indicates that the center of Pisces–Perseus is still far from full dynamical relaxation. Analysis of the central subclusters shows that most of them have internal dynamical crossing timescales longer than the Hubble age of the Universe, also suggesting an incompletely virialized state (Saslaw and Haque-Copilah, 1998). Within finite, bound regions satisfying the virial theorem (13.16), the GQED would not apply, for the reasons discussed in Chapter 24.

The result that relatively few Pisces–Perseus galaxies are needed to provide a fair sample for the distribution function in this region suggests how it formed. To maintain the form of the GQED, clusters would have had to collect in a fairly quiescent manner, in fact in quasi-equilibrium. This implies a roughly homologous form of accumulation, without clusters tidally shredding each other, nor many galaxies migrating between them.

Other superclusters have not yet been so extensively studied. Eventually we might find one that reveals a clear nongravitational clue to its origin and changes the present simple picture. Or the use of more powerful statistics for Pisces–Perseus may detect subtler clues.

33.3.6 Distribution Functions Are Similar for Galaxies over a Wide Range of Individual Properties

In the simulations of Section 31.1 there was some dependence of the distribution functions on galaxy mass, particularly when the masses ranged widely. But this dependence was not overwhelming. The reason is that most of the later relaxation is collective, so that the masses of individual galaxies are only of secondary importance. Masses are related to the morphology, luminosity, and sizes of galaxies, although this relation may be complex and involve many other aspects of galaxy formation and evolution. Thus even if the distribution functions of different types of galaxies differ in detail, we would expect them to have the same basic form for a wide range of galaxies.

Hubble (1936a,b) knew that ellipticals cluster more strongly than spirals and his results have often been refined, especially for different types within clusters. But a more general analysis, encompassing all levels of clustering, awaited large samples and more subtle statistics. First the two-point correlation functions showed higher amplitudes for ellipticals than for spirals (see Chapters 14 and 20). Then distribution functions confirmed this impression (Lahav and Saslaw, 1992).

Morphological information in the UGC and ESO catalogs has made it possible to test the dependence of distribution functions on morphology. Future catalogs will naturally be even better. Figure 33.11 shows the different distribution functions for the combined elliptical and SO galaxies and for the combined spiral Sa and Sb galaxies in the UGC and ESO catalogs. These are for cells of size $\Delta l^{II} \times \Delta \sin b^{II} = 6° \times 0.1 \approx 34.4$ square degrees at galactic latitudes greater than 30° from the plane. The fits to the GQED (solid smooth lines) are reasonably good in all cases, with the best values of b shown on each panel. These cells are large enough to represent the asymptotic regime for b. The resulting value of b for all morphological types in the UGC catalog is 0.62; for the ESO catalog it is 0.58 (Lahav and Saslaw, 1992).

Smaller values of b for these samples compared with those of the Pisces–Perseus or Zwicky catalogs arise partly from criteria (e.g., angular diameter limits) that lead to fairly random selection of a smaller fraction of all galaxies for the UGC and ESO samples. Within each catalog, we can check whether the value of b for a particular subsample represents a random selection from the entire catalog, according to (28.24). For Sa + Sb galaxies in the UGC catalog, $b_{random}(p = 0.26) = 0.37$ while $b_{observed} = 0.33$; this is consistent with the early spirals being distributed essentially randomly. For the ESO catalog, $b_{random}(p = 0.33) = 0.37$ and $b_{observed} = 0.49$ and so random selection seems less certain here, but it is difficult to track down the origin of such small discrepancies. The situation is much different for the ellipticals and SOs. In the UGC, $b_{random}(p = 0.16) = 0.28$ while $b_{observed} = 0.44$. In the ESO, $b_{random}(p = 0.18) = 0.26$ while $b_{observed} = 0.52$. There is no doubt that the

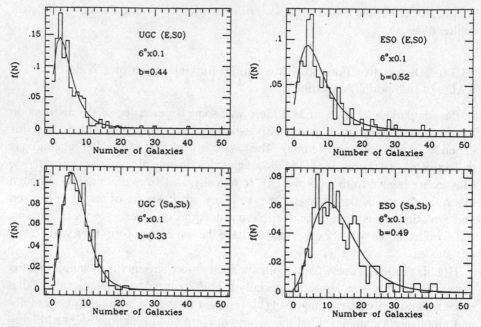

Fig. 33.11. Distribution functions of counts-in-cells for spiral (Sa, Sb) and elliptical (E, SO) galaxies in the UGC and ESO samples. The smooth curve is the GQED with the indicated value of b. (From Lahav & Saslaw, 1992.)

E + SO subsamples are clustered more strongly than would be the case for random selection. Although earlier surveys found these properties within samples of clusters or field galaxies, distribution functions can characterize differential properties across the whole range of clustering from the most underdense regions to the richest cluster.

We can use this last feature to estimate the average probability $p(M)$ for finding an elliptical in a region containing M galaxies. The technique is based on how bias affects distribution functions (discussed in Section 28.10), and the result (Lahav and Saslaw, 1992) for the ESO and UGC catalogs is that $p(M) \propto M^{0.3}$ for ellipticals. A similar analysis for the spirals gives no appreciable dependence on M.

This rather weak dependence of $p(M)$ on M for ellipticals, throughout the whole range of clustering, has a rather interesting interpretation. If spirals generally turned into ellipticals by a process such as gas stripping, which involves interaction with intergalactic gas, and if the average density of intergalactic gas were roughly proportional to the galaxy number density, then we would expect $p(M) \sim M$. Alternatively, if ellipticals were produced primarily by merging, they would have $p(M) \propto M^2$, which is even less in agreement with observations. It is possible that more complicated models that depend more weakly on these physical processes could be made more compatible with the data, but they would need an independent physical justification.

These results, of course, do not preclude individual instances of stripping and merging, and many are known. However, they are unlikely to be generic effects that dominate the difference between ellipticals and spirals at small redshifts. This interpretation is subject to the caveat that it refers to the current galaxy distribution. Earlier merging, for example, might have depleted the excess density around ellipticals. It would then be necessary to show that subsequent evolution leads to the observed distribution of satellite galaxies remaining in regions around ellipticals. Going back to even earlier times, initial formation conditions and dark matter halos may have affected the present distributions for different galaxy types.

Gravitational segregation is inevitable and is probably the simplest explanation for galaxy segregation. Simulations such as those outlined in Chapter 31 show that for heavier particles b is greater than it would be if these particles were selected at random. If the average mass associated with elliptical galaxies is several times that of spirals, it might help explain the weak dependence of $p(M)$ on M. Thus it is important to measure individual masses for large representative samples of clustering spirals and ellipticals.

33.3.7 Abell Clusters and Radio Sources

Radio sources and rich clusters of galaxies sample the distribution of matter over very large scales, but in a more dilute manner. Although the Abell catalog of rich clusters has uncertainties from incompleteness and from entries of small superimposed clusters (see Chapter 7), one can estimate how these uncertainties affect its distribution function (Coleman and Saslaw, 1990). Treating each cluster as a point, a sample of 460 Abell clusters above galactic latitude 30° has the distribution functions $f_N(R)$ and $f_R(N)$ shown in Figure 33.12 for projected cells of angular radius R minutes of arc. Agreement with the GQED is fairly good, considering the relatively small sample. For cells of radius $\sim 7°$ the value of $b = 0.43$. Randomly removing one third of the sample reduces b to 0.37 on this scale, randomly adding one third reduces b to 0.35, and combining both random uncertainties gives $b \approx 0.34$. All these "bootstrap" modifications have worse fits than the actual observed catalog.

Distribution functions clearly indicate that Abell clusters participate in the gravitational clustering of all galaxies on scales of tens of megaparsecs. Their two-point correlation function (Section 14.3) provides another measure of this clustering. What is not so clear is the physical interpretation of all these measures of clustering, since they apply to predefined collections of galaxies, rather than to galaxies themselves, as the fundamental units. It may be that rich clusters form around peaks of the perturbed initial density, leading to their own enhanced clustering (see Section 18.3). Or, if clusters contain most of the total inhomogeneous light and dark mass in their neighborhoods, their mutual attraction may promote their relative clustering over a Hubble time. We need to disentangle the roles of initial conditions and of subsequent clustering for a wide range of possibilities, wider than those examined so far.

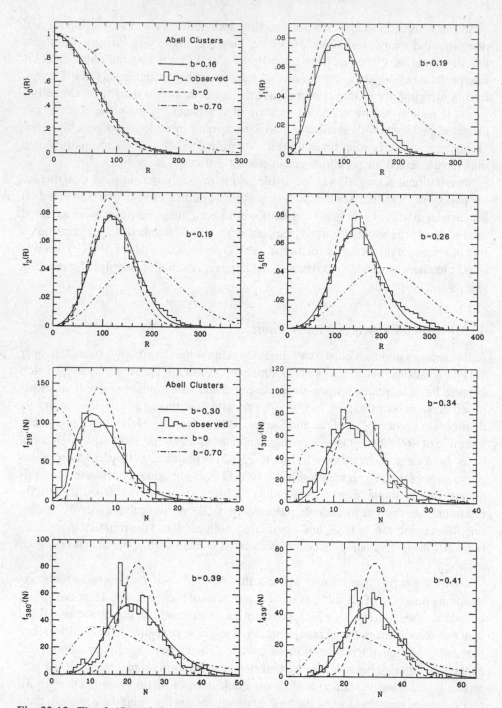

Fig. 33.12. The $f_N(R)$ and $f_R(N)$ distributions for an Abell sample of 460 clusters for $N = 0$ (voids), 1, 2, 3 and $R = 219'$, $310'$, $380'$, and $439'$. The GQEDs (solid curves) are shown for the best fit values of b along with comparisons for $b = 0$ (Poisson, higher dashed curves) and $b = 0.70$ (lower dashed curves). (From Coleman & Saslaw, 1990.)

Radio galaxies present an even more difficult story. Here the basic problem is that flux-limited samples, which may have millions of radiogalaxies, contain appreciable numbers of faint nearby sources with $z \lesssim 0.1$ as well as many bright sources at very high redshifts. Without measured redshifts, sources are projected over vastly greater distances than in most optical samples. This dilutes any statistical measure of structure. Of course, one can model projection with plausible estimates of luminosity functions and evolution. However, at best this smooths out the actual statistic one is trying to measure.

The other problem (also present but less important in optical samples where clustering is stronger) is that inhomogeneities in sensitivity or sampling can mimic weak correlations. This applies particularly to statistics such as dipoles and higher harmonics (e.g., Baleisis et al., 1998).

Naturally these problems have not prevented astronomers from searching for structure in radio catalogs as they came along. Early analyses (e.g., Webster, 1976) revealed random distributions. More recently, there have been suggestions (e.g., Loan, Wall, and Lahav, 1997) of a very weak $W(\theta) \approx 0.01$ two-point angular correlation at $\theta = 1°$ in large area surveys reasonably complete above 50 mJy. With very sensitive new catalogs restricted to low-redshift samples, it should become possible to detect faint radio emission from many nearby optical galaxies. These should then show radio correlations consistent with those in the visual.

Using a very sensitive survey complete to 5 mJy (at 8σ above the noise level), over a $4° \times 4°$ area of the sky, gave a value of $b = 0.01 \pm 0.01$ for counts in cells, in agreement with a Poisson distribution (Coleman and Saslaw, 1990). This is consistent with random dilution if the probability for detecting a radio source in an optical galaxy is $p \approx 4 \times 10^{-3}$ in this sample. It is also consistent with negligible intrinsic average clustering of galaxies on scales $\gtrsim 100$ Mpc. To distinguish between these two possibilities requires yet larger, fainter, homogeneous radio catalogs with reasonable distance information.

Thus, on the whole, much has been learned about the galaxies' spatial distribution. Many of these basic questions are well on their way to being resolved, and the way to future refinements looks clear. Less is known about peculiar velocity distributions, and we turn to them next.

34

Observed Peculiar Velocity Distribution Functions

no one can bottle a *breeze*
Auden

Velocities are the current frontier, as spectrographs with fiber optic cameras scramble to produce larger catalogs. Some galaxies will have good secondary distance indicators and be useful for peculiar velocities. Catalogs whose peculiar velocities are homogeneously sampled can provide direct comparisons with theoretical predictions.

So far, most catalogs of peculiar velocities are for restricted samples such as spirals in clusters, or for isolated field galaxies. This again involves finding a suitable definition of a cluster or field galaxy. Relating convenient morphological definitions of clusters to the underlying physics of their formation is often quite difficult. It is simpler, and perhaps less model dependent, to consider the combined peculiar velocity distribution for all clustering scales and degrees from isolated field galaxies and small groups to the richest dense clusters. The corresponding predictions of gravitational quasi-equilibrium clustering are given for $f(v)$ by (29.4) and for the observed radial velocity distribution $f(v_r)$ by (29.16). Chapter 32 describes their agreement with cosmological many-body simulations.

The GQED predictions are for local peculiar velocities averaged over a large and varied system. They do not include the motion of the system itself, which corresponds to regional bulk flow produced by distant, rare, and massive attractors. There is general agreement that such bulk flows exist, though their detailed nature is subtle and still controversial. An analogy illustrating the relation between bulk and local flows is a room with a gentle breeze blowing through an open window. There is an overall flow, but within each small region of a few cubic centimeters the velocity distribution is nearly Maxwellian. Observationally, the bulk and local peculiar velocities are linearly superimposed. Therefore the bulk flow can be subtracted off in every direction.

Soon after the GQED predictions for $f(v_r)$ were made, I asked several observers if they could be observationally tested. Responses fell into two categories. The optimists said "Come back in five years." The pessimists said "Don't come back." The pessimistic view was dominated by difficulties of removing bulk flow and finding sufficiently accurate secondary distance indicators.

As it turned out, the pessimists were too pessimistic and the optimists were closer to the mark, but for a reason that was then unsuspected. While several long-term projects were slowly accumulating suitable data, Somak Raychaudhury discovered that a subset of a sample of peculiar velocities in the literature, though made for other purposes, could already be used for a first determination of $f(v_r)$ (see Section

452

15.5 for a summary). As additional more accurate peculiar velocities are measured, we expect important refinements.

This useful sample of Mathewson, Ford, and Buchorn (1992) contains peculiar velocities of 1,353 spirals with redshifts $<7,000$ km s^{-1}. It covers about onefourth of the sky. Distances to the spirals are estimated using the I-band Tully–Fisher relation, described in Section 15.5. All late spirals (Sb–Sd) in the ESO catalog with major diameter $>1'.7$, inclination $>40°$ (to facilitate measurement of interior rotation velocities), and galactic latitude $>11°$ were included in their list. Their list, however, also included some galaxies observed for neutral hydrogen rotation velocities by Haynes and Giovanelli plus a "sprinkling" of spirals beyond 7,000 km s^{-1}, many of which have smaller diameters and lower inclinations.

The key idea (Raychaudhury and Saslaw, 1996) was to extract a more homogeneous, well-defined subsample from the Mathewson et al. (1992) catalog. Since the catalog was originally designed to test for bulk gravitational infall into the back of a large supercluster – the "great attractor," which might be responsible for much of the nearby bulk flow on scales $\gtrsim 50$ Mpc – galaxies were not selected with regard to their clustering. This is the catalog's main virtue for determining $f(v_r)$. Although spirals dominate, many of them are in groups and clusters and reasonably represent the overall distribution. The subsample was chosen, again without regard for clustering, to include only those galaxies with major diameter $\geq 1'$ and disk inclination $\geq 35°$, so that the inclination corrections to magnitudes and rotational velocities are relatively small. Then the subsample was spatially restricted to those parts of the sky where the sampling was most uniform. The resulting subsample had 825 galaxies.

Putting this subsample into a suitable state for analysis requires correcting the total I-band magnitudes for internal extinction and for galactic extinction. Then the Tully–Fisher relation of Figure 15.9, with a standard uniform Malmquist bias correction, gives relative distances of individual galaxies accurate to 15–20% (see Raychaudhury and Saslaw, 1996 for details). This is good enough for a first determination of $f(v_r)$, though it needs to be improved in the future.

Removal of bulk flow is the most serious and uncertain correction. This may be done in several ways for comparisons. Figure 15.10 showed $f(v_r)$ found for the 825 galaxies whose Tully–Fisher distances correspond to velocities $\leq 5,000$ km s^{-1} ($H_0 = 75$ km s^{-1} Mpc^{-1}). Bulk velocities of 600 km s^{-1} toward $l = 312°, b = 6°$ found by Dressler et al. (1987), probably the most relevant ones for this region, were subtracted off. The histogram is the mean of 10^4 histograms, each calculated from a randomly selected subsample consisting of two thirds of the entire sample. The error bars are standard deviations for each bin among the 10^4 subsamples, and the mean equals that of the entire sample. Subtracting instead the bulk flow of Courteau et al. (1993) of 360 km s^{-1} toward $l = 294°, b = 0°$ resulted in $f(v_r)$ being somewhat asymmetric around $v_r = 0$, a value of $b = 0.68$, and a generally poorer fit to the GQED. This is probably because the Courteau et al. (1993) flow is measured over a region less relevant to the Mathewson et al. (1992) subsample. A third method of removing bulk flow subtracts an extra Hubble expansion from the observed peculiar velocities so that $\langle v_r - rH \rangle = 0$. This required an effective local value of $H = 92$ km s^{-1}Mpc^{-1} (which is not the true value of H_0 for uniform expansion on large scales). It gave a very good fit to the GQED with $b = 0.80$, $\beta = 2.3$, and $\langle v_r^2 \rangle = 0.7$. The results that one value of H works well for all velocities and that $\beta \approx 3\langle v_r^2 \rangle$ suggest that the distribution is quite symmetric around $v_r = 0$.

All these methods for removing bulk flow leave a decidedly non-Maxwellian velocity distribution function. The peak around $v_r = 0$ is higher, mostly representing field galaxies whose peculiar velocities decay as the Universe expands. The high-velocity tails are also higher than the best-fit Maxwellian; they represent the galaxies in rich clusters. Observed values of b for $f(v_r)$ tend to be a few percent larger than for the spatial distribution function. This is because velocities involve interactions on all scales, including those larger than the spatial cells that determine $f(N, V)$.

Analogous to the different spatial distributions for different types of galaxies, it is also possible to make a preliminary estimate of $f(v_r)$ for ellipticals (Raychaudhury and Saslaw, 1996). A sample (Burstein et al., 1987; Lynden-Bell et al., 1988) of 449 elliptical galaxies chosen from all over the sky has distances estimated from the D_n–σ relation between the diameter within which the galaxy's corrected mean surface brightness is 20.75 B magnitude arcsec^{-2} and the internal central velocity dispersion σ. Their dispersion in the logarithm of estimated distance is 0.21. A catalog of 376 ellipticals (nearly all from this sample) with $D < 5,000$ km s^{-1}, comparable to the Mathewson et al. spiral galaxy distances, gives the radial velocity distribution in Figure 34.1. Apart from being more irregular because of its smaller number of galaxies, the tails of this distribution are significantly skewed toward negative velocities. These properties may be caused partly by preferential clustering of ellipticals (Section 33.3.6) and partly by chance fluctuations in the number, positions, and relative velocities of rich clusters in this small sample. Again there are clear departures from a Maxwell–Boltzmann distribution, and these departures occur in the same sense as for the spiral galaxies.

Many uncertainties affect analyses of $f(v_r)$. These include 1. samples limited by redshift instead of distance, 2. different bulk flow subtractions, 3. modified Tully–Fisher relations for distances, 4. random errors in peculiar velocity, 5. random errors in distance measures, 6. systematic errors in distance measures, and 7. inhomogeneous Malmquist bias. Most of these uncertainties can be modeled and, fortunately, for reasonable ranges of values they do not substantially change $f(v_r)$ (Raychaudhury and Saslaw, 1996). Departures from Maxwell–Boltzmann distributions remain clear, with values of $0.75 \lesssim b \lesssim 0.9$.

While the effects of most of these uncertainties are best treated using computer models for particular samples, it may be useful to discuss random errors, in distance measurements in somewhat more detail, again following Raychaudhury and Saslaw (1996). This is because their effects are often misunderstood. Frequently it is thought

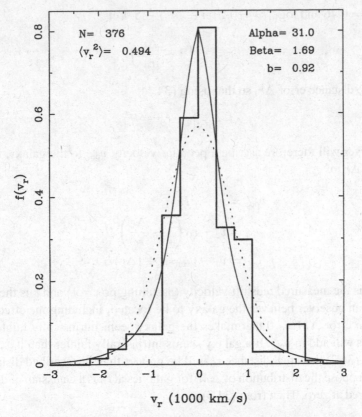

Fig. 34.1. The distribution of radial peculiar velocities for 376 elliptical galaxies with distances less than 5,000 km. The dotted line is the Maxwell–Boltzmann distribution with the same velocity dispersion as the sample. The solid line is the best fitting GQED whose parameters are in the upper right-hand corner. (From Raychaudhury & Saslaw, 1996.)

that because distances are exponentially related to magnitudes, a Gaussian distribution of magnitude errors will produce a lognormal distribution of distance errors and therefore of peculiar velocity errors. Such errors, it is thought, would then produce departures from an intrinsic Maxwell–Boltzmann distribution making it look more like the observations.

The actual situation is more complicated, however, and a lognormal distribution for distance errors departs significantly from a Gaussian only when the fractional distance errors become large. To see this, consider a simple model that illustrates how observational errors ΔM in the inferred absolute magnitude M_0 of the galaxies affect the distribution $f(v_r)$ for galaxies at a given actual distance r_0. The distance r (in Mpc), which an observer will ascribe to a galaxy of apparent magnitude m, is

$$r = 10^{-5} \times 10^{0.2[m-(M_0+\Delta M)]}$$

$$= r_0 \exp(-a\Delta M), \tag{34.1}$$

where $a = 0.46$ and $\log_{10} r_0 = 0.2 (m - M_0) - 5$. Let

$$r = r_0 \left(1 - \frac{\Delta r}{r_0}\right) \qquad (34.2)$$

define the distance error Δr, so that using (34.1)

$$\Delta r = r_0[1 - e^{-a\Delta M}]. \qquad (34.3)$$

An observer will therefore ascribe a peculiar velocity v_{pec} to the galaxy, including the error Δr, of

$$v_{pec} = v_z - rH$$
$$= v_z - r_0 \left(1 - \frac{\Delta r}{r_0}\right) H$$
$$= v_z - v_{Hubble} + (\Delta r)H. \qquad (34.4)$$

Since v_z is the measured redshift velocity (assuming no error) and r is the distance at which an observer believes the galaxy to be located, including the effect of ΔM on Δr, an error $\Delta M > 0$ that makes the galaxy seem intrinsically fainter than it actually is will add to v_{pec}. If a galaxy seems intrinsically fainter than it is, $\Delta r > 0$, and from (34.2) its actual distance r_0 will be greater than the ascribed distance r.

Now suppose the distribution of ΔM for galaxies at r_0 is a Gaussian of dispersion σ_M^2 centered at zero. Then from (34.3)

$$\Delta M = -\frac{1}{a} \ln \left(1 - \frac{\Delta r}{r_0}\right), \qquad (34.5)$$

so that the absolute value of the Jacobian is

$$\left| \frac{\partial(\Delta M)}{\partial(\Delta r)} \right| = \frac{1}{a (r_0 - \Delta r)} \qquad (34.6)$$

for $\Delta r < r_0$, and thus

$$f(\Delta r) = \frac{1}{\sqrt{2\pi \sigma_M^2 a^2 r_0^2}} \frac{1}{[1 - (\Delta r/r_0)]} \exp\left[-\frac{1}{2\sigma_M^2 a^2} \ln^2 \left(1 - \frac{\Delta r}{r_0}\right)\right]. \qquad (34.7)$$

This is indeed a lognormal distribution, but in $[1 - (\Delta r/r_0)]$ rather than in $\Delta r/r_0$ directly. Since $\Delta v_{pec} = v_{pec} - (v_z - v_{Hubble}) = H\Delta r$ from (34.4), and the actual Hubble velocity is $v_0 = v_{Hubble} = Hr_0$, we have

$$\frac{\Delta v_{pec}}{v_0} = \frac{\Delta r}{r_0}. \qquad (34.8)$$

Therefore $f(\Delta v_{pec}/v_0)$ is lognormal in $[1 - (\Delta v_{pec}/v_0)]$.

The most probable value of $\Delta v_{pec}/v_0 = \Delta r/r_0 \approx a\sigma_M \approx 0.2$ for $a = 0.46$ and adopting $\sigma_M = 0.4$ as a conservative estimate for the Mathewson et al. subsample. Then expanding (34.7) for $f(\Delta v_{pec})$ for small $\Delta v_{pec}/v_0$ gives

$$f(\Delta v_{pec}) = \frac{1}{\sqrt{2\pi \sigma_M^2 a^2 v_0^2}} \left[1 + \frac{\Delta v_{pec}}{v_0} + \left(\frac{\Delta v_{pec}}{v_0}\right)^2 + \cdots \right]$$
$$\times \exp\left\{ -\frac{1}{2\sigma_M^2 a^2} \left(\frac{\Delta v_{pec}}{v_0}\right)^2 \left[1 + \frac{\Delta v_{pec}}{v_0} + \frac{11}{12}\left(\frac{\Delta v_{pec}}{v_0}\right)^2 + \cdots \right] \right\}.$$

(34.9)

Therefore, if the errors in the Tully–Fisher magnitudes for galaxies at r_0 produce errors in the peculiar velocities that are small compared to the Hubble velocity $r_0 H$, then the distribution of these peculiar velocity errors is essentially Gaussian (Maxwell–Boltzmann).

As the peculiar velocity errors become larger, the influence of the $\Delta v_{pec}/v_0$ term makes the error distribution asymmetric. There does not seem to be any significant asymmetry of this type in the observed $f(v_{pec})$ in Figure 15.10. Moreover, if the true peculiar velocity distribution were Gaussian, it would convolve with a Gaussian error distribution to give an observed Gaussian peculiar velocity distribution, which is, however, not observed. This suggests that the observed departures from a Gaussian in Figures 15.10 and 34.1 are probably not dominated by errors in Tully–Fisher magnitudes. More detailed models, in which the errors are non-Gaussian, or become distance dependent, confirm that the actual distribution function departs intrinsically from a Maxwell–Boltzmann distribution. Models also suggest that observational errors may imply that the true $f(v_r)$ is somewhat more peaked than the observed one.

The seven sources of uncertainty mentioned earlier may also interact to change $f(v_r)$ in opposing ways. Their net effect is unclear but does not seem to dominate the presently available relevant data. Naturally, the best solution is more accurate data. Accumulating accurate data is often slow but seldom deters pioneering explorers. After all, Hubble's Hubble constant was nearly ten times the value accepted today.

A fundamental check of these results, possible with future catalogs, could be made with the velocity–edgelength correlation mentioned in Chapter 10 for observed minimal spanning trees (Krzewina and Saslaw, 1996). Consider the dimensionless average edgelength, Λ_i, associated with any particular node i of degree M:

$$\Lambda_i = (M\bar{x})^{-1} \sum_{j=1}^{m} x_{ij},$$

(34.10)

where x_{ij} is the edgelength of the edge connecting nodes i and j, and $\bar{x}$ is the average edgelength of the entire minimal spanning tree. Next, for each node i, calculate the dimensionless quantity v_i from $v_i = v_{ri}/\bar{v}_r$, where v_{ri} is the magnitude of the peculiar radial velocity of node i, and $\bar{v}_r$ is the average of all v_{ri}. Then we can investigate radial velocity-position correlations by plotting Λ_i versus v_i. If there are many points, it is convenient to bin with respect to v_i and plot the average value, $\bar{\Lambda}$, for each bin.

Since galaxies in clusters have short edgelengths connecting them, they necessarily have small $\bar{\Lambda}$. Similarly, galaxies outside clusters have large $\bar{\Lambda}$. Thus $\bar{\Lambda}$ is a quantitative measure of the average surrounding density. The higher average peculiar velocities of galaxies in clusters implies that v_i decreases as $\bar{\Lambda}$ increases.

We can estimate this relation for cosmological many-body clustering quite simply. The local average density $\bar{n}$ of a given number of galaxies is inversely proportional to the volume V and approximately proportional to the square of the average peculiar velocity of galaxies within V. Since $V \propto x^3$ for edgelength x and $\bar{n} \propto x^{-3} \propto v_r^2$, we obtain $x \propto v_r^{-2/3}$. With $\bar{\Lambda} \propto x$, we conclude that

$$\bar{\Lambda} = \bar{\Lambda}_0 \, (v_r/\bar{v}_r)^{-2/3} . \tag{34.11}$$

Naturally this will not hold for very small $v_r/\bar{v}_r$ since the low peculiar velocities of the most isolated field galaxies are dominated by the adiabatic expansion of the Universe. However, comparison (Krzewina and Saslaw, 1996, Figure 8.2) with the 10,000-body, $\Omega_0 = 1$, initially Poisson simulation after an expansion factor of 8 (Model S of Itoh et al., 1993; see Section 31.1) shows that (34.11) describes the relation between $\bar{\Lambda}$ and v_r very well for $1 \lesssim v_r/\bar{v}_r \lesssim 7$ with $\bar{\Lambda}_0 = 0.85$. So it is worth seeing if the observations also provide this indication of many-body clustering.

Our current overall conclusion is that the observed nonlinear velocity and spatial distributions of galaxies are consistent with each other and with cosmological many-body clustering.

35

Observed Evolution of Distribution Functions

I will to be and to know,
facing in four directions,
outwards and inwards in Space,
observing and reflecting,
backwards and forwards in Time,
recalling and forecasting.

Auden

The past is our future frontier. Distribution functions at high redshift have not yet been observed. Therefore this chapter will be very short.

Here and there we have glimpses of how galaxy clustering may have developed. These are from observations of two-particle correlations, Lyman alpha clouds, merging protogalaxies, and rich clusters, all at great distances. Eventually, when the half-dozen or so high-redshift catalogs now being started accumulate complete and well-chosen samples, they will yield up the distribution functions of the past. Insofar as these agree with the GQED, their evolution is represented by the changing of b. Thus there is time for genuine predictions, such as those of (30.12) and (30.19)–(30.20) shown in Figure 30.5. These catalogs will also test the importance of merging, which would alter $\bar{N}$'s conservation (see Chapter 36).

At high redshifts the value of b for gravitational quasi-equilibrium evolution depends quite strongly on Ω_0. Figures 30.5, 31.12, and 31.13 indicate that as we look further into the past, b will decrease more slowly for lower Ω_0. This is essentially because the clustering pattern is "frozen" at higher redshifts for lower Ω_0. Equivalently, for higher Ω_0 most of the evolution occurs more recently. Zhan (1989) and Saslaw and Edgar (1999) give useful diagrams to show how this can help determine Ω_0 and the redshift at which galaxies start clustering.

All observations of past evolution will be complicated by distance-dependent selection effects, unknown initial conditions, and evolving luminosity functions, as well as by the value of Ω_0. It is possible to simulate complications in any model. Until very wide ranges of possible models are strongly restricted by independent observations, large uncertainties will remain. This is especially true for clustering of objects like Lyman alpha clouds and X-ray clusters whose properties depend on models of their formation and internal astrophysical evolution. Although both these objects clearly cluster at moderate redshifts, the detailed form of their clustering remains controversial. Their distribution functions have not yet been obtained. And, of course, the matter distribution may be biased, in a complicated way, relative to the objects we study. Sorting everything out is a great challenge. Only the problem's importance guarantees its solution.

To glimpse a solution, consider the simple fact that superclusters, such as Pisces–Perseus described in Section 33.3.5, exist. In the cosmological many-body problem, unaided by strong initial large-scale perturbations, these superclusters are the last structures to form. How many of them would we expect?

For a rough estimate, consider the respective volumes of a supercluster containing N galaxies and of the observable universe to be $V_{sc} \approx (3 \times 10^{24} L_{Mpc} h^{-1})^3$ and $V_u \approx (10^{28} h^{-1})^3$, where h is Hubble's constant in units of 100 km s^{-1} Mpc^{-1} and L is the supercluster's characteristic lengthscale. Then the number of such superclusters observable in a large fair sample volume tesselated with cells of size V_{sc} is approximately

$$N_{sc}(N) \approx \frac{V_u}{V_{sc}} f_{V_{sc}}(N) = 3.7 \times 10^{10} L_{Mpc}^{-3} f_{V_{sc}}(N). \tag{35.1}$$

We can use Pisces–Perseus as an example. It appears consistent with the GQED and so (27.24) gives $f_{V_{sc}}(N)$. Since galaxies remain the fundamental units of this distribution, superclusters are far out on the tail at large N. This makes their mere existence very sensitive to the values of b and $\bar{N}$ (with similar sensitivity to different parameters for other models).

Next, we need to define the Pisces–Perseus region in terms of N and V. The volume occupied by the 4,501 galaxies in Figure 33.8 is $V_{p-p} = (128 \times 70 \times 140)h^{-3} = 1.25 \times 10^6 h^{-3}$ Mpc (Saslaw and Haque-Copilah, 1998). Suppose the overall galaxy distribution has a value of $b_0 = 0.8$ at present. To estimate $\bar{n}$, we note that the Pisces–Perseus region is essentially complete to about the same apparent magnitude as the Zwicky catalog, which, over a much larger area of the sky, averages about 2.2 galaxies per square degree (Saslaw and Crane, 1991). Taking the effective depth of the Zwicky catalog for these galaxies to be 200 Mpc and considering any conical volume to this distance gives $\bar{n} = 2.7 \times 10^{-3}$ galaxies Mpc^{-3}. A more precise analysis for the SSRS II catalog with a generally similar galaxy population gives $\bar{n} = 2.6 \times 10^{-3}$ Mpc^{-3} (Benoist et al., 1998) for a representative sample within 91 Mpc. (On smaller scales, sampling inhomogeneities are important and on larger scales the 15.5 apparent magnitude cutoff reduces the completeness significantly.) This gives an expected number of galaxies in the Pisces–Perseus region of $\bar{N} \approx 3,400$. Hence the entire region is about 32% overdense. What makes this relatively improbable is the large scale and the large number of galaxies involved. Its resulting probability from gravitational clustering in the GQED is $f_{p-p}(N) = 6.1 \times 10^{-4}$. Then (35.1) gives $N_{p-p} \approx 18$.

"But," I hear you say, "we are not just interested in overdense regions with exactly the specifications of Pisces–Perseus, rather in a range around it, especially those at least as dense or denser." Quite so. Just sum (35.1) over the values of V and N for any range you fancy.

In fact we can estimate this sum simply for volumes of any given size containing N or more galaxies by noting that for large N the ratio $f(N+1)/f(N) \approx be^{1-b} \approx$

0.977 for $b = 0.8$ from (27.24). Then the sum of (35.1) just becomes a geometrical progression and

$$\sum_{k=0}^{N_1} f(N+k) = \sum_{k=0}^{N_1} f(N)\beta^k = f(N)\frac{1 - \beta^{N_1}}{1 - \beta}, \qquad (35.2)$$

where

$$\beta \equiv be^{1-b} < 1. \qquad (35.3)$$

Even if β is close to unity, the second term in the numerator of (35.2) becomes unimportant for values of $N_1 \gtrsim 10^2$, which are reasonable in this context, and the sum essentially encompasses all overdense regions of volume V with N or more galaxies. To find the corresponding sum for a specific range of overdensity $N_0 \leq N \leq N_1$, just replace $f(N)$ by $f(N_0)$ in (35.2). Thus the number of supercluster regions is

$$N_{sc}(N_0 \leq N \leq N_1) \approx \frac{V_u}{V_{sc}} \frac{(1 - \beta^{N_1})}{(1 - \beta)} f_{V_{sc}}(N_0). \qquad (35.4)$$

Note that (35.4) also applies to any overdense or underdense regions satisfying $Nb \gg \bar{N}(1 - b)$. If this condition is not satisfied, the series for (35.2) must be summed more exactly.

Returning to our example, (35.4) gives about 10^3 regions of at least the Pisces–Perseus region overdensity in the Universe. If instead we consider regions of 20% or more overdensity, this is multiplied by a factor $f(N_0 = 4,080)/f(N = 4,501) \approx 30$ to give $\sim 3 \times 10^4$ such regions. For a Poisson distribution of superclusters, the expectation value of the distance (e.g., GPSGS, Equation 33.22 with $a = 0$) from a random place to the nearest Pisces–Perseus type region is $0.55\,n^{-1/3} \approx 150$ Mpc, and thus we are somewhat closer than expected in this model. For all regions of this volume with more than 20% overdensity the expected distance ~ 50 Mpc, illustrating its sensitivity to the threshold used to define a Pisces–Perseus type region. These probabilities are also quite sensitive to the values of $\bar{N}$ and b.

This last sensitivity holds out a hope (one of many) for finding Ω_0. Because smaller values of b produce much less clustering on the high-N tail of the GQED, at moderate redshifts there might not be any Pisces–Perseus type overdense regions. The redshift at which this happens indicates Ω_0. Continuing our simple illustration, consider a previous redshift when $b = 0.7$. Then $f(N = 4,501)$ is reduced by a factor $\sim 10^3$ and we expect only of order one Pisces–Perseus or denser region within the same comoving volume of the earlier visible universe.

Now go back to the past. Suppose $\Omega_0 = 1$. Then from (30.19)–(30.20) and Section 16.2, as illustrated in Figure 30.5 for $b_0 = 0.75$ the value of $b = 0.7$ occurs at $z \approx 0.45$ and we would not expect to see regions more overpopulated than Pisces–Perseus beyond that redshift. In contrast, if $\Omega_0 = 0.3$, then $b = 0.7$ at $z \approx 0.6$. This illustrates how detection of clusters at high redshifts can put an upper limit on Ω_0.

A similar analysis applies to the discovery (Steidel et al., 1998) of small groups of galaxies at $z \approx 3$, where b is more sensitive to Ω_0 and preliminary indications suggest $\Omega_0 \lesssim 0.3$ (Saslaw and Edgar, 1999).

General clustering is a rather more powerful technique for determining Ω_0 than using the evolution of predefined types of clusters (e.g., Bahcall et al., 1997), even though both methods are model dependent. As this illustration for Pisces–Perseus type superclusters shows, the results from predefined clusters are very sensitive to their definitions, to complete sampling, and to the precise inferred values of N, $\bar{n}$, and b. It will probably be more accurate to determine $b(z)$ and thus Ω_0 directly from the entire $f_V(N)$ distribution in large future high-redshift surveys.

PART VII

Future Unfoldings

Finally, four vignettes of the future. Some basic questions whose true understanding awaits new ideas and new observations. They follow in order of their solution's remoteness.

> Nature that framed us of four elements,
> Warring within our breasts for regiment,
> Doth teach us all to have aspiring minds:
> Our souls whose faculties can comprehend
> The wondrous architecture of the world,
> And measure every wandering planet's course,
> Still climbing after knowledge infinite,
> And always moving as the restless spheres,
> Will us to wear ourselves and never rest
>
> Marlowe

36

Galaxy Merging

E pluribus unus

Virgil

In hierarchical models of galaxy production, many protogalaxies merge at high redshifts. Each merger results in a spectacular wreck, which gradually restructures itself into a more unified system. Collisions engender vast conflagrations of stars as unstable gas clouds collapse and ignite thermonuclear fires. This process repeats and repeats until galaxies form as we know them today.

We see many galaxies still merging at present. In hierarchical models these late mergers are all that remain of earlier more active combining, or they result from encounters in recent dense groups. The long history of merging changes the observed galaxy distribution by destroying the conservation of galaxy numbers and by modifying the luminosity function. The first of these is easier to model; the second, at present, is really a guess. Galaxy luminosities depend on their unknown stellar initial mass functions, their subsequent star formation by merging or other violent activity, their nonthermal radiation, their production and distribution of dust, and their stellar evolution. Of these factors, only stellar evolution is reasonably well understood. On the other hand, number nonconservation depends on galaxy velocities, densities, and collision cross sections. These too must be modeled, but they seem more straightforward.

By describing the evolution of both the luminosity and the spatial distribution functions with one common formalism, we can use their mutual self-consistency to help constrain free parameters. Master equations are a particularly useful way to do this. We start with the master equation for the galaxy mass function evolving by mergers, which can be connected to the evolving luminosity function. Then we find a related master equation for the evolution of counts in cells as galaxies merge, and we follow some of its implications (Fang and Saslaw, 1997).

The mass function $N(m, t)dm$ is just the number density of galaxies within the mass range $m \rightarrow m + dm$ at time t. It evolves with time because each merged galaxy contains the sum of the masses of its parents, and the total number density of galaxies decreases as well. Let $\Lambda(m_j, m - m_j)$ be a typical volume in which one pair of galaxies with masses m_j and $m - m_j$ can merge per unit time. It is related to the merger cross section S by $\Lambda = Sv_0$, where v_0 is the relative speed of merging galaxies. The master equation describing the evolution of $N(m, t)$ is

$$\frac{\partial N(m, t)}{\partial t} = \frac{1}{2} \sum_{m_j=1}^{m-1} \Lambda(m_j, m - m_j)N(m_j, t)N(m - m_j, t)$$

$$-N(m, t) \sum_{\substack{m_j=1 \\ m_j \neq m}}^{m_c} \Lambda(m, m_j)N(m_j, t) - 2\Lambda(m, m)N^2(m, t), \qquad (36.1)$$

465

where m is in units of the smallest galaxy mass, and m_c is the largest galaxy mass in these same units. The first term in (36.1) sums up all possible mergers that yield galaxy mass m, so it contributes positively to $N(m, t)$. The factor 1/2 takes the symmetry of $N(m_j, t)$ and $N(m - m_j, t)$ into account. The negative terms represent mergers of galaxies of mass m with others of all possible masses. The sums become integrals for a continuous mass spectrum of galaxies. For evolution of $N(m, t)$ in finite time steps, (36.1) becomes a difference equation for ΔN and Δt. This assumes that there are no additional galaxy formation or destruction processes.

Next, consider the analogous master equation for the merger evolution of the counts-in-cells distribution function. Suppose the total number of cells is constant in time. Then the normalized probability that a cell contains N galaxies at time t is $f(N, t)$. For $N > 1$, $f(N, t)$ changes when galaxies merge, as well as when the general clustering measured by $b(t)$ increases. Considering just the effects of mergers will leave the number of cells containing one or no galaxy constant. A cell with $N + N_j$ galaxies ($N_j > 0$ and $N \geq 1$) at t has a certain probability to become a cell containing N galaxies at $t + dt$ as a result of merging. We denote this transition probability per unit time by $\Gamma(N + N_j, N)$. Then the equation describing the merger evolution of $f(N, t)$ is

$$\frac{\partial f(N, t)}{\partial t} = \sum_{N_j > 0} \Gamma(N + N_j, N) f(N + N_j, t) - \sum_{N_j > 0}^{N-1} \Gamma(N, N - N_j) f(N, t).$$

$$(36.2)$$

Here again there are no other galaxy formation, destruction, or clustering processes. Note that $\Gamma \Delta t$ has a different physical meaning than $\Lambda \Delta t$ since (36.2) deals with properties of cells rather than of galaxies.

Master equations are just the bare bones of a model. To give them physical content, we need to specify Λ and Γ. Suppose for simplicity that two galaxies of masses m_1 and m_2, and geometrical sizes with average diameter r_*, merge with isotropically distributed collision velocities. They are initially separated by large enough distances that their initial gravitational potential energy is much less than their kinetic energy. Let them have an initial relative speed v_0, impact parameter p, and a final relative speed v at their closest approach. Then the collision is identical to one in which a galaxy with reduced mass $\mu = m_1 m_2/(m_1 + m_2)$ collides with a galaxy of mass $m = m_1 + m_2$. As soon as $r \leq r_*$ so that the galaxies (including their dark matter halos) overlap, they are assumed to merge. If the merger conserves energy and angular momentum, then it has the standard cross section (e.g., GPSGS, Section 17.2):

$$S = \pi p^2 = \pi r_*^2 \left(1 + \frac{Gm}{r_* v_0^2}\right).$$

$$(36.3)$$

The first term on the right-hand side is just the geometric cross section. The second term dominates when v_0 is small and is the main contribution to merging. An important condition for merging is that the galaxies approach each other sufficiently slowly. We can estimate this from the relevant timescales. The timescale for a star to cross a typical colliding galaxy is $\tau_{cross} \approx (r_*/\sigma)$, where σ is the stellar velocity dispersion inside the galaxy. On the other hand, the timescale of the galaxy encounter is typically $\tau_{collide} \approx (p/v)$. If $v \gg \sigma$, the star cannot cross the galaxy during the collision when the impact parameter $p \approx r_*$. Galaxy mergers, however, are inelastic processes that substantially change the internal energy and stellar distributions of the colliding galaxies. Energy transfer from galaxy orbits to their internal motions causes these changes. Therefore encounters whose relative approaching speed is comparable with their internal velocity dispersion are the ones most likely to merge. When $v \gg \sigma$, so that v_0 is large, the geometrical cross section dominates and the encounter can be studied analytically using an "impulse approximation" (cf. Section 65 of GPSGS). Under somewhat more general conditions (cf. Fang and Saslaw, 1997), we may write the merging cross section as

$$S = \eta \left(\frac{2\pi Gm}{v_0^2} \right)^\mu r_*^\nu. \tag{36.4}$$

We use the values $\mu = 1$, $\nu = 1$, and $1 \lesssim \eta \lesssim 10$ as simple illustrations consistent with (36.3) for the low-velocity case.

Since merging changes the galaxies' mass distribution directly, we need to assume a relation between galaxy luminosity and mass to obtain the luminosity function. Typically, a galaxy's local mass-to-light ratio varies between 10 and 10^2 in solar units, depending on the galaxy's internal properties such as morphology, starbursts, and dust formation, as well as on the clustering environment. Merging naturally complicates the situation. In spite of all these uncertainties, it is customary to assume a constant mass/luminosity ratio for illustration, until the details are understood better.

The luminosity function of nearby galaxies has been studied extensively (e.g., Ellis et al., 1996) and can be described well by a gamma-distribution (Schechter, 1976). If all galaxies have a constant mass-to-light ratio α, this gives the form

$$N(m)\, dm = \phi_* \left(\frac{m}{m_*} \right)^\beta e^{-m/m_*} \frac{dm}{m_*} \tag{36.5}$$

with slope β, normalization ϕ_*, and $\alpha = m_*/L_*$. Estimates of these observed parameters from the Autofib survey (Ellis et al., 1996) at redshifts of $z \sim (0.7, 0.0)$ respectively give $\phi_* = (0.0355, 0.0245)h^3$ Mpc^{-3}, $\beta = (-1.45, -1.16)$, and $M_* = (-19.38, -19.30)$ for the absolute magnitude corresponding to L_*. Choosing α then determines m_*. Of the observed parameters, the normalization ϕ_* is particularly uncertain, as is the slope of the faint end of the distribution. Low normalization at $z = 0$ would imply that a strong evolutionary decrease occurred in the large numbers of faint blue galaxies found at intermediate redshifts.

These results enable us to follow the evolution of the luminosity function. With (36.4), we find that in (36.1)

$$\Lambda(m_j, m - m_j) = \eta \frac{2\pi G m}{v_0} r_*$$
(36.6)

and

$$\Lambda(m, m_j) = \eta \frac{2\pi G (m + m_j)}{v_0} r_*.$$
(36.7)

Thus the summation terms in (36.1) both have a common factor

$$\Theta \equiv \eta \pi r_* (G m_* \phi_*^2 / v_0) \Delta t,$$
(36.8)

which will also appear in the evolution of the counts-in-cells distribution function. Therefore the luminosity function and the evolution of $f(N, t)$ provide a double observational constraint.

To evolve luminosity functions $\phi(M)$, starting with the previous parameters for $z = 0.7$ in (36.5), it is necessary to extrapolate from the observed absolute magnitude limit at $M_B \approx -18 + 5 \log h$ to a fainter limit at $M_B \approx -17 + 5 \log h$. This allows for mergers, which produce the faint observed end of $\phi(M)$. Figure 36.1 illustrates

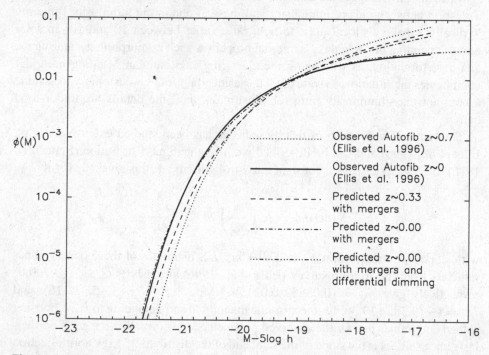

Fig. 36.1. Merger driven evolution of the galaxy luminosity function as described in the text. (From Fang & Saslaw, 1997.)

this evolution, based on (36.1) and (36.4)–(36.8). The solid line represents the local luminosity function at $z = 0$ found by Ellis et al. (1996). The dotted line is the luminosity function observed at $z = 0.7$ and linearly extrapolated one magnitude further. The resulting galaxy merging evolution at $z < 0.7$ is constrained by these two curves, for this illustration. At $z \approx 0$, it is the dot-dashed line, and at $z \approx 0.33$ it is the dashed line. The bright ends of both these calculated evolved luminosity functions are consistent with the observations of Ellis et al. (1996) for $0.15 < z < 0.35$.

Although the calculated bright end ($M \lesssim M_* \approx -19.30$) reproduces the observations quite well, the faint end does not. The simplest way to reconcile these results would be to drop the assumption of a constant universal mass to light ratio. But if the fainter $M \gtrsim M_*$ merged galaxies become less bright in a uniform way, they still would not match the observed luminosity function at $z = 0$. Somehow the less luminous galaxies would have to become differentially fainter than the more luminous galaxies for the evolved distribution to fit the observed faint-end slope of the local galaxies. To model the process, one might suppose that the absolute magnitude M of a galaxy fainter than M_* increases by an amount proportional to 10^M after merging. The luminosity distribution resulting from this ad hoc "differential dimming" is the dot-dot-dot-dashed line in Figure 36.1, and it matches the observations quite well over the entire range. Such differential dimming may be related to a relative lack of starburst activity in smaller merging galaxies, perhaps because they lose a larger fraction of their gas than in mergers of more massive galaxies. This could happen on timescales much shorter than the Hubble time and lead to faster general dimming of the smaller galaxies.

The best fit of these models to the $z = 0$ observed luminosity function gives a value of $\Theta \approx 4.5 \times 10^{-4} h^3$ Mpc^{-3}. A reasonably consistent set of values for the quantities on which Θ depends in (36.8) would be $\eta \approx 5.5$, $v_0 \approx 10^2$, $\Delta t \approx 5 \times 10^9 h^{-1}$ yr, and with the observationally determined values of ϕ_* and L_* used earlier, an average mass/light ratio $\alpha \approx 10^2$ gives $r_* \approx 10^2 h^{-2}$ kpc. This average galaxy radius includes any dynamically important dark matter halo. It will be interesting to see how well these estimates agree with future more accurate observations.

Now we can turn to the evolution of the galaxy spatial distribution function. Again using the parameterized cross section (36.4), the number of mergers, $\Gamma \Delta t$, in cells containing N galaxies during Δt is the interger just less than

$$N\, n\, S v_0 \Delta t = \eta N\, n \frac{2\pi G m r_*}{v_0} \Delta t = 4(1 + \beta)! \frac{N}{\phi_*} \Theta, \qquad (36.9)$$

where n is the average number density of galaxies and m, r_*, and v_0 all represent the average properties of the galaxies as a simple approximation. A Maxwellian distribution for the galaxy peculiar velocities, for example, would give $N_j = \eta N n [2\pi G m r_* / (\pi \sigma_v)^{1/2}] \Delta t$, where σ_v is the one-dimensional velocity dispersion of the galaxies. Different assumptions for the peculiar velocity distribution would introduce different numerical factors into σ_v. Since neither the velocity distribution nor its dispersion for distant galaxies is well known, we simply denote the denominator by v_0, which has the same order of magnitude as σ_v. The average mass density follows from $\int_0^\infty m\, N(m)\, dm = (1+\beta)!\, m_*\phi_*$ using (36.5).

Notice that the same value of Θ obtained from observations of the evolved luminosity function, along with the parameters of the observed luminosity function, determines the number of mergers that occur in cells containing N galaxies and therefore how the counts-in-cells distribution function evolves. This property of relating two major observational results to the same basic physical parameter is an attractive feature of the master equation description for the two different evolution processes.

The simplest way to start the evolution, lacking more specific knowledge, is to assume a GQED for $f(N)$ at $z = 0.7$ with $b(z = 0.7) = 0.65$ from (30.12) or Figure 31.12 for $\Omega_0 = 1$. If Ω_0 is smaller, $b(z = 0.7)$ is closer to its present value of about 0.75 (see Figure 30.5). This enables us to see directly how merging would modify the adiabatic evolution of $f(N)$ given by (30.12).

Galaxies continue to cluster while they merge, and as their clustering amplitude increases they move into different cells. Adding realistic clustering terms to the master equation (36.2) would be complicated because clustering and merging are coupled. Decoupling them is a reasonable simplifying approximation. First let the clustering of galaxies evolve for a timestep according to (30.12) for $\Omega_0 = 1$, or according to b_{pattern} from (30.19) for $\Omega_0 < 1$. At the end of the step, let the galaxies merge suddenly, with the number of mergers calculated from (36.2) and (36.9) for this time span. The same process repeats in the next timestep. After the evolution has gone as far as desired, check whether the GQED still describes $f(N)$. If so, then the evolution of a merging distribution can be followed by just $b(t)$, which will differ from the $b(t)$ for simple point mass clustering. To do this reasonably realistically, the parameters m, r_*, and v_0 in (36.9) for the transition probabilities $\Gamma\Delta t$ should be represented by distributions rather than just by their average values. Since these distributions are not well determined observationally, they are usually assumed to be Gaussian.

As an illustration, consider an $\Omega_0 = 1$ model in which galaxy halo diameters have a Gaussian distribution whose dispersion is equal to their average value r_*. Starting at $z = 0.67$ with a GQED $f_V(N)$ having $b = 0.65$ for cells whose expectation value of N is $\langle N \rangle = \bar{N} = 30$, Figure 36.2 compares the results of evolution with and without mergers. The dot-dot-dot-dashed line is the simple adiabatic evolution without merging, which leads to $b = 0.80$ at present. Adding merging gives the solid line. It does indeed fit the GQED very well, but with $b = 0.74$ on this scale at present. Moreover, $\langle N \rangle$ is also smaller for the case with mergers, since mergers deplete the high-N tail and enhance $f_V(N)$ for small N. These essential results also apply to models with more continuous merging and with distributions of other galaxy parameters. Incorporating a weakly varying mass/luminosity ratio such as $m/L \propto m^{0.2}$ found from the D_n–σ relation for ellipticals by Oegerle and Hoessel (1991) does not affect these results significantly, but a strong mass dependence related to "differential dimming" could.

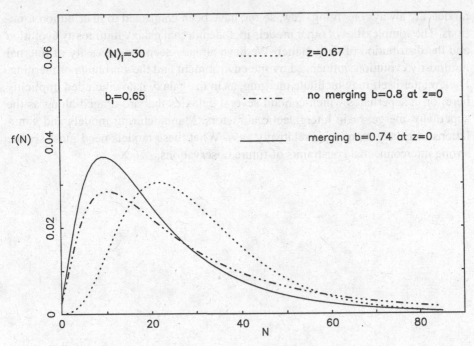

Fig. 36.2. The evolution of the spatial distribution function for the combined effects of galaxy merging and adiabatic clustering for the $\Omega_0 = 1$ model as described in the text. (From Fang & Saslaw, 1997.)

In most cases when there is a broad probability distribution for the number of mergers N_j at given N, the merger-driven evolved distribution fits the GQED very well. This suggests that galaxy merging can regain its quasi-equilibrium state soon after clustering, but with a reduced value of b. This probably implies that the centers of mass of the merging galaxy pairs are also described reasonably well by the GQED. The cells illustrated here have $\langle N \rangle = 30$ and are large enough to represent the asymptotic value of b until very late stages of merging evolution. Smaller cells, in which most of the merging occurs across cell boundaries, may lead to differences from the GQED.

The lower value of b that merging produces may be understood approximately as a result of randomly selecting the galaxies that merge. The ratio of the number of galaxies left after merging to the number available initially gives the value of p in (28.24) or (28.86). The corresponding value of b and the near-GQED that result agree reasonably well with these types of models. In fact too much merging will decrease b below its observed value. This puts limits on the numbers of major mergers – in which two similar high-mass galaxies combine – in such models. Present indications are that fewer than about one fourth of all galaxies were major mergers more recently than $z \lesssim 0.7$.

Here I have emphasized a particularly simple model of merging and its connection with the luminosity and spatial distribution functions. Many more complicated models are always emerging. Few, so far, have been connected to distribution functions. The complexities of other models include internal galaxy luminosity evolution and the distribution of dark matter. We have already seen the necessity of internal luminosity evolution, influenced by the environment and the starbursts of merging. Dark matter itself may facilitate merging, as in the galaxy halos included implicitly here; or in superhalos which contain several galaxies that are draged along as the superhalos merge; or in intergalactic attractors. Manufacturing models and simulations of galaxy merging is relatively easy. What these models need most are the strong interconnected constraints of future observations.

37

Dark Matter Again

No light, but rather darkness visible

Milton

How to observe the dark matter of our Universe remains a primary puzzle of astronomy. Many searches over the radiative spectrum from radio to X rays have uncovered much of great interest, but not enough to close the Universe. Many laboratory experiments have searched directly for exotic weakly interacting massive particles (WIMPS) streaming through space, but to no avail. Gravitational lensing of more distant sources reveals dark matter in clusters along the line of sight; its amount and detailed location are quite model dependent. So far, the main direct evidence for dark matter comes from the gravitational motions of stars within galaxies and galaxies within clusters. All of it adds up to only $\Omega_0 \lesssim 0.3$. Its nature, form, and distribution still are unknown.

Close agreement between the form of the GQED and the observed spatial and velocity distributions of galaxies suggests methods for constraining dark matter. It is relatively easy to start with the cosmological many-body model and formulate dark matter variations around it. These variations should not destroy the observed agreement unduly, nor attract epicycles.

As an illustration (Fang and Saslaw, 1999), consider the peculiar velocity distribution function $f(v)$. This is especially sensitive to local dark matter. In the simplest case, the dark matter is closely associated with each galaxy (e.g., in its halo) as it clusters, and (29.4) describes the velocities. This is consistent with observations. If dark matter is present outside galaxy halos, it may influence the velocities through a generalization of (29.1), which adds the galactic and dark matter together in the form

$$\alpha v^2 = Gm \left\langle \frac{1}{r} \right\rangle N + GD \left\langle \frac{1}{r} \right\rangle M_d(N). \tag{37.1}$$

Here D describes the ensemble average of r^{-1} for the distribution of dark matter in terms of its value $\langle r^{-1} \rangle$ for a Poisson distribution of galaxies as in (29.1). The amount of dark matter, $M_d(N)$, in a region may be related to the number, N, of galaxies there. D is effectively an average over orbits and thus represents a configurational "form factor" for the dark matter. So the velocities depend on the combination DM_d rather than on M_d alone.

The case $M_d \propto N$ is physically plausible and represents a simple version of "linear bias" between the dark and luminous matter. This does not change the form of the GQED $f(v)$. The constant of proportionality between M_d and N can be absorbed into D, and we again obtain (29.4) with α replaced by $\alpha/(1+D)$, recalling the natural units $G = m = R = 1$. All that changes is the interpretation of α to include the dark matter, and this can be incorporated instead into a rescaling of the

473

velocities v and β since α always occurs with them in a dimensionless combination. Physically this arises because, on average, the peculiar velocities are caused by the same linear combination of dark and luminous matter everywhere.

Other relations between M_d and N introduce a spatially "nonlinear bias" and generally alter the form of $f(v)$. Since the observed spatial distribution of galaxies conforms closely to the GQED, it is reasonable to retain the form (29.3) for $f(N)$ and see how (37.1) (which continues to assume that the average kinetic and modified potential energy fluctuations are proportional) modifies $f(v)$. Because there is no currently accepted physical theory for $M_d(N)$, we will just consider two cases that straddle $M_d \propto N$ as illustrations.

The first case is $M_d =$ constant, independent of N. This may represent a situation in which dark matter congeals into superhalos and the number of galaxies inside each superhalo is unrelated to its mass. If all superhalos have the same mass, then (37.1) gives

$$N = \alpha v^2 - DM_d > 0, \tag{37.2}$$

with M_d in units of m, which are unity, and α rescaled by $\langle \frac{1}{r} \rangle^{-1}$. Figure 37.1 shows the

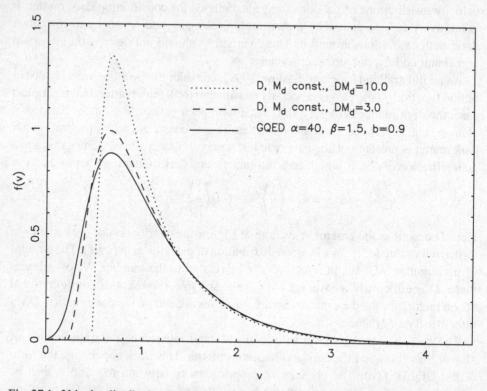

Fig. 37.1. Velocity distribution functions with dark matter of constant mass DM_d added to the observed spatial clustering. Two cases with $DM_d = 10.0$ and 3.0 are compared with the GQED for the same values of α, β, and b. (From Fang & Saslaw, 1999.)

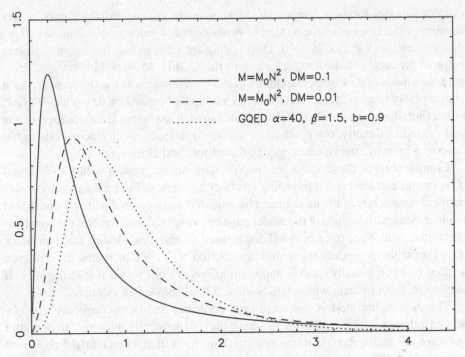

Fig. 37.2. Velocity distribution functions with dark matter of mass $M_d = M_0N^2$ added to the observed spatial clustering. Two cases with $DM_0 = 0.1$ and 0.01 are compared with the GQED for the same values of α, β, and b. (From Fang & Saslaw, 1999.)

resulting $f(v)$ derived analogously to (29.4) for several values of DM_d. The sharp low-velocity cutoff for large M_d results from the inequality in (37.2). More realistically, if a distribution of halo masses were predicted theoretically, DM_d could be treated as a stochastic variable and $f(v, DM_d)$ could be compounded with the probability $P(DM_d)$ to obtain $f[v, P(DM_d)]$ and then compared with observations.

The second illustrative case is $M_d = M_0N^2$. The amount of dark matter now depends quite strongly on the local number of galaxies. Figure 37.2 shows some sample distribution functions. They are quite sensitive, especially in the high-velocity tail, to the value of DM_0. These long tails are produced by the excess dark matter in regions of strong clustering.

A wide variety of other dark matter models can be conceived. However, to be taken seriously they must be shown to be consistent with dynamical evolution, with the observed galaxy spatial distribution function, and with the observed velocity distribution function, as well as with other statistical properties such as the measured low-order correlations and topology. This is a tall order. It is difficult to meet even with the additional parameters that dark matter introduces. Consider some examples (Fang and Saslaw, 1999). Simple models that try to fit the observed radial peculiar velocity distribution function with a double Gaussian, representing strongly and weakly clustered components, have difficulty finding an observationally acceptable ratio of galaxies for each component. More specific models, in which dark matter dominates the cores of clusters, tend to produce radial velocity distributions $f(v_r)$ that are too steep around $v_r = 0$. Distributing the dark matter throughout clusters helps to produce a shape for $f(v)$ close to the GQED. Models with massive great attractors made of dark matter randomly distributed throughout the Universe produce unacceptably large peculiar velocities after a couple of expansion timescales if $\Omega_0 \gtrsim 0.3$. If (hot) dark matter were uniformly distributed, it would not influence local orbits and velocities directly, but it would be harder to reconcile with fluctuations in the cosmic microwave background, galaxy formation, and clustering.

Comparisons of these and other specific dark matter models with the observed $f(v_r)$ require at least two intervening levels of discussion. First, all the possible dynamical parameters in the model must be adjusted and explored in a self-consistent manner. Second the results of the model must be convolved with the observational uncertainties, which are often not well determined themselves. Models that reproduce $f(v_r)$ must also reproduce the spatial distribution, $f(V, N)$, including the observed scaling of $b(r)$. Usually models and simulations are not tested this stringently. If they were, most of them would fail. Section 33.3.3 showed an example.

The dark matter models that seem least contrived and most consistent with observed galaxy positions and velocities have their dynamically important dark matter in or around each galaxy as it clusters. Looking for it there or as tidal debris in rich clusters would seem the best current bet.

38

Initial States

We shall not cease from exploration
And the end of all our exploring
Will be to arrive where we started
And know the place for the first time.
Through the unknown, remembered gate
When the last of earth left to discover
Is that which was the beginning

T. S. Eliot

Why is it so difficult for current observations to determine the initial state of galaxy clustering and even earlier of galaxy formation? The answer, in a word, is dissipation. Much energy changed as its entropy gained, first as galaxies formed and then as they clustered.

To see the magnitude of this transformation, imagine a cube now a hundred megaparsecs across in a universe with $\Omega_0 = 1$ as an example. If the matter in these million cubic megaparsecs had not condensed at all as the universe expanded, if its temperature had decayed adiabatically $\propto R^{-2}$ since decoupling so that now $T \approx 3 \times 10^{-3}$ K, then the total random kinetic energy in this volume would be about 3×10^{56} erg. Gravitational condensation produces dissipation. In gaseous condensation, much of the energy exits as radiation, and some leaves as hot particles. If the dissipation is mostly particulate, as in many-body clustering, escaping orbits carry energy away. The remaining part of the system condenses into a deepening gravitational well and acquires the increased random kinetic energy it needs for quasi-stability. The magnitude of this kinetic energy, $K \approx |W|/2 \approx -E_{total}$, provides an estimate of dissipation.

Consider the energy that would be dissipated if all the matter in this cube were to condense into galaxies, stars, and clusters of galaxies. If all the matter condenses into about 10^6 galaxies, each of $\sim 2 \times 10^{11}$ solar masses (including any dark matter halos) with average stellar velocities ~ 100 km s^{-1}, then their total internal kinetic energy is about 2×10^{64} erg. This is $\sim 10^8$ times the kinetic energy that a uniform particle distribution would have had at $z = 0$. Next, if all the matter within the galaxies condensed into solar type stars, they would have an internal kinetic energy of $\sim 2 \times 10^{65}$ erg. (This would be reduced by the total amount of dark matter outside stars.) Finally, we estimate the kinetic energy of galaxy clustering for a radial peculiar velocity dispersion ~ 700 km s^{-1} (Raychaudhury and Saslaw, 1996) as in the velocity distribution function of Figure 15.10, which includes the rich clusters. Assuming isotropic velocities relative to the cosmic expansion, shows that about 4×10^{66} ergs are involved in the kinetic energy of galaxy clustering. All these numbers are for a representative million cubic megaparsecs.

Gravitational dissipation clearly dominates over simple extrapolation of the uniform universe to the present, and it dominates on all scales from stars to clusters of galaxies. For clusters of galaxies this dominance is greatest – ten orders of magnitude. Looking back into even a simple adiabatic past, for which $K \propto R^{-2}$, the redshift at which the uniform universe would have contained as much local kinetic energy as was subsequently dissipated was $z_{gal} \approx 10^4$ for galaxies, $z_{star} \approx 3 \times 10^4$ for stars, and $z_{cluster} \approx 10^5$ for clustering. All these inhomogeneities have actually formed at much smaller redshifts, indicating that their initial conditions have faded away and dissolved.

Faced with this wall of dissipation, how are we ever to find the initial conditions behind it? The usual way has been to make an educated guess, then try to evolve its initial conditions through the wall and into the present, and see if its results resemble observations. If not, try again. One's educated guess depends on one's education. Thus fluid dynamicists and plasma physicists have looked for instabilities that might arise naturally from chance fluctuations during and after decoupling. Particle physicists have devised a series of increasingly complex inflationary cosmologies that suggest various spectra for primordial perturbations. Other particle physicists have proposed that galaxy formation and clustering started with cosmic strings and textures left over from phase transitions in the very early universe. Astrophysicists have postulated a population of primitive objects whose massive explosions impelled galaxies to form and perhaps also to cluster. Believers in primordial black holes have promoted their virtues as seeds for starting the accretion of galaxies.

While there have been years when one or another approach such as these has held out high hopes for our understanding, in the end all have led to inconsistencies, overwrought complexities, ambiguities, or disagreements with observations. Eventually one may succeed so convincingly and uniquely that the search stops. For now, the search continues.

A less ambitious search also continues for relics of the past, harbored in the present, that may have escaped complete dissipation. The obvious place to look is on scales so large that little has happened. But on these scales, where time is too short for change to develop, observational signals are lost in the linear noise. At the edge of linearity, where initial perturbations have grown but not yet evanesced, prime clues may hover long enough to be seen. These clues will differ from model to model, occurring on scales where two-point correlations approach unity.

The difficulty in all these searches is not in imagining new multiparameter models, which but for inflation would be "a dime a dozen." Rather it is in making clear predictions by which the models stand or fall. And when they fall, as fall they must until we know the final answer, why then we continue the chase in another direction. Past decades have seen the excitement of many pursuits, but the quarry has always escaped.

39

Ultimate Fates

The eternal silence of these infinite spaces terrifies me.

Pascal

That's all there is, there isn't any more.

Ethyl Barrymore

Suppose that cosmological many-body clustering runs on forever. What will happen in the infinite future?

Standard Einstein–Friedmann universes suggest three main possibilities. If the Universe is closed ($\Omega_0 > 1$, $k = +1$) and recollapses into a singularity, all large-scale structure will eventually be destroyed in the big crunch. Whether anything can be resurrected from surviving seeds if the crunch is incomplete (Saslaw, 1991) is unknown. Oscillating universes are possible, though in practice we do not know if the physical requirements for repeated oscillations are consistent with reasonable equations of state. Oscillations whose amplitudes were too small to produce equilibrium would accumulate the debris of previous cycles. Quite apart from the question of increasing entropy, such models would probably require especially fine tuning to produce our observable Universe.

If the Universe is open and expands forever with negative curvature ($\Omega_0 < 1$, $k = -1$), it will expand so rapidly after redshifts $z \lesssim \Omega_0^{-1}$ (see 30.13) that new larger structures will generally cease to form, and the largest scale patterns at $z \approx \Omega_0^{-1}$ will be essentially frozen. These patterns then tend to expand homologously, becoming increasingly stretched and dilute in physical space: Pascal's nightmare. In models with a cosmological constant, the expansion may pause. But it will have to be carefully tuned, so the quasi-stationary period does not produce overclustering, and also satisfy other constraints.

The infinite flat critical density universe ($\Omega_0 = 1$, $k = 0$), which has zero total energy and asymptotically reaches $R \rightarrow \infty$ with $\dot{R} = 0$, is the most interesting with regard to ultimate clustering. Its gravitational clustering may continue forever. The largest linear perturbation or inhomogeneity at any time begins to grow with a timescale slightly less than the then current Hubble timescale $\tau_H = 2R(t)/3\dot{R}(t)$. Given an infinite time, all the inhomogeneities accrete around the region of maximum density in the universe. The rest is empty: Pascal's nightmare again. The discrete and collective dynamical relaxation processes among the galaxies redistribute their orbital energies until the system becomes approximately isothermal – supposing the ultimate stability of matter itself.

What might such a universe look like (Saslaw, Maharaj, and Dadhich, 1996)? Unlike the spherical, homogeneous Robertson–Walker models (Section 16.2) in which there is no center (or as Nicholas of Cusa pointed out, in which every spatial point is an equivalent center (see Chapter 2), or in which, as Gertrude Stein put it, "There is no there there"), the isothermal model singles out the point with the highest density as its center. Asymptotically the universe is static, and so it satisfies the general static, spherically symmetric line element

$$ds^2 = e^\nu dt^2 - e^\lambda dr^2 - r^2(d\theta^2 + \sin^2\theta\, d\phi^2), \tag{39.1}$$

where $\nu = \nu(r)$ and $\lambda = \lambda(r)$. The Einstein field equations are

$$R_{ab} - \frac{1}{2}Rg_{ab} = -8\pi T_{ab} \tag{39.2}$$

with the usual energy–momentum tensor

$$T_{ab} = (p + \rho)u_a u_b - pg_{ab} \quad \text{and} \quad u^a u_a = 1 \tag{39.3}$$

for a perfect fluid.

Isothermal metrics are characterized by a pressure gradient that balances the mutual self-gravity of its constituent particles (considered here as usual to be idealized point galaxies). The dispersion of the particles' peculiar velocities is independent of position, with a simple equation of state

$$p = \alpha\rho \tag{39.4}$$

for the pressure p and density ρ. This is independent of temperature, and α is a constant satisfying $0 < \alpha \leq 1$. Galaxy peculiar velocities are nonrelativistic and ρ is the total energy density including the rest mass energy.

Under these conditions, (39.1)–(39.4) have the exact solution

$$e^\nu = A r^{4\alpha/(1+\alpha)}, \tag{39.5}$$

$$e^\lambda = 1 + \frac{4\alpha}{(1+\alpha)^2} \tag{39.6}$$

for the metric where A is an arbitrary constant. This provides the energy density

$$8\pi\rho = \frac{4\alpha}{4\alpha + (1+\alpha)^2} \frac{1}{r^2} \tag{39.7}$$

and a corresponding relation from (39.4) for the pressure.

This form of the density is remarkably simple and closely resembles that of a Newtonian isothermal sphere (which applies approximately to rich relaxed clusters of galaxies such as Coma; cf. Zwicky, 1957). In the Newtonian case, the central boundary condition requires $\partial\rho/\partial r \to 0$, whereas the cosmological solution has a central singularity $\rho \propto r^{-2}$. This is of little consequence, however, since the mass interior to any radius r, which is the physically important quantity, remains finite, and $M(r) \to 0$ as $r \to 0$. The result (39.7) is an example of a class of spherical similarity solutions that has also been discussed in other contexts (e.g., Chandrasekhar, 1972; Wesson, 1989; Carr, 1997). Other energy density distributions $\rho(r)$ can lead to more complicated isothermal solutions. Moreover, departures from isothermality can also lead to self-consistent static cosmological models, although they may be mechanically stable everywhere only in their isothermal limit. Thus an isothermal sphere has a consistent interpretation as a cosmological model.

Is such a universe stable? If not, the ultimate cluster would be an isothermal sphere only if its instabilities eventually lead to a self-similar state. Although there are no rigorous proofs yet, general physical arguments suggest stability against a range of alterations.

First, in an isothermal sphere of galaxies the pressure gradient adjusts itself to hydrostatic equilibrium with the gravitational force, and so any Jeans type of instability (Section 16.3) caused by local linear perturbations would damp. The critical Jeans length is essentially the size of the sphere – in this case infinite. If local instabilities were to develop, their nonlinear interactions with surrounding galaxies would ultimately rethermalize the distribution. This could change the value of α in (39.4), but the result would still be an isothermal sphere. The form of the density distribution $\rho \propto r^{-2}$ need not change significantly if the temperature changes.

Second, as mentioned earlier, there is a more general class of exact solutions for the spherically symmetric static field equations. These allow α in (39.4) to be an arbitrary positive nonzero constant (so that $p/\rho > 0$ as $r \rightarrow 0$). The solutions depend on an additional arbitrary constant and have a more complicated $\rho(r)$ dependence, which reduces to the isothermal solution (39.7) when this new constant is zero. For this more general class of models to be mechanically stable requires $(\partial p/\partial \rho)_T > 0$ at all radii r. This is not generally true for these models and only holds in their isothermal limit. Therefore, the isothermal models are the mechanically stable subset of this and perhaps other yet undiscovered more general classes of solutions.

Third, consider a more detailed dynamical instability that has long been known to apply to an isothermal sphere. This is its tendency to form a core–halo structure (reviewed, e.g., in GPSGS, Chapters 43–45). In a finite system such as a globular cluster containing N stars, the timescale for this density redistribution is approximately of order $N(G\bar{\rho})^{-1/2}$, which itself becomes infinite for the Einstein–de Sitter universe as $\bar{\rho} \rightarrow 0$. (However, smaller clusters of galaxies will tend to evaporate.) In the central part of the isothermal universe, galaxies may redistribute energy and alter the density by forming binaries that exchange energy with their neighbors. This process is self-limiting for two reasons. First, as the binaries become smaller and more deeply bound, their effective interaction cross section with their neighbors decreases. Second, after the separation of the binary decreases to the radius of its larger member, the binary merges and ceases to transfer energy to its neighbors. This creates some very massive galaxies at the center, perhaps forming black holes. The resulting departures from an r^{-2} density profile, however, would be relatively local.

Fourth, even though detailed galaxy interactions may alter the density profile, the result may still belong to a class of isothermal solutions satisfying (39.4) but without the specific $\rho \propto r^{-2}$ density profile.

Fifth, since the particle horizon of the Einstein–de Sitter universe expands as $3ct$ (where c is the velocity of light), and the universe itself expands as $t^{2/3}$, all the galaxies will have been able to interact in the limit $t \rightarrow \infty$ and $\bar{\rho} \rightarrow 0$, and thus new instabilities cannot enter the horizons.

Sixth, could perturbations in the density or equation of state cause the type of instability found in the classical Einstein or the Lemaître models? The important point to appreciate is that in both these other models equilibrium is attained because the cosmological constant provides a repulsive force. This equilibrium depends crucially on a constant that cannot respond to perturbations. This is the basic cause of their instability. But the static isothermal models are different. Because they are in hydrostatic equilibrium (independent of a global property such as the cosmological constant) any perturbations in density will produce corresponding changes in pressure so as to maintain the equilibrium. Thus the isothermal models will be stable in this respect as well.

Since the isothermal equilibrium state forms only asymptotically, even if these instabilities were important they would need an infinite time to develop. However, the stability of the isothermal state strengthens the tendency toward its establishment.

How would the universe turn into a static isothermal sphere from its earlier expanding homogeneous Robertson–Walker metric? As in other many-body systems where fundamental symmetries change from one state to another, the answer is: through a phase transition. The universe could fluctuate and undergo a phase transition from a statistically homogeneous state of translational and rotational symmetry around every point, to a state of rotational symmetry around one point only, and translational symmetry nowhere.

Let us consider whether the Einstein–de Sitter universe can evolve into an isothermal end state continuously. We can roughly imagine three stages: first, the Einstein–de Sitter expansion, which tends asymptotically to stationarity; second, the condensing stage that leads to the development of cosmologically significant pressure; and finally the isothermal sphere with $p = \alpha\rho \propto r^{-2}$. The first two stages are essentially pressure free, while the last has a nonzero cosmological pressure. To match two spacetimes such as that of (39.1), (39.5), and (39.6) for the isothermal universe with that of (16.3) and (16.7) for the Einstein–de Sitter universe, the junction conditions require continuity of pressure across a specific hypersurface $r = $ constant. Comparison of these two metrics shows no physical hypersurface where this matching might occur. Therefore, we must treat the discontinuous changes as phase transitions: first from the uniform expanding phase to the centrally condensed inhomogeneous state, and later from the pressure-free condensation in which the cosmological importance of galaxy motions can be neglected to an isothermal equilibrium with a nonzero cosmologically significant pressure. In most cosmological models, the metric is fundamental and configurations of matter evolve within it. In the last stages of an

Einstein–de Sitter universe, however, the evolving distribution of matter may alter the symmetry of spacetime's global structure.

With one proviso. If matter itself is inherently unstable and turns into radiation, which becomes more uniformly distributed, it might prevent this phase transition. Or, if it decays after the isothermal universe forms, then the energy redistribution might induce yet another phase transition, perhaps instituting collapse.

Naturally, the application of an isothermal model to the eventual state of our own Universe is very speculative and will remain so for much of the future. If our Universe were to evolve into an isothermal cosmology, it would represent the ultimate astrophysical prediction.

40

Epilogue

But Time, the Domain of Deeds,
calls for a complex grammar
with many Moods and Tenses
and prime the Imperative.

Auden

The search for the structure of our Universe and our position within it never will cease. As we answer each question, others arise with even greater insistence. And the context of our questions is ever changing. From the mythological background of Babylon, to the mechanical clockwork of Newton, through the opening of our minds to prodigous swarms of distant galaxies, to the mathematical models of general relativity and gravitational clustering within them, each new context inspires new questions. Nor is there reason to suppose that the present context will bring this search to a close.

Throughout the roughhewn matrix of our understanding, dark matter weaves threads of uncertainty. Its amount governs the flight of the galaxies and the fate of the Universe. Many models undertake to confine it to various distributions and forms. So far, dark matter has resisted all but gravitational attempts at detection, leaving the models to flicker and shift in the ebb and flow of theoretical fashion.

Nor do observations always provide simple truths. Most are so riddled with selection and filtered with theory that their interpretation is seldom straightforward. Simple ideas like filaments and voids, walls and clusters, become much more complex when closely examined. Their simple grammar often remains suitable mainly for slogans. All good observers know this in their bones. Results, regardless, can still be astounding.

Where then will the search at the start of the new millennium lead? Astronomy's history usually shows that only incremental predictions come true. But the most effective promoter of science is often surprise. As the millennium turns, sights of unforeseen paths should beckon most strongly. Our prime imperative is to create new connections and audacious discoveries.

Bibliography

Aarseth, S. J., 1985. "Direct methods for N-body simulations." In *Multiple Time Scales*, ed. J. V. Brackbill and B. I. Cohen (New York: Academic Press), p. 377.

Aarseth, S. J., 1994. "Direct methods for N-body simulations." In *Galactic Dynamics and N-Body Simulations*, ed. G. Contopoulos, N. K. Spyrou, and L. Vlahos (Berlin: Springer-Verlag), p. 277.

Aarseth, S. J. and Inagaki, S., 1986. "Vectorization of N-body codes." In *The Uses of Supercomputers in Stellar Dynamics*, ed. P. Hut and S. McMillan (Berlin: Springer-Verlag), p. 203.

Aarseth, S. J. and Saslaw, W. C., 1972. "Virial mass determinations of bound and unstable groups of galaxies," *Astrophys. J.* **172**, 17.

Aarseth, S. J. and Saslaw, W. C., 1982. "Formation of voids in the galaxy distribution," *Astrophys. J. Lett.* **258**, L7.

Aarseth, S. J. and Zare, K., 1974. "A regularization of the three-body problem," *Celest. Mech.* **10**, 185.

Abbe, C., 1867. "On the distribution of the nebulae in space," *Mon. Not. R. Astron. Soc.* **27**, 257.

Abel, N. B., 1826. Crelle **1**, 159.

Abell, G. O., 1958. "The distribution of rich clusters of galaxies," *Astrophys. J. Suppl.* **3**, 211.

Abell, G. O., 1965. "Clustering of galaxies," *Annu. Rev. Astron. Astrophys.* **3**, 1.

Abramowitz, M. and Stegun, I. A., 1964. *Handbook of Mathematical Functions* (Washington, DC: National Bureau of Standards).

Adler, R. J., 1981. *The Geometry of Random Fields* (Chichester: Wiley).

Ahmad, F., 1996. "Two-particle correlation function and gravitational galaxy clustering," IUCAA preprint 8/96.

Ambartsumian, V. A., 1958. "On the evolution of galaxies." In *La Structure et L'evolution de L'Universe*, ed. by R. Stoops (Brussels: Institut International de Physique Solvay), p. 241.

Anesaki, M., 1930. *History of Japanese Religion* (London: Kegan Paul).

Antonov, V. A., 1962. "Most probable phase distribution in spherical stellar systems and conditions of its existence," *J. Leningrad Univ.* **7**, Issue 2.

Arp, H. C., 1987. *Quasars, Redshifts and Controversies* (Berkeley: Interstellar Media).

Baade, W., 1963. *Evolution of Stars and Galaxies* (Cambridge, MA: Harvard Univ. Press).

Bahcall, N. A., Fan, X., and Cen, R., 1997. "Constraining Ω with cluster evolution," *Astrophys. J. Lett.* **485**, L53.

Baier, F. W., Fritze, K., and Tiersch, H., 1990. "The asymmetry of the Coma cluster of galaxies," *Astron. Nachr.* **311**, 89.

Bak, P., 1996. *How Nature Works* (New York: Springer-Verlag).

Baleisis, A., Lahav, O., Loan, A. J., and Wall, J. V., 1998. "Searching for large scale structure in deep radio surveys," *Mon. Not. R. Astron. Soc.*, **297**, 545.

Balescu, R., 1975. *Equilibrium and Nonequilibrium Statistical Mechanics* (New York: Wiley).

Balian, R. and Schaeffer, R., 1989. "Scale-invariant matter distribution in the universe. I. Counts in cells," *Astron. Astrophys.* **220**, 1.

Bardeen, J. M., Bond, J. R., Kaiser, N., and Szalay, A. S., 1986. "The statistics of peaks of Gaussian random fields," *Astrophys. J.* **304**, 15.

Barnard, E. E., 1906. "Groups of small nebulae," *Astron. Nachr.* **173**, 118.

Barnes, J. and Hut, P., 1986. "A hierarchical O (N log N) force-calculation algorithm," *Nature* **324**, 446.

Barrow, J. D., and Bhavsar, S. P., 1987. "Filaments: What the astronomer's eye tells the astronomer's brain," *Quart. J. R. Astron. Soc.* **28**, 109.

Barrow, J. D., Bhavsar, S. P., and Sonoda, D. H., 1985. "Minimal spanning trees, filaments and galaxy clustering," *Mon. Not. R. Astron. Soc.* **216**, 17.

Beaky, M. W., Sherrer, R. J., and Villumsem, J. V., 1992. "Topology of large-scale structure in seeded hot dark matter models," *Astrophys. J.* **387**, 443.

Beckwith, M., 1940. *Hawaiian Mythology* (New Haven: Yale Univ. Press).

Beers, T. C., Flynn, K., and Gebhardt, K., 1990. "Measures of location and scale for velocities in clusters of galaxies – a robust approach," *Astron. J.* **100**, 32.

Benoist, C. et al., 1996. "Biasing in the galaxy distribution," *Astrophys. J.* **472**, 452.

Benoist, C. et al., 1997. "Biasing and high-order statistics from the SSRS2," preprint.

Bergmann, A. G., Petrosian, V., and Lynds, R., 1990. "Gravitational lens models of arcs in clusters," *Astrophys. J.* **350**, 23.

Bernardeau, F., 1992. "The gravity-induced quasi-Gaussian correlation hierarchy," *Astrophys. J.* **392**, 1.

Bernstein, G. M., 1994. "The variance of correlation function estimates," *Astrophys. J.* **424**, 569.

Bertschinger, E. and Dekel, A., 1989. "Recovering the full velocity and density fields from large-scale redshift-distance samples," *Astrophys. J. Lett.* **336**, L5.

Bertschinger, E. and Gelb, J. M., 1991. "Cosmological N-body simulations," *Computers in Physics* **5**, 164.

Bertschinger, E., 1998. "Simulations of structure formation in the Universe," *Annu. Rev. Astron. Astrophysics.* **36**, 599.

Bettis, D. G. and Szebehely, V., 1972. "Treatment of close approaches in the numerical integration of the gravitational problem of N-bodies." In *Gravitational N-Body Problem*, ed. M. Lecar (Dordrecht: Reidel), p. 388.

Bhavsar, S. P. and Cohen, J. P., 1990. "The statistics of bright galaxies," *Mon. Not. R. Astron. Soc.* **247**, 462.

Bhavsar, S. P. and Ling, E. N., 1988a. "Large-scale distribution of galaxies: Filamentary structure and visual bias," *Publ. Astron. Soc. Pacific* **100**, 1314.

Bhavsar, S. P. and Ling, E. N., 1988b. "Are the filaments real?," *Astrophys. J. Lett.* **331**, L63.

Binney, J. and Tremaine, S., 1987. *Galactic Dynamics* (Princeton: Princeton Univ. Press).

Bok, B., 1934. "The apparent clustering of external galaxies," *Harvard College Obs. Bull.* **895**, 1.

Bond, J. R., Cole, S., Efstathiou, G., and Kaiser, N., 1991. "Excursion set mass functions for hierarchial Gaussian fluctuations," *Astrophys. J.* **379**, 440.

Bondi, H., 1960. *Cosmology*, 2nd ed. (Cambridge: Cambridge University Press).

Bonometto, S. A. et al., 1993. "Correlation functions from the Perseus–Pisces redshift survey," *Astrophys. J.* **419**, 451.

Bonnor, W. B., 1956. "Boyle's law and gravitational instability," *Mon. Not. R. Astron. Soc.* **116**, 351.

Bonnor, W. B., 1957. "Jeans' formula for gravitational instability," *Mon. Not. R. Astron. Soc.* **117**, 104.

Bonnor, W. B., 1958. "Stability of polytropic gas spheres," *Mon. Not. R. Astron. Soc.* **118**, 523.

Borgani, S., 1993. "The multifractal behaviour of hierarchial density distributions," *Mon. Not. R. Astron. Soc.* **260**, 537.

Börner, G. and Mo, H. J., 1989a. "Geometrical analysis of galaxy clustering: Dependence on luminosity," *Astron. Astrophys.* **223**, 25.

Börner, G. and Mo, H. J., 1989b. "Percolation analysis of cluster superclustering," *Astron. Astrophys.* **224**, 1.

Bouchet, F. R. and Hernquist, L., 1992. "Gravity and count probabilities in an expanding universe," *Astrophys. J.* **400**, 25.

Bouchet, F. R., Shaeffer, R., and Davis, M., 1991. "Nonlinear matter clustering properties of a cold dark matter universe," *Astrophys. J.* **383**, 19.

Bouchet, F. R. et al., 1993. "Moments of the counts distribution in the 1.2 Jansky *IRAS* galaxy redshift survey," *Astrophys. J.* **417**, 36.

Bowyer, S. and Leinert, C., eds., 1990. *The Galactic and Extragalactic Background Radiation.* IAU Symp. 139 (Dordrecht: Kluwer).

Brecher, K. and Silk, J., 1969. "Lemaitre Universe, galaxy formation and observations," *Astrophys. J.* **158**, 91.

Brillouin, L., 1963. *Science and Information Theory* (New York: Academic Press).

Brown, S. C., 1979. *Benjamin Thompson, Count Rumford* (Cambridge, MA: The MIT Press).

Buchdahl, H. A., 1966. *The Concepts of Classical Thermodynamics* (Cambridge, UK: Cambridge University Press).

Burland, C., 1965. *North America Indian Mythology* (Feltham: Paul Hamlyn).

Burstein, D. et al., 1987. "Spectroscopy and photometry of elliptical galaxies. III. UBV aperture photometry, CCD photometry, and magnitude-related parameters," *Astrophys. J. Supplement* **64**, 601.

Callen, H. B., 1960. *Thermodynamics* (New York: Wiley).

Callen, H. B., 1985. *Thermodynamics and an Introduction to Thermostatics* (New York: Wiley).

Cantor, G., 1883. "Veber unendliche, lineare punktmannich faltigkeiten," *Math Ann.* **21**, 545.

Carpenter, E. F., 1931. "A cluster of extra-galactic nebulae in Cancer," *Publ. Astron. Soc. Pacific* **43**, 247.

Carpenter, E. F., 1938. "Some characteristics of associated galaxies. I. A density restriction in the metagalaxy," *Astrophys. J.* **88**, 344.

Carr, B. J., 1997. "Spherically symmetric similarity solutions and their applications in cosmology and astrophysics." In *Proceedings of the 7th Canadian Conference on General Relativity and Relativistic Astrophysics*, ed. D. Hobill.

Chandrasekhar, S., 1960. *Principles of Stellar Dynamics* (New York: Dover).

Chandrasekhar, S., 1972. "A limiting case of relativistic equilibrium." In *General Relativity*, ed. L. O'Raifeartaigh (Oxford: Clarendon Press).

Chandrasekhar, S. and Münch, G., 1952. "The theory of the fluctuations in brightness of the Milky Way. V," *Astrophys. J.* **115**, 103.

Charlier, C. V. L., 1908. "Wie eine unendliche welt aufgebout sein kann," *Arkiv Matematick Astronomi Fysik* **4** (24).

Charlier, C. V. L., 1922. "How an infinite world may be built up," *Arkiv Matematick Astronomi Fysik* **16** (22).

Charlier, C. V. L., 1927. "On the structure of the universe," *Publ. Astron. Soc. Pacific* **37**, 53, 115, 177.

Clerke, A. M., 1890. *The System of the Stars* (London: Longmans, Green) p. 368.

Cole, S., Ellis, R., Broadhurst, T., and Colless, M., 1994. "The spatial clustering of faint galaxies," *Mon. Not. R. Astron. Soc.* **267**, 541.

Coleman, P. H. and Pietronero, L., 1992. "The fractal structure of the universe," *Physics Reports* **213**, 311.

Coleman, P. H. and Saslaw, W. C., 1990. "Structure in the universe on scales of 10–100 Mpc: Analysis of a new deep radio survey and of the Abell cluster catalog," *Astrophys. J.* **353**, 354.

Coles, P. and Lucchin, F., 1995. *Cosmology, the Origin and Evolution of Cosmic Structure* (Chichester: Wiley).

Consul, P. C., 1989. *Generalized Poisson Distributions: Properties and Applications* (New York: Dekker).

Courteau, S. et al., 1993. "Streaming motions in the local universe: Evidence for large-scale, low-amplitude density fluctuations," *Astrophys. J. Lett.* **412**, L51.

Couchman, H. M. P., 1991. "Mesh-refined P3M: A fast adaptive N-body algorithm," *Astrophys. J.* **368** L23.

Cox, D. R. and Miller, H. D., 1965. *The Theory of Stochastic Processes* (London: Chapman and Hall).

Crane, P. and Saslaw, W. C., 1986. "How relaxed is the distribution of galaxies?," *Astrophys. J.* **301**, 1.

Crane, P. and Saslaw, W. C., 1989. "Gravitational clustering of galaxies in the CfA slice." In *Dark Matter*, ed. J. Audouze and J. Tran Thanh Van (Gif-sur-Yvette: Editions Frontiers).

Curtis, H. D., 1919. "Modern theories of spiral nebulae," *J. Washington. Acad. Science* **9**, 217. Also see *J. Roy. Astron. Soc. Canada* **14**, 317 (1920).

Curtis, H. D., 1921. "Dimensions and structure of the galaxy," *Bull. Natl. Research Council of the Natl. Acad. Sci.* **2**, 194.

Cusa, Nicolas of, 1954. *Of Learned Ignorance.* Transl. by Fr. G. Heron (New Haven: Yale Univ. Press).

da Costa, L. N. et al., 1988. "The southern sky redshift survey," *Astrophys. J.* **327**, 544.

da Costa L. N. et al., 1994. "A complete southern sky redshift survey," *Astrophys. J. Lett.* **424**, L1.

Daley, D. J. and Vere-Jones, D., 1988. *An Introduction to the Theory of Point Processes* (New York: Springer-Verlag).

Dalton, G. B., Efstathiou, G., Maddox, S. J., and Sutherland, W. J. 1992. "Spatial correlations in a redshift survey of APM galaxy clusters," *Astrophys. J. Lett.* **390**, L1.

Davis, M. and Peebles, P. J. E., 1977. "On the integration of the BBGKY equations for the development of strongly nonlinear clustering in an expanding universe," *Astrophys. J. Suppl.* **34**, 425.

de Fontenelle, B., 1769. *A Conversation on the Plurality of Worlds* (London: Daniel Evans) p. 126.

Degerle, W. R. and Hoessel, J. G., 1991. "Fundamental parameters of brightest cluster galaxies," *Astrophys. J.* **375**, 15.

de Groot, S. R. and Mazur, P., 1962. *Non-Equilibrium Thermodynamics* (Amsterdam: North Holland).

Dekel, A., 1994. "Dynamics of cosmic flows," *Annu. Rev. Astron. Astrophys.* **32**, 371.

Dekel, A. and Aarseth, S. J., 1984. "The spatial correlation function of galaxies confronted with theoretical scenarios," *Astrophys. J.* **283**, 1.

Dekel, A., Blumenthal, G. R., Primak, J. R., and Oliver, S., 1989. "Projection contamination in cluster catalogs: Are the Abell redshift sample correlations significant?," *Astrophys. J. Lett.* **338**, L5. See also *Astrophys. J.* **356**, 1.

Dekel, A. and West, M. J., 1985. "On percolation as a cosmological test," *Astrophys. J.* **288**, 11.

de Lapparent, V., Geller, M. J., and Huchra, J., 1986. "A slice of the universe," *Astrophys. J. Lett.* **302**, L1.

de Lapparent, V., Geller, M. J., and Huchra, J. P., 1991. "Measures of large-scale structure in the CfA redshift survey slices," *Astrophys. J.* **369**, 273.

Deprit, A., 1983. "Monsignor Georges Lemaitre." In *The Big Bang and Georges Lemaitre*, ed. A. Berger (Dordrecht: Reidel).

de Sitter, W., 1917a. "On the curvature of space," *Proc. Akad. Amsterdam* **20**.

de Sitter, W., 1917b. "On Einstein's theory of gravitation and its astronomical consequences," *Mon. Not. R. Astron. Soc.* **78**, 3.

de Vaucouleurs, G., 1970. "The case for a hierarchical cosmology," *Science* **167**, 1203.

de Vaucouleurs, G., 1971. "The large scale distribution of galaxies and clusters of galaxies," *Publ. Astron. Soc. Pacific* **83**, 113.

de Vaucouleurs, G., 1989. "Who discovered the local supercluster of galaxies?," *Observatory* **109**, 237.

Drazin, P. G., 1992. *Nonlinear Systems* (Cambridge, UK: Cambridge Univ. Press).

Dressler, A. et al., 1987. "Spectroscopy and photometry of elliptical galaxies: A large-scale streaming motion in the local universe," *Astrophys. J. Lett.* **313**, L37.

Dreyer, J. L. E., 1905. *A History of the Planetary Systems from Thales to Kepler* (Cambridge, UK: Cambridge Univ. Press).

Duhem, P., 1985. *Medieval Cosmology*. Transl. by R. Ariew (Chicago: Univ. Chicago Press).

Easton, C., 1904. "La distribution des nebuleuses et leurs relation avec le system galactique," *Astron. Nachr.* **166**, 131.

Eddington, A. S., 1930. "On the instability of Einstein's spherical world," *Mon. Not. R. Astron. Soc.* **90**, 668.

Eddington, A. S., 1931a. "The expansion of the universe," *Mon. Not. R. Astron. Soc.* **91**, 412.

Eddington, A. S., 1931b. "The end of the world: From the standpoint of mathematical physics," *Nature* **127**, 447 (Supplement).

Efstathiou, G., Davis, M., Frenk, C. S., and White, S. D. M., 1985. "Numerical techniques for large cosmological N-body simulations," *Astrophys. J. Suppl.* **57**, 241.

Efstathiou, G., Bond, J. R., and White, S. D. M., 1992. "*COBE* background radiation anisotropies and large scale structure in the universe," *Mon. Not. R. Astron. Soc.* **258**, 1p.

Einasto, J. and Saar, E., 1987. "Spatial distribution of galaxies: Biased galaxy formation, supercluster-void topology, and isolated galaxies." In *Observational Cosmology*, ed. A. Hewitt et al., IAU Symp. 124 (Dordrecht: Reidel), p. 349.

Einstein, A., 1917. "Kosmologische betrachtungen zur allgemeinen relativistatstheorie," *Sitzungsber* Berlin. Feb. 8, 142.

Ellis, D. H. R., 1964. *Gods and Myths of Northern Europe* (London: Penguin).

Ellis, R. S., Colless, M., Broadhurst, T. J., Heyl, J., and Glazebrook, K., 1996. "Autofib redshift survey – I. Evolution of the galaxy luminosity function," *Mon. Not. R. Astron. Soc.* **280**, 235.

Epstein, R. I., 1983. "Proto-galactic perturbations," *Mon. Not. R. Astron. Soc.* **205**, 207.

Evrard, A. E., 1988. "Beyond N-body: 3D cosmological gas dynamics," *Mon. Not. R. Astron. Soc.* **235**, 911.

Eyles, C. J. et al., 1991. "The distribution of dark matter in the Perseus cluster," *Astrophys. J.* **376**, 23.

Faber, S. M. and Gallagher, J. S., 1979. "Masses and mass to light ratios of galaxies," *Annu. Rev. Astron. Astrophys.* **17**, 135.

Faber, S. M. and Jackson, R. E., 1976. "Velocity dispersions and mass-to-light ratios for elliptical galaxies," *Astrophys. J.* **204**, 668.

Fabian, A. C. and Barcons, X., 1991. "Intergalactic matter," *Rep. Prog. Phys.* **54**, 1069.

Fabian, A. C., Nulsen, P. E. J., and Canizares, C. R., 1991. "Cooling flows in clusters of galaxies." *Astron. Astrophys. Rev.* **2**, 191.

Fairall, A., 1998. *Large-Scale Structures in the Universe* (Chichester: Wiley).

Fairall, A. P. et al., 1990. "Large-scale structure in the Universe: Plots from the updated *Catalogue of Radial Velocities of Galaxies and the Southern Redshift Catalogue*," *Mon. Not. R. Astron. Soc.* **247**, 21 p.

Fall, S. M. and Saslaw, W. C., 1976. "The growth of correlations in an expanding universe and the clustering of galaxies," *Astrophys. J.* **204**, 631.

Fang, F. and Saslaw, W. C., 1995. "Some implications of the radial peculiar velocity distribution of galaxies." In *Wide-Field Spectroscopy and the Distant Universe*, ed. S. J. Maddox and A. Aragón-Salamanca (Singapore: World Scientific).

Fang, F. and Saslaw, W. C., 1997. "Effects of galaxy mergers on their spatial distribution and luminosity function," *Astrophys. J.* **476**, 534.

Fang, F. and Saslaw, W. C., 1999, in preparation.

Fang, F. and Zou, Z., 1994. "Structure in the distribution of southern galaxies described by the distribution function," *Astrophys. J.* **421**, 9.

Feder, J., 1988. *Fractals* (New York: Plenum).

Feller, W., 1957. *An Introduction to Probability Theory and Its Applications*, Vol. I, 2nd ed. (Vol. II 1966) (New York: Wiley).

Field, G. B. and Saslaw, W. C., 1971. "Groups of galaxies: Hidden mass or quick disintegration?," *Astrophys. J.* **170**, 199.

Fisher, K. B., Davis, M., Strauss, M. A., Yahil, A., and Huchra, J., 1994. "Clustering in the 1.2-Jy *IRAS* galaxy redshift survey – I. The redshift and real space correlation functions," *Mon. Not. R. Astron. Soc.* **266**, 50.

Flin, P., 1988. "Early studies of the distribution of the nebulae," *Acta Cosmologica* **15**, 25.

Flory, P. J., 1941. "Molecular size distribution in three dimensional polymers. II. Trifunctional branching units," *J. Am. Chem. Soc.* **63**, 3091.

Fong, R., Stevenson, P. R. F., and Shanks, T., 1990. "The orientation of galaxies and clusters from an objectively defined catalog," *Mon. Not. R. Astron. Soc.* **242**, 146.

Fong, R., Hale-Sutton, D., and Shanks, T., 1992. "Correlation function constraints on large-scale structure in the distribution of galaxies," *Mon. Not. R. Astron. Soc.* **257**, 650.

Fowler, R. H., 1936 also 1966. *Statistical Mechanics* (Cambridge, UK: Cambridge University Press).

Friedmann, A., 1922. "Uber die krümmung des raümes," *Z. Phys.* **10**, 377.

Fry, J. N., 1985. "Cosmological density-fluctuations and large-scale structure – From N-point correlation-functions to the probability-distribution," *Astrophys. J.* **289**, 10.

Fry, J. N. and Gaztañaga, E., 1994. "Redshift distortions of galaxy correlation functions," *Astrophys. J.* **425**, 1.

Fry, J. N., Giovanelli, R., Haynes, M., Melott, A., and Scherrer, R. J., 1989. "Void statistics, scaling, and the origins of large-scale structure," *Astrophys. J.* **340**, 11.

Gamow, G., 1954. "On the formation of protogalaxies in the turbulent primordial gas," *Proc. Natl. Acad. Sci.* **40**, 480.

Gamow, G. and Teller, E., 1939. "The expanding universe and the origin of the great nebulae," *Nature* **143**, 116.

Geller, M. J. and Huchra, J. P., 1989. "Mapping the universe," *Science* **246**, 897.

Giovanelli, R. and Haynes, M. P., 1993. "A survey of the Pisces–Perseus supercluster. VI. The declination zone $+15.5°$ to $+21.5°$," *Astron. J.* **105**, 1271.

Glass, L., 1969. "Moiré effect from random dots," *Nature* **223**, 578.

Glass, L. and Perez, R., 1973. "Perception of random dot interference patterns," *Nature* **246**, 360.

Goodstein, D. L., 1985. *States of Matter* (New York: Dover).

Gott, J. R. et al., 1989. "The topology of large-scale structure III. Analysis of observations," *Astrophys. J.* **340**, 625.

Grad, H., 1958. "Principles of the kinetic theory of gases." In *Handbuch der Physik*, ed. S. Flugge, Vol. XII (Berlin: Springer-Verlag), p. 205.

Gramann, M., 1990. "Structure and formation of superclusters – XI. Voids, connectivity and mean density in the CDM universes," *Mon. Not. R. Astron. Soc.* **244**, 214.

Gramann, M., Cen, R., and Bahcall, N. A., 1993. "Clustering of galaxies in redshift space: Velocity distortion of the power spectrum and correlation function," *Astrophys. J.* **419**, 440.

Grant, M., 1962. *Myths of the Greeks and Romans* (London: Weidenfield & Nicholson).

Graves, R., 1948. *The Greek Myths* (London: Penguin).

Gregory, S. A. and Thompson, L. A., 1978. "The Coma/A1367 supercluster and its environs," *Astrophys. J.* **222**, 784.

Grey, G., 1965. *Polynesian Mythology* (London: Whitcomb & Tombs).

Grimmett, G., 1989. *Percolation* (New York: Springer-Verlag).

Gunn, J. E. and Gott, J. R. III., 1972. "On the infall of matter into clusters of galaxies and some effects on their evolution," *Astrophys. J.* **176**, 1.

Hamilton, A. J. S., Saslaw, W. C., and Thuan, T. X., 1985. "Thermodynamics and galaxy clustering: Analysis of the CfA catalog," *Astrophys. J.* **297**, 37.

Hamilton, A. J. S., Gott, J. R., and Weinberg, D., 1986. "The topology of the large-scale structure of the universe," *Astrophys. J.* **309**, 1.

Hamilton, A. J. S., Kumar, P., Lu, E., and Matthews, A., 1991. "Reconstructing the primordial spectrum of fluctuations of the universe from the observed nonlinear clustering of galaxies," *Astrophys. J.* **374**, L1.

Hammer, F., 1991. "Thin and giant luminous arcs: A strong test of the lensing cluster mass distribution," *Astrophys. J.* **383**, 66.

Hammer, F. and Rigaut, F., 1989. "Giant luminous arcs from lensing: Determination of the mass distribution inside distant cluster cores," *Astron. Astrophys.* **226**, 45.

Hardcastle, J. A., 1914. "Nebulae seen on Mr. Franklin-Adams, plates," *Mon. Not. R. Astron. Soc.* **74**, 698.

Hansel, D., Bouchet, F. R., Pellot, R., and Ramani, A., 1986. "On the closure of the hierarchy of galaxy correlation function in phase space," *Astrophys. J.* **310**, 23.

Harrison, E. R., 1965. "Olbers' paradox and the background radiation in an isotropic homogeneous universe," *Mon. Not. R. Astron. Soc.* **131**, 1.

Harrison, E. R., 1985. *Masks of the Universe* (New York: Macmillan).

Harrison, E. R., 1986. "Newton and the infinite universe," *Physics Today* **39**, 24.

Harrison, E. R., 1987. *Darkness at Night* (Cambridge, MA: Harvard Univ. Press).

Hartwick, F. D. A., 1981. "Groups of spiral galaxies around the Coma cluster and upper limits to its mass," *Astrophys. J.* **248**, 423.

Haud, V., 1990. "Are the groups of galaxies flat?," *Astron. Astrophys.* **229**, 47.

Hausdorff, F., 1919. "Dimension and ä usseres mass," *Math. Ann.* **79**, 157.

Henning, W. B., 1951. *Zoroaster* (Oxford: Oxford Univ. Press).

Hernquist, L., 1987. "Performance characteristics of tree codes," *Astrophys. J. Suppl.* **64**, 715.

Hernquist, L. and Katz, N., 1989. "TREESPH: A unification of SPH with the hierarchical tree method," *Astrophys. J. Suppl.* **70**, 419.

Herschel, W., 1784. "Account of some observations tending to investigate the construction of the heavens," *Phil. Trans.* **LXXIV**, 437.

Herschel, W., 1785. "On the construction of the heavens," *Phil. Trans.* **LXXV**, 213.

Herschel, W., 1811. "Astronomical observations relating to the construction of the heavens, arranged for the purpose of a critical examination, the results of which appears to throw some new light upon the organization of the celestial bodies," *Phil. Trans. R. Soc.* **101**, 269.

Herschel, J., 1847. "Results of astronomical observations made during the years 1834, 5, 6, 7, 8 at Cape of Good Hope, being the completion of a telescopic survey of the whole surface of the visible heavens, commenced in 1825," *Phil. Trans.* **137**, 1.

Hertsprung, E., 1913. "Über die raüm liche verteilung der veränderlichen vom δ Cephei-Typus," *Astron. Nachr.* **196**, 201.

Hesiod, ca. 800 B.C. *Theogony*. Transl. by M. L. West, 1988 (Oxford: Oxford Univ. Press).

Hetherington, N. S., 1988. *Science and Objectivity* (Ames: Iowa State Univ. Press).

Hickson, P., 1997. "Compact groups of galaxies," *Annu. Rev. Astron. Astrophys.* **35**, 357.

Hill, T. L., 1956. *Statistical Mechanics: Principles and Selected Applications* (New York: McGraw-Hill).

Hill, T. L., 1960. *An Inroduction to Statistical Thermodynamics* (Reading: Addison-Wesley).

Hill, T. L., 1986. *An Introduction to Statistical Thermodynamics* (New York: Dover).

Hill, T. L., 1987. *Statistical Mechanics* (New York: Dover).

Hinks, A. R., 1914. "Note," *Mon. Not. R. Astron. Soc.* **74**, 706.

Hockney, R. W. and Eastwood, J. W., 1981. *Computer Simulation Using Particles* (New York: McGraw-Hill).

Holmberg, E., 1940. "On the clustering tendencies among the nebulae I," *Astrophys. J.* **92**, 200.

Holmberg, E., 1941. "On the clustering tendencies among the nebulae II," *Astrophys. J.* **94**, 385.

Horwitz, G. and Katz, J., 1978. "Steepest descent technique and stellar equilibrium statistical mechanics. III. Stability of various ensembles," *Astrophys. J.* **222**, 941.

Hoskin, M. A., 1963. *William Herschel and the Construction of the Heavens* (London: Oldbourne).

Hoskin, M. A., 1987. "John Herschel's cosmology," *J. Hist. Astron.* **18**, 1.

Huang, K., 1987. *Statistical Mechanics* (New York: Wiley).

Hubble, E., 1925a. "Cepheids in spiral nebulae," *Publ. Am. Astron. Soc.* **5**, 261.

Hubble, E., 1925b. "Cepheids in spiral nebulae," *Observatory* **48**, 139.

Hubble, E., 1925c. "N.G.C. 6822 a remote stellar system," *Astrophys. J.* **62**, 409.

Hubble, E., 1926. "A spiral nebula as a stellar system Messier 33," *Astrophys. J.* **63**, 236.

Hubble, E., 1929a. "A spiral nebula as a stellar system Messier 31," *Astrophys. J.* **69**, 103.

Hubble, E., 1929b. "A relation between distance and radial velocity among extra-galactic nebulae," *Proc. Natl. Acad. Sci.* **15**, 168.

Hubble, E., 1931. "The distribution of nebulae," *Publ. Astron. Soc. Pacific* **43**, 282.

Hubble, E., 1934. "The distribution of extra-galactic nebulae," *Astrophys. J.* **79**, 8.

Hubble, E., 1936a. *The Realm of the Nebulae* (New Haven: Yale Univ. Press).

Hubble, E., 1936b. "Effects of red shifts on the distribution of nebulae," *Astrophys. J.* **84**, 517.

Hudson, M. J. and Lynden-Bell, D., 1991. "Diameter functions of UGC and ESO galaxies," *Mon. Not. R. Astron. Soc.* **252**, 219.

Huggins, W., 1864. "On the spectra of some nebulae," *Phil. Trans.* **154**, 437.

Hughes, J. P., 1989. "The mass of the Coma cluster: Combined X-ray and optical results," *Astrophys. J.* **337**, 21.

Inagaki, S., 1976. "On the density correlations of fluctuations in an expanding universe. II," *Publ. Astron. Soc. Jpn.* **28**, 463.

Inagaki, S., 1991. "On the validity of the quasi-linear theory for the cosmological three-point correlation function," *Publ. Astron. Soc. Jpn* **43**, 661.

Inagaki, S., Itoh, M., and Saslaw, W. C., 1992. "Gravitational clustering of galaxies: Comparison between thermodynamic theory and N-body simulations: III. Velocity distributions for different mass components," *Astrophys. J.* **386**, 9.

Ions, V., 1967. *Indian Mythology* (Feltham: Paul Hamlyn).

Ions, V., 1968. *Egyptian Mythology* (Feltham: Paul Hamlyn).

Irvine, W. M., 1961. *Local Irregularities in a Universe Satisfying the Cosmological Principle* (Harvard University Thesis).

Itoh, M., 1990. "Quantitative descriptions of nonlinear gravitational galaxy clustering," *Publ. Astron. Soc. Jpn.* **42**, 481.

Itoh, M., Inagaki, S., and Saslaw, W. C., 1988. Gravitational clustering of galaxies: Comparison between thermodynamic theory and N-body simulations," *Astrophys. J.* **331**, 45.

Itoh, M., Inagaki, S., and Saslaw, W. C., 1990. "Gravitational clustering of galaxies: Comparison between thermodynamic theory and N-body simulations: II. The effects of different mass components," *Astrophys. J.* **356**, 315.

Itoh, M., Inagaki, S., and Saslaw, W. C., 1993. "Gravitational clustering of galaxies: comparison between thermodynamic theory and N-body simulations. IV. The effects of continuous mass spectra," *Astrophys. J.* **403**, 476.

James, E. O., 1958. *Myth and Ritual in the Ancient Near East* (London: Thames & Hudson).

Jeans, J. H., 1902a. "The stability of a spherical nebula," *Phil. Trans.* 199A, 1.

Jeans, J. H., 1902b. "The stability of a spherical nebula," *Phil. Trans.* 199A, 49.

Jeans, J. H., 1913. "On the kinetic theory of star-clusters," *Mon. Not. R. Astron. Soc.* **74**, 109.

Jeans, J. H., 1929. *Astronomy and Cosmogony* (Cambridge, UK: Cambridge Univ. Press).

Jedamzik, K., 1995. "The cloud-in-cloud problem in the Press–Schechter formalism of hierarchial structure formation," *Astrophys. J.* **448**, 1.

Kaiser, N., 1984. "On the spatial correlations of Abell clusters," *Astrophys. J.* **284**, L9.

Kaiser, N., 1987. "Clustering in real space and in redshift space," *Mon. Not. R. Astron. Soc.* **227**, 1.

Kang, H. et al., 1994. "A comparison of cosmological hydrodynamic codes," *Astrophys. J.* **430**, 83.

Kant, I., 1755. "Universal natural history and the theory of the heavens," transl. by W. Hastie, in *Kant's Cosmogony*, Glasgow, 1900.

Katz, L. and Mulders, G. F. W., 1942. "On the clustering of nebulae II," *Astrophys. J.* **95**, 565.

Kauffmann, G. and Fairall, A. P., 1991. "Voids in the distribution of galaxies: An assessment of their significance and derivation of a void spectrum," *Mon. Not. R. Astron. Soc.* **248**, 313.

Kendall, M. and Stuart, A., 1977. *The Advanced Theory of Statistics*, Vol. I, 4th ed. (Vol. 2, 1973; Vol. 3, 1976) (London: Griffin).

Kiang, T., 1967. "On the clustering of rich clusters of galaxies," *Mon. Not. R. Astron. Soc.* **135**, 1.

Kiang, T. and Saslaw, W. C., 1969. "The distribution in space of clusters of galaxies," *Mon. Not. R. Astron. Soc.* **143**, 129.

Kim, K. -T., Kronberg, P. P., Dewdney, P. E., and Landecker, T. L. 1990. "The halo and magnetic field of the Coma cluster of galaxies," *Astrophys. J.* **355**, 29.

Kim, K. -T., Tribble, P. C., and Kronberg, P. P., 1991. "Detection of excess rotation measure due to intracluster magnetic fields in clusters of galaxies," *Astrophys. J.* **379**, 80.

Kirk, G. S., Raven, J. E., and Schofield, M., 1983. *The Presocratic Philosophers*, 2nd ed. (Cambridge, UK: Cambridge Univ. Press).

Kirkwood, J. G., 1935. "Statistical mechanics of fluid liquids," *J. Chem. Phys.* **3**, 300.

Kirshner, R. P., Oemler, A., Schechter, P. L., and Schectman, S. A., 1981. "A million cubic megaparsec void in Boötes?," *Astrophys. J. Lett.* **248**, L57.

Kirshner, R. P., Oemler, A., Schechter, P. L., and Schectman, S. A., 1987. "A survey of the Boötes void," *Astrophys. J.* **314**, 493.

Klinkhamer, F. R., 1980. "Groups of galaxies with large crossing times," *Astron. Astrophys.* **91**, 365.

Kodaira, K., Doi, M., Ichikawa, S., and Okamura, S., 1990. "An observational study of Shakhbazyan's compact group of galaxies. II. SCGG 202, 205, 223, 245 and 348." *Publ. Natl. Astron. Obs. Jpn.*, **1**, 283.

Kofman, L. A., Pogosyan, D., and Shandarin, S., 1990. "Structure of the universe in the two-dimensional model of adhesion," *Mon. Not. R. Astron. Soc.* **242**, 200.

Kolmogorov, A. N., 1958. "A new invariant of transitive dynamical systems," *Dokl. Akad. Nauk. USSR* **119**, 861.

Kreyszig, E., 1959. *Differential Geometry* (Toronto: University of Toronto Press).

Krzewina, L. and Saslaw, W. C., 1996. "Minimal spanning tree statistics for the analysis of large-scale structure," *Mon. Not. R. Astron. Soc.* **278**, 869.

Lacey, C. and Cole, S., 1994. "Merger rates in hierarchial models of galaxy formation. II. Comparison with N-body simulations," *Mon. Not. R. Astron. Soc.* **271**, 676.

Lahav, O. and Saslaw, W. C., 1992. "Bias and distribution functions for different galaxy types in optical and *IRAS* catalogs," *Astrophys. J.* **396**, 430.

Lahav, O., Rowan-Robinson, M., and Lynden-Bell, D., 1988. "The peculiar acceleration of the Local Group as deduced from the optical and *IRAS* flux dipoles," *Mon. Not. R. Astron. Soc.* **234**, 677.

Lahav, O., Itoh, M., Inagaki, S., and Suto, Y., 1993. "Non-Gaussian signatures from Gaussian initial fluctuations: Evolution of skewness and kurtosis from cosmological simulations in the highly nonlinear regime," *Astrophys. J.* **402**, 387.

Lambas, D. G., Nicotra, M., Muriel, H., and Ruiz, L., 1990. "Alignment effects of clusters of galaxies," *Astron. J.* **100**, 1006.

Lambert, J. H., 1800. *The System of the World*. Transl. by J. Jacque (London).

Landau, L. D. and Lifshitz, E. M., 1969. *Statistical Physics*, 2nd ed. (Reading: Addison-Wesley).

Landau, L. and Lifshitz, E. M., 1979. *Fluid Mechanics* (New York: Pergamon Press).

Landau, L. and Lifshitz, E. M., 1980. *Statistical Physics* (Oxford: Pergamon Press).

Landsberg, P. T., 1961. *Thermodynamics* (New York: Interscience).

Landsberg, P. T., 1972. "What Olbers might have said." In *The Emerging Universe*, ed. W. C. Saslaw and K. C. Jacobs (Charlottesville: Univ. Virginia Press).

Landy, S. D. and Szalay, A. S., 1993. "Bias and variance of angular correlation functions," *Astrophys. J.* **412**, 64.

Lang, A., 1884. *Custom and Myth* (London: Longmans, Green).

Lauberts, A., 1982. *The ESO–Uppsala Survey of the ESO(B) Atlas* (Garching: ESO).

Layzer, D., 1956. "A new model for the distribution of galaxies in space," *Astron. J.* **61**, 383.

Layzer, D., 1963. "A preface to cosmogony. I. The energy equation and the virial theorem for cosmic distributions," *Astrophys. J.* **138**, 174.

Lemaître, G., 1927. "A homogeneous universe of constant mass and increasing radius accounting for the radial velocity of extra-galactic nebulae," *Annales de la Société Scientifique de Bruxelles XLVII A part 1*. Translated in *Mon. Not. R. Astron. Soc.***91**, 483 (1931).

Lemaître, G., 1931. "The beginning of the world from the point of view of quantum theory," *Nature* **127**, 706.

Lemaître, G., 1950. *The Primeval Atom* (New York: Van Nostrand).

Lemaître, G., 1961. "Exchange of galaxies between clusters and field," *Astron. J.* **66**, 603.

Leavitt, H., 1912. "Periods of 25 variable stars in the Small Magellanic Cloud," *Harvard College Obs. Circ. No. 173*.

Lifshitz, E., 1946. "On the gravitational stability of the expanding universe," *J. Physics* (USSR) **X**, 116.

Limber, D. N., 1953. "The analysis of counts of the extragalactic nebulae in terms of a fluctuating density field," *Astrophys. J.* **117**, 134.

Limber, D. N., 1954. "Analysis of counts of the extragalactic nebulae in terms of a fluctuating density field II," *Astrophys. J.* **119**, 655.

Limber, D. N., 1957. "Analysis of counts of the extragalactic nebulae in terms of a fluctuating density field III," *Astrophys. J.* **125**, 9.

Loan, A. J., Wall, J. V., and Lahav, O., 1997. "The correlation function of radio sources," *Mon. Not. R. Astron. Soc.* **286**, 994.

Long, A. A. and Sedley, D. N., 1987. *The Hellenistic Philosophers* (Cambridge, UK: Cambridge Univ. Press).

Loveday, J., Efstathiou, G., Peterson, B. A., and Maddox, S. J., 1992. "Large-scale structure in the universe: Results from the Stromlo-APM redshift survey," *Astrophys. J. Lett.* **400**, L43.

Lucey, J. R., 1983. "An assessment of the completeness and correctness of the Abell catalogue," *Mon. Not. R. Astron. Soc.* **204**, 33.

Lundmark, K., 1920. "On the relation of the globular clusters and spiral nebulae to the stellar system," *Kungl. Svenska Vetenskapskakademiens Handlingar*, **60**, (8), 1.

Lundmark, K., 1921. "The spiral nebula Messier 33," *Publ. Astron. Soc. Pacific* **33**, 324.

Luquet, G. H., 1959. "Oceanic mythology." In *New Larousse Encyclopedia of Mythology* (London: Hamlyn Publishing) p. 460

Lynden-Bell, D., 1967, "Statistical mechanics of violent relaxation in stellar systems," *Mon. Not. R. Astron. Soc.* **136**, 101.

Lynden-Bell, D. and Wood, R., 1968. "The gravo-thermal catastrophe in isothermal spheres and the onset of red giant structure for stellar systems," *Mon. Not. R. Astron. Soc.* **138**, 495.

Lynden-Bell, D. et al., 1988. "Spectroscopy and photometry of elliptical galaxies. V. Galaxy streaming toward the new supergalactic center," *Astrophys. J.* **326**, 19.

Lynds, R. and Petrosian, V., 1989. "Luminous arcs in clusters of galaxies," *Astrophys. J.* **336**, 1.

Mackie, G., Visvanathan, N., and Carter, D., 1990. "The stellar content of central dominant galaxies. I. CCD surface photometry," *Astrophys. J. Suppl.* **73**, 637.

Maddox, S. J. and Aragón-Salamanca, A., eds., 1995. *Wide Field Spectroscopy and the Distant Universe* (Singapore: World Scientific).

Maddox, S. J., Efstathiou, G., Sutherland, W. J., and Loveday, J., 1990. "Galaxy correlations on large scales," *Mon. Not. R. Astron. Soc.* **242**, 43p.

Maddox, S. J., Efstathiou, G. P., and Sutherland, W. J., 1996. "The APM galaxy survey – III. An analysis of systematic errors in the angular correlation function and cosmological implications," *Mon. Not. R. Astron. Soc.* **283**, 1227.

Maia, M. A. G. and da Costa, L. N., 1990. "On the properties of galaxies in groups," *Astrophys. J.* **352**, 457.

Makino, J., Kokubo, E., and Taiji, M., 1993. "HARP: A special-purpose computer for N-body problem," *Publ. Astron. Soc. Jpn.* **45**, 349.

Mandelbrot, B. B., 1982. *The Fractal Geometry of Nature* (San Francisco: Freeman).

Martínez, V. J. and Jones, B. J. T., 1990. "Why the universe is not a fractal," *Mon. Not. R. Astron. Soc.* **242**, 517.

Martínez, V. J., Jones, B. J. T., Dominguez-Tenreiro, R., and van de Weygaert, R., 1990. "Clustering paradigms and multifractal measures," *Astrophys. J.* **357**, 50.

Mather, J. C. et al., 1990. "A preliminary measurement of the cosmic microwave background spectrum by the *Cosmic Background Explorer (COBE)* satellite," *Astrophys. J. Lett.* **354**, L37.

Mathewson, D. S., Ford, V. L., and Buchhorn, M., 1992. "A southern sky summary of the peculiar velocities of 1355 spiral galaxies," *Astrophys. J. Suppl.* **81**, 413.

Matsubara, T., 1994. "Peculiar velocity effect on galaxy correlation functions in nonlinear clustering regime," *Astrophys. J.* **424**, 30.

Matsubara, T. and Suto, Y., 1994. "Scale dependence of three-point correlation functions: Model predictions and redshift-space contamination," *Astrophys. J.* **420**, 497.

Maurogordato, S., Schaeffer, R., and da Costa, L. N., 1992. "The large-scale galaxy distribution in the southern sky redshift survey," *Astrophys. J.* **390**, 17.

Mayall, N. U., 1934. "A study of the distribution of extra-galactic nebulae based on plates taken with the Crossley reflector," *Lick Obs. Bull.* **458**, 177.

McGill, C. and Couchman, H. M. P., 1990. "Detecting clusters in galaxy catalogs," *Astrophys. J.* **364**, 426.

McLaughlin, D. B., 1922. "The present position of the island universe theory of the spiral nebulae," *Popular Astron.* **30**, 286.

McMillan, S. L. W. and Aarseth, S. J., 1993. "An O ($N \log N$) integration scheme for collisional stellar systems," *Astrophys. J.* **414**, 200.

Melott, A. L. et al., 1989. "Topology of large-scale structure IV. Topology in two dimensions," *Astrophys. J.* **345**, 618.

Melott, A. L. and Shandarin, S., 1990. "Generation of large-scale cosmological structures by gravitational clustering," *Nature* **346**, 633.

Messier, C., 1781. "Catalogue des nebuleuses et des amas d'etoiles." In *Connoissance des Temps Pour l'Annee Bissexile* 1784, Paris, p. 263.

Milne, E. A., 1935. *Relativity Gravitation and World Structure* (Oxford: Clarendon Press).

Mikkola, S. and Aarseth, S. J., 1993. "An implementation of N-body chain regularization," *Celest. Mech. Dyn. Astron.* **57**, 439.

Mo, H. J., 1988. "Probability functions and systematics of large-scale clustering," *Int. J. Mod. Phys. A* **3**, 1373.

Mo, H. J. and Börner, G., 1990. "Percolation analysis and void probability functions of galaxies with different morphological type," *Astron. Astrophys.* **238**, 3.

Mo, H. J., Jing, Y. P., and Börner, G., 1992. "On the error estimates of correlation functions," *Astrophys. J.* **392**, 452.

Mo, H. J., McGaugh, S. S., and Bothun, G. D., 1994. "Spatial distribution of low-surface-brightness galaxies," *Mon. Not. R. Astron. Soc.* **267**, 129.

Moran, P. A. P., 1968. *An Introduction to Probability Theory* (Oxford: Oxford University Press).

Mowbray, A. G., 1938. "Non-random distribution of extragalactic nebulae," *Publ. Astron. Soc. Pacific* **50**, 275.

Murray, M. A., 1913. *Ancient Egpytian Legends* (London: John Murray).

Newton, I., 1692. In *The Correspondence of Isaac Newton, Vol.* III, ed. H. W. Turnbull, 1961 (Cambridge, UK: Cambridge Univ. Press).

Neyman, J., 1962. "Alternative stochastic models of the spatial distribution of galaxies." In *Problems of Extra-Galactic Research*, ed. G. C. McVittie (London: Macmillan) p. 294.

Neyman, J. and Scott, E. L., 1952. "A theory of the spatial distribution of galaxies," *Astrophys. J.* **116**, 144.

Neyman, J. and Scott, E. L., 1955. "On the inapplicability of the theory of fluctuations to galaxies," *Astron. J.* **60**, 33.

Neyman, J. and Scott, E. L., 1959. "Large scale organization of the distribution of galaxies." In *Encyclopedia of Physics*, ed. S. Flügge (Berlin: Springer-Verlag) **53**, 416.

Neyman, J., Scott, E. L., and Shane, C. D., 1953. "On the spatial distribution of galaxies: A specific model," *Astrophys. J.* **117**, 92.

Neyman, J., Scott, E. L., and Shane, C. D., 1956. *Proc. Third Berkeley Symp. Math. Statist. Probability* **3**, 75.

Nichol, J. P. 1848. *The Stellar Universe; Views of Its Arrangements, Motions, and Evolutions* (Edinburgh: J. Johnstone).

Nilson, P., 1973. *Upsalla General Catalogue of Galaxies*, Vol. 6 (Uppsala Astron. Obs.).

Oegerle, W. R., Fitchett, M. J., and Hoessel, J. G., 1989. "The distribution of galaxies in Abell 1795," *Astron. J.* **97**, 637.

Oegerle, W. R. and Hoessel, J. G., 1991. "Fundamental parameters of brightest cluster galaxies," *Astrophys. J.* **375**, 15.

Öpik, E., 1922. "An estimate of the distance of the Andromeda nebula," *Astrophys. J.* **55**, 406.

Ore, O., 1963. *Graphs and Their Uses* (New York: Mathematical Association of America).

Padmanabhan, T., 1993. *Structure Formation in the Universe* (Cambridge, UK: Cambridge University Press).

Park, C., 1990. "Large N-body simulations of a universe dominated by cold dark matter," *Mon. Not. R. Astron. Soc.* **242**, 59p.

Park, C. and Gott, J. R., 1991. "Dynamical evolution of topology of large-scale structure," *Astrophys. J.* **378**, 457.

Park, C. et al., 1992. "The topology of large-scale structure VI. Slices of the universe," *Astrophys. J.* **387**, 1.

Partridge, R. B., 1995. *3 K: The Cosmic Microwave Background Radiation* (Cambridge, UK: Cambridge University Press).

Peebles, P. J. E., 1980. *The Large Scale Structure of the Universe* (Princeton: Princeton Univ. Press).

Peebles, P. J. E., 1993. *Principles of Physical Cosmology* (Princeton: Princeton Univ. Press).

Pellegrini, P. S. et al., 1990. "On the statistical properties of the galaxy distribution," *Astrophys. J.* **350**, 95.

Penzias, A. A. and Wilson, R. W., 1965. "A measurement of excess antenna temperature at 4080 Mcs," *Astrophys. J.* **142**, 419.

Pincherle, S., 1904. *Acta. Math.* **28**, 225.

Plionis, M., Barrow, J. D., and Frenk, C. S., 1991. "Projected and intrinsic shapes of galaxy clusters," *Mon. Not. R. Astron. Soc.* **249**, 662.

Poignant, R., 1968. *Oceanic Mythology* (Feltham: Paul Hamlyn).

Press, W. H. and Schechter, P., 1974. "Formation of galaxies and clusters of galaxies by self-similar gravitational condensation," *Astrophys. J.* **187**, 425.

Proctor, R. A., 1869. "Distribution of the nebulae," *Mon. Not. R. Astron. Soc.* **29**, 337.

Ramella, M., Geller, M. J., and Huchra, J. P., 1992. "The distribution of galaxies within the "Great Wall," *Astrophys. J.* **384**, 396.

Raphaeli, Y. and Saslaw, W. C., 1986. "Correlated statistical fluctuations and galaxy formation," *Astrophys. J.* **309**, 13.

Raychaudhury, S. and Saslaw, W. C., 1996. "The observed distribution function of peculiar velocities of galaxies," *Astrophys. J.* **461**, 514.

Renyi, A., 1970. *Probability Theory* (Amsterdam: North Holland).

Reynolds, J. H., 1920. "The galactic distribution of the large spiral nebulae," *Mon. Not. R. Astron. Soc.* **81**, 129.

Reynolds, J. H., 1923. "The galactic distribution of the spiral nebulae with special reference to galactic longitude," *Mon. Not. R. Astron. Soc.* **83**, 147.

Reynolds, J. H., 1924. "The galactic distribution of the small spiral and lenticular nebulae," *Mon. Not. R. Astron. Soc.* **84**, 76.

Rhee, G. F. R. N., van Haarlem, M. P., and Katgert, P., 1991a. "A study of the elongation of Abell clusters. II. A sample of 107 rich clusters," *Astron. Astrophys. Suppl.* **91**, 513.

Rhee, G. F. R. N., van Haarlem, M. P., and Katgert, P., 1991b. "Substructure in Abell clusters," *Astron. J.* **97**, 637.

Richardson, L. F., 1961. "The problem of contiguity," an appendix to *Statistics of Deadly Quarrels. General Systems Yearbook*, **6**, 139.

Rood, H. J. and Williams, B. A., 1989. "The neighbourhood of a compact group of galaxies," *Astrophys. J.* **339**, 772.

Rose, J. A. and Graham, J. A., 1979. "Mass-to-light ratios of two compact groups of galaxies," *Astrophys. J.* **231**, 320.

Rosse, Third Earl (William Parsons) 1850. "Observations of the nebulae," *Phil. Trans.* **140**, 499.

Rowan-Robinson, M., Saunders, W., Lawrence, A., and Leech, K. L., 1991. "The QMW *IRAS* galaxy catalogue: a highly complete and reliable *IRAS* 60-μm galaxy catalogue," *Mon. Not. R. Astron. Soc.* **253**, 485.

Roxburgh, I. W. and Saffman, P. G., 1965. "The growth of condensations in a Newtonian model of the steady-state universe," *Mon. Not. R. Astron. Soc.* **129**, 181.

Ruamsuwan, L. and Fry, J. N., 1992. "Stability of scale-invariant nonlinear gravitational clustering," *Astrophys. J.* **396**, 416.

Rubin, V. C., 1954. "Fluctuations in the space distribution of the galaxies," *Proc. Natl. Acad. Sci. Washington* **40**, 541.

Rubin, V. C., Hunger, D. A., and Ford, W. J. Jr., 1990. "One galaxy from several: The Hickson compact group H31," *Astrophys. J.* **365**, 86.

Russell, B., 1945. *A History of Western Philosophy* (New York: Simon and Schuster) p. 472.

Sanford, R. F., 1917. "On some relation of the spiral nebulae to the Milky Way," *Lick Bull.* No. 297 Vol. IX, p. 80.

Saslaw, W. C., 1968. "Gravithermodynamics I. Phenomenological equilibrium theory and zero time fluctuations," *Mon. Not. R. Astron. Soc.* **141**, 1.

Saslaw, W. C., 1969. "Gravithermodynamics II. Generalized statistical mechanics of violent agitation," *Mon. Not. R. Astron. Soc.* **143**, 437.

Saslaw, W. C., 1972. "Conditions for rapid growth of perturbations in an expanding universe," *Astrophys. J.* **173**, 1.

Saslaw, W. C., 1985a. *Gravitational Physics of Stellar and Galactic Systems* (GPSGS) (Cambridge, UK: Cambridge University Press).

Saslaw, W. C., 1985b. "Galaxy clustering and thermodynamics: Relaxation of N-body experiments," *Astrophys. J.* **297**, 49.

Saslaw, W. C., 1986. "Galaxy clustering and thermodynamics: Quasi-static evolution of $b(t)$," *Astrophys. J.* **304**, 11.

Saslaw, W. C., 1989. "Some properties of a statistical distribution function for galaxy clustering," *Astrophys. J.* **341**, 588.

Saslaw, W. C., 1991. "Black holes and structure in an oscillating universe," *Nature* **350**, 43.

Saslaw, W. C., 1992. "The rate of gravitational galaxy clustering," *Astrophys. J.* **391**, 423.

Saslaw, W. C., 1995. "Gravitational constraints on the distribution of intergalactic matter." In *The Physics of the Interstellar Medium and Intergalactic Medium*, ed. A. Ferrara et al. (San Francisco: Astron. Soc. Pacific).

Saslaw, W. C., Antia, H. M., and Chitre, S. M., 1987. "New limits on the amount of dark matter in the universe," *Astrophys. J. Lett.* **315**, L1.

Saslaw, W. C., Chitre, S. M., Itoh, M., and Inagaki, S., 1990. "The kinetic evolution and velocity distribution of gravitational galaxy clustering," *Astrophys. J.* **365**, 419.

Saslaw, W. C. and Crane, P., 1991. "The scale dependence of galaxy distribution functions," *Astrophys. J.* **380**, 315.

Saslaw, W. C. and Edgar, J., 1999. "High redshift galaxy clustering and Ω_0" in preparation.

Saslaw, W. C. and Fang, F., 1996. "The thermodynamic description of the cosmological many-body problem," *Astrophys. J.* **460**, 16.

Saslaw, W. C. and Hamilton, A. J. S., 1984. "Thermodynamics and galaxy clustering: Non-linear theory of high order correlations," *Astrophys. J.* **276**, 13.

Saslaw, W. C. and Haque-Copilah, S., 1998. "The Pisces–Perseus supercluster and gravitational quasi-equilibrium clustering," *Astrophys. J.* **509**, 595

Saslaw, W. C., Maharaj, S. D., and Dadhich, N., 1996. "An isothermal universe," *Astrophys. J.* **471**, 571.

Saslaw, W. C., Narasimha, D., and Chitre, S. M., 1985. "The gravitational lens as an astronomical diagnostic," *Astrophys. J.* **292**, 348.

Saslaw, W. C. and Sheth, R. K., 1993. "Non-linear properties and time evolution of gravitational galaxy clustering," *Astrophys. J.* **404**, 504.

Sathyaprakash, B. S., Sahni, V., Munshi, D., Pogosyan, D., and Mellott, A. L., 1995. "Gravitational instability in the strongly non-linear regime: A study of various approximations," *Mon. Not. R. Astron. Soc.* **275**, 463.

Saunders, W., Rowan-Robinson, M., and Lawrence, A., 1992. "The spatial correlation function of *IRAS* galaxies on small and intermediate scales," *Mon. Not. R. Astron. Soc.* **258**, 134.

Saunders, W., Rowan-Robinson, M., Lawrence, A., Efstathiou, G., Kaiser, N., Ellis, R. S., and Frenk, C. S., 1990. "The 60 μm and far-infrared luminosity functions of *IRAS* galaxies," *Mon. Not. R. Astron. Soc.* **242**, 318.

Schechter, P., 1976. "An analytic expression for the luminosity function for galaxies," *Astrophys. J.* **203** 297.

Schombert, J. M. and West, M. J., 1990. "A morphology–density relation for clusters of galaxies," *Astrophys. J.* **363**, 331.

Scott, E. L., 1962. "Distribution of galaxies on the sphere: Observed structures (groups, clusters, clouds)." In *Problems of Extragalactic Research*, ed. G. C. McVittie (London: Macmillan), p. 269.

Seeliger, H., 1895. "Ueber das Newton'sche gravitationsgesetz," *Astron. Nachr.* **137**, No. 3273.

Shane, C. D., 1956. "The distribution of extragalactic nebulae, II," *Astron. J.* **61**, 292.

Shane, C. D. and Wirtanen, C. A., 1954. "The distribution of extragalactic nebulae," *Astron. J.* **59**, 285.

Shane, C. D. and Wirtanen, C. A., 1967. "The distribution of galaxies," *Publ. Lick Obs.* **23**, Part 1.

Shane, C. D., Wirtanen, C. A., and Steinlin, U., 1959. "The distribution of extragalactic nebulae, III," *Astron. J.* **64**, 197.

Shapley, H., 1917. "Note on the magnitudes of novae in spiral nebulae," *Publ. Astron. Soc. Pacific* **29**, 213.

Shapley, H., 1919. "On the existence of external galaxies," *Publ. Astron. Soc. Pacific* **31**, 261.

Shapley, H., 1921. "Evolution of the idea of galactic size," *Bull. National Research Council of the Natl. Acad. Sci.* **2**, 171.

Shapley, H., 1932. "Note on the distribution of remote galaxies and faint stars," *Harvard College Obs. Bull.* **890**, 1.

Shapley, H., 1933. "On the distribution of galaxies," *Proc. Natl. Acad. Sci.* **19**, 389.

Shapley, H., 1934a. "A first search for a metagalactic gradient," *Harvard College Obs. Bull.* **894**, 5.

Shapley, H., 1934b. "On some structural features of the metagalaxy," (George Darwin Lecture) *Mon. Not. R. Astron. Soc.* **94**, 792.

Shapley, H., 1938a. "A metagalactic density gradient," *Proc. Natl. Acad. Sci.* **24**, 282.

Shapley, H., 1938b. "Second note on a metagalactic density gradient," *Proc. Natl. Acad. Sci.* **24**, 527.

Shapley, H. and Shapley, M., 1919. "Studies based on the colors and magnitudes in stellar clusters," *Astrophys. J.* **50**, 107.

Sheth, R. K., 1995a. "Press–Schechter, thermodynamics and gravitational clustering," *Mon. Not. R. Astron. Soc.* **274**, 313.

Sheth, R. K., 1995b. "Merging and hierarchial clustering from an initially Poisson distribution," *Mon. Not. R. Astron. Soc.* **276**, 796.

Sheth, R. K., 1996a. "Random dilutions, generating functions, and the void probability distribution function," *Mon. Not. R. Astron. Soc.* **278**, 101.

Sheth, R. K., 1996b. "The distribution of pairwise peculiar velocities in the non-linear regime," *Mon. Not. R. Astron. Soc.* **279**, 1310.

Sheth, R. K., 1996c. "The distribution of counts in cells in the non-linear regime," *Mon. Not. R. Astron. Soc.* **281**, 1124.

Sheth, R. K., 1996d. "Galton–Watson branching processes and the growth of gravitational clustering," *Mon. Not. R. Astron. Soc.* **281**, 1277.

Sheth, R. K., Mo, H. J., and Saslaw, W. C., 1994. "The distribution of *IRAS* galaxies on linear and non-linear scales," *Astrophys. J.* **427**, 562.

Sheth, R. K. and Saslaw, W. C., 1994. "Synthesizing the observed distribution of galaxies," *Astrophys. J.* **437**, 35.

Sheth, R. and Saslaw, W. C., 1996. "Scale dependence of nonlinear gravitational clustering in the universe," *Astrophys. J.* **478**, 78.

Slipher, V. M., 1913. "The radial velocity of the Andromeda nebula," *Lowell Obs. Bull.* **2**, 56.

Slipher, V. M., 1915. "Spectroscopic observations of nebulae," *Popular Astron.* **23**, 21.

Smart, W. M., 1922. "An infinite universe," *Observatory* **45**, 216.

Smith, R. M., Frenk, C. S., Maddox, S. J., and Plionis, M., 1991. "The reality of large-scale filamentary structure in the Lick map." In *Large-Scale Structures and Peculiar Motions in the Universe*, ed. D. W. Latham and L. A. Nicolaci da Costa (San Francisco: Astron. Soc. Pacific), p. 225.

Stauffer, D., 1979. "Scaling theories of percolation clusters," *Physics Reports* **54**, 1.

Stauffer, D., 1985. *Introduction to Percolation Theory* (London: Taylor & Francis).

Steidel, C. C. et al., 1998. "A large structure of galaxies at redshift $z \sim 3$ and its cosmological implications," *Astrophys. J.* **492**, 428.

Stiefel, E. L. and Scheifele, G., 1971. *Linear and Regular Celestial Mechanics* (Berlin: Springer-Verlag).

Stratonoff, W., 1900. "Etudes sur la structure de l'Universe," *Publ. Tashkent* **2**, 1.

Struble, M. F., 1990. "Position angle statistics of the first and second brightest galaxies in a sample of Coma-like Abell clusters," *Astron. J.* **99**, 743.

Suto, Y., Itoh, M., and Inagaki, S., 1990. "A gravitational thermodynamic approach to probe the primodial spectrum of cosmological density fluctuations," *Astrophys. J.* **350**, 492.

Suto, Y. and Suginohara, T., 1991. "Redshift-space correlation functions in the cold dark matter scenario," *Astrophys. J.* **370**, L15.

Szapudi, I. and Szalay, A. S., 1993. "High order statistics of the galaxy distribution using generating functions," *Astrophys. J.* **408**, 43.

Thirring, W., 1970. "Systems with negative specific heat," *Z. Physik* **235**, 339.

Tifft, W. and Gregory, S. A., 1976. "Direct observations of large-scale distribution of galaxies," *Astrophys. J.* **205**, 696.

Tolman, R. C., 1934. *Relativity, Thermodynamics and Cosmology* (Oxford: Oxford University Press).

Tolman, R. C., 1938. *The Principles of Statistical Mechanics* (Oxford: Oxford University Press).

Totsuji, H. and Kihara, T., 1969. "The correlation function for the distribution of galaxies," *Publ. Astron. Soc. Jpn.* **21**, 221.

Tully, R. B. and Fisher, J. R., 1977. "A new method of determining distances to galaxies," *Astron. Astrophys.* **54**, 661.

Tully, R. B. and Fisher, J. R., 1988. *Nearby Galaxies Catalog* (Cambridge, UK: Cambridge University Press).

Ueda, H. and Yokoyama, J., 1996. "Counts-in-cells analysis of the statistical distribution in an N-body simulated universe," *Mon. Not. R. Astron. Soc.* **280**, 754.

Ulam, S., 1954. "Infinite models in physics," Los Alamos Scientific Laboratory manuscript.

Ulmer, M. P., McMillan, S. L. W., and Kowlaski, M. P., 1989. "Do the major axes of rich clusters of galaxies point toward their neighbors?," *Astrophys. J.* **338**, 711.

Uson, J. M., Boughn, S. P., and Kuhn, J. R., 1990. "The central galaxy in Abell 2029: An old supergiant," *Science* **250**, 539.

Valdarnini, R., Borgani, S., and Provenzale, A., 1992. "Multifractal properties of cosmological N-body simulations," *Astrophys. J.* **394**, 422.

Vallée, J. P., 1990. "A possible excess rotation measure and large scale magnetic field in the Virgo supercluster of galaxies," *Astron. J.* **99**, 459.

van Albada, G. B., 1960. "Formation and evolution of clusters of galaxies," *Bull. Astron. Inst. Neth.* **15**, 165.

van Albada, G. B., 1961. "Evolution of clusters of galaxies under gravitational forces," *Astron. J.* **66**, 590.

van Kampen, E. and Rhee, G. F. R. N., 1990. "Orientation of bright galaxies in Abell clusters," *Astron. Astrophys.* **237**, 283.

von Humboldt, Alexander, 1860. *Cosmos: A Sketch of a Physical Description of the Universe.* Transl. by E. C. Otte & B. H. Paul (New York: Harper & Bros.).

Vorontsov-Vel'yaminov, B. A., Dostal', V. A., and Metlov, V. G., 1980. "The most compact nest of galaxies VV644," *Sov. Astron. Lett.* **6**, 217.

Vorren, O. and Manker, E., 1962. *Lapp Life and Customs* (Oxford: Oxford Univ. Press).

Walke, D. G. and Mamon, G. A., 1989. "The frequency of chance alignments of galaxies in loose groups" *Astron. Astrophys.* **225**, 291.

Warren, M. S., Quinn, P. J., Salmon, J. K., and Zurek, W. H., 1992. "Dark halos formed via dissipationless collapse: I. Shapes and alignment of angular momentum," *Astrophys. J.* **399**, 405.

Waters, S., 1873. "The distribution of the clusters and nebulae," *Mon. Not. R. Astron. Soc.* **33**, 558.

Webster, A., 1976. "The clustering of radio sources – II. The 4C, GB and MC1 surveys," *Mon. Not. R. Astron. Soc.* **175**, 71.

Weinberg, D. H., Gott, J. R., and Mellot, A. L., 1987. "The topology of large-scale structure I. Topology and the random phase hypothesis," *Astrophys. J.* **321**, 2.

Weinberg, D. H. and Gunn, J. E., 1990. "Simulations of deep redshift surveys," *Astrophys. J.* **352**, L25.

Werner, E. T. C., 1922. *Myths and Legends of China* (London: Harrap).

Wesson, P. S., 1989. "A class of solutions in general relativity of interest for cosmology and astrophysics," *Astrophys. J.* **336**, 58.

West, M. J. and Bothum, G. D., 1990. "A reanalysis of substructure in clusters of galaxies and their surroundings," *Astrophys. J.* **350**, 36.

Westfall, R. S., 1980. *Never at Rest. A Biography of Isaac Newton* (Cambridge, UK: Cambridge Univ. Press).

White, R. J., 1965. *Dr Bentley: A Study in Academic Scarlet* (London: Eyre & Spottiswoode).

White, S. D. M., 1979. "The hierarchy of correlation functions and its relation to other measures of galaxy clustering," *Mon. Not. R. Astron. Soc.* **186**, 145.

Whittaker, E. T. and Watson, G. N., 1927. *A Course in Modern Analysis*, 4th ed. (Cambridge, UK: Cambridge University Press).

Wright, M. R., 1995. *Cosmology in Antiquity* (New York: Routledge).

Wright, T., 1750. *An Original Theory or New Hypothesis of the Universe.* (New York: Elsevier, 1971).

Yano, T. and Gouda, N., 1997. "Scale-invariant correlation functions of cosmological density fluctuations in the strong-clustering regime," *Astrophys. J.* **487**, 42.

Yepes, G., Domínguez-Tenreiro, R., del Pozo-Sanz, 1991. "Luminosity segregation as an indication of dynamical evolution in rich clusters of galaxies," *Astrophys. J.* **373**, 336.

Yoshii, Y., Petersen, B., and Takahara, F., 1993. "On the angular correlation function of faint galaxies," *Astrophys. J.* **414**, 431.

Zaehner, R.C., 1961. *The Rise and Fall of Zoroastrianism* (London: Weidenfield & Nicholson).

Zeldovich, Ya. B., 1965. "Survey of modern cosmology." In *Advances in Astronomy and Astrophysics*, Vol. 3, ed. Z. Kopal, p. 241 (New York: Academic Press).

Zeldovich, Ya. B., 1970. "Gravitational instability: An approximate theory for large density perturbations," *Astron. Astrophys.* **5**, 84.

Zeldovich, Ya. B., Einasto, J., and Shandarin, S. F., 1982. "Giant voids in the universe," *Nature* **300**, 407.

Zhan, Y., 1989. "Gravitational clustering of galaxies: The probability distribution function," *Astrophys. J.* **340**, 23.

Zhan, Y. and Dyer, C.C., 1989. "An integral constraint on the N-point correlation function of galaxies," *Astrophys. J.* **343**, 107.

Zhao, J., Huang, S., Pan, R., and He, Y., 1991. "Studies of the Coma cluster III. The virial mass and kinematics," *Chin. Astron. Astrophys.* **15**, 260.

Zwicky, F., 1937. "On the masses of nebulae and of clusters of nebulae," *Publ. Astron. Soc. Pacific* **86**, 217.

Zwicky, F., 1938. "On the clustering of nebulae," *Publ. Astron. Soc. Pacific* **50**, 218.

Zwicky, F., 1957. *Morphological Astronomy* (Berlin: Springer-Verlag).

Zwicky, F., 1967. "Compact and dispersed cosmic matter, Part I," *Adv. Astron. Astrophys.* **5**, 268.

Zwicky, F., Herzog, E., and Wild, P., 1961. *Catalogue of Galaxies and Clusters of Galaxies* (Pasadena: Calif. Inst. of Technol.) and succeeding volumes.

Index

Index-learning turns no student pale,
Yet holds the eel of science by the tail.
Alexander Pope

adhesion approximation, 232
adiabatic decay
  in evolution of $b(t)$, 161
  and heat flow into regions, 373–4, 376–8
  of initial peculiar velocities, 391
  as universe expands, 269, 276, 416–21

bank robbers, 141
Bayes's theorem, 250–1
BBGKY hierarchy
  derived from Liouville's equation, 205–7
  growth of two-particle correlations in,
    207–15
  scaling in, 218–19
Bentley, Richard, 11–12
bias, 131, 221–3, 355–61, 447–9
bulk velocity flow
  effects on peculiar velocity distribution
    functions, 452–8
  nature of, 166–7
  produced by dark matter, 476

Cantor's set, 91–2
catalogs of galaxies
  Abell, 49–50, 90
  Herschel, J., 22, 23
  Herschel, W., 17–22
  Hubble, 27–8
  Lick, 28, 53
  Messier, 17
  modern, 135, 242–5, 430–45, 453
  Shapley, 28
  Zwicky, 50, 162–3, 430
clusters of galaxies
  Abell, clustering of, 449–50
  binding of, 100–2
  controversy over uniformity, 27–31,
    50–1
  correlations of, 51, 135

discoveries of, 17, 18, 22, 25, 49–50
evolution of b for, 381–5
multiplicity function, 361–4, 378–81
Neyman-Scott models, 41–2, 363–4
observed properties, 107–12
probability for finding, 460–2
physical processes in, 112–20
Pisces-Perseus, 442–6, 460–2
topology of Abell cluster distribution, 90
comoving equations of particle motion, 182,
  228, 268–72, 307, 315–6
comoving potential, 306–8, 315
conditional and unconditional density,
  121–2
confusion, utter, 279
correlation dimension, 98–9
correlation energy, 147, 224, 270, 286,
  310–11
correlation functions, 65–6, 121–40
  angular, 130–1, 133–4
  BBGKY hierarchy growth, 207–15
  bias, 131, 221–3
  of clusters, 135–6
  and distribution functions, 262–3
  in Einstein–de Sitter model, 185
  evolution of, 171–215, 381–5
  and fluctuations, 126, 131, 321–2
  for GQED, 353–4, 372
  higher order, 127, 128
  magnitude scaling, 131
  normalization of, 126–8
  observation of, 133–6, 242–5
  in redshift space, 130, 135, 219–21
  see also two-point correlations
cosmic blackbody background, 51
cosmic energy equation, 225
  conditions for validity, 271
  derivation from adiabatic evolution,
    374–5